Plant Resistance to Herbivores and Pathogens

EDITED BY

Robert S. Fritz
and
Ellen L. Simms

Plant Resistance to Herbivores and Pathogens

Ecology, Evolution, and Genetics

The University of Chicago Press

Chicago and London

ROBERT S. FRITZ is associate professor in the
Department of Biology at Vassar College. ELLEN
L. SIMMS is assistant professor in the Department
of Ecology and Evolution at the University of
Chicago.

The University of Chicago Press, Chicago 60637
The University of Chicago Press, Ltd., London

© 1992 by The University of Chicago
All rights reserved. Published 1992
Printed in the United States of America

01 00 99 98 97 96 95 94 93 92   5 4 3 2 1

ISBN (cloth): 0–226–26553–6
ISBN (paper): 0–226–26554–4

Library of Congress Cataloging-in-Publication Data

Plant resistance to herbivores and pathogens : ecology, evolution, and genetics / edited by Robert
    S. Fritz and Ellen L. Simms.
        p.   cm.
    Includes bibliographical references and index.
    ISBN 0-226-26553-6 — ISBN 0-226-26554-4 (pbk.)
    1. Plants—Disease and pest resistance.   2. Plants—Disease and pest resistance—Genetic
aspects.   3. Herbivores—Food.   4. Plant ecological genetics.   5. Coevolution.   I. Fritz,
Robert S.   II. Sims, Ellen Louise, 1955–   .
    SB750.P56   1992
    581.29—dc20
                                                                                                91-5060
                                                                                                CIP

⊗The paper used in this publication meets the minimum requirements of the American National
    Standard for Information Sciences—Permanence of Paper for Printed Library Materials, ANSI
    Z39.48–1984.

Contents

Preface

The study of plant resistance has been conducted along many fronts. The coordination among these fronts is sometimes impressive, but at other times seems nonexistent. In this volume we have tried to integrate the broad field of plant resistance by soliciting chapters from a diversity of authors who represent the variety of approaches and points of intersection between applied and basic research, herbivore and pathogen interactions with plants, and empirical and theoretical studies.

Of course, no volume on plant resistance would be complete without emphasizing the agricultural traditions that so fundamentally contribute to our knowledge and shape our perceptions of plant resistance. Concerns about plant resistance to herbivores and pathogens must date from the beginnings of agriculture, and today selection and breeding for pest-resistant crops is a major focus of agricultural research. The vast conceptual and empirical literature from managed systems is a rich lode that has been successfully (but perhaps too selectively) mined by workers in natural systems. Against a backdrop provided by this strong tradition, evolutionary biologists and ecologists studying coevolution of plants and phytophages have recently focused attention on the role of genetic variation in plant resistance in natural systems. Empirical studies have documented the widespread existence of genetic variation in plant resistance in natural populations and we are beginning to explore its effects on many aspects of the ecology and evolution of organisms affected by it. Now agronomists, entomologists, and plant pathologists are incorporating into their work the contributions from this outpouring of work on herbivores and pathogens in natural systems.

A synthesis of these two bodies of literature was first made by Denno and McClure in their important volume, *Variable Plants and Herbivores in Natural and Managed Systems*. We have built upon this sturdy foundation by both narrowing and broadening our focus. Whereas Denno and McClure considered all levels of variation in resistance, from among species to within a single individual, we have chosen to concentrate on variation among individuals within a population. This level of variation is interesting because it determines in part the potential for evolution of resistance traits in a plant population. In addition to focusing on one level of variation, our recognition of the many similarities between plant-herbivore and plant-pathogen interactions led us to expand our coverage into the latter field.

There are numerous points of intersection between the lines of applied and basic research on the interactions of plants with herbivores and pathogens in managed and natural systems. Evolutionary biologists have produced models to explain the evolution of plant resistance and the maintenance of genetic variation in resistance in natural populations, to understand the evolution of virulence in phytophages in both natural and managed systems, and to determine how the evolution of virulence to crop resistance can be prevented in crop pests. Many empiricists are actively testing the assumptions and predictions of these models, which will soon permit a newer, more refined generation of models. Ecologists are also beginning to understand some of the effects of genetic variation in plant resistance on the population biology of phytophages and their natural enemies. Progress on this front has been made both by people working on crop plants and by those studying natural plant populations.

This volume was given impetus by a symposium on the genetics of plant resistance at the 1987 meeting of the Ecological Society of America. At that meeting it became apparent that an increasing number of ecologists and evolutionary biologists were studying the evolution of plant resistance in natural populations and its ecological effects on herbivores, pathogens, and natural enemies. Since then, many more people have joined the fray. In this book, we explore a variety of approaches to the study of plant resistance and attempt some synthesis of these viewpoints. We hope that presenting ideas and information about interactions of plants with both herbivores and pathogens and from both applied and managed systems will stimulate new directions of research in each of these fields. We also hope to show how research in these fields, by integrating genetic with evolutionary and ecological methods, can contribute to a new evolutionary synthesis.

Part 1 of this volume emphasizes the inheritance and mechanisms of plant resistance, and methods of studying resistance traits in natural populations. The evolutionary responses of herbivores and pathogens to within-population resistance variation in natural and managed systems comprises part 2. In part 3, the effects of resistance on population biology of herbivores and grazers, on interactions between herbivore species, and on interactions between organisms on the second and third trophic levels are considered. Selection on resistance by herbivores and pathogens, the effects of plant population structure, and defense theory and costs of resistance are the themes of part 4.

Various drafts of the chapters in this book were reviewed by other authors and/or anonymous reviewers. The entire manuscript was reviewed by two reviewers for the University of Chicago Press. We are immensely grateful for their careful comments and advice. We specifically want to think Deane Bowers, Jeremy Burdon, Phyllis Coley, Robert Denno, Fred Gould, Marvin Harris, John Lawton, Svata Louda, Nancy Moran, Deborah Roach, Richard Roush, Don Stratton, and Don Strong. We wish to thank Nora Murphy and Daniella Scott for clerical assistance. Bonnie Milne worked with us tirelessly in preparing the bibliography and with numerous other details needed to complete this volume. We are most grateful for her assistance. We are grateful to our respective spouses, Chris Fritz and Tom Colton, for their moral and material support.

Robert S. Fritz
Ellen L. Simms

1

Ecological Genetics of Plant-Phytophage Interactions

Robert S. Fritz and Ellen L. Simms

In any terrestrial community, photosynthesis is the first step in an extensive array of trophic relationships involving many species of herbivores, pathogens, and their natural enemies. Because of their ability to fix light energy, plants play a central role in community interactions and therefore profoundly influence the evolution, population dynamics, and community structure of herbivores and pathogens. Moreover, characteristics of individual plants and the physical and chemical structure of plant communities directly and indirectly affect the evolution and ecology of predators and parasitoids of herbivores (Price et al. 1980; Boethel and Eikenbary 1986) and parasites of pathogens.

Plant species and communities in turn are influenced directly by herbivores and pathogens and indirectly by parasites, parasitoids, and predators. These organisms may affect the population dynamics of plants by altering growth, reproductive success, seedling establishment, range, and habitat occupancy (Kulman 1971; Harper 1977; Crawley 1983). Furthermore, because herbivores and pathogens consume plants, they likely reduce plant fitness and therefore should selectively favor any plant traits that minimize phytophagy (i.e., plant resistance).

If resistance improves plant fitness in the presence of phytophages, and if such traits are genetically variable, then plant populations should evolve to maximal levels of resistance, and in the process lose genetic variation for those traits. Nevertheless, numerous studies have documented significant levels of genetic variation in resistance in natural plant populations. Explaining the presence of high levels of genetic variation in phenotypic traits such as plant resistance is a goal of many evolutionary biologists. This book focuses on the mechanisms by which resistance traits evolve in plant populations, the factors responsible for their variability, and the ecological and evolutionary implications of this variability for phytophages and their natural enemies.

1

Organization and Emphasis of Chapters

Analysis and Inheritance of Resistance Variation

Much of our knowledge of the genetics of plant resistance to herbivores and pathogens comes from agricultural research. It is therefore appropriate that the first section of the book begin with chapter 2, in which Kennedy and Barbour selectively review what is known about the inheritance of resistance in crop plants (the literature is vast). These authors take several approaches to their subject. First, they consider the resistance of specific crops to herbivorous insects and mites and briefly review what is known about resistance in several natural systems. This review reveals striking differences between natural and managed populations in the mode of inheritance and extent of genetic and environmental variation, and in their potential for evolutionary change. Kennedy and Barbour continue by discussing a number of putative resistance traits, reviewing the evidence that each plays a causal role in resistance. Furthermore, they enumerate the environmental factors that can influence the expression of plant resistance.

Kennedy and Barbour also contrast various definitions of resistance used in managed and natural systems. There are perhaps as many definitions of resistance as there are scientists studying the phenomenon. Kennedy and Barbour define host-plant resistance as including genetically controlled traits (antibiosis and nonpreference) possessed by some subset of a plant species (individuals, clones, populations, races, or varieties) that result in the members of that subset being less damaged by a particular herbivore species compared with members of other subsets of the same plant species. This concern for defining resistance is echoed in chapter 14 by Alexander. She reviews definitions of resistance to pathogens and then suggests how these definitions can be operationalized to measure resistance in experimental situations. Both chapters remind us that resistance is both relative and context dependent. For example, a plant genotype is resistant only relative to less resistant genotypes, and its level of resistance will depend on its environment and the species of phytophage being considered. Usually excluded from definitions of resistance are nonhost immunity (Kennedy and Barbour, chapter 2) and tolerance (Alexander, chapter 14).

The remaining chapters of the first section introduce valuable techniques for analyzing plant resistance variation. In chapter 3, Simms and Rausher describe how quantitative genetic methods may be used to study the evolution of plant resistance. Although their hypothetical examples involve resistance to herbivory, the methods they describe are equally suitable for studying the evolution of resistance to pathogens. They begin by enumerating the various factors that contribute to phenotypic variation in resistance and briefly describing how quantitative genetic methods can be used to estimate the relative contribution of these factors to resistance variation in a particular population. Simms and Rausher then consider the evolution of quantitative traits, with special emphasis on how genetic correlations influence evolutionary trajectories. The latter part of the chapter presents a more technical discussion of the various experimental methods for estimating evolutionary parameters, including ways to test hypotheses and avoid pitfalls inherent in the indiscriminant use of these techniques.

With the widespread interest in chemical mediation of plant-insect interactions and the often-assumed coevolutionary relationship between plant chemistry and phytophagy, it is essential that the inheritance of chemically mediated resistance mechanisms be carefully documented in natural systems. This documentation has been attempted in only a few natural systems. The technical problems of quantifying chemical variation among plants (both among genotypes for one chemical variation, and measuring plant fitness can be daunting. In chapter 4, Berenbaum and Zangerl review these issues, pointing out the difficulties as well as potential solutions. The care needed to produce both genetic and chemical analyses of plant-herbivore coevolution might suggest that it is an unprofitable line of research for ecologists, but Berenbaum and Zangerl chart a navigable course for determining the usefulness of particular systems for answering questions about chemical coevolution.

Evolutionary Responses of Herbivores and Pathogens to Variation in Plant Resistance

Phenotypic diversity within host-plant populations affects the distribution, abundance, and fitness of phytophages. When host plants vary in their level of resistance, selection should favor phytophages that can overcome such resistance and utilize the previously underexploited resource. When phytophages vary genetically for preference and/or performance on different host phenotypes, then this selection should cause evolution in these traits. The chapters in part 2 of this book explore how genotypic variation among host plants influences phytophage distribution and population structure (Pilson, chapter 6; Weis, chapter 7; Groth and Christ, chapter 8), the evolution of host use and virulence (Wilhoit, chapter 5; Pilson, chapter 6; Groth and Christ, chapter 8), and the evolutionary options herbivores have for responding to plant variation (Weis, chapter 7).

Wilhoit (chapter 5) explores theoretical models of the evolution of herbivore virulence to mixtures of resistant and susceptible varieties of crops. He focuses on aphids as herbivores and reviews the available empirical observations relevant to these models. Although most of the chapter concentrates on agricultural issues, the predictions of the models have numerous implications for the evolutionary responses of herbivores in natural populations of plants that vary in resistance.

Pilson (chapter 6) considers the influence of genetic variation in natural plant populations on herbivore distributions. She discusses what is known from empirical studies in natural systems and reviews models of the evolution of host use, oviposition preference, and herbivore performance. Pilson considers both direct and indirect effects of resistance variation on herbivore preference and performance. The indirect effects she describes include the potential that herbivores have for altering the phenotype of their preferred host plant (simply by using it), which then influences its level of resistance as perceived by later herbivores.

Weis (chapter 7) examines the evolution of herbivore performance in response to host-plant phenotypic variation. Herbivores experience plant phenotypic variation as environmental variation, and variable performance of an insect genotype on different plant phenotypes constitutes genotype-environmental interaction (plasticity). Weis models the evolution of plasticity in performance characters of herbivores using a multivariate selection model, and reevaluates two important underlying assumptions: that "a jack-of-all-trades is a master of none," and that "specialization is an evolutionary dead end."

Groth and Christ (chapter 8) describe the influence of host-plant characteristics on plant-pathogen diversity. After reviewing the literature on both agricultural (disturbed) and natural (presumed to be coevolved) systems, they conclude that with only a few exceptions, very little is known about the extent of genetic diversity in plant-pathogen populations. Variation in resistance or compatibility genes among host-plant individuals, populations, and species evidently affects pathogen diversity in agricultural systems, but only in one case was there evidence of a relationship between host diversity and pathogen diversity in a natural system. They argue that the use of other markers (e.g., restriction-fragment-length polymorphisms, RFLPs) will permit much more detailed characterizations of pathogen diversity and pave the way for a better understanding of how host-plant traits contribute to maintaining genetic diversity with pathogen populations.

Population and Community Responses to Plant Resistance Variation

The widespread presence of genetic variation in plant populations is leading to the conclusion that it may have a pervasive influence on the distribution, abundance, associations, and interactions of phytophages. This is becoming a productive focus of current research, where genetics and ecology are becoming integrated and where new patterns of herbivore distributions, community patterns, and herbivore interactions are being revealed. The chapters in this section address much of what we currently know about these interactions and focus attention on several future directions.

In early studies designed to understand how host plants influence herbivore fitness and population size, emphasis was placed on plant phenotypic variation due to environmental factors such as water, light, and nutrients rather than to genetic variation in resistance. More recent research has corrected this imbalance and raised two major questions. First, relative to environmental factors, how important a role does plant genetic variation play in causing phenotypic variation in host plants? Second, how do genotype and environment interact to produce host-plant phenotypic effects on herbivore populations? In chapter 9 Karban reviews this recent literature and concludes that plant genotype is at least as important to herbivore populations as other environmental factors. Furthermore, although environmental and genetic effects can interact to influence herbivore populations, the effects of these interactions are usually small relative to the main effect of host-plant genotype.

Grazers differ from other types of herbivores in many ways that can influence their evolutionary and ecological interactions with plants. In chapter 10 Pollard examines how intraspecific variation in plant resistance may influence grazer-plant interactions. Because many grazers are large and consume numerous host plants in their lifetimes, individuals may not be adversely affected by occasional resistant plants. Thus, plant variation may not affect grazers to the extent that it can influence more sedentary herbivores. However, unlike small sedentary herbivores, which some authors have claimed have little impact on host-plant fitness, grazers could be responsible for strong selection pressures on plants. Pollard shows that phenotypic (and genetic) variation in plant chemical and physical traits associated with resistance to grazers appears to be widespread and does affect grazing pressure. Moreover, these traits appear to be correlated with plant fitness, lending credence to the idea that grazers may strongly influence plant evolution. Pollard goes on to critique the continuing contentious debate as to whether coevolution between grazers and host plants has led to a mutualistic interaction. He suggests that the potential for coevolutionary relationships between grazers and their food plants exists, but that evidence that plants benefit from grazing and should evolve to become more palatable is lacking.

The impact on phytophages of genetic variation among plants is not limited to population-level effects. Plant variation can also affect community-level interactions among plant pathogens, herbivores, and their natural enemies. Fritz (chapter 11) examines the influence of phenotypic variation in plant resistance on structuring herbivore communities on plants and in creating variable competitive interactions among herbivores among conspecific host plants. Variation among plants in resistance to a diverse group of herbivores can influence the covariances of herbivore abundances across host plants. Ives (1988) has noted that these covariances can affect interspecific competitive interactions and thus determine the outcome of competition (e.g., coexistence versus exclusion). The effects of variation in plant quality on the dynamics of herbivore competition are only recently receiving attention (Fritz et al. 1986; Karban 1986), and more knowledge of the role of plant genetic variation in affecting this fine scale of herbivore interactions will complement our growing appreciation of subtle and indirect competitive interactions among herbivores (Faeth 1987, 1988).

In addition to influencing competitive interactions among herbivores, genetic variation in plant resistance traits can also affect the interactions between herbivores and their natural enemies. Hare (chapter 12) reviews studies that examine how plant genetic variation influences interactions across trophic levels. He provides many examples of parasitism or predation on herbivores varying among crop varieties, but finds that few studies have demonstrated variable enemy impact on herbivores among plant genotypes in natural systems. Hare describes several ways that plant resistance can influence the impact of natural enemies on herbivores. Resistance may enhance enemy impact by facilitating the search for prey or host, or resistance may impede parasitism or predation and therefore oppose enemy impact. Finally, predator or parasite searching may be unaffected by plant resistance traits. An important outstanding question is how the direct effects of plant resistance on herbivores interact with indirect effects mediated through natural enemies to influence herbivore population dynamics and community structure.

Evolutionary Responses of Plants

Coevolutionary theory has played a preeminent role in directing both the theoretical and empirical study of plant-phytophage ecology. As a result, empiricists have directed much of their work toward answering the basic questions enumerated below: (1) Is there heritable variation for resistance to phytophages in natural plant populations? (2) Can herbivores and pathogens affect plant fitness, and therefore influence the evolution of resistance traits? (3) Does resistance to one phytophage (species or phenotype) hamper resistance to others, or can resistance traits operate generally against many types of phytophages? (4) Is a model of multispecies coevolution more realistic than a pairwise model?

Marquis (chapter 13) and Alexander (chapter 14) address the first two questions for herbivores and pathogens, respectively. Marquis reviews the effects of herbivores on plant fitness and asks whether herbivory is selective. Few studies to date have demonstrated a connection between differential herbivory and plant fitness, and more research is needed in this area. Furthermore, Marquis points out that we know very little about how herbivory acts at certain stages in a plant's life history (such as the seedling stage) or about how the genetic correlations of fitness with other plant characters (such as life-history traits) may in turn be correlated with herbivory.

It is often taken as self-evident that plants in natural populations vary genetically in resistance to herbivores, but as Alexander notes in chapter 14, we have little empirical evidence with which to support this assumption. Despite the extensive literature on pathogens in managed forest and agricultural systems, studies of resistance to pathogens in natural plant populations are rare, leaving us unable to answer even the most basic questions: Do plants vary in disease resistance? Is resistance variation heritable? Does disease resistance enhance fitness? Alexander reviews the limited evidence available and then provides an important perspective on empirical methods for answering each of these major questions. The approach she promotes will also be useful for those studying insect herbivores and for those wishing to consider pathogens and herbivores simultaneously.

Parker (chapter 15) addresses question (3) above by comparing equilibrium and nonequilibrium models of disease resistance polymorphism in plants. He uses these as a context in which to examine how selection on other plant traits that are in disequilibrium with resistance alleles might be responsible in part for maintaining resistance variation. Parker also reviews models that explore the potential for disease to favor the evolution of sexual reproduction in plants. This important theory has sparked extensive theoretical studies of host-parasite and host-pathogen interactions, but has rarely been tested empirically. In an attempt to evaluate the empirical validity of this theory, Parker explores the effects of various plant reproductive patterns on susceptibility to and transmission of disease.

Coevolutionary models of plant-phytophage interactions indicate that costs of plant resistance traits will determine their rates of evolution, their evolutionary equilibrium values, and their equilibrium levels of genetic variation. Consequently, determining the cost of resistance traits has been an important empirical endeavor. Three chapters (Parker, chapter 15; Zangerl and Bazzaz, chapter 16; and Simms, chapter 17) describe in more detail the various reasons why it is important to know whether resistance traits are costly, and they evaluate the empirical evidence for costs of resistance.

Resource-allocation theory greatly influenced the direction of early research on the evolution of plant resistance to insect herbivores, and continues to play an important role in the field. Since plant defenses are assumed to be costly, this theory predicts that plants that allocate resources optimally to defend valuable tissues at appropriate times in development will be most fit. Zangerl and Bazzaz evaluate this theory and then examine the empirical data on allocation patterns of defenses and the evidence for direct and indirect costs of resistance. Finally, they present a defense allocation model and relate it to the resource availability model (Coley et al. 1985).

Simms (chapter 17) briefly reviews the importance of resistance costs to coevolutionary theory and optimal allocation theory. She then describes a number of mechanisms by which the evolution of maximal levels of resistance might be constrained, which provides a broader definition of costs than that discussed by Zangerl and Bazzaz. The variety of mechanisms by which resistance traits might involve costs dictates that more than one method be used for detecting costs. Simms describes these methods and then reviews the empirical evidence for costs of resistance gleaned from both the agricultural and ecological literature. This review reveals that in at least some cases resistance traits do not appear to involve costs, a result that leads Simms to consider how costs themselves might evolve in response to selection.

Antonovics (chapter 18) wraps up the book by suggesting that combining demographic and ecological genetic analyses in the study of two or more interacting populations leads us toward a new area of biology that he christens "community genetics." This development may be a necessary outgrowth of interest in coevolution and in the evolution of species interactions (Thompson 1988d). Antonovics illustrates this approach with evidence from a study of the interactions between anther smut disease and its host *Silene alba*. He calls for a holistic approach to studying how interactions among species influence and are influenced by trophic diversity and genetic diversity, and discusses the pitfalls and potential advantages of the integrative field he calls community genetics.

Conclusion

Phenotypic variation in plant resistance has numerous implications for the ecology and evolution of plant-phytophage interactions. Resistance variation may influence the evolution of many phytophage characteristics, including the ability to survive and reproduce on particular host genotypes, and in the case of insect herbivores, the ability to select those hosts that will maximize fitness. Population dynamics of herbivores and pathogens can also be influenced by intrapopulational variation in resistance among plants. Variation in resistance should reduce the likelihood of insect outbreaks (Barbosa and Schultz 1987) and slow the development and spread of disease epidemics (Burdon 1987b). Differential resistance of individual plants causes variation among plants in their herbivorous insect communities (Fritz and Price 1988) and probably also creates microvariation in pathogen communities. Moreover, the effects of variation in plant resistance extend up the trophic structure of a community to affect predators, parasitoids, and parasites of phytophages (Price 1981; Price et al. 1980; Boethel and Eikenbary 1986). Finally, when variation in plant resistance is heritable and affects plant fitness, then natural selection, which can be imposed by herbivores and pathogens, will produce evolutionary change in these traits.

As emphasized by Antonovics, a more complete understanding of the multiplicity and complexity of ecological and evolutionary interactions among plants, phytophages, and organisms at higher trophic levels will require the integration of genetics with other more traditional ecological methods. Endler (1986) has argued that a poor understanding of the modes of inheritance of ecologically important traits has kept ecology and genetics as distinct fields. The research reviewed by the authors in this book, however, illustrates the accelerating integration of these two once-disparate fields. Recent studies of plant-phytophage interactions have been using increasingly sophisticated genetic methodologies, and it is clear that this trend will continue. While on the one hand a reductionist course, this path also leads to a more synthetic understanding of ecological and evolutionary interactions.

Some important gaps remain in the synthesis, however. With a few notable exceptions (e.g., Kearsley and Whitham 1989), ecologists and evolutionary biologists studying plant-phytophage interactions have generally ignored the potentially important contribution of plant ontogeny to resistance variation. Another realm that remains largely unexplored is the impact that other selective factors in the environment might have on the evolution of plant resistance traits. Plant physiological ecologists and others have discussed these factors (Olney 1968; Whittaker and Feeny 1971; Lee and Lowry 1980; Lieberman and Lieberman 1984; Rhoades 1977), but little has been done to understand how selection from these factors might interact with selection by phytophages in molding plant resistance characters. Other gaps exist, but we would be remiss if we failed to note one other important lacuna; the paucity of research on the interactions among herbivores and pathogens on host plants in natural populations. Although work by Clay (1987a) and colleagues has begun to fill this gap for at least one ecosystem, more work is needed.

Increased interest in and understanding of the genetic mechanisms underlying ecological interactions will enhance our ability to address these gaps in our knowledge and will facilitate our exploration of better-studied aspects of plant-phytophage interactions. Furthermore, new techniques from molecular genetics, immunogenetics, and systematics are opening more doors to understanding the ecological and evolutionary interactions among organisms in natural communities.

PART ONE

Analysis and Inheritance of Resistance Variation

2

Resistance Variation in Natural and Managed Systems

George G. Kennedy and James D. Barbour

There is abundant literature documenting variation in arthropod and pathogen resistance in plants. Most of this literature has focused on agriculturally important plants, and has emphasized the exploitation of genetically based resistance variation, through plant breeding, for the development of pest-resistant crop cultivars (Painter 1951; Harris 1979; Maxwell and Jennings 1980; Fraser 1985; Smith 1989). In contrast, research on plants in natural systems has emphasized documentation of resistance variation, with relatively few studies partitioning that variation into its genetic and environmental components (Denno and McClure 1983). It is thus impossible to interpret the evolutionary significance of much of the resistance variation reported from natural systems.

The literature on genetic mechanisms of resistance to plant pathogens is extensive and has been reviewed by several authors (see van der Plank 1984; Leonard and Fry 1986; Wolfe and Caten 1987). Our focus is on resistance to anthropods. Our objective is to provide a perspective on the nature and extent of genetically based resistance variation in plants of natural and managed systems. We recognize that there exists a continuum in the degree to which systems are managed that ranges from pristine natural systems to intensively managed agricultural systems. Because of limitations of space and available information, we have restricted our treatment of managed systems to agricultural systems.

A knowledge of the genetic mechanisms of resistance in plants of natural systems should provide a basis for understanding the role of genetically based defenses in the coevolution of plants and herbivores and the community and population processes that operate in those systems. This, in turn, may provide a basis for designing more efficient and durable agricultural systems. Such knowledge should also enhance our ability to estimate the extent of genetically based resistance variation in natural systems and to devise and implement strategies to conserve that variation.

Resistance genes operating in agricultural systems were originally derived

13

from wild or ancestral forms of crop plants or their relatives. Therefore, the extensive data base pertaining to genetic mechanisms of resistance in agriculturally important plants should prove of value in helping us to understand the genetic mechanisms of resistance operating in natural systems. However, before making inferences regarding the nature of genetic mechanisms of resistance operating in one system based on information derived from the other, it is important to be aware of differences between agricultural and natural systems that might bias the data.

The Definition of Resistance is Context Dependent

Before attempting to compare genetic mechanisms of resistance in natural and managed systems, it is important to know how resistance is perceived by scientists working in each of these systems. The focus of our discussion is on host resistance as distinct from nonhost immunity. Nonhost immunity refers to the array of qualities possessed by a plant species that place it outside the host range of potential herbivore species with which it is spatially and temporally sympatric. Nonhost immunity could result from either the presence of genes conditioning plant qualities that actively interfere with an herbivore's ability to recognize or utilize the nonhost, or the absence of genes conditioning qualities necessary for an herbivore to recognize or utilize the plant. In the former case, genes conferring nonhost immunity might be available for transfer to other plant species. In the latter, nonhost immunity cannot be transferred genetically to other plants.

In contrast, host resistance refers to those genetically controlled qualities possessed by some individuals, clones, populations, races, or varieties of a plant species that result in their being less damaged by a particular herbivore species than other individuals, clones, populations, races, or varieties of the same plant species within the host range of the herbivore. The distinction is important in an evolutionary context. Herbivores that do not feed upon or otherwise utilize any individuals of a given plant species cannot impose any direct selection pressure on that species. However, any mutation in the nonhost conferring host status for a particular herbivore could lead to selection against the mutation, if it conferred no other beneficial qualities and the herbivores were a significant selective force. Alternatively, genetic changes in an herbivore allowing it to exploit a nonhost could lead to selection of the new host for resistance.

Host resistance may result from the expression of genes conditioning the presence of chemical or physical attributes that interfere with the ability of an herbivore to utilize a plant compared to a plant not expressing those attributes. Resistance may also result from the absence of qualities essential for full utilization of a host plant by an herbivore. These relationships are illustrated by the effects of cucurbitacins on cucumber beetles (*Diabrotica* spp.) and two-spotted spider mites (*Tetranychus urticae* Koch). Cucurbitacins are tetracyclic triterpenoid compounds that act as powerful feeding stimulants for cucumber beetles (see Chambliss and Jones 1966). Cucurbitacins also confer antibiosis resistance to spider mites. The absence of cucurbitacins in cucumber (*Cucumis sativus* L.) (Cucurbitaceae) is controlled by the recessive bi gene. Cucumber plants lacking cucurbitacins (bibi) are highly susceptible to mites and resistant to beetles, whereas those possessing cucurbitacins (Bibi or BiBi) are resistant to mites and susceptible to beetles (DaCosta and Jones 1971; but see dePonti and Garretsen 1980; dePonti et al. 1983; and discussion below).

Host Resistance is Relative

Resistance is the manifestation of an interaction between plant, herbivore, and environment. Consequently, expression of the resistant phenotype depends not only on the genotypes of both the plant and the herbivore, but also on the environment in which the plant and herbivore interact (Gallun and Khush 1980; Tingey and Singh 1980; Gould 1983; Diehl and Bush 1984; Futuyma and Peterson 1985; Smith 1989). Even in environments conducive to expression of the resistant phenotype and in the presence of herbivore populations affected by the resistance, we can recognize resistant phenotypes only by comparison of herbivore performance (developmental time, feeding, survival, or reproduction) on or plant damage to the resistant phenotype with that on or to a more susceptible phenotype.

Differences between Managed and Natural Systems Result in Different Operational Definitions of Resistance

Crop protection capabilities in primitive agricultural systems were minimal and dependability of yield was paramount. Under these conditions, although humans exerted selection for particular characteristics that suited their needs, natural selection was a powerful determinant of the characteristics of the plant varieties or populations that were grown as crops (Harlan 1979). In modern agriculture, production systems and crop habitats are defined, established, and maintained by humans through the use of technology and the expenditure of energy. Modern crop cultivars and hybrids are selected for attributes that conform to the requirements of these production systems. Modern crop breeding programs emphasize agronomic characters that enhance crop value; these include crop uniformity, harvestability, and yield, as well as storage life and marketable qualities of those plant parts that will be used for food, fiber, or animal feed. In many instances these agronomic characters have little or no relationship to reproductive fitness. In some crops, such as seedless grapes, oranges, cucumbers, and watermelons, the two are clearly incompatible.

Selection on crop species results in greatly reduced phenotypic and genetic diversity and a high degree of adaptation to the intensely managed environment of agricultural systems. Selection for resistance or tolerance to abiotic and biotic stresses is focused on specific stresses that are limiting factors in production (e.g., drought and salt tolerance, resistance to pathogens and arthropods). Selection for agronomic characters in the development of modern crop cultivars frequently occurs in plots protected from herbivory through the use of insecticides. This "insecticide umbrella" can obscure differences in resistance that exist between plant genotypes. As a result, this resistance may be lost during subsequent selection for other characters (dePonti 1983).

By contrast, plants growing in natural systems are subjected to selection by a broad array of abiotic and biotic forces in an environment unbuffered (or minimally buffered) by human intervention. Unlike plants of agricultural systems for which selection is specific, uniform, and consistent from generation to generation, the types and intensity of selective forces operating on plants in natural systems are highly diverse both spatially and temporally.

Strategies for Resistance in Managed and Natural Systems

There are four basic strategies by which plants can reduce damage by herbivores: (1) association with other species; (2) escape in space and/or time; (3) accommodation of the herbivore (tolerance); or (4) confrontation (use of physical or chemical defenses, including both antibiosis and nonpreference resistance) (Painter 1951; Atsatt and O'Dowd 1976; Feeny 1976; Rhoades and Cates 1976; Harris 1979).

In natural systems all of these strategies can contribute to the total defense of any particular plant. However, particular strategies or combinations of strategies effective in natural systems may be precluded in agricultural systems because of production and marketing requirements. For example, in natural systems, the options for escape from herbivory in time may be limited only by the onset of unfavorable abiotic conditions or the occurrence of a second biotic agent. In agricultural systems, market requirements impose additional major restrictions on when a crop can be produced profitably in a particular area. Similarly, in natural systems, options for avoidance in space may be limited only by the plant's mating system and the availability of suitable habitat, whereas in agricultural systems production efficiency requires that large numbers of uniformly mature plants be produced in close proximity. Typically, avoidance in space in agricultural systems takes the form of crop rotation to escape pest species that have long generation times and are not highly mobile (National Research Council 1969). Avoidance in agricultural systems also includes the production of crop species in areas in which they are not indigenous.

Accommodation, or tolerance, is a viable defense strategy for plants in both managed and natural systems. Some forms of tolerance, however, are not compatible with the requirements of production agriculture. Ability to produce a second seed crop following destruction of the first by an herbivore, for example, is of little or no value in a crop that is destructively harvested when the seeds are immature (e.g., green beans) or that must meet a temporally restricted market. Similar limitations apply in natural systems. The ability to produce a second crop may be useful for morning glories in North Carolina but not for morning glories in Massachusetts.

Confrontation is also a viable defense strategy in both natural and managed systems (Ehrlich and Raven 1964; Whittaker and Feeny 1971; Levin 1976; Swain 1977; Rosenthal and Janzen 1979; Norris and Kogan 1980; Harborne 1972, 1988; Green and Hedin 1986). However, in agricultural systems, biochemical and physical plant characters providing confrontational resistance must be compatible with the production requirements of the crop as well as its intended use. For example, high levels of glycoalkaloids in the foilage of *Solanum* spp. (Solanaceae) are associated with resistance to the Colorado potato beetle (*Leptinotarsa decemlineata* Say) (Coleoptera: Chrysomelidae). In potato (*S. tuberosum* L.), glycoalkaloid levels in the foilage are often correlated with those in the tubers. Since these glycoalkaloids are toxic to humans, selection for insect resistance attributable to high levels of foliar glycoalkaloids is feasible only to the extent that it does not result in unacceptably high glycoalkaloid levels in the tubers (Sinden et al. 1984; Tingey 1984).

There is circumstantial evidence that selection of crop species for improved agricultural value has been associated with increased herbivore susceptibility and reduced levels of particular plant secondary chemicals. Genetic drift in populations maintained by plant breeders as well as negative genetic correlations between desirable agronomic qualities and plant characters conferring resistance are undoubtedly both involved (see Way 1988).

An example of negative genetic correlations between agricultural value and plant chemicals conferring resistance is seen in potato. Efforts to transfer high levels of foliar glycoalkaloids and associated resistance to Colorado potato beetle to *S. tuberosum* by crossing it with the resistant, wild species *S. chacoense* and *S. demissum* showed promise in the early generations of selection. However, continued selection for improved agronomic qualities resulted in a decrease in foliar glycoalkaloids and resistance (Tingey 1984). Whether this is a result of the cost of alkaloid production or of linkage disequilibrium among the traits involved is an important but unanswered question. Conversely for lupines (*Lupinus albus* and *L. mutabilis*), selection for plants devoid of quinolizidine alkaloids for use as forage was successful but resulted in plants highly susceptible to a variety of herbivores and pathogens (Wink 1988).

Negative genetic correlations between plant characters of defensive value and fitness can also limit the types and levels of resistance in plants of natural systems. Berenbaum et al. (1986), for example, found a significant negative genetic correlation between two different furanocoumarin variables conferring resistance in parsnip (*Pastinaca sativa*) (Umbelliferae) to the parsnip webworm (*Depressaria pastinacella* Duponchel) (Lepidoptera: Oecophoridae) such that selection for increased levels of one furanocoumarin variable would result in decreased levels of the other. Negative genetic correlations between several furanocoumarin variables and seed production were also found (see also Coley 1986; but see Simms and Rausher 1987, and Simms, this volume, for counterexamples).

Resistance in Agricultural Systems

Because the objective of agriculture is the production of plant (or animal) products for human use, resistance has been defined for agricultural plants in terms of its consequences for plant value. Agricultural workers generally accept Painter's (1951) definition of resistance as "the relative amount of heritable qualities possessed by the plant that influences the ultimate degree of damage done by the insect. In practical agriculture, resistance represents the ability of a certain variety to produce a larger crop of good quality than do ordinary varieties at the same level of infestation."

Development of arthropod-resistant crop cultivars is a time-consuming and expensive process, and emphasis has focused on achieving high levels of resistance to arthropod pests of major importance. Although low and moderate levels of resistance can provide significant crop protection benefits in some systems, concerted efforts to develop crop cultivars expressing reduced levels of resistance have not been undertaken unless the pest has been a major limiting factor in production or the resistance promises to provide nearly complete and dependable pest control (see Kennedy, Gould et al. 1987 for a discussion of important considerations in the development and deployment of arthropod-resistant germplasm for crop protection).

Emphasis on high levels of resistance is further dictated by the difficulties associated with reliably selecting plants for low to moderate levels of resistance. Unless selection is conducted under carefully controlled conditions, methodological difficulties and variation inherent in working with insect populations typically result in the ratio of environmental to genetic variation being greater for low than for high levels of resistance. The degree of control necessary to maintain environmental variation at levels permitting efficient selection for low levels of resistance is frequently impractical in large breeding programs.

Resistance in Natural Systems

In contrast to agricultural systems, options for resistance in natural systems are not restricted by anthropocentric considerations. Resistance in plants of natural systems can be defined strictly in terms of the selective advantage it provides relative to less resistant phenotypes. Thus, in plants of natural systems, resistance represents those heritable qualities that confer enhanced fitness relative to other members of the population or species growing sympatrically with the target herbivore. To the extent that they represent heritable qualities, the defensive strategies of escape and association are encompassed in this definition, although they were categorized separately as pseudoresistance by Painter (1951).

Herbivore resistance in plants of natural systems is seldom measured in terms of its direct effects on fitness. Instead, parameters thought to be correlated with fitness, but more amenable to measurement, are used as indices of resistance. These parameters include amount (leaf area or biomass) of plant material lost to herbivory, amount of seed or flower injury, and plant effects on herbivore growth, survival, and reproduction, or population size. Other frequently measured indices of resistance include the quantity and/or quality of putative resistance factors such as various allelochemicals or morphological characters (Harborne 1972; Norris and Kogan 1980; Crawley 1983; Krischik and Denno 1983).

If we are to understand resistance in natural plant populations, the relationship between the parameters measured, presumed resistance, and plant fitness must be unambiguously established (see discussions by Harper 1977; Hartl 1980; Falconer 1981). Currently, limitations in our knowledge of these relationships represent a major constraint on our understanding of the evolution and genetic mechanisms of resistance in plants of natural systems.

Genetic Mechanisms of Resistance in Agricultural Plants

Although there is an abundant literature documenting the genetic control of arthropod resistance and resistance variation in agricultural systems (Painter 1951; Panda 1979; Maxwell and Jennings 1980; Gould 1983; Fery and Kennedy 1987; Smith 1989), the extent of this information is quite limited with respect to the universe of crops and arthropods available for study. Because of the large and diverse nature of this literature, we have made no attempt to provide a comprehensive review. Rather, we have attempted to provide an overview of genetic variation for resistance in crop species followed by a brief discussion of selected examples to illustrate the types of genetic mechanisms that have been documented.

The task of developing resistant crop cultivars is considerably easier when the sources of resistance genes are plant types already adapted to agricultural production. It is much more time-consuming when resistance must be transferred from agriculturally unadapted or wild plant types. For this reason, the quest for potentially useful sources of resistance generally begins with a survey of cultivars presently and previously grown in the area of interest. This is followed by surveys of exotic cultivars, land races (plant varieties cultivated and maintained by local farming), and wild types of the same species. Currently, it is only as a last resort that related species are considered as potential sources of resistance genes. This may change, however, with further improvement of our ability to identify and transfer genes among species.

Variation in reistance to many important arthropod pests species has been repeatedly demonstrated among crop cultivars not intentionally selected for resistance (table 2.1). These cultivars represent distinct genotypes. Hence, much of this variation can be assumed to have a genetic basis, even in those instances where heritability has not been documented. However, the potential contribution of genotype-by-environment interactions should not be overlooked.

In addition to variation among cultivars, significant variation in resistance has been found to exist within cultivars. For example, alfalfa (*Medicago sativa:* Fabaceae) cultivars represent progeny from open pollinated crossings among several (usually four or more) lines selected for uniformity in characters of interest (Busbice et al. 1972). Sources of resistance to the spotted alfalfa aphid (*Therioaphis maculata* Buckton) (Homoptera: Aphididae), pea aphid (*Acyrthosiphon pisum* Harris), and blue alfalfa aphid (*A. kondoi* Shinji), used in the subsequent development of resistant cultivars, were found within commercial alfalfa cultivars (Nielson and Lehman 1980).

In carrot (Apiaceae), another open pollinated crop, Scott (1970) documented significant differences in mortality of *Lygus* spp. (Hemiptera: Lygaeidae) among cultivars and among plants within cultivars. More recent work demonstrated that levels of resistance could be increased by selecting and selfing plants within cultivars (Scott 1977).

Significant variation in resistance is less likely to be observed within culti-

TABLE 2.1

Some examples of crops found to possess resistance to arthropod pests even though they were not artificially selected for resistance

Crop	Pest	References
Peach *Prunus persica* (L)	Peach tree borer *Samminoidea exitosa* Say	Chapin and Schneider 1975
Raspberry *Rubus* spp.	Large Raspberry Aphid *Amphorophora agathonica* Hottes	Daubeny 1973
Sweet Potato *Ipomoea battalas*	Complex of root-feeding soil insects	Cuthbert and Davis 1970
Common bean *Phaseolus vulgaris* L	Potato leafhopper *Empoasca fabae* (Harris)	Chalfant 1965
Cabbage *Brassica oleracea*	Onion thrips *Thrips tabaci* Lindeman	Shelton et al. 1983
Carrot *Daucus carota*	*Lygus hesperus*	Scott 1970, 1977
Tomato *Lycopersicon* *esculentum*	Tomato fruitworm *Helicoverpa zea* Potato aphid *Macrosiphum euphorbiae*	Fery and Cuthbert 1974b Gentile and Stoner 1968
Alfalfa *Medicago sativa*	Spotted alfalfa aphid *Therioaphis maculata* Buckton Blue alfalfa aphid *Acyrthosiphon kondoi*	Nielson and Lehman 1980

vars or hybrids of less diverse crops. However, spontaneous mutation within cultivars has given rise to valuable morphological, physiological, and biochemical characters even in highly inbred crops such as tomato (Rick 1986). Although spontaneous mutations affecting insect resistance undoubtedly arise in the absence of severe insect infestations, these mutations are less readily identified than those affecting highly visible plant characters such as leaf shape, fruit size or color, or plant growth form.

The complexity of the genetic mechanisms of resistance to arthropods varies greatly, as does our level of understanding. The following discussion is presented to illustrate the kinds of information available on the genetic mechanisms of resistance and some of the complexities associated with resistance in different crop species. Although we attempted to provide a balanced treatment, the relative amount of information available on the genetic mechanisms of resistance is greater for some crops than for others. To the extent that this is true, our discussion is biased.

Alfalfa (Medicago sativa)

Pea aphid (*A. pisum*) resistance in alfalfa involves antibiosis and acts to suppress population development; it is conditioned by one or a few dominant genes inherited tetrasomically (involving two extra chromosomes of the same type) (Jones et al. 1950; Ortman et al. 1960; Glover and Stanford 1966). In contrast, spotted alfalfa aphid (*T. maculata*) resistance, which involves both tolerance and antibiosis, is under polygenic control (Harvey et al. 1960; McMurtry and Stanford 1960).

Resistance to potato leafhopper (*Empoasca fabae* [Harris]) (Homoptera: Cicadellidae) involves both tolerance, which inhibits yellowing in response to leafhopper feeding, and antibiosis, which results in lower leafhopper populations; both are under polygenic control. A diallel analysis of resistance in ten selected alfalfa clones indicated narrow-sense heritabilities of 0.31 for tolerance and 0.56 for antibiosis (Soper et al. 1984). In another diallel analysis, Elden et al. (1986) reported a low but significant negative correlation between leaf pubescence and antibiosis resistance with additive and nonadditive gene action for leaf and stem pubescence.

Beans (Phaseolus spp.)

An evaluation of 185 lima bean (*P. lunatus* L.) accessions for resistance to the leafhopper *E. kraemeri* identified four that were resistant on the basis of damage scores (Lyman and Cardona 1982). The resistance was associated with high densities of foliar trichomes (see also Pillemer and Tingey 1976). Lyman and Cardona (1982) found that trichome densities of resistant by susceptible F_1 hybrids did not differ from those of the low-density, susceptible parent. F_2 progeny in their study showed continuous variation in trichome density with mean densities intermediate between parental means but lower than the midparent mean;

no high-density F_2 progeny were observed. These results may be explained by either (or both) incomplete penetrance of the high-density trait in a genetic background different from the high-density parent or by the quantitative inheritance and/or environmental lability of trichome density with dominance of low density.

Corn (Zea mays)

In whorl-stage corn, larvae of the European corn borer (*Ostrinia nubilalis* Hubner) (Leptidoptera: Pyralidae) feed on leaf tissue. Resistance to *O. nubilalis* involves antibiosis attributable to elevated levels of DIMBOA (2,4-dihydroxy-7-methoxy-2H-1, 4-benzoxazin-3-[4H]-one) in the leaf tissue. Inheritance of DIMBOA-mediated resistance involves additive gene action (Tseng et al. 1984). A diallel analysis of 11 inbred lines revealed highly significant correlations between DIMBOA concentration in plant tissues and leaf feeding, for inbreds ($r = -0.89$) and single crosses ($r = -0.74$). Genetic effects due to general and specific combining ability accounted for 84% of the variation in resistance ratings for the 91% of the variation in DIMBOA concentration (Klun et al. 1970).

Using reciprocal translocations to map genes conferring resistance to leaf feeding, Scott et al. (1966) found the resistant inbred lines C131A and B49 possessed genes contributing to resistance on the short arms of chromosomes 1, 2, and 4, and on the long arms of 4 and 6. B49 also possessed an additional gene(s) on the long arm of chromosome 8. This represents an estimate of the minimum number of genes involved, since in this translocation analysis linked genes would appear as one, no recessive genes could be detected, and only those genes with sufficient potency to be measured in the heterozygous condition could be detected (Scott et al. 1966).

The DIMBOA content of foliage declines as corn matures and leaf feeding resistance is lost (Klun and Robinson 1969). European corn borer larvae feed primarily in the leaf sheath and collar region of reproductive-stage corn. Resistance in reproductive-stage corn is distinct from that in whorl-stage corn and is referred to as sheath/collar resistance. Scott et al. (1967) studied sheath/collar resistance in 45 diallel crosses and among 10 inbreds. They concluded that resistance involved mostly additive gene effects. However, there was some evidence for epistatic effects, since crosses involving one resistant inbred (Oh43) with other resistant and susceptible inbreds gave larger differences in resistance than did crosses involving the other resistant inbreds. Jennings et al. (1974b) reported sheath/collar resistance of inbred B52 to be partially dominant when crossed with inbred Oh13. A translocation analysis to map genes for sheath/collar resistance by Onukogu et al. (1978) indicated that inbred B52 has genes for resistance on the long arms of chromosomes 1, 2, 4, and 8 and on the short arms of 1, 3, and 5. They noted that three of the seven chromosome arms carrying genes for sheath/collar resistance in inbred B52 also contained genes for leaf feeding resistance in inbreds B49 and CI31A. For the most part, genes conditioning leaf feeding resistance and sheath/collar resistance are different, although some genes appear to contribute to both kinds of resistance (Jennings et al. 1974a).

Cowpea (Vigna unguiculata [L])

Cowpea curculio adults, *Chalcodermis aeneus* Boheman (Coleoptera: Cucurlionidae), feed on cowpea pods and oviposit on the developing seeds within the pods. Three genetically distinct types of resistance to *C. aeneus* are known: antixenosis, which inhibits adult feeding on the pods; a pod factor, which inhibits penetration of the pod wall and oviposition; and antibiosis, which prolongs development and increases mortality among larvae feeding on the seeds (Cuthbert et al. 1974). Because the frequency of damaged seed is not reduced by the antibiosis, emphasis in developing resistant cultivars for crop protection has focused on the antixenosis and pod factors. Despite low heritability estimates (broad-sense ≤ 0.191), when cowpea curculio populations were low, selection for the antixenosis resistance based on low levels of adult feeding injury resulted in an increase in the frequency of resistant plants (Fery and Cuthbert 1978). The pod factor was shown to be simply inherited (few genes) with additive gene action. The narrow- and broad-sense heritability estimates for the pod factor were 0.475 and 0.490, respectively (Fery and Cuthbert 1975).

Muskmelon (Cucumis melo L.)

Resistance in *C. melo* (Cucurbitaceae) to the melon aphid, *Aphis gossypii* Glover, involves both tolerance and antibiosis. The tolerance consists of two genetically distinct components. One component involves the absence of leaf curling in response to aphid feeding and is controlled by a single dominant gene (Bohn et al. 1973). The other involves reduced stunting in response to aphid infestation and is a more complex character. F_1 progeny of crosses between tolerant (reduced stunting) and susceptible parents showed levels of tolerance similar to the tolerant parent, whereas F_2 and F_1 backcross progeny manifested frequency distributions with single broad peaks. The authors concluded that reduced stunting was dominant but quantitatively inherited.

Antibiosis resistance in *C. melo* accessions is inherited independently of tolerance. It operates by interfering with the ability of *A. gossypii* to recognize its normal feeding site, the phloem sieve elements (Kennedy et al. 1978). The consequences of this for the aphid are reduced fecundity and greater morality. *A. gossypii* populations on susceptible plants are more than 40 times greater than those on resistant plants (Kennedy and Kishaba 1976). Studies of aphid reproduction on F_1, F_2, and F_1 backcross hybrids from crosses between resistant and susceptible parents indicated that antibiosis was conferred by a single dominant, major gene, the expression of which was modified by additional genes with minor effects (Kishaba et al. 1976).

Sorghum (Sorghum bicolor [L])

There are at least two distinct sources of resistance to the sorghum midge, *Contarinia sorghicola* (Coquillet) (Diptera: Cecidomyiidae), which feeds on developing seeds of sorghum. The genetics of resistance as measured by reduced damage to seed heads (blasting) is complex. Based on segregation patterns in F_2, F_3, and backcross hybrids, the resistance from both sources is inherited recessively and involves more than two gene pairs. In crosses with susceptible lines, parents representing the two resistance sources give different segregation patterns; however, the choice of susceptible parent also influences the inheritance of resistance (Boozaya-Angoon et al. 1984). When resistant accession PI 383856 was used as a male parent in crosses with susceptible sorghum, the resistance was recovered in the F_2 and subsequent generations; but when used as a female parent the resistance was either greatly reduced or not recovered at all in the F_2 and subsequent generations. In contrast, when plants of resistant accession SGIRL-MR-L were used as resistant parents, differences in resistance among reciprocal crosses were not observed (Widstrom et al. 1984). The exact basis for the differences is now known; nor is it known if the resistance sources share any resistance genes in common.

Sunflower (Helianthus annuus L.)

In sunflower (Asteraceae), the presence of a phytomelanin layer in the seed restricts penetration of the seed by young larvae of the sunflower moth *Homoeosoma electellum* (Hulst) (Lepidoptera: Pyralidae). The presence of the phytomelanin layer is controlled by a single dominant gene and explains most of the resistance to the sunflower moth. However, additional factors apparently contribute to the resistance, since some F_2 plants lacking a phytomelanin layer possess a low level of resistance (Johnson and Beard 1977).

Tomato (Lycopersicon esculentum Mill)

Arthropod resistance in *Lycopersicon* spp. (Solanaceae) is complex, involving a number of plant characters. Many accessions of the wild tomato species *L. hirsutum* f. *glabratum* are highly resistant to a number of arthropod species that are pests of the cultivated tomato *L. esculentum* (Kennedy et al. 1985; Kennedy 1986; Kennedy, Nienhuis et al. 1987). Since the two species cross readily when the wild species is the pollen source, *L. hirsutum* f. *glabratum* provides a valuable source of resistant germplasm for use in tomato breeding programs. Resistance in accessions PI 134417 of *L. hirsutum* f. *glabratum* to the tobacco hornworm, *Manduca sexta* (L.) (Lepidoptera: Sphyngidae), and *L. decemlineata* causes high levels of mortality among neonates. The resistance results from high levels of the toxic methyl ketone, 2-tridecanone, in the tips of type VI (sensu Luckwill 1943) glandular trichomes that abound on the foliage (Williams et al. 1981; Kennedy 1986; Fery and Kennedy 1987). Neonates of these species rupture the membranes of the type VI trichome glands, contacting lethal quantities of this toxin.

Analysis of F_1, F_2, and F_1 backcross generations from crosses between PI 134417 and *L. esculentum* or *L. hirsutum* (2-tridecanone absent) indicated that *M. sexta* and *L. decemlineata* resistance and high levels of 2-tridecanone followed the same patterns of segregation and were controlled by several recessive genes. In these studies, the type VI trichome densities on the resistant and susceptible parents differed by only twofold; nonetheless, type VI trichome density accounted for a significant, although small, amount of total variation in *M. sexta* resistance in several segregating populations. Since 2-tridecanone occurs primarily in the tips of type VI trichomes, it is not surprising that trichome density affects the expression of resistance by influencing the availability of 2-tridecanone over the plant surface (Fery and Kennedy 1987; Sorenson et al. 1989).

A restriction-fragment-length polymorphism (RFLP) analysis of loci associated with 2-tridecanone levels in F_2 progeny of a cross between PI 134417 and *L. esculentum* revealed three different linkage groups containing RFLP loci associated with the expression of 2-tridecanone. One of the RFLP loci appeared to be primarily associated with the expression of type VI trichome density. This analysis indicated additive by additive, epistatic interactions among linked quantitative loci important in the expression of 2-tridecanone levels. In addition, it indicated the likely involvement of one or more quantitative loci not identified as linked to the RFLP loci identified in this analysis (Nienhuis et al. 1987).

PI 134417 contains additional genes conferring a second form of resistance to *L. decemlineata* that does not involve 2-tridecanone (Kennedy et al. 1985; Sorenson et al. 1989). Whereas 2-tridecanone is located in the tips of type VI trichomes, kills early instar larvae, and is inherited recessively, this second form of resistance is associated with the leaf lamellae, causes chronic effects leading to an accumulation of mortality throughout larval and pupal development, and is not inherited recessively. A genetic analysis of this lamellar-based resistance has not been completed. However, F_1 hybrids are intermediate in their resistance between the theoretical midparent and the resistant parent (PI 134417) from which the type VI trichome glands, and hence 2-tridecanone, had been eliminated (Sorenson et al. 1989).

In that potato beetles are killed as neonates on foliage possessing high levels of 2-tridecanone, the lamellar-based resistance, which kills larvae later in their development, is only apparent in PI 134417 when the trichomes have been removed. This illustrates one of the difficulties inherent in understanding the genetic basis of resistance in complex systems. At the level of the gene product, genes responsible for the specific characters conferring both high levels of 2-tridecanone–mediated and lamellar-based resistance are being expressed inde-

TABLE 2.2
**Mortality of Colorado potato beetle due to 2-tridecanone and lamellar-based resistances
from *L. hirsutum* f. *glabratum***

Basis of resistance	Mortality during 1st instar	Mortality during pupal stage[a]	Level of defoliation
2-Tridecanone	High	High	None
Lamellar	None	High	Low to Moderate
Both	High	High	None
Neither	None	Low	High

[a]Level of mortality observed when recorded as 100 − (percentage of original cohort surviving to adult emergence).

pendently. However, when viewed as the phenotypic expression of resistance as measured by reduced survival of beetles to the adult stage or by reduced defoliation, expression of lamellar-based resistance is masked by the presence of 2-tridecanone, which kills the larvae before the lamellar factor exerts its effects (table 2.2). Since these two resistances are genetically separate, it is possible (even likely) that a genetic analysis involving crosses among resistant plants, each possessing only one of the resistance types, and susceptible plants could lead to a conclusion that the expression of resistance as measured by reduced defoliation or survival of beetles to adulthood involves an epistatic interaction among genes from the two resistant parents. However, in the same analysis, if the specific plant characters responsible for resistance were measured directly or if age-specific mortality of the beetles were measured, there would be no evidence for epistasis and the two characteristics would be considered to be completely independent. Thus, the way in which the phenotypes are measured may affect conclusions regarding the genetic mechanisms of resistance and can have important consequences for our attempts to understand the evolution of resistance or for the development of plant breeding strategies.

Wheat (Triticum aestivum L.)

Resistance in wheat, *Triticum aestivum* (Gramineae) to several insect pests has been reviewed by Gallun and Khush (1980). The genetic bases of resistance to biotypes of greenbug (*Shizaphis graminum* [Homoptera: Aphididae]) and the Hessian fly (*Mayetiola destructor* [Diptera: Cecidomyiidae]) are of special interest.

Resistance to greenbug biotype C is due to the action of a single dominant gene present on a section of a rye chromosome transferred into wheat. This was the first example of selection in intergenic crossings being used to successfully incorporate resistance against an insect (Sebasta 1980, cited in Gallun and Khush 1980).

Resistance in wheat to the Hessian fly is the best-documented case of a gene-for-gene interaction for resistance in a plant/insect system. The gene-for-gene concept was originally formulated by Flor (1942, 1956; see also Person et al. 1952) for the flax, *Linum usitatissimum* (Linaceae), flax rust, *Melampsora lini,* system. Basically, a gene-for-gene interaction results when for every major gene for resistance in a host (plant) species, there exists in a pathogen/herbivore species a corresponding gene for overcoming resistance (i.e., a virulence gene). When a pathogen or herbivore lacking a virulence allele encounters a host-plant species or variety bearing an allele for resistance at the "matching" locus, the plant is phenotypically resistant and the pathogen/herbivore phenotypically avirulent. When a pathogen/herbivore species or biotype bearing a virulent allele at the locus corresponding to the resistance locus encounters the same plant species or variety, the plant is phenotypically susceptible and the pathogen/herbivore phenotypically virulent. Ten biotypes of Hessian fly have been described based on how various combinations of four Hessian fly genes for the ability to overcome resistance interact with four genes for resistance in wheat. The ability of Hessian fly to overcome resistance conferred by a particular gene for resistance in wheat is controlled by a specific gene for virulence that is recessive to its corresponding allele for avirulence. The details of these interactions have been reviewed by Gallun (1977), Gallun and Khush (1980), and Smith (1989).

Genetic Mechanisms of Resistance in Plants of Natural Systems

The study of genetically based resistance variation has only recently received the serious attention of scientists studying plant-herbivore interactions in natural systems (Gould 1983; Fritz and Price 1988). In his review of the literature on genetically based resistance variation in natural and managed plant populations, Gould (1983) presented three studies documenting genetically based resistance variation in plants from natural systems. Currently, studies examining the causes and effects of variation in resistance for plants of natural systems are becoming an increasingly important part of the ecological literature. Since many of these studies are examined in some detail in the various chapters of this volume, the discussion that follows is intended to illustrate briefly the nature of the evidence for genetic variation in resistance, and the extent to which the genetic mechanisms of resistance are known for natural plant populations.

Variation in resistance of goldenrod (*Solidago* spp.: Asteraceae) clones, grown in common garden or greenhouse environments, to various aphid species has been attributed to genetic differences among the goldenrod clones (Moran 1981; Service 1984b; Maddox and Cappuccino 1986). Maddox and Root (1987) examined resistance in 18 *Solidago altissima* clones to 16 insect species. Resistance among clones was found to vary for 15 of the 16 insect species. Parent-offspring regression and sib-correlation indicated significant heritabilities (h^2) for 10 of these instances of resistance.

In cocklebur, *Xanthium strumarium* L. (Asteraceae), Hare (1980) found genetically based variation in burr size to be a major factor conditioning variation in the intensity of seed predation by the insects *Euaresta aequalis* Loew (Diptera: Tephritidae) and *Phaneta imbridana* (Lepidoptera: Tortricidae). Coley (1986), in both field and greenhouse studies, found that variation in herbivory by *Spodoptera latifascia* (Fernald) (Lepidoptera: Noctuidae) was negatively correlated with tannin content of *Cecropia peltata* (Moraceae). Variation in tannin content of greenhouse-grown *C. peltata* was shown to have a genetic basis. Differences in herbivory on four clones of *Piper arieianum* (Piperaceae) grown at a common site were also attributed to genetic differences among clones (Marquis 1984).

The genetic variation in flowering time and several furanocoumarin variables associated with resistance in wild parsnip, *P. sativa,* to the parsnip webworm *D. pastinacella* (Berenbaum et al. 1986) has been discussed above. Additive genetic variation was a significant component of the total variation in resistance of the annual morning glory, *Ipomoea purpurea* Roth (Convolvulaceae) to the flea beetle, *Chaetocnema confinis* (Crotch) (Coleoptera: Chrysomelidae) (Simms and Rausher 1987). However, the heritability (h^2) of that resistance was low. Since no differences in seed production were detected in the presence or absence of *C. confinis,* the authors concluded that there were no fitness costs associated with the levels of resistance studied.

Other investigations have focused on the effects of genetically based resistance on patterns of insect abundance and community structure. Wainhouse and Howell (1983) suggested that infestation patterns of the beech scale, *Cryptococcus fagisuga* Lind. (Homoptera: Coccidae), were influenced by genetically based variation in the susceptibility of beech, *Fagus sylvatica* (Fagaceae), to *C. fagisuga.* Similarly, Fritz and Price (1988) concluded that genetic variation in resistance was a major factor influencing the density and relative proportion of each of four sawfly (Hymenoptera: Tenthredinidae) species among willow (*Salix lasiolepis:* Salicaceae) clones (see also Kinsman 1982 for a somewhat similar example in *Oenothera biennis* [Onagraceae]).

Taken together, these studies provide strong evidence for the existence of heritable variation in resistance of plants from natural systems. However, information relating to the genetic mechanisms operating in natural systems is virtually absent, limiting our ability to compare directly the genetic mechanisms of resistance operating in natural and managed systems.

Genetic Mechanisms of Putative Resistance Characters

In the absence of additional information on the genetics of resistance in natural systems, we offer the following discussion of the genetics of plant characters known to contribute to arthropod resistance in plants of agricultural importance. To the extent that these and similar traits also contribute to resistance in natural plant populations, a knowledge of their genetics in particular crop species may facilitate the formulation of hypotheses regarding the genetic mechanisms of resistance in natural plant populations.

We have selected for discussion an array of plant characters that have been correlated with resistance in several plant/arthropod systems. The quality of the direct evidence for a causal relationship with resistance is variable, although in most cases the weight of evidence for a role in resistance is substantial.

Biochemical Mechanisms

Biochemical mechanisms of resistance may involve inorganic compounds, such as silica or calcium carbonate incrustations, and primary metabolites (Norris and Kogan 1980). However, the overwhelming bulk of evidence deals with the presence or absence of secondary plant compounds (Denno and McClure 1983; Rosenthal and Janzen 1979).

Alkaloids

Waller and Nowacki (1978) reviewed the literature on alkaloid genetics and found evidence for simple (few genes) inheritance of alkaloids implicated in insect resistance in lupines and tobacco. Alkaloid presence in several lupine species was controlled by a single dominant gene at a single locus, whereas absence was the result of a recessive gene, with many allelic forms, that disrupted alkaloid synthesis. In crosses between two tobacco species, nicotine presence was controlled by a single recessive gene regulating nicotine degradation.

Crosses between wild and cultivated *Lycopersicon* species revealed that variation in tomatine content was controlled by two codominant alleles (Juvick and Stevens 1982a). Tomatine content, however, is subject to considerable variation owing to environment (Sinden et al. 1978; Barbour and Kennedy 1991; Hare 1987).

In potato (*Solanum* spp.), the total glycoalkaloid content appears to be under polygenic control and subject to considerable environmental variation (Sinden et al. 1984). Total glycoalkaloid content is a measure of the combined amounts of several chemically distinct glycoalkaloids present. Thus, it is conceivable that the presence or absence of individual component alkaloids is under simple genetic control. Since the individual sugar moieties of glycoalkaloids are themselves under simple genetic control, the potential exists for the production of new glycoalkaloids in crosses between different potato genotypes (Sinden et al. 1984; Dimock and Tingey 1985).

Terpenoids

Segregation patterns of monoterpenes in intra- and interspecific crosses among three closely related *Hedeoma* spp. (Lamiaceae) indicated that the inheritance of individual monoterpenes was under simple genetic control but provided some evidence for epistasis and modifying genes (Irving and Adams 1973).

Variation in quantity and quality of monoterpenes in some *Mentha* species (Murray 1960a, 1960b; Murray et al. 1980), yerba buena (*Satureja douglasii*) (Lincoln and Langenheim 1981), and some conifers (Forde 1964; Hanover 1966a; Zavarin et al. 1969; Zavarin and Cobb 1970) also appears to be controlled by relatively few genes. However, Hanover's (1966b) study suggested multigenic control of most monoterpenes present in *Pinus monticola*. In many of these studies, environmental effects contributed significantly to both qualitative and quantitative variation in monoterpene content, suggesting the potential for strong genotype-by-environment interaction effects on monoterpene composition (see Lincoln and Langenheim, 1976, 1978, 1979; Krischik and Denno 1983).

Cucurbitacins

Cucurbitacins are a class of tetracyclic triterpenoids implicated in arthropod resistance of cucurbits (Nath and Hall 1965; Chambliss and Jones 1966; Sharma and Hall 1971). The absence of the biologically active cucurbitacin components (cucurbitacins E and B) appears to be controlled by a recessive gene in cucumber (*Cucumis sativus* L.) (Andeweg and DeBruyn 1959), squash (*Cucurbita pepo* L.) (Nath and Hall 1965; Sharma and Hall 1971), and watermelon (*Citrullus vulgares* Schrad.) (Chambliss and Jones 1966). However, it is not known whether the same gene controls cucurbitacin presence in all crops. More recent work in cucumber indicates the involvement of a number of additively acting genes that affect cucurbitacin concentration in plants (dePonti and Garretsen 1980).

Although the inheritance of cucurbitacins has been clearly established, the segregation of *Diabrotica* resistance depends on how it is measured. Chambliss and Jones (1966) found that resistance in F_1, F_2, and backcross populations from crosses between low- and high-cucurbitacin watermelon lines appeared to be under polygenic control if insect counts were used to measure resistance. However, on the basis of feeding damage to cotyledons, resistance appeared to be controlled by the segregation of "a relatively few" genes with a strong dominance component. In a similar study with squash, resistance to *Diabrotica* (as measured by feeding damage) appeared to be under polygenic control, although cucurbitacin level was controlled by segregation of a single gene (Sharma and Hall 1971). The weight of the evidence indicates that although cucurbitacin level, which is under relatively simple genetic control, is a factor in the resistance of cucurbits to *Diabrotica*, other minor factors, inherited polygenetically or subject to large environmental effects, contribute to variation in host suitability.

Morphological Mechanisms

A number of morphological plant characters have been implicated in arthropod resistance in a number of systems. Although plant morphological characters are thought, in general, to be under polygenic control (Wright 1978), there is evidence that at least some of those implicated in arthropod resistance are controlled by relatively few genes.

Plant Trichomes

Plant trichomes are probably the best studied of the morphological traits associated with arthropod resistance in plants of both natural and managed systems. However, studies of the genetics of trichome presence or variation in trichome density as a resistance phenomenon are limited to economically important plant species.

The inheritance of glandular trichomes in solanaceous crop species has been particularly well studied. Gibson (1979) reported clinal variation in secreting activity of type B trichomes in the wild potato species *Solanum berthaultii* and *S. tarijense*. Intra- and interspecific crosses among plants with secreting and nonsecreting trichomes indicated that the presence or absence of secreting activity was controlled by a single dominant gene. However, in crosses between *S. berthaultii* and *S. phureja*, two or more recessive genes appeared to be involved in the inheritance of secreting activity. In a subsequent study involving crosses between *S. berthaultii*, *S. tuberosum*, and hybrids between *S. phureja* and *S. tuberosum*, Mehlenbacker et al. (1983) found that segregation for droplet size on the tips of type B trichomes indicated additive gene action. In addition, in those crosses, density of both type A and B trichomes seemed to be controlled by a small number of genes. Although none of the trichome characters measured were highly correlated with resistance to green peach aphid (*Myzus persicae* [Sulzer]) in this study, other studies have documented a role for these trichomes in resistance to *M. persicae* and shown that the resistance involves complex chemical and environmental interactions (Tingey et al. 1982; Mehlenbacker et al. 1983, 1984; Gregory et al. 1986).

In tobacco (*Nicotiana* spp.: Solanaceae), glandular trichomes have been implicated in resistance to several insect species (see Roberts et al. 1981). Two apparently distinct genetic mechanisms controlling the presence or absence of trichome glands have been identified from different tobacco lines (Burke et al. 1982; Nielsen et al. 1982). The absence of trichome glands in tobacco accession TI 1112 is controlled by partially dominant alleles at three loci with the presence of a recessive allele at any of the loci resulting in the presence of glands. In contrast, the glandless condition of accession TI 1401 is conditioned by recessive alleles at two loci (Johnson et al. 1988).

In tomato, *Lycopersicon* spp., both type IV and type VI glandular trichomes (sensu Luckwill 1943) play a role in arthropod resistance (Snyder and

Carter 1984; Kennedy 1986; Kennedy, Nienhuis et al. 1987). In crosses between *L. esculentum* and *L. pennellii,* the presence or absence of type IV trichomes was found to be controlled by two unlinked genes, while density of type VI trichomes appeared to be inherited quantitatively. The inheritance of type VI trichome density in these crosses appeared more complex and subject to large environmental influences. Heritability estimates for type IV trichome density were much higher than those for type VI density (Lemke and Mutshler 1984).

Pubescence in cotton has been shown to condition resistance to several insect species including leafhoppers, thrips, and boll weevil (Norris and Kogan 1980). Four genes at three loci are involved in control of the glabrous or smooth-leaf character in cotton. The genes Sm1 and Sm2 are associated with different loci but are expressed similarly. In the absence of other smooth leaf alleles, each of these genes, when homozygous, results in a fringe of trichomes on the leaf margins. In contrast, Sm3 and sm3, when homozygous and in the absence of other smooth leaf alleles, confine trichomes to stems, petioles, and leaf margins. Sm3 is dominant and sm3 is recessive to the normally pubescent condition. In a given genetic background, Sm3 results in smoother leaves than sm3 (Lee 1968). Pubescence in cotton above that controlled by the smooth leaf genes is governed by two major genes, H_1 and H_2 (Ramey 1962). H_1 and H_2 are incompletely dominant and a complex of modifier genes exerts substantial effect on the expression of pubescence governed by these genes. Although genes affecting pubescence may contribute significantly to the resistance of wild cottons in natural systems, they are of limited value in domesticated cottons because of their association with undesirable agronomic qualities (Simpson 1947; Ramey 1962; Lee 1971).

Solid Stem Character in Wheat

Stem solidness in wheat (*Triticum aestivum*), important in the resistance of wheat to the stem sawfly, *Cephus cinctus* Norton (Hymenoptera: Cephidae), is controlled by three genes. One allele exerts a major influence in that the dominant allele for susceptibility is epistatic to the other genes when homozygous; the recessive allele for resistance is twice as effective as the other two genes in conditioning resistance (McKenzie 1964).

The Gossypol Gland Character in Cotton

The presence of the polyphenolic gossypol in cotton contributes resistance to the tobacco budworm (Lee 1971; Wilson and Shaver 1973). Gossypol is found in epidermal, pigment glands in the foliage and seeds of cotton. Alleles at three different loci (G1, G2, and G3) are involved in inheritance of gossypol level (McMichael 1960; Roux 1960). However, alleles at only two of these loci are required for resistance. The alleles at G2 and G3 control gland number on the cotyledons, leaves, stems, petioles, and carpel walls. The allele at G1 affects gland number on the stems, petioles, and carpels; its presence is not required for resistance (McMichael 1969). The recessive alleles g2 and g3, when homozygous, produce a glandless phenotype virtually devoid of foliar gossypol (high gossypol levels can occur in the roots of both glanded and glandless cotton [Singh and Weaver 1972]). The dominant alleles G2 and G3 result in a normal glanded phenotype. Additive effects account for more than 90% of the variation in gossypol level, although small but significant epistatic effects occur (Lee 1973).

The examples discussed above illustrate the diversity of genetic mechanisms controlling plant characters contributing to resistance. However, in attempting to derive from these examples insights regarding the genetic mechanisms of resistance in natural systems attributable to similar or related plant characters, it is important to recognize that many, if not most, available studies primarily address genetic control of the presence or absence of the character in question. Relatively few studies exist detailing the genetic mechanisms responsible for variation in the levels of resistance characters. Frequently, where levels have been considered, complex genetic mechanisms and significant environmental effects were observed.

Environmental Influences on the Expression of Resistance

Environmental effects on the suitability of plants as hosts for insects have been well documented and a number of specific environmental factors affecting resistance have been identified, especially for plants of managed systems (Tingey and Singh 1980; Heinricks 1988; Smith 1989). The following discussion is presented to illustrate briefly the types of environmental factors known to influence the expression of resistance.

Biotic Factors

Plant Age

Plant age has been shown to influence the level of resistance to arthropods in a number of crops. For example, two-week-old seedlings of the tomato accession PI 134417 manifest no resistance to the tobacco hornworm and Colorado potato beetle; five-week-old seedlings are highly resistant to both species (Kennedy et al. 1985). Conversely, resistance in sorghum to the aphid *Rhopalosiphum maidis* (Fitch), and the plant-hopper *Peregrinus maidis* (Ashm.) is greatest in young plants and decreases with plant age (Fish 1978).

Plant Injury

Plant damage due to herbivory or mechanical wounding can induce elevated levels of resistance to some herbivores (Ryan 1979; Kogan and Paxton 1983). For example, mechanical injury or feeding by two-spotted spider mite on cotton plants induces elevated levels of resistance to spider mites (Karban and Carey 1984; Karban 1985) and similar effects have been observed in a number of other plant/herbivore systems (see Smith 1989). Induced resistance of this type can alter or mask the expression of constitutively expressed resistance.

Plant Disease

Infection of a plant by pathogen can alter its suitability as a host, in some instances by improving host quality and in others by reducing it (Hammond and Hardy 1988). Aphid resistance in sugar beet was lost when plants were infected with yellows virus (Baker 1960; see Kennedy 1951 for other examples), but infection by cucumber mosaic virus reduced the suitability of three other plant species as hosts for the green peach aphid (*Myzus persicae* Sulzer) (Lowe and Strong 1963; see also Hare and Dodds 1987).

Infection of some plants with endophytic fungi has been shown to enhance the level of herbivore resistance. For example, ryegrass foliage infected with the fungal endophyte *Acremonium loliae* is more resistant than uninfected foliage to several arthropod species, including the weevils *Listronotus bonariensis* (Kuschel) and *Sphenophorus parvulus* Gyllenhal, and the armyworms *Spodoptera frugiperda* (J. E. Smith) and *S. eridania* Cramer (Gaynor and Hunt 1983; Hardy et al. 1985; Ahmad et al. 1986; Ahmad et al. 1987).

Planting Arrangement

Plant resistance is often inferred on the basis of measurements of insect numbers or amount of feeding injury per plant. However, these measurements may be influenced by such factors as planting density and whether or not the planting is arranged so that the subject insects are free to select the most suitable plants.

Planting density has been shown to influence profoundly the amount of plant damage that occurs. Relative differences among cucumber lines in *Diabrotica* beetle feeding injury were observed to decrease as plant density increased; maximum differences among lines were observed with single hill plots (Quisumbing 1975). The reasons for this observation were not investigated but may have involved simply a dilution effect whereby the number of insects per plant is much smaller in the high-density plots. In contrast, in tomato and in cowpea, plant damage by *H. zea* and *C. aeneus* was shown to be positively correlated with plant density (Fery and Cuthbert 1972, 1974a).

In diverse plantings in which insects are able to move freely among plants, plant families, or plant accessions, the insects are able to select those plants most suitable for oviposition or feeding. In such choice situations, differences in insect numbers or feeding damage among plants (or entries) reflect the insects' relative preference for the various entries as well as the inherent resistance of the entries. To the extent that differences among entries reflect relative preference rather than inherent suitability as a host, the differences in insect attack observed in choice situations will disappear when the insects are exposed to each entry in isolation. To illustrate, differences in striped cucumber beetle feeding injury among varieties of both squash and muskmelon were readily detected in field and greenhouse experiments that involved a choice, but not when the element of choice was removed (Wiseman et al. 1961). Similarly, in a study involving spotted cucumber beetle, Overman and MacCarter (1972) found the magnitude of differences in feeding injury among muskmelon culti-vars (genotypes) to vary depending on whether the beetles were allowed to choose among cultivars. In this experiment, however, the same cultivars ap-peared most resistant under choice and nonchoice conditions.

In agricultural systems characterized by large monocultures, the element of choice may be limited for all but the most highly mobile species. However, in other more diverse agricultural systems, the element of choice may be a significant determinant of the patterns of host utilization that exist even among suitable host plants (Stinner 1979; Kennedy and Margolies 1985). In natural systems characterized by high vegetational diversity, patterns of plant utiliza-tion will often be influenced to a great extent by the array of potential alterna-tive suitable hosts available. This is especially so for highly mobile polypha-gous species. However, even monophagous species may be expected to manifest preferences among plants in populations that are variable for particular characteristics as a result of genetic and environmental effects, but which would be equally suitable if encountered in isolation.

Factors Related to the Insect Population

Population Size

The size of the arthropod population can have a profound effect on our ability to recognize resistant plant phenotypes (Harris 1979). At low arthropod popu-lation densities, many susceptible phenotypes may appear resistant because they escape damaging infestations. At very high densities, even highly resistant phenotypes may succumb to prolonged arthropod attack.

Variation in Arthropod Population

Variation in a number of arthropod population attributes such as age struc-ture, sex ratio, and prior host-plant experience can influence patterns of host-plant utilization (Jesiotr 1979, 1980; Bernays and Wege 1987; Smith 1989) and thus contribute to phenotypic variation in the expression of resistance. To the extent that they affect utilization of plant genotypes differentially (e.g., Smith et al. 1976, 1979) arthropod population attributes will result in genotype-by-environment interaction.

Genetic variation in arthropod populations for host adaptation is well doc-umented in both natural and managed systems and can be an important source of genotype-by-environment interactions in the expression of resistance (Gould 1983; Diehl and Bush 1984, Service 1984b; Futuyma and Peterson 1985; Hare and Kennedy 1986; Smith 1989).

Abiotic Factors

The effects of a variety of abiotic factors on plant suitability and the expression of resistance have been reviewed recently (Tingey and Singh 1980; Heinricks 1988; Smith 1989). Light, temperature, moisture, relative humidity, plant nutrient status, and air pollutants have all been shown to alter resistance expression in some plant/arthropod systems. The effects of abiotic factors on host-plant suitability have been investigated more extensively in crop plants than in natural systems and are illustrated by the effects of day length, light intensity, and soil fertility on the expression of resistance of *Lycopersicon hirsutum* f. *glabratum* to *Manduca sexta, Leptinotarsa decemlineata* and *Helicoverpa zea.*

In experiments involving the production of plants under different light regimes, mortality of *Manduca sexta* and *Leptinotarsa decemlineata* on susceptible *Lycopersicon esculentum* was unaffected by light regime, but on *L. hirsutum* f. *glabratum* the level of mortality was significantly lower on plants grown under short photophase. This effect was mediated primarily by reduced density of type VI trichomes but may also have involved slightly reduced levels of 2-tridecanone per trichome tip. In the case of *H. zea*, which is attributable to other factors in addition to 2-tridecanone, resistance in *L. hirsutum* f. *glabratum* was affected by light intensity when the plants were grown under long but not short photophase regimes. Light intensity over the range tested had no effect on suitability of *L. esculentum* as a host (Dimock 1981; Kennedy et al. 1981; Kennedy unpublished data).

High levels of NPK fertilization can completely eliminate the expression of resistance in *L. hirsutum* f. *glabratum* to all three species. In the case of *M. sexta* and *L. decemlineata* resistance, the effects are mediated by both reduced densities of type VI trichomes and reduced concentrations of 2-tridecanone per trichome tip at high fertilization rates. The loss of resistance to *H. zea* is attributable to changes in factors associated with the leaf lamellae (Barbour et al. 1991).

Genotype-by-Environment Interactions

The concept of environmental variation in relation to genetic variation is discussed by Antonovics et al. (1988), who point out the importance of differentiating between the external environment of an organism, the ecological environment, which is the component of the external environment that influences the organisms's contribution to population growth, and the selective environment. The latter is the component of the external environment that influences the differential contribution of genotypes to subsequent generations. These distinctions are important in understanding the expression and evolution of resistance in natural systems and our ability to select and utilize pest resistance in crop plants.

Although the occurrence of genotype-by-environment interactions affecting the expression of resistance is common in crop plants, the need for dependability in expression of the resistant phenotype necessitates restricting the resistances exploited for crop protection to those that show a minimum of variation across the range of environments in which the resistant crops are to be grown, or that can be selected through breeding for minimal sensitivity to environmental variation.

Relatively few studies clearly document the occurrence of genotype-by-environment interactions as they affect resistance in natural systems (but see Karban, this volume). Given the diversity of factors known to alter the expression of resistance to arthropods and the high degree of environmental heterogeneity that exists in most natural systems, there can be little doubt that environmental effects contribute significantly to the variation in resistance observed among plants of natural systems. Given that in natural systems, within- and among-population genetic diversity in plants is generally high, the frequency of genotype-by-environment interactions may also be high. To the extent that this is true, in the absence of carefully designed genetic analyses it becomes difficult to determine how much of the observed variation in host-plant utilization by herbivores in natural systems is attributable to genetic variation, environmental variation, or genotype-by-environment interactions (see Fritz et al. 1987b; Maddox and Root 1987). The relative contributions of each, however, must be determined if we are to understand the role of genetically based resistance in mediating plant/herbivore interactions and its evolution (Fritz 1990a).

Conservation of Resistant Germplasm

Future demands placed on agricultural production by a burgeoning world population combined with the effects of increasing restrictions on pesticide use will almost certainly lead to a greatly increased demand for development and use of pest-resistant crop cultivars as a key element in crop protection. If this demand is to be met, resistant germplasm will have to be identified, genetically characterized, transferred to agriculturally acceptable plant types, and deployed in a manner designed to minimize the selection of resistance-breaking pest races (see Gould 1983; Kennedy, Gould et al. 1987; and Wilhoit, this volume).

Genetic variation for resistance to arthropods is clearly widespread. There are few, if any, crop species for which extensive searches have failed to identify significant levels of resistance against particular pests. Oftentimes, this resistance has been identified in land races, wild and unadapted plant types, and related species, all of which are extremely important as potential sources of resistant germplasm. The continued existence of these sources of genetic diversity is in great jeopardy owing to the effects of habitat destruction (Myers 1980,

1981; Namkoong 1983; Ingram and Williams 1988), and the increasingly wide-spread adoption of modern crop cultivars by farmers in developing countries. The latter results in a loss of locally adapted and highly diverse land races of crop species that have been important sources of crop diversity and pest resistance in the past (Harlan 1979; Harlan and Starks 1980; Brown et al. 1988).

The issue of germplasm conservation involves philosophical and logistical components, both of which are currently being debated. The threats to world genetic resources have been well documented, as has the need to preserve genetic diversity of all types, including genetic variation in resistance (Muhammed et al. 1970; Jain and Mehra 1982; Frankel and Brown 1984; Fehr 1987; Brown et al. 1988). Foremost among the logistical issues are questions regarding the content, extent, and organization of germplasm to be maintained in germplasm collections or to be preserved in situ in habitat preserves. Biologically sound answers to these questions require a knowledge of the nature and extent of genetically based resistance in natural populations and in land races of economically important species and their near relatives. A more thorough understanding of existing patterns of resistance variation is needed; including the relative importance of intra- and interpopulation as well as clinal genetic variation for resistance (Gibson 1979; Hare 1980, 1983; Gould 1983). There is also a need for more information concerning the relative importance of environmental variation and of genotype-by-environment interactions with respect to the expression of resistance in plant species of demonstrated or potential interest. Additionally, it is important that we understand the extent to which existing genetic variation in resistance represents the result of coevolution between host plant and herbivore.

Assumptions as to the evolutionary origins of resistance variation can profoundly influence decisions regarding how and where to look for genetically based variation in resistance. Thus, these assumptions may accidently bias the samples of germplasm selected for preservation in collections. In this regard, Leppik (1970) has argued on theoretical grounds that the search for pest-resistant germplasm should concentrate in areas of sympatry between the plant and the insect, with particular emphasis on centers of origin or diversity of the crop species in question. Implicit in this approach is the assumption of a coevolutionary relationship between plant and pest resulting in an accumulation of resistance genes in the plant population over evolutionary time. Leppik's (1970) focus was on resistance to plant pathogens with an emphasis on major (single) gene resistance of a high level. The numerous examples of agriculturally useful levels of resistance that exist in a number of crop-pest systems indicate that sources of resistance can be found in areas of pest/plant sympatry (see reviews by Harris 1975; Harlan and Starks 1980). It is, however, by no means certain that all such resistances represent the result of coevolutionary interactions.

There are also numerous instances in which high levels of genetically based resistance to arthropod species have been documented from regions in which the plant and the arthropod are not sympatric. Harris (1975) coined the term "allopatric resistance" to describe this phenomenon and argued that because there has been no coevolutionary interaction between the plant and arthropod species, allopatric resistance is more likely to be polygenic (see also Jermy 1976). Allopatric resistance could result from the effects of genes conferring resistance to arthropods present in the native habitat of the plant. DIMBOA (2,4-dihydroxy-7-methoxy-2H-1, 4-benzoxazin-3-[4H]-one), responsible for antibiosis resistance in corn to the European corn borer, is effective against other insect (e.g., the corn leaf aphid, *Rhopalosiphum maidis;* Long et al. 1976) and fungal (e.g., *Helminthosporium turcicum;* Toldine 1984) pests of corn. Allopatric resistance could also result from the pleiotropic effects of genes conditioning characters of selective value, but not associated with resistance, in the plant's native habit. For example, solid stem in wheat could be selected for by abiotic factors. Alternatively, allopatric resistance could result from genes present owing to processes such as drift and immigration, but that are selectively neutral in the plant's native habitat.

In his recent review, Smith (1989) lists examples of resistance to 13 arthropod species in 10 crop species for which there was no known prior association between the resistance source and the resisted arthropod. Of these, at least three involve major genes for resistance with independent effects (wheat/Hessian fly, *Rubus*/aphid, Rice/ green rice leafhopper); one involves three genes with epistasis (wheat/wheat stem sawfly); and at least one involves polygenic, additive genetic variation (corn/European corn borer). The nature of the genetic variation has not been elucidated for the remaining examples.

At present, the evidence is inadequate to permit a rigorous testing of the Leppik and Harris hypotheses, both of which have important implications for the development of germplasm samples for preservation in working collections and repositories.

Concluding Remarks

Genetic variation for resistance has been clearly documented in plants of natural and managed systems, but the genetic mechanisms are more thoroughly documented for plants of agricultural systems. The data for agriculturally important plant species reveal the full spectrum of genetic mechanisms for resistance.

Resistance to a particular insect species can result from the presence of one or more genetically distinct resistance traits (e.g., cowpea resistance to cowpea curculio [Cuthbert et al. 1974]). In agricultural systems, resistance is usually targeted against a particular arthropod pest. However, there exist instances in which resistance to more than one arthropod species is conferred by a single genetic mechanism (e.g., resistance to cotton fleahopper, tarnished plant bug, and pink bollworm conferred by the nectariless character in cotton). There also exist examples in which the resistance of a particular plant line to several arthropod species is attributable to the simultaneous occurrence of several distinct genetic mechanisms, each conditioning resistance to a different arthropod species (e.g., resistance in alfalfa to the spotted alfalfa aphid and pea aphid). Levels of resistance that are so high that populations of the resisted arthropod are completely unable to use the resistant plant as a host are rare (e.g., resistance in raspberry to the aphid *Amphorophora agathonica* Hottes [Kennedy et al. 1973]). In most cases, resultant plants will be utilized to some degree by the resisted arthropod species.

The degree to which the genetic mechanisms for resistance useful in agriculturally important plants are fully representative of the genetic mechanisms for resistance operational in natural systems is not clear. Differences in the attributes of natural and managed systems result in emphasis on those types and levels of resistance that are readily recognized, environmentally stable, compatible with the requirements of the agricultural production system, and capable of providing an economically significant level of crop protection. Thus, emphasis in agricultural systems has generally been on the more potent resistance types. There is a relatively high frequency of monogenic control among the resistance traits known to exist in agricultural plants and their wild relatives. This may reflect the greater potency of such traits and the greater ease of working with them in breeding programs, since highly potent polygenic resistances are also documented (see discussion by Gould 1983).

The significant differences that exist between natural and managed systems clearly bias the types of resistance studied in plants of agricultural importance. Whereas resistance in plants of natural systems most likely results from an accumulation of traits conferring resistance that may involve independent, epistatic, and pleiotropic effects, and thus in toto are polygenic, the resistance selected to use in agriculture often represents only a portion of the total heritable resistance (Harris and Frederickson 1984). One consequence of this may be that the gene-for-gene relationships observed between resistance genes in plants and resistance-breaking genes in pest biotypes (Gallun and Khush 1980; Gould 1983; Smith 1989) may be, in part, an artifact of the presentation of resistance genes in vastly different plant genetic backgrounds and in more simplified, managed systems.

It is essential that we understand the relationship that exists between genetic mechanisms of resistance in natural and managed systems if we are to devise scientifically sound approaches to germplasm conservation and techniques for deploying resistant germplasm to minimize the development of resistance-breaking pest biotypes. This type of knowledge is essential if we are to take advantage of the simplified nature of managed systems to experimentally address important questions regarding the evolution and population genetics of plant/herbivore interactions (Snaydon 1980; Gould 1983; Harris and Rogers 1988). An understanding of these relationships will require collaboration among basically oriented evolutionary biologists and population geneticists and agricultural scientists working at the applied level. The fruits of such collaboration will nourish both areas of science.

Acknowledgments

The authors would like to thank R. Fritz, F. Gould, M. Harris, E. Simms, and J. Walgenbach for their comments on an earlier draft of this manuscript.

3

Uses of Quantitative Genetics for Studying the Evolution of Plant Resistance

Ellen L. Simms and Mark D. Rausher

Because genetic variation is a prerequisite for evolutionary change, understanding the evolution of plant resistance to herbivores and pathogens requires that we understand the origin and dynamics of genetic variation for resistance traits within natural populations. These dynamics depend largely on three major factors: (1) the extent to which a trait is genetically controlled and heritable, (2) the probability with which chance effects can change the distribution of a trait (genetic drift), and (3) the nature of natural selection that may act on a trait. Quantitative genetics is a methodology for elucidating the role of each of these factors in the evolution of resistance.

Overview of Quantitative Genetics and Evolution

Quantitative Characters and Their Variances

Quantitative genetics is the study of inheritance of characters that are distributed continuously (Crow 1986; Falconer 1981; Bulmer 1980). Unlike qualitative traits, for which phenotypes can be placed into discrete categories, quantitative traits differ among individuals by degree. Many plant resistance traits reported in the literature exhibit continuous variation (Jokela 1966; Russo and Tricerri 1976; Shehata et al. 1983; Notteghem 1985; Berenbaum et al. 1986; Maddox and Cappuccino 1986; McCrea and Abrahamson 1987; Simms and Rausher 1987, 1989; Bjarko and Line 1988a, 1988b; Fritz and Price 1988; Nash and Gardner 1988; Geiger and Heun 1989). Consequently, quantitative genetics will often be an appropriate framework for examining the evolution of resistance.

Typically, the inheritance of quantitative traits is complex. A basic premise of quantitative genetics is that the phenotypic expression of continuous characters is influenced by genes of individually small and roughly equal effect, as well as by various environmental factors. Furthermore, quantitative differences are usually, although not necessarily always, influenced by many loci. Despite this complexity, the genes that determine continuous characters are assumed to behave in the same way as the genes of major effect that control discrete characters (Crow 1986; Falconer 1981; Hedrick 1985), making quantitative genetics an extension of Mendelian genetics (Falconer 1981).

In addition to considering the small, additive effects of many genes, quantitative genetics also presumes that the phenotypic expression of a trait is modifiable by the interactions of alleles at each locus (dominance), the interactions of genes at different loci (epistasis), effects of genes on multiple traits (pleiotropy), mutations, and effects of the environment. Because of the complexity of the inheritance of quantitative traits, statistical procedures are used to infer the existence and properties of genes controlling them. In particular, these procedures are used to partition the total amount of variation among individuals for a particular trait, termed phenotypic variance, into various components that reflect causes of variation, which are called causal components. These components are in turn used to infer or predict how the trait will evolve.

The two primary causal components of, or influences on, the phenotype are genotype and environment. These causes are reflected in the standard partitioning of the total phenotypic variance (a measure of the total variation among individuals in a population) into a sum of environmental and genetic variances:

(1) $$V_P = V_G + V_E.$$

Here V_P is the total phenotypic variance, V_G is the genotypic variance component, which is due to variation among genotypes, and V_E is the environmental variance, which is due to the variation among environments.

The ultimate purpose of this partitioning is to derive an estimate of the genetic variance (or, more exactly, of the portion of the genetic variance termed the additive genetic variance), for this quantity determines the response of a character to selection, that is, how the character evolves. A complete understanding of this partitioning and what it tells us about evolutionary change, however, requires an explanation of all components of V_P.

To illustrate the meaning of the terms in equation (1), imagine the following experiment. A number of individuals of an alkaloid-containing plant are sampled and the concentration of alkaloids in the foliage of each plant is measured. This measured concentration is an individual's *phenotypic value, P.* The total phenotypic variance of the population, V_P, is simply the variance of the phenotypic values in the experimental population (fig. 3.1a).

Now, suppose that each individual is cloned and fifty ramets of each clone are planted into a field. At some later date, the concentration of alkaloids in the foliage of each plant is measured. The average alkaloid concentration across the ramets of a particular clone estimates the *genotypic value, G,* of alkaloid concentration for that clone. (This estimate becomes more precise as the number of

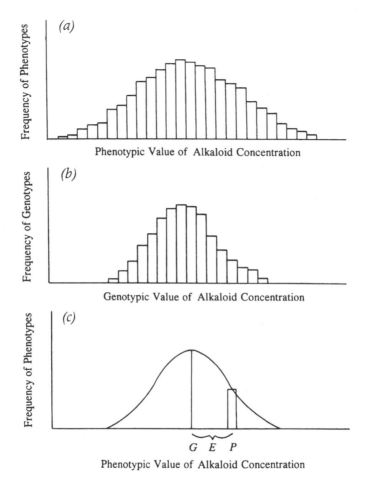

Fig. 3.1. Graphical illustration of several quantitative genetic parameters. (a) Frequency histogram of phenotypic values, P, of alkaloid concentration measured in a random sample of individuals from a hypothetical plant population. The sample mean estimates the mean alkaloid concentration phenotype for the population, P. (b) Frequency histogram of genotypic values of alkaloid concentration measured in a random sample of genotypes. Genotypic values, G, are measured as described below. The sample variance estimates the genotypic variance, V_G, for alkaloid concentration in the population. (c) Frequency histogram of phenotypic values, P, of alkaloid concentration measured in a population of ramets cloned from a single genotype and planted randomly across the environment. The mean alkaloid concentration calculated across ramets estimates the genotypic value, G. The difference $G - P$ equals E, the environmental deviation of the alkaloid concentration of a particular ramet from the genotypic value.

ramets per clone increases without bound.) *G is the contribution to an individual's phenotypic value that is determined by the* individual's genetic makeup. The genetic variance, V_G, for alkaloid production is then simply the variance of the genotypic values in the population (fig. 3.1b).

Not all plants of a clone will have an alkaloid concentration equal to that clone's average or genetic value. Individual plants will deviate from this average because, for instance, they grow in pockets of soil with greater or less than the average amount of soil nutrients, or because they are more or less shaded. In other words, an individual's phenotypic value deviates from this genotypic value because of unique environmental effects on alkaloid production (for examples of environmental effects on resistance factors, see Maddox and Cappuccino 1986; Craig et al. 1988; Bryant et al. 1988). This difference between the phenotypic and genotypic values is termed the *environmental deviation, E* (fig. 3.1c). The variance of the environmental deviations of all individuals in the population is the environmental variance, V_E.

From these definitions,

$$(2) \qquad\qquad P = G + E.$$

Assuming G and E are uncorrelated, equation (1) follows from taking the variance of each side of equation (2).

The genotypic value of an individual may be broken down into two major components: the additive and nonadditive genetic values, symbolized by A and I respectively:

$$(3) \qquad\qquad G = A + I.$$

The additive genetic value is best conceptualized by the following thought experiment, in which we temporarily assume little effect of environment on phenotype (i.e., $V_E = 0$). Take one of the alkaloid-producing plants discussed previously and mate it as a sire to a large number of other plants chosen randomly from the base population (i.e., the population from which the clones were previously drawn). From each mating, pick one seed, grow it in the field, and measure the alkaloid content of the foliage. Then twice the average value, measured as a deviation from the population mean, of the alkaloid content in the offspring of an individual estimates a quantity known as the *breeding value* of that individual. Because the breeding value of a trait in an individual is measured by the value of that trait in the individual's offspring, it indicates the portion of the individual's genotypic value that can be passed on to the next generation. Breeding value and additive genetic value are synonymous, and the additive genetic variance of a trait is simply the variance of the breeding values in a population.

The nonadditive genetic value is simply the difference between an individual's genotypic value and breeding value. In effect, I represents the portion of an individual's genotypic value that cannot be passed on to the next generation.

Taking the variances on each side of equation (3) yields

$$(4) \qquad\qquad V_G = V_A + V_I,$$

since it can be shown that additive and nonadditive genotypic values are uncorrelated (Falconer 1981). In other words, the total genetic variance for a trait can be partitioned into an additive genetic variance and a nonadditive genetic variance.

Although quantitative genetics does not provide a mechanism for doing so, standard ecological experiments could allow one to further subdivide the environmental component of phenotypic variance within a population into a portion caused by variation in the physical environment, V_{Ep}, and a portion caused by variation in the biotic environment, V_{Eb}, such that $V_E + V_{Ep} + V_{Eb}$. This division is useful because the biotic environment differs from the physical environment in its potential to evolve in response to changes in plant variance traits.

Phenotypic variation caused by the physical environment is well documented for plant traits because plants are sessile and thus reflect small-scale variation in the environment in a way that many mobile organisms cannot. For example, fine spatial heterogeneity in water or nutrient availability can affect the concentrations of secondary compounds in plant tissues, and thus influence their suitability for herbivores and pathogens (Colhoun 1973; Mattson 1980; Mihaliak and Lincoln 1985; Louda 1986; Louda et al. 1987; Bryant et al. 1988). The patchy distribution of sunlight in some communities can also cause individual variation in foliage quality (Read 1968; Pandey and Wilcoxson 1970; Lukens and Mullany 1972; Lincoln and Langenheim 1979; Lincoln and Mooney 1984).

The biotic environment is also an important cause of phenotypic variation among host plants. Prior herbivory can produce changes in host plants that affect their suitability for later herbivores (Schultz and Baldwin 1982; Rhoades 1983; Haukioja and Hanhimaki 1985). Moreover, pathogen attack can induce the production of phytoalexins (Deverall 1972) and other changes, which may affect the suitability of infected plants for herbivores (Kingsley et al. 1983; Dowd 1989; Wicklow 1988) and for other pathogens (Cruickshank and Perrin 1964). Competition among neighboring plants may reduce the amount of resources available, causing changes in resistance and lowering the nutritional quality of host plants. Furthermore, variation in plant density and diversity can also affect the rate of spread of disease organisms (Burdon 1987a, 41–48) and the ability of herbivores to locate host plants (Kareiva 1983). Such variation in plant community structure also influences the vulnerability of herbivores to attack by natural enemies (Price et al. 1980; Schultz 1983a).

A special type of biotic environmental effect arises because siblings all share an environment created by the same maternal parent. A common maternal environment tends to make siblings more similar, which tends to increase the variance between groups of siblings produced by different mothers and thus, as will be described below, tends to increase the apparent amount of genetic variation. Such maternal effects can be caused by variation among mothers in the availability of resources that can be allocated to offspring. For example, the maternal environment can influence the nutritive value of seeds, and hence their probability of being consumed by predators (Sosulski et al. 1963). Other mechanisms can also produce maternal effects. The physical proximity of dispersed seeds to their female parent alters their probability of being eaten by herbivores (Janzen 1970, 1971; O'Dowd and Hay 1980) or infected by pathogens (Augspurger 1983, 1984). Recently, maternal transmission of nongenomic genetic information has become another topic of interest. Viruses can often be transferred from mother to offspring (Shepherd 1972), and may be an important cause of maternal effects (Burdon 1987a, 15).

The Evolution of Quantitative Traits

One common approach to understanding the evolution of quantitative traits is to adopt the standard equations used by breeders to predict changes in characters subjected to artificial selection:

$$(5) \qquad \Delta \bar{z} = V_A V_P^{-1} s = h^2 s$$

(Kempthorne 1969; Falconer 1981). In this equation, $\bar{z}$ is the population mean of the character under selection; h^2 is a quantity known as the heritability of the character, equal to the ratio V_A/V_P, and s is the selection differential.

The selection differential represents the intensity with which selection acts on the character. Directional selection, i.e., the tendency for selection to favor an increase or decrease in the mean of the trait, occurs when there is a correlation between the value of the trait and fitness. Consequently, the selection differential is defined as the covariance between the trait and relative fitness:

$$(6)^* \qquad s = \text{cov}(w, z).$$

When selection operates primarily by differential survival of individuals with different phenotypes, the selection differential defined by this equation is simply the difference between the mean value of the trait in the surviving individuals and the mean value of the trait in the population before selective mortality occurs (fig. 3.2).

A great number of studies, as well as the practical experience of animal

---

*As a reminder, the covariance between two traits x and y and the correlation between those traits, r_{xy}, are related by the equation $\text{cov}(x,y) = r_{xy}V(x)V(y)$.

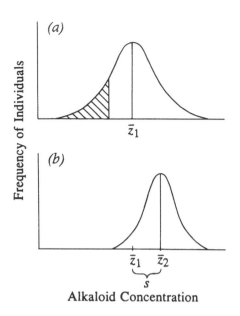

Fig. 3.2. Directional selection on alkaloid concentration. (*a*) Distribution of phenotypes in the parental population, with mean $\bar{z}_1$. Suppose individuals in the shaded region have low alkaloid concentration, making them vulnerable to herbivory, which kills them. (*b*) Then, $\bar{z}_2$ represents the mean alkaloid concentration in surviving members of the parental population. The difference $\bar{z}_2 - \bar{z}_1$ equals *s*, the selection differential on alkaloid concentration.

and plant breeders, have demonstrated that equation (5) predicts short-term evolutionary change in single characters reasonably accurately under simplified laboratory or agricultural conditions, in which selection is imposed on a single trait by an investigator (e.g., Falconer 1953; Clayton et al. 1957). It is thus tempting to use this equation to understand the evolution of traits in natural populations. In particular, equation (5) suggests that if an investigator finds significant heritability for a trait ($h^2 > 0$) and a nonzero selection differential, he or she could reasonably conclude that the trait was evolving in the population examined. If this inference were valid, demonstrating and quantifying evolution in natural populations would be a simple matter of measuring heritabilities and selection differentials.

Unfortunately, life is not so simple. One important difference between selection experiments and selection in nature is that in nature selection may act on several characters simultaneously. If these characters are correlated with each other, then evolutionary change in one character may be brought about not only by selection acting directly on that character, but also indirectly by selection acting on correlated characters. A more general theoretical framework is therefore needed to describe the evolution of a complex phenotype composed of a suite of correlated characters. This framework is the subject of the next two sections. The first section introduces the concept of an additive genetic covariance between two characters, while the second section discusses how such covariance affects the response of correlated characters to selection.

Correlations among Characters

Suppose that in the experiment described above involving a population of alkaloid-producing plants we had measured the concentrations of two different alkaloids, which we will label A and B. These measurements are illustrated in figure 3.3. The simple correlation of the concentrations of A and B calculated across all plants in the experiment is known as the phenotypic correlation of alkaloids A and B. Similarly, the simple covariance of concentrations is known as the phenotypic covariance.

Just as the phenotypic variance of a trait can be partitioned into genetic and environmental components, so can the phenotypic covariance between two traits. In particular,

$$(7) \qquad\qquad \text{cov}_P = \text{cov}_G + \text{cov}_E,$$

which says simply that the phenotypic covariance between two traits is equal to the sum of the genetic and environmental covariances. The importance of this partitioning will become apparent below, when we consider the evolutionary dynamics of correlated characters.

A genetic covariance is the covariance of genotypic values of two traits. In our example, it could be estimated by calculating the covariances between clone means for the concentrations of the two alkaloids, since the clone means estimate genotypic values. Of particular interest, however, is a component of the genetic covariance known as the additive (genetic) covariance. Conceptually, the additive covariance is simply the covariance of the breeding values of two traits (fig. 3.4). For alkaloids A and B, we described above how the breeding value of each trait in an individual could be estimated from the mean

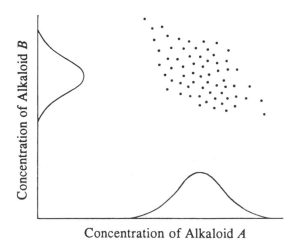

Fig. 3.3. A phenotypic correlation between concentrations of two alkaloids. Each point in space represents the phenotypic values of concentrations of alkaloids A and B measured on an individual plant. The frequency histogram along each axis describes the phenotypic distribution of concentrations of the alkaloid represented by that axis. The phenotypic correlation shown here is negative.

Concentration of Alkaloid B

Concentration of Alkaloid A

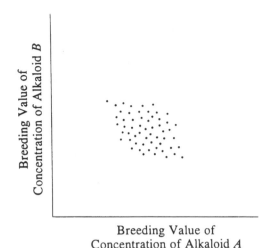

Breeding Value of
Concentration of Alkaloid *A*

Fig. 3.4. Additive genetic covariance between concentrations of two alkaloids. Each point in space represents the breeding values of concentrations of alkaloids *A* and *B* measured on a particular genotype.

of that trait in a set of offspring of different females but sired by a common male. The additive covariance between these two alkaloids could then be estimated by calculating the covariance of the offspring means. (As noted above, this is strictly correct only if the number of matings per sire is very large.)

Genetic covariances in general, and additive covariances in particular, arise for two reasons: pleiotropy and linkage disequilibrium. Pleiotropy occurs when variation at a genetic locus affects more than one trait. If alleles at a pleiotropic locus affect two traits in the same direction, that locus will contribute toward a positive covariance. By contrast, if alleles at a locus affect two traits in opposite directions, the locus will contribute toward a negative genetic covariance. Assuming that all loci contribute roughly equally to determining the genotypic (or breeding) value of each trait, the overall genetic covariance will be determined by which type of locus is more prevalent.

Linkage disequilibrium occurs between two or more loci if there is a statistical association between a particular allele at one locus and a particular allele at a second locus. Such an association may arise from the mixing of populations with different allele frequencies at these loci, or because selection favors certain combinations of alleles over others. To the extent that the two loci influence the values or levels of different traits, the association between alleles will produce a corresponding association or correlation between the values of the two traits.

The environmental covariance between two traits is related in a straightforward way to the environmental deviations for those traits: the environmental covariance is simply the covariance of the environmental deviations of the traits across all individuals in the population. Environmental covariances thus arise because an environmental factor that causes one trait in an individual to deviate in a particular direction from its genotypic value also tends to cause the second trait to deviate in the same or the opposite direction (depending upon the sign of the covariance).

In terms of the alkaloid-producing plant we used in our example, consider the ramets of a single clone distributed across a field (fig. 3.5). Ramets that grow in pockets of soil poor in nitrogen might produce less than the clone-average amounts of both alkaloids A and B, since availability of nitrogen is often thought to limit the production of alkaloids in plants (Robinson 1974). In these sites, the environmental deviation for both alkaloids would be negative. By contrast, plants growing in soil pockets rich in nitrogen might be expected to produce greater than average amounts of both alkaloids, and the environmental deviation for both alkaloids would be positive. In such a situation, there would be within the clone a positive covariance between the concentrations of the two alkaloids. And if all clones responded to variation in soil nitrogen in the same way, there would be a positive environmental covariance.

The Evolution of Correlated Traits

The importance of additive correlations among traits is that they alter the predicted course of evolution compared to that expected when each trait is considered separately (e.g., equation 5). The analog of equation (5) for a set of correlated characters is

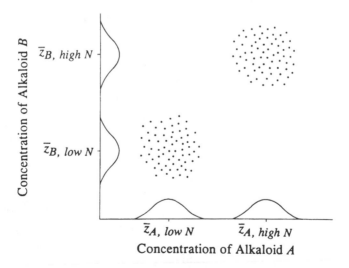

Fig. 3.5. A positive environmental correlation between two alkaloids. Each point in space represents the phenotypic values of concentrations of alkaloids A and B in an individual ramet. All ramets are cloned from the same genotype. Ramets in the upper right cluster grew in a soil high in available nitrogen, whereas those in the lower left cluster occupied a low-nitrogen soil.

(8) $$\Delta \bar{z} = GP^{-1}s = G\beta$$

(Lande 1979). Here $\bar{z}$ is a vector of mean values of the traits, G is a matrix of genetic variances and covariances for those traits, P is the corresponding matrix of phenotypic variances and covariances, and s is the vector of selection differentials for the traits. Lande (1979) showed that the selection gradient β, which is equal to $P^{-1}s$, represents the magnitude of directional selection acting on the set of characters. In particular, each element, β_i, represents the magnitude of directional selection acting directly on character i, with the effects of all other characters held constant.

The effect of genetic correlations is most easily seen by considering the joint evolution of two characters. From equation 8, the change in the mean of each trait due to selection is

(9a) $$\Delta \bar{z}_1 = V_{A1}\beta_1 + \text{cov}_a\beta_2$$

(9b) $$\Delta \bar{z}_2 = \text{cov}_a\beta_1 + v_{A2}\beta_2,$$

where V_{Ai} is the additive genetic variance of trait i and cov_a is the additive genetic covariance between the traits. If there is no genetic covariance, and hence no genetic correlation, between traits 1 and 2, the equations (9a) and (9b) each reduce to equation (5), that is, each trait evolves independently of selection on the other trait. However, if cov_a is not zero, then selection on trait 2, manifested by β_2, contributes to the change in the mean of trait 1 via the term $cov_a\beta_2$. Similarly, selection on trait 1 contributes to the change in the mean of trait 2. In our example, trait 1 is the concentration of alkaloid A and trait 2 is the concentration of alkaloid B, and equation (9a) appears as

$$\overline{\Delta \text{alkaloid } A} = V_{\text{alkaloid } A}\beta_A + \text{cov}_{AB}\beta_B.$$

Equation (9b) appears as

$$\overline{\Delta \text{alkaloid } B} = V_{\text{alkaloid } B}B_B + cov_{AB}\beta_A.$$

Negative genetic correlations are often thought to constitute constraints on evolution (Rausher 1984b; Berenbaum et al. 1986; Futuyma and Philippi 1987; Loeschcke 1987). The way in which such a constraint may operate can be visualized by considering two traits and their associated adaptive landscape. The adaptive landscape is simply the relationship between the vector of mean values of the traits, z, and the mean fitness in a population. For two characters, such as concentrations of alkaloids A and B, the adaptive landscape can be represented as a series of fitness contours in a plane having mean concentration of alkaloid A as one axis and mean concentration of alkaloid B as the second axis (fig. 3.6). The selection gradient, β, at any point in the adaptive landscape is then a vector representing the rate of change of fitness as one moves perpendicular to the fitness contours at that point (fig. 3.6; technically, the selection gradient is the gradient of the natural logarithm of fitness). It indicates the direction selection is "pushing" the population. The projection of the gradient vector on an axis represents the magnitude of selection acting on the character corresponding to that axis.

In the absence of genetic covariance between two traits, equation (9) predicts that selection will tend to cause the population to evolve toward the selective optimum (peak of adaptive landscape) following a path perpendicular to the fitness contours (fig. 3.7a). By contrast, if there is a genetic correlation of $^-1$ between the traits, the same equation predicts that selection may move the population toward an equilibrium away from the optimum (fig. 3.7b). In essence, the population is constrained from reaching the optimum because one trait cannot evolve toward the optimum without a negative correlated response in the other trait causing that trait to evolve away from the optimum. Such a situation might occur if a plant could allocate a given amount of nitrogen to

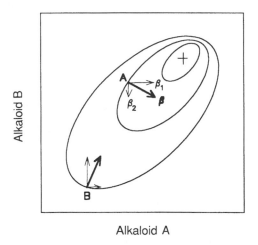

Alkaloid A

Fig. 3.6. Hypothetical adaptive landscape for the concentrations of two alkaloids. Contours represent combinations of values of alkaloids A and B that have equal fitness. The "$+$" indicates the combination of values with the highest fitness, and is thus the adaptive peak or optimum. Points A and B represent the locations of two hypothetical populations on the adaptive landscape. The vector of selection gradients, β, for either point A or B is shown as a thick arrow. This vector points in the direction of steepest assent up the contours. The thin horizontal (β_1) and vertical (β_2) arrows are the orthogonal components of β and represent the selection gradients for traits 1 (alkaloid A) and 2 (alkaloid B), respectively. Note that the selection gradient vectors do not necessarily point to the optimum.

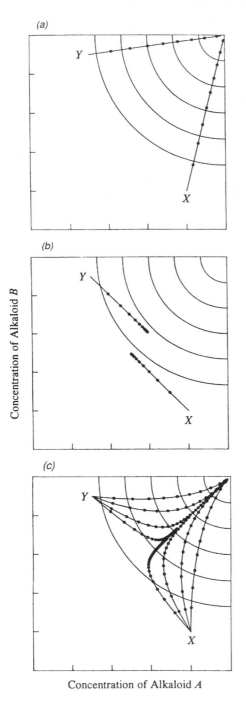

Fig. 3.7. Evolutionary trajectories for two correlated traits, concentrations of alkaloids A and B. The optimal combination of traits occurs at the peak at the upper right corner of the figure. Contours of the adaptive landscape are symmetrical about the optimum, but for convenience only one quadrant of the landscape is depicted. Solid lines represent evolutionary trajectories from initial points X and Y. Each segment between the dots represent 50 generations. In all figures, $V_{a1} + V_{a2} = 1$. (a) Trajectories for $cov_a = 0$. Note that all trajectories converge towards the optimum, indicating an absence of constraint. (b) Trajectories for $cov_a = 1$. Trajectories converge at right angles to a line perpendicular to the contour lines, and then stop. So long as the covariance persists, the trajectories will never reach the optimum. (c) Trajectories for $^-1 < cov_a < 0$. Trajectories eventually converge on the optimum, but do so slowly for cov_a close to $^-1$.

Concentration of Alkaloid B

Concentration of Alkaloid A

either alkaloid A or B, but not to both. In that case, evolving to higher concentrations of A would cause simultaneous evolution to lower concentrations of B, and vice versa. Neither alkaloid could reach its optimal value and a compromise level would be attained.

For genetic correlations between $^-1$ and 0, evolution toward the optimum is slowed (compared to the absence of a correlation), but the population eventually is expected to reach the optimum, assuming that genetic variation for the two traits is not depleted (fig. 3.7c). The rate at which evolution proceeds is reduced substantially only by negative correlations close to $^-1$. Consequently, negative genetic correlations must be strong in order to constrain evolution significantly, at least when the number of correlated characters is small.

For more than two correlated traits contributing to fitness, the constraining effects of negative genetic correlations are more complicated. Dickerson (1955) argued that as the number of correlated characters increases, the average negative genetic correlation needed to prevent attainment of a joint optimum decreases. Specifically, an average correlation of $^-1/(N\text{-}1)$, where N is the number of characters, can effectively constrain evolution.

This result should not be taken as necessarily indicating that as N becomes large, weak correlations can potentially constitute a significant constraint to evolutionary change. In most cases, it is likely that at least some correlations must be fairly strong in order for such a constraint to exist. For example, it is commonly observed that positive genetic correlations among fitness traits are at least as common as negative correlations (Gould 1979; Rausher 1984; Via 1984; Futuyma and Philippi 1987). Under these conditions, the average value of the negative correlations, r_{neg}, must satisfy

$$(10) \qquad |r_{neg}| > |r_{pos}| + 2/(N\text{-}1),$$

where r_{pos} is the average value of the positive correlations. With ten correlated traits and a relatively low value of r_{pos} of 0.25, the average negative correlation must exceed 0.47 in absolute value. In other words, at least some of the negative correlations must be large to constrain evolution. If most negative correlations are small, it is probably unlikely that equation (10) will be satisfied.

On the other hand, the absence of a strong genetic correlation between two traits cannot be taken as evidence for lack of a genetic constraint on their evolution. As noted by Pease and Bull (1988) and developed more fully by Charlesworth (1990), if several characters are involved in plant resistance to herbivory, then the pattern of additive genetic correlations between the pairs of traits will be only indirectly related to the functional trade-offs that constrain their evolution. Consequently, a positive genetic correlation could occur between two traits within a suite of resistance characters, even if the evolution of each trait in the pair is constrained by the other via a complex structure of benefits and trade-offs with other characters in the suite. Nevertheless, within the suite of characters, some negative genetic correlations must exist in order for their joint evolution to be constrained (Charlesworth 1990).

Measuring Evolutionary Parameters

Equation (8) represents the best paradigm available to evolutionary biologists for understanding the evolution of quantitative traits. Empirical biologists who rely on this paradigm to interpret the evolution of resistance in plant populations need methods for estimating evolutionary parameters such as genetic variances and covariances for resistance, heritability, and the strength and direction of selection. In this section we present a more technical discussion of the methods available for estimating these parameters.

Heritability and Genetic Variance

As indicated in equation (5), estimating the heritability of a trait involves measuring both the phenotypic and additive genetic variances for the trait. As described above, the phenotypic variance is easily measured as the variance of phenotypic values. By contrast, the additive genetic variance of a trait is estimated from the degree of resemblance among relatives. Various statistical methods may be used to achieve this goal. Most partition the total observed variance into two components: within groups of related individuals and among such groups. From the among-group component, the additive genetic variance may then be estimated based on the expected resemblance among relatives within the groups due to shared genes.

The usual procedure is to identify or generate a population of individuals with known relationships. Although many types of breeding designs are employed to produce such populations (e.g., Kearsey 1965), two major types of relationships are most commonly used. The first type consists of various combinations of offspring and parents. This method can be illustrated as follows: mate an alkaloid-producing plant with another plant chosen at random from the population. Plant three offspring from this cross and measure the alkaloid concentration of each parent and of the offspring. Repeat this procedure for a large number of plants; then perform a regression of the mean alkaloid concentration in the offspring on the alkaloid concentration of the sire (or the average of both parents, termed the midparent value) (fig. 3.8). The slope of this line estimates the heritability (in the case of a regression on the midparent value) or half the heritability (if regressed on one parent) of alkaloid concentration (Falconer 1981). Computational details using least-squares methods are described for various permutations of this design by Hill (1970), Hill and Thompson (1977), and Becker (1984, 133–138).

The second type of relationship involves families of siblings sharing one (half-sib family) or both (full-sib family) parents in common. For example, mate an alkaloid-containing plant as a sire with many other plants chosen at

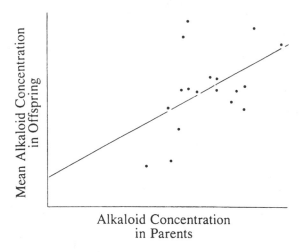

Fig. 3.8. Parent-offspring regression for alkaloid concentration. The x-axis represents the phenotypic value of alkaloid concentration measured in each parent and the y-axis represents the mean of the phenotypic values of alkaloid concentration measured in the offspring of that parent. The equation $y_i = mx_i + e_i$ describes the regression of the mean offspring phenotype on the parental phenotype, where x_i is the observation of the ith parent (usually the sire), y_i is the mean of the offspring of the ith parent, m is the regression of y on x, and e_i is the error associated with an individual offspring. If the parent is a sire, the narrow-sense heritability is described by the equation $h^2 = 2m$.

random from the population. Plant several offspring from each cross and measure their alkaloid concentrations. The offspring from one cross (the sire mated to one dam) represent a full-sib family. Because they all share the same father, the offspring from all the crosses represent a paternal half-sib family. The mean alkaloid concentration, averaged over all his offspring, is called the paternal half-sib family mean and is an estimate of the sire's breeding value for alkaloid concentration. If this procedure is repeated many times (mating each sire with many randomly chosen dams), the variance among paternal half-sib family means estimates one-fourth of the additive variance (V_A) of alkaloid concentration in the population. Thus, V_A is estimated by multiplying the component of variance due to paternal half-sib family by four. The heritability is calculated as usual, $h^2 = V_A/V_P$, where V_P is estimated by the total variance in alkaloid concentration calculated over all offspring from all crosses. In the cross described above, each sire is mated with a different set of randomly chosen dams, so dams are nested within sires, producing a nested mating design.

Numerous sib designs have been published. The most commonly encountered designs include full (Hayman 1954a, 1954b, 1958; Griffing 1956a, 1956b; Kempthorne 1956) and partial (Kempthorne and Curnow 1961; Fyfe and Gilbert 1963) diallels, factorials (Turner and Young 1969; Cockerham and Weir 1977), and nested (Becker 1984) designs. The important characteristic that these designs share is the production of an experimental population consisting of half-sib families (maternal, paternal, or both) each of which is comprised of several full-sib families. Analysis of variance is used to partition the total variance in the experimental population into components found within and among the various types of families. The relationships among and within the families are then used to partition these variance components further into those presented in equations (1) and (4), which may be used to calculate the heritability of the trait. Although the general theory underlying calculation of heritabilities by sib analysis is well described by Falconer (1981), some of the references cited above are better sources of information about actual computation.

Many modifications of the basic diallel and factorial mating designs were published prior to the mid-1970s (e.g., Youden and Connor 1953; Mandel 1954; Fyfe and Gilbert 1963). Their virtue lay mainly in the ease with which they could be analyzed on primitive calculating devices. With the advent of computers and sophisticated statistical packages (e.g., SAS Institute 1985 and SPSS 1986) this is no longer a crucial issue in breeding design. As long as a design is balanced (Searle 1971, 138–140), it can be analyzed by least-squares methods with any of the aforementioned statistical packages. Furthermore, if maximum-likelihood (ML) methods are employed (R. G. Shaw 1987; Thompson and Shaw 1990), even balance is not essential. An additional virtue of ML methods is that parent-offspring and sib designs can be combined in the same analysis. These combinations may provide more efficient estimates of genetic parameters than either element used alone (R. G. Shaw 1987).

The choice of breeding design depends upon the circumstances. Two important factors to consider in this decision are precision and bias in the resulting estimates of genetic parameters. In general, closer relationships produce more precise estimates (Falconer 1981). Estimates of genetic parameters are notoriously imprecise (Falconer 1981; Mitchell-Olds and Rutledge 1986), making precision a valuable commodity. However, bias is introduced to estimates of genetic parameters by covariance among relatives sharing the same environment and, in the case of full-sibs, by dominance effects (Falconer 1981). Maternal effects are often a particularly intractable source of bias (Mitchell-Olds and Rutledge 1986). If enough kinds of relatives are available (e.g., a full diallel or reciprocal factorial design), bias due to dominance or maternal effects can be estimated and removed by subtraction (e.g., Cockerham and Weir 1977). However, unless one is particularly interested in measuring nonadditive genetic effects, simpler designs (e.g., dams nested within sires) more efficiently estimate V_A and consequently heritability.

One modification of sib mating designs that frequently appears in the quantitative genetics literature is the use of inbred lines as parents in the breeding design. Cockerham (1963) provides a lucid discussion of the effects of using inbred parents in mating designs. Essentially, the estimates of genetic parameters become more precise with greater inbreeding of the parents. However, the estimates no longer reflect the true values of the genetic parameters as they exist in the original population.

In addition to considerations of precision and bias, there may be practical limitations to the types of relatives available. For example, if artificial pollination is not possible, then mother-offspring regression or analysis of variance within and among maternal half-sib families must be used. Both of these designs are liable to bias by inestimable maternal effects. Moreover, in the latter case, it is not always possible to ensure that members of a family collected from a single mother have different fathers; and in the case of self-compatible plants, offspring may even be self-sibs. These uncharacterized families introduce further complications to the analysis of maternal-family designs and should be avoided where possible.

Although not actually a breeding design, clone analysis (described above) is another method commonly used to partition the genetic component of variance from phenotypic variance. This method is favored by those working on long-lived perennial plants, especially when the traits of interest are not expressed by juveniles. Because this method fails to partition additive genetic variance from dominance and interaction effects, it estimates broad-sense heritability as

(11) $$h_{bs}^2 = V_G/V_P.$$

If the population is outbreeding, dominance and interaction effects can be important components of the total genetic variance, and may cause broad-sense heritabilities to be much higher than narrow-sense heritabilities. For example, in a study of the inheritance of resistance to rust in poplars, Jokela (1966) found that broad-sense heritabilities estimated from clone analysis were 12–100% greater than narrow-sense heritabilities estimated from maternal half-sib families (which were themselves likely to be biased by maternal effects). This result was likely due to the fact that dominance and epistatic effects caused 40–50% of V_G.

One might argue that clonal estimates are adequate because evolutionary biologists are generally interested only in determining whether heritability is equal to or different from zero. However, contributions of nonadditive genetic effects to the numerator of broad-sense heritability could significantly bias predictions regarding the evolutionary responsiveness of a character to natural selection. For predicting responses to selection, clonally derived heritability estimates should therefore be viewed with caution.

On the other hand, if the intent is to predict the effect of natural selection among clonally propagated ramets, broad-sense heritability may be an appropriate parameter for predicting the within-generation response to selection.

This method has been employed by foresters for predicting gain expected from selection of clonally propagated trees (e.g., Foster et al. 1984). Because this type of selection program does not involve sexual reproduction, recombination will not disrupt extant genotypes. Hence V_D and V_I should be included in the numerator of the heritability estimate used to predict the outcome of selection among clones.

Even for this limited purpose, however, clonally derived heritability estimates are fraught with problems. Estimates of V_G are subject to large biases introduced by the environment of the ortet (the source of the clonal progeny). This bias, called the c-effect, is likely to inflate estimates of V_G. Burdon and Shelbourne (1974) suggested subcloning (taking a second set of cuttings from the primary clones of the ortet) as a method of estimating c-effects, thus eliminating them from the estimate of V_G. Another problem with clones is that the developmental stage of the part of the ortet from which a cutting is taken influences the morphology and physiology of the resulting ramet. By increasing the variability among ramets from a single ortet, this phenomenon, known as topophysis (Olesen 1978), is likely to cause underestimation of V_G. These considerations indicate that heritability estimates calculated from clonal designs should be interpreted cautiously.

An exciting field now being explored is the estimation of genetic parameters from natural populations (as opposed to experimental populations generated by crosses). Two different approaches are being taken. Thompson and Shaw (1990) use ML methods to develop estimators of genetic parameters from inferred relationships among individuals in a natural population (e.g., Meagher 1986; Meagher and Thompson 1987; Devlin et al. 1988). In contrast, Ritland (1990) has developed a method for determining heritability based on "gene identity disequilibrium" (Weir and Cockerham 1969). In this method, the level of relatedness between pairs of individuals is inferred from whether or not members of each pair share the same allele at a marker locus. This information, in conjunction with an estimate of gene identity disequilibrium between the marker locus and the collection of quantitative trait loci (those loci actually contributing to the phenotype), is used to estimate the probability that pairs with different levels of relatedness will share alleles at quantitative trait loci. These probabilities are then used to estimate the heritability of the quantitative trait.

Genetic Covariance and Correlation

The complex nature of the phenotypic covariance dictates that the magnitude and the sign of the genetic covariance cannot be determined from the phenotypic covariance alone (Falconer 1981). However, the two parameters will be similar if genetic and environmental covariances are similar and if the traits have similar heritabilities (Barton and Turelli 1989). Cheverud (1988) argued that an estimate of the phenotypic variance-covariance matrix will adequately predict the genetic matrix. He compared phenotypic and genetic variance-covariance matrices reported in 41 studies and found that in studies with larger sample sizes, genetic and phenotypic matrices were similar (Cheverud 1988). He concluded that empirically observed differences between the two matrices are often due to inaccurate estimates of genetic variance-covariance matrices taken from inadequate sample sizes. In Cheverud's study, however, sample size was often positively correlated with heritability. Environmental variation is a less important contributor to phenotypic variance in traits with high heritability and thus may be less likely to alter significantly phenotypic covariances among such traits from their genetic covariances. Furthermore, most of the studies reported by Cheverud measured genetic parameters on organisms raised in laboratories or other controlled environments in which environmental variance was likely to have been less than that observed in nature and thus less likely to have influenced the structure of phenotypic variance-covariance matrices. Finally, other studies have failed to find agreement between estimates of phenotypic and genetic variance-covariance matrices (Bell and Koufopanou 1986; Roff and Mosseau 1987). Estimating genetic variance-covariance matrices is expensive, as will become apparent below, and substitution of phenotypic matrices would be welcome. However, until further surveys are completed, conflicting empirical results counsel the more conservative approach.

Two methods available for estimating the genetic covariance are analogous to those used for estimating the genetic variance, while a third method directly estimates the genetic correlation. The first method is the multivariate analog of parent-offspring regression for determining heritability. The individual phenotype is treated as a vector; the length of the vector reflects the number of traits included in the multivariate phenotype. In our hypothetical example, this vector consists of two values: the concentrations of alkaloids A and B. Multivariate regression is used to regress the mean value of the offspring phenotype vector on the mean phenotype vector of one or both parent(s). The slope of the regression of the mean concentration of alkaloid A in the offspring on the parent mean concentration of alkaloid B estimates the genetic covariance of alkaloids A and B (Becker 1984, 133–138).

In sib analysis, multivariate ANOVA is used to partition the phenotypic variance-covariance matrix into within- and among-family components. The relationships of these families are then used to repartition the covariance into components due to various environmental and genetic effects (Becker 1984, 113–132). Like its variance-component analog, the least-squares methods of estimating covariance components via sib analysis require a balanced mating design.

A third method is to estimate the genetic correlation directly, and involves calculating the Pearson product-moment coefficient of correlation (Sokal and Rohlf 1981, 565–582) between mean values of pairs of characters averaged over families. In other words, measure the concentrations of alkaloids A and B

in the offspring of several paternal half-sib families. Then, calculate the mean concentration of each alkaloid for each family. Finally, use the family means to calculate a simple correlation of alkaloid A with alkaloid B. Called a family-mean correlation (where the type of family dictates the types of biases in the estimate), this parameter only approximates the true genetic correlation because each term contains not only the variance or covariance among families, but also a fraction of the within-family variance or covariance (Arnold 1981):

$$(12) \qquad \text{cov}_M = \text{cov}_{\text{among}} + 1/n \, \text{cov}_{\text{within}}$$

where n is the average number of sibs per family (Via 1984). As n increases, the family mean correlation approaches the value of the component correlation for the relevant type of family.

The trade-offs between bias and precision described for the estimation of additive genetic variances from sib analysis also apply to the latter two methods of estimating additive genetic covariances or correlations. Again, crosses between inbred lines enhance the precision of covariance or correlation estimates, but inbreeding depression could produce positive genetic covariances between traits (such as major fitness components) in the sample population that are actually negatively correlated in the original population (Rose 1984). Another drawback is that crosses between inbred lines set up very strong linkage disequilibria, which render estimates of genetic correlations useless for predicting the evolutionary response to selection on correlated characters (Clark 1987).

In addition to providing information on the response of a multivariate phenotype to selection (e.g., equation [8]), genetic correlations can also be used to measure genotype-by-environment ($G \times E$) interactions. Falconer (1952) noted that the expression of the same trait in two different environments may involve different sets of genes, and thus can be treated as two traits that are genetically correlated. This relationship permits estimation of $G \times E$ interactions and indicates the importance of genetic correlations for understanding evolution in heterogeneous environments (Via 1984; Via and Lande 1985, 1987). Genetic correlations as estimates of the $G \times E$ interaction are particularly valuable for examining the effects of spatial and temporal variability in herbivore and pathogen population sizes and compositions on the evolution of plant resistance traits. For example, some models of plant-pathogen coevolution assume that plants resistant to one pathogen strain will necessarily be less resistant to another and that plants resistant to the second strain will be less resistant to the first (e.g., Clarke 1976). One criterion for validating such models is to determine whether these types of trade-offs are common in nature. If different pathogen resistance types are considered as different environments, such trade-offs can be detected as crossing type $G \times E$ interactions (Falconer 1981, 122–124). Furthermore, measurement of genetic correlations between resistance traits expressed in different environments, where variables such as soil fertility, water availability, and sun exposure have been experimentally manipulated, will increase our understanding of how environmental heterogeneity may constrain their evolution. Several authors have highlighted this problem as an important gap in our current understanding of the evolution of plant resistance traits (Bryant et al. 1988; Craig et al. 1988).

Yamada (1962) first provided a method for calculating the genetic correlation between expressions of the same trait in two different environments, which involves a univariate linear model treating environment as a fixed effect and genotype as a random effect and includes the interaction term. However, Yamada's procedure is inappropriate for unbalanced designs (Fernando et al. 1984) and in such cases should be replaced by another method such as those provided by Schaeffer et al. (1978), Wiggans et al. (1980), and Fernando et al. (1984). Finally, Eisen and Saxton (1983) have provided a method for calculating genetic correlations among traits influenced by more than one environmental factor.

Selection

As noted earlier, because in nature selection can act on more than one character, evolution in one character may be influenced by selection on characters with which it is genetically correlated. These facts complicate our attempts to understand the role of natural selection in natural populations. Nevertheless, Lande and Arnold (1983) developed a conceptually and computationally simple method of calculating the selection gradients for a set of correlated characters. In particular, they showed that β_i, the directional selection gradient, which represents the magnitude of directional selection acting directly on character i (with all other characters held constant), can be estimated by performing a linear multiple regression of relative fitness on the phenotypic values, z, using the model

$$(13) \qquad w = \sum_{i=1}^{n} \beta_i \, z_i + \text{error}.$$

The stabilizing/disruptive selection gradient, γ_{ij}, can be obtained from the quadratic partial regression of relative fitness on phenotypic values using the model

$$(14) \qquad w = \sum_{i=1}^{n} \beta_i' z_i + \sum_{i=1}^{n} \sum_{j>i}^{n} (1 - 1/2 \, \delta) \, v_{ij} \, z_i z_j + \text{error},$$

where $\delta = 1$ if $i = j$ and 0 otherwise.

To illustrate the method, let's return to the hypothetical population of alkaloid-producing plants. Suppose each individual in a wild population is sampled to determine its concentration of alkaloids A and B. Furthermore, at the end of the year the seeds produced by each plant (assume they are annuals) are counted to estimate its fitness. Directional selection on alkaloid concentrations can then be estimated by fitting the following regression equation:

$$\text{seed number} = \text{intercept} + \beta_A(\text{conc. } A) + \beta_B(\text{conc. } B) + \text{error}$$

The slope of the regression of seed number on alkaloid A concentration (while alkaloid B is held at constant concentration) is β_A. The sign and magnitude of this parameter estimates the direction and strength of directional selection acting directly on the concentration of alkaloid A.

Stabilizing or disruptive selection on alkaloid concentration can be estimated by fitting this quadratic regression equation:

$$
\begin{aligned}
\text{seed number} = {} & \text{intercept}' + \beta_A'(\text{conc. } A) + \beta_B'(\text{conc. } B) \\
& + 1/2\,\gamma_{AA}(\text{conc. } A)^2 + 1/2\gamma_{BB}(\text{conc. } B)^2 \\
& + \gamma_{AB}(\text{conc. } A)(\text{conc. } B) + \text{error}
\end{aligned}
$$

The shape of the relationship between seed number and concentration of alkaloid A (while the concentration of alkaloid B is held constant) is indicated by γ_{AA}. The sign of this parameter indicates whether selection is stabilizing ($\gamma_{AA} < 0$) or disruptive ($\gamma_{AA} > 0$). The cross-product coefficient (γ_{AB}) indicates whether the evolutionary equilibrium occurs at a peak, a saddle, a ridge, or a valley (Phillips and Arnold 1989; Simms 1990).

Although the Lande-Arnold method has been adopted readily by field biologists (e.g., Kalisz 1986; Conner 1988; Campbell 1989; Rausher and Simms 1989), there are several limitations in its use (see also Mitchell-Olds and Shaw 1987). One major qualification is that the estimates of selection gradients do not account for correlations between measured and unmeasured traits. A second problem is that environmental correlations between the measured characters and fitness may render selection gradients calculated by the Lande-Arnold method unreliable (Price et al. 1988). When selection is measured on offspring from half-sib families, these two problems can be addressed by another method, described below. An important statistical problem inherent to linear regression methods is that multicollinearity, caused by subsets of highly correlated characters, can distort parameter estimates (Mitchell-Olds and Shaw 1987). Hence, adding more characters to the analysis, in an attempt to avoid the problems of unmeasured characters, can reduce the reliability of the estimates of selection gradients. Finally, estimating the selection surface by multiple quadratic regression produces a simple, unimodal surface. Because certain models of evolution (e.g., Wright 1970) posit multiple adaptive peaks, more flexible nonparametric methods for estimating the selection surface have been developed (e.g., Schluter 1988). Despite these limitations, the Lande-Arnold method can help generate hypotheses about selection pressures which can then be tested with manipulative experiments to demonstrate the operation of selection on particular characters.

When the genetic relationships among individuals in a population are known (such as when the individuals are derived from controlled crosses), it is possible to measure selection acting directly on genetic variation. It is well-known that for a single quantitative trait, an evolutionary change in the mean value from one generation to the next is determined by the additive genetic covariance between that trait and fitness, i.e.,

$$(15) \qquad \Delta \bar{z} = \text{cov}_g(W, z)$$

(Crow and Nagylaki 1976). This relationship holds regardless of genetic correlations between characters. Consequently, the net selection on a character can be estimated by measuring the genetic correlation between that character and fitness using methods described earlier.

To separate the contributions of indirect selection on correlated characters and direct selection as causes of evolutionary change in a character, an analysis similar to that described by Lande and Arnold can be used. In particular, Rausher (in preparation) shows that for a set of characters, the predicted response to selection is

$$(16) \qquad \Delta \bar{z} = GB,$$

where $B = G^{-1}\text{cov}_g(W, z)$ is the vector of partial regression coefficients of the breeding value of fitness on the breeding values of the characters measured. The elements of B are thus analogous to the selection gradients in the Lande-Arnold analysis, and hence indicate the direct force of selection on each character.

Estimation of B can be relatively straightforward. Methods described in the previous section for calculating the genetic correlations between traits can be used to estimate the individual elements of both G and $\text{cov}_g(W, z)$, from which B can be calculated. Alternatively, since twice the deviation of the average half-sib offspring value of a trait estimates the breeding value of the parent for that trait asymptotically as family size increases (Arnold 1981), B can sometimes be estimated approximately as the regression coefficients resulting from a multiple family-mean regression of fitness on the measured characters (e.g., Rausher and Simms 1989).

Testing Hypotheses and Estimating Confidence Intervals

Estimates of genetic parameters are usually sought for purposes of testing a specific hypothesis about evolution of the traits being studied. For these cases, estimating the parameters is not an end in itself; confidence intervals must also be estimated so that hypotheses may be tested. Unfortunately, most genetic parameters are complex values, often ratios, for which the underlying distribution is unknown. In a few instances (such as specific parameters estimated from particular crossing designs) parametric methods have been used to define estimators of confidence intervals (Robertson 1959; Boygo and Becker 1963; Graybill 1976; Knapp et al. 1985; Knapp 1986; Knapp and Bridges 1987). However, in most cases, parametric interval estimators are not available. Furthermore, Arvesen (1969) and Miller (1974a) recommend against such estimators because they are not robust to deviations from normality in the data (Box 1953; Scheffé 1959). Instead, confidence intervals can be generated by resampling schemes such as the jackknife or bootstrapping (Miller 1974a, 1974b; Efron 1979; Efron and Gong 1983; Wu 1986; Knapp and Bridges 1988). Widespread availability of fast computers has only recently facilitated the use of nonparametric methods of this type, and they are likely to increase in importance in many statistical applications. Finally, whenever many different genetic parameters are compared within a single study, the increased probability of committing a type I error (falsely rejecting the null hypothesis; Sokal and Rohlf 1981) should be considered (Rice 1989).

Limitations of Quantitative Genetics

Although the methods of quantitative genetics provide valuable techniques for studying the evolution of continuously distributed traits such as plant resistance, we should also understand the limitations and assumptions of these techniques. First, it should be understood that the heritability of a trait is not some fundamental constant (Barker and Thomas 1987). As an observed ratio between additive and phenotypic variance, and because the latter value depends upon both environmental and genetic variance, the heritability of a trait may differ among populations of the same species and among samples of the same population exposed to different environments. It is especially important to recognize that the heritability of a trait measured in a laboratory-reared population is likely to differ from the heritability of the same trait in the same population in a natural environment. These same cautionary notes hold for genetic correlations between traits; they are specific to the particular population and circumstances in which they are measured.

Quantitative genetics is constrained by the same limitations and assumptions as its statistical procedures. Many of these assumptions are reviewed in detail by Mitchell-Olds and Rutledge (1986) and more recently by Barton and Turelli (1989). To fulfill the Gaussian criteria of quantitative genetic models, the phenotypic expression of a trait must be normally distributed. The phenotypic distribution of most traits can be rendered normal with the appropriate transformation (Powers 1950), although no single transformation is suitable for all situations (Barton and Turelli 1989). Fisher (1918) proposed that such a normal distribution occurs because phenotypic expression of a quantitative trait is determined by both the environment and a large number of independently segregating loci, each with effects that are equal, additive, and small relative to the environmental effect. It is this explanation for the Gaussian distribution of a trait that motivates and justifies application of statistical descriptions of polygenic evolution (Barton and Turelli 1989). In general, however, the nature of quantitative traits usually precludes confirmation of these underlying assumptions. Even observation of a normal phenotypic distribution is not sufficient, because normality can be quite robust to violation of the assumption of many genes, each with small effect (e.g., Hammond and James 1970). Turelli (1988) and Slatkin (1987) have recently begun exploring the implications of violations of these assumptions and are constructing less restrictive models. More theoretical work is needed in this area.

Empirical studies are also needed. Several studies have attempted to estimate the number of loci determining various quantitative traits (called quantitative trait loci, or QTLs). Two methods have been employed. The Wright-Castle method uses biometrical techniques to estimate an effective number of loci from a cross between two populations (Castle 1921; Wright 1968, 381–391; 1978, 336–345; Lande 1981). The second method employs genetic markers to identify chromosome regions containing QTLs. In the past, visible markers were used for this purpose (e.g., Shrimpton and Robertson 1988). More recently, molecular methods have greatly increased the number of available markers, and QTLs are now sought using enzyme markers (e.g., Tanksley et al. 1982; Weller 1987) and restriction-fragment-length polymorphisms (RFLPs) (e.g., Edwards et al. 1987; Steuber et al. 1987; Nienhuis et al. 1987; Paterson et al. 1988). Unfortunately, because both methods count the number of alleles that contribute to phenotypic differences between populations or inbred lines, and because such differences have often been produced by artificial selection, they are biased toward finding alleles of major effect (Lande 1981). Thus, we still do not even know whether the number of QTLs responsible for genetic variation in quantitative traits is small (5 to 20) or large (> 100) (Barton and Turelli 1989).

Assumptions about the mode of gene action are also problematic. All models of quantitative genetics are strictly applicable only when all genetic variance is additive (Falconer 1981). It is not clear how components of dominance and epistatic variance will influence the long-term evolution of a trait in response to selection. Bryant et al. (1986) demonstrated that genetic drift, caused by passage of laboratory housefly populations through experimental bottlenecks, can increase the additive variance of traits. This effect could be due to an increase in epistatic variance and conversion of epistatic variance into additive variance during genetic drift (Goodnight 1987, 1988). It is not inconceivable that selection could have similar effects on genetic variance under certain circumstances. Although we do not know how frequently genetic variation in quantitative traits is determined by additive gene action, there is plenty of empirical evidence that dominance, epistasis, and pleiotropy are common (Kearsey and Kojima 1967; Robertson 1967; Wright 1978; McKenzie et al. 1982; McKenzie and Purvis 1984; Kinghorn 1987; Lenski 1988a, 1988b). More empirical studies are needed before we can make general statements as to the relative importance of various forms of gene action in the expression of continuously distributed traits.

Another important assumption underlying the application of quantitative genetic techniques is that the population being examined is at Hardy-Weinberg equilibrium (Falconer 1981). One important consequence of this assumption is that these techniques can be used to examine patterns of variation only within, and not between, populations. Another consequence is that genetic parameters estimated from small populations evolving by genetic drift or populations that are currently evolving in response to strong selection may be inaccurate.

In addition to these fundamental problems with quantitative genetics theory, there are also practical problems. Interpretation and estimation of quantitative genetic parameters is exquisitely sensitive to the statistical models used to produce them (Ayers and Thomas 1990; Fry, in press). Researchers should understand the implications of various statistical models and seek the assistance of statisticians in utilizing them. Furthermore, the standard errors of genetic parameters are often distressingly large (Mitchell-Olds and Rutledge 1986), which necessitates heroic sample sizes to provide acceptable levels of statistical power. Statistical problems inherent to analyses of selection (Lande and Arnold 1983) were discussed above, where those methods were presented.

Despite the limitations enumerated here, quantitative genetics is an important tool for biologists seeking to understand evolution in natural populations. We must not, however, depend solely on this methodology. As Lewontin (1974a) cautioned, the analysis of phenotypic variance and its partitioning into "causal" components does not truly analyze causes. Such an analysis provides only a statistical description of the action of putative underlying causes. As such, it is an imperfect but nonetheless useful tool for predicting short-term evolutionary change. Manipulative selection experiments can complement this tool by testing what factors influence the accuracy of its evolutionary predictions. Quantitative genetics also plays a valuable role as a theoretical framework for formalizing various evolutionary hypotheses. Finally, it may provide a theoretical structure for research into gene structure and regulation, gene action and interaction, and investigations into the mechanisms by which genotype and environment interact during development to produce the phenotype.

Acknowledgments

The authors gratefully acknowledge support for exploring these methods and preparation of this chapter from the NSF (BSR-8507359 to ELS and MDR, BSR-8817899 to MDR, and BSR-9196188 to ELS).

4

Quantification of Chemical Coevolution

May R. Berenbaum and Arthur R. Zangerl

It was probably some nameless neolithic farmer who first recognized that, within a species, individual plants vary in their ability to resist damage inflicted by herbivorous insects. Although most likely unaware of the genetic basis for such resistance, early agriculturists no doubt profited from genetically based resistance traits by virtue of vegetative propagation; indeed, many of the oldest known domesticated crop plants from southeastern Asia (the "cradle of earliest agriculture"), such as taro, sago palm, bamboo, sugarcane, and yams, were propagated by cuttings or clones (Sauer 1952). New World aboriginal cultures, particularly in the Caribbean and South America, also employed vegetative propagation extensively to maintain such staples as sweet potato and yuca. Even for those crops propagated by seed, agriculturists most likely engaged in selective planting of seeds produced by resistant individuals, probably at times out of necessity, if not by design.

For over eight thousand years, selective planting of resistant genotypes was a hit-or-miss affair. However, almost two hundred years ago, Chapman (1826) had a revelation inspired by events surrounding the introduction of the Hessian fly (*Mayetiola destructor*) into the United States in 1776. Within five years of its introduction, "the important fact became accidentally discovered, certain varieties of wheat are capable of withstanding its attacks. In the year 1781, a prize schooner loaded with wheat, was taken in the Delaware River, and carried into New York, whence the cargo was sent to the mill of Isaac Underhill, near Flushing, Long Island, to be ground. Mr. Underhill's own crop of the previous year having been so entirely destroyed that he had no grain for seed, he took what he required for sowing from this cargo, and reaped therefrom upwards of twenty bushels per acre, whilst few of his neighbors for miles around had any to reap, so calamitous were the operations of the fly. To his praise be it recorded, he distributed his entire crop . . . at a moderate price, among his neighbors, for seed; and all who made use of it were similarly successful. The 'Underhill wheat' at once became noted, for effectually resisting the attacks of the fly, and for many years subsequently . . . was eagerly sought for and success-fully cultivated, where all other varieties of this grain failed." Based on the signal success of Underhill wheat, Chapman was among the first to advocate deliberate plant breeding to facilitate discovery of resistant varieties.

Artificial selection techniques, whereby plant populations were exposed to insect infestation and surviving individuals used for subsequent plantings, dem-onstrated repeatedly that differences in resistance to insect damage can be in-herited. In a review of the subject, Harris and Frederiksen (1984) cited evidence for heritability of resistance to 33 species of insects on 14 crop species. In this review, resistance was dominant for 17 species of insect, recessive for 1 spe-cies, quantitative for 5 species, and assumed more than one form for 10 insect species. For 23 of 33 pest species, resistance involved more than one gene.

Recognition of the fact that host plants can have a selective impact on the insects that eat them was contingent upon the development not only of evolu-tionary theory but also of insect taxonomy. Observing the host shift of *Rhago-letis pomonella* from its native *Crataegus* host to introduced *Malus* species (and its attendant shift from noneconomic to economic status), Walsh (1867) was among the first to describe host races within a species and to attribute them to evolution, resulting from "attachment" to different host plants. Just barely 12 decades after this great insight, there is still a paucity of evidence in support of the genetic bases for adaptation to host plants. Curiously, it was the interaction between Hessian fly and wheat (which helped to inspire the practice of plant breeding for resistant varieties) that provided the first suggestive evidence of genetically based resistance in insect populations to plant varieties. Painter (1930) noted that certain wheat varieties were prone to different patterns of Hessian fly infestation when planted in different parts of the state of Kansas; these patterns were reproducible in a common greenhouse environment, using flies from different parts of the state. In rearing experiments, flies differing in infestation capacity maintained differences over three generations; moreover, a mass selection experiment led to differential infestation capacity within a fly population. These results led Painter to conclude that "these biological or phys-iological strains of fly are genetically distinct," arising as a result of natural selection.

As for the nature of the genetic changes associated with host shifts, perhaps the most thoroughly documented are those of host-associated populations of none other than *Rhagoletis pomonella*, the insect whose host shift first led to Walsh's speculations about host race formation. Feder et al. (1988) and Mc-Pheron et al. (1988) demonstrated gene frequency (electrophoretic) differences between sympatric host races and attributed them to differential selection by and restricted gene flow between two host species; at least part of the restricted gene flow may be attributable to heritable differences in postdiapause eclosion time associated with differences in host-plant phenology (Smith 1988) or to genetic differences in host-fruit acceptance behavior (Prokopy et al. 1988).

Considerable (and at times acrimonious) debate has raged over the relationship between the evolution of resistance to insects in plants and adaptation to plant hosts by insects, particularly in natural systems. At least two schools of thought on the matter exist. According to Jermy (1984), insects do not constitute a major selective force in plant evolution; consequently, insect adaptation to host plants occurs subsequently to or independently of plant evolution. In contrast to such "sequential evolution," Ehrlich and Raven (1964) proposed that insects and plants coevolve—that is, act as reciprocal selective pressures. While Ehrlich and Raven (1964) based their description of the process on macroevolutionary patterns of host-plant utilization and butterfly relationship, Janzen (1980b) explicitly redefined the term to refer to microevolutionary processes at the population level: "an evolutionary change in a trait of the individuals in one population in response to a trait of the individuals of a second population, followed by an evolutionary response by the second population to the change in the first."

Although these schools differ dramatically with regard to many details, they part company most dramatically when considering the role of plant chemistry in the evolution of insect–host-plant associations. The first suggestion that phytochemical differences between plants could result from depredations of herbivorous invertebrates appeared in the literature over a hundred years ago, when Stahl (1888) and Errera (1887) (cited in Berenbaum 1986) raised the possibility that the multitudes of plant chemicals with no known physiological function may be important in protecting plants against their enemies. Subsequently, in this century, literally hundreds of studies have been conducted that demonstrate physiological or behavioral effects of plant chemicals on insects—and indeed, even prior to the suggestion of a natural function for these compounds, plant products were in widespread use for insect control on crop plants (Smith and Secoy 1981). Fraenkel (1959) suggested that insects were largely responsible for maintaining phytochemical diversity in angiosperms, and in fact that insect herbivory was the "raison d'être" for these compounds in plants. Ehrlich and Raven (1964) further proposed that increasing chemical diversity in plants leads in turn to increasing insect diversity.

As hundreds of studies demonstrating physiological and behavioral effects of plant chemical constituents on insects were being published, however, hundreds more studies were simultaneously being published documenting the effects of these phytochemicals on bacteria, fungi, viruses, other plants, other animals, and even humans. Thus, those disinclined to believe that insects can act as selective agents on plants in the first place were unwilling to believe that insects are selective agents capable of effecting changes in plant chemistry. To quote Jermy (1976), "These substances . . . turned out to be important information carriers in plant-plant interactions (allelopathy) and probably even more important as factors of resistance against diseases . . . Therefore, resistance to insect attacks cannot be the 'raison d'être' of these substances."

Empirical Estimates of Selection

In order to distinguish between the putative processes of sequential evolution and coevolution, and, accordingly, to determine the importance of plant chemistry in the evolutionary outcome of insect-plant interactions, it is first necessary to find ways to measure the selective impact of insects on their host plants and the selective impact of plants on their insect associates. The methods of quantitative genetics are well suited to this enterprise. Among other reasons, in natural situations, selection is likely to affect many characters at the same time, rather than single characters in isolation, and the methods of quantitative genetics are most appropriate for evaluating such selection (Lande and Arnold 1983). Even in agricultural situations, in which selection may be so intense as to favor the evolution of monogenic resistance (Day 1974), there are many examples of quantitative resistance in crop plants to insect pests (tables 4.1 and 4.2). By the same token, adaptation to new host plants in all probability results from simultaneous selection on a range of phenotypic characters (Thompson et al. 1990).

Quantitative genetics methodologies are conveyed in detail above in chapter 3 by Simms and Rausher; for the purpose of this discussion, suffice it to say that the introduction of these techniques to the ecological literature (in a landmark paper by Lande and Arnold published in 1983) allowed the design of empirical tests of the impact of insect herbivory on plant chemical traits in particular and plant fitness in general and allowed assessment of the impact of plant chemistry on insect adaptation in particular and insect fitness in general. Despite the existence of an experimental framework, designing experimental details has continued to be a daunting prospect. Essential elements (crucial in testing for selection responses) such as "fitness," "plant chemistry," and "resistance" have proved distressingly refractory to simple quantification. Yet another difficulty is in identifying appropriate variables to examine; regressions of characters on fitness are extremely sensitive to the number and type of characters

TABLE 4.1
Quantitative genetic resistance to insects in crop plants

Crop	Insect	Reference
Alfalfa	Spotted alfalfa aphid	Gallun and Khush 1980
Corn	Corn earworm	Gallun and Khush 1980
	European corn borer	Gallun and Khush 1980
	Fall armyworm	Gallun and Khush 1980
Potato	Green peach aphid	Smith 1989
Rice	Brown planthopper	Smith 1989
Sorghum	Chilo partellus	Smith 1989
	Shoot fly	Gallun and Khush 1980
Soybean	Mexican bean beetle	Smith 1989
Squash	Squash bug	Gallun and Khush 1980
Wheat	Cereal leaf beetle	Gallun and Khush 1980

TABLE 4.2
Estimates of heritability of resistance to insects in natural (N) and managed (M) ecosystems

Plant	Insect	h^2
Brassica oleracea (M)	Plutella xylostella	.75–.88[a]
Lycopersicon esculentum (M)	Heliothis armigera	.57–.90[b]
Medicago sativa (M)	Empoasca fabae	.31–.56[c]
Vigna unguiculata (M)	Lygus hesperus	.49–.72[d]
Pastinaca sativa (N)	Depressaria pastinacella	.94[e]
Solidago altissima (N)	Philaenus spumarius	.62[f]
	Uroleucon spp.	.49–.52[f]
	Exema canadensis	.66[f]
	Microrhopala vittata	.78[f]
	Trirhabda spp.	.68[f]
	Epiblema spp.	.5–1.5[f]
	Rhopalomyia solidaginis	1.34[f]
	Eurosta solidaginis	1.12[f]
Salix lasiolepis (N)	Euura sp.	.28–.30[g]
	Phyllocolpa sp.	.48–.51[g]
	Pontania sp.	.09–.11[g]

[a]Lin et al. 1984. [b]Kalloo et al. 1989. [c]Soper et al. 1984. [d]Bosque-Perez et al. 1987. [e]Berenbaum et al. 1986. [f]Maddox and Root 1987. [g]Fritz and Price 1988; broad-sense heritability.

entered into the regression in the first place (Pease and Bull 1988). Even if variables are identified and quantified, there is limited consensus on interpreting quantitative genetics data; statistical distributions are unknown for many calculated variables (Via 1984).

Such obstacles would seem to suggest that time and effort might be more profitably spent in other pursuits but for the fact that ascertaining the genetic basis of chemical coevolution, or reciprocal adaptive evolution as mediated by chemicals, can be tremendously useful. In an applied context, such understanding can provide new insights into plant breeding for resistance, an occupation still carried out largely as it was eight thousand years ago. In a theoretical context, establishing the genetic basis for chemically mediated plant resistance and physiological and behavioral insect adaptations can lead to restructuring of the framework in which plant-insect interactions are currently examined. Thus, a discussion of the requirements for appropriate quantification of chemical coevolution seems a constructive enterprise.

Measuring Plant Chemical Traits

How Many Chemicals?

A chemical trait is not as easily recognized in a plant as a morphological trait; half the battle in measuring a chemical trait is defining it. Thus, it is not surprising that plant chemistry as a possible source of resistance was not recognized until early this century, at least a hundred years after people freely acknowledged the importance of morphology and phenology in conferring resistance. One of the earliest suggestions came from Collins and Kempthorne (1917) who, in breeding for resistance in sweet corn to corn earworm, postulated that "at least a part of the immunity is the result of chemical differences, perhaps the presence of some volatile substance distasteful alike to the moths and larvae."

Given that a single plant can contain hundreds of constituents in its essential oil alone (Guenther 1948), it is virtually impossible a priori to guess which particular compounds are likely to be important with respect to insect colonization or growth. This embarrassment of chemical richness presents the investigator with several options. One approach is to quantify collectively all constituents belonging to a single class of compounds, defined either biosynthetically (sharing a common precursor), structurally (sharing a common chemical configuration), or operationally (sharing a common chemical, physical, or biological property). This approach has been enormously popular among ecologists in the last two decades—investigations of total alkaloids (or total Dragendorf-positive constituents) and total phenolics (total Folin-Denis positive constituents) are commonplace in the ecological literature (Swain 1979; Robinson 1979).

An alternative to quantifying whole groups of chemicals is isolating and quantifying a single chemical, preferably one with known behavioral or physiological significance to the interacting insect species. There is indeed evidence that single compounds can have profound effects on herbivorous insects. The glucosinolate sinigrin from cruciferous plants, for example, is a spectacular feeding stimulant for *Pieris rapae,* the cabbageworm, a crucifer specialist (Feeny 1977), as are individual cucurbitacins for diabroticite beetles feeding on cucurbits (Metcalf et al. 1982).

Neither of these approaches however, is ideal, in that as a rule herbivorous insects in general use neither approach in evaluating or processing host plants. There is very little evidence that insects are physiologically or neurologically capable of "summing up" total concentrations of a class of compounds; there is at the same time a large body of evidence that individual compounds are rarely sufficient in eliciting a full behavioral or physiological response to a host plant (Stadler and Buser 1984; Feeny et al. 1989; Berenbaum and Neal 1985; Renwick 1988).

Thus, yet another approach to evaluating the chemical profile of the plant is to quantify many different chemicals. The question, however, remains: how many chemicals are enough? Factors to consider include not only the number of individual compounds, which may exert effects individualistically or synergistically (Berenbaum 1985), but also the number of classes of compounds, since the effects of synergistic interactions between biosynthetically distinct chemicals on insect behavior and physiology are well documented (Stadler and Buser 1984; Feeny et al. 1989; Berenbaum and Neal 1985). The only definitive source for an answer to that question is the insect of interest. Correlative associations between chemical phenotypes and insect herbivory in field situations (table 4.3) can be used not only to determine whether chemistry can account for variation in herbivory patterns but also to indicate directions for more definitive investigation. Laboratory bioassays (confirming correlative associations in field observations) are a necessary element in conducting any study purporting to document the selective impact of plant chemistry on herbivorous insects.

To illustrate this approach, we can cite our own work on *Pastinaca sativa*, the wild parsnip. *P. sativa* produces at least five furanocoumarins, compounds with broadly biocidal properties (see Zangerl and Bazzaz, this volume, for details). The principal insect associate of *P. sativa* throughout eastern North America is *Depressaria pastinacella* (Lepidoptera: Oecophoridae), the parsnip webworm, which feeds on developing buds, flowers, and fruits. Resistance to parsnip webworm in a common garden of 20 maternal half-sib families of *P. sativa* was, as determined by stepwise multiple regression, associated with high levels of two furanocoumarins, bergapten and sphondin (Berenbaum et al. 1986). However, furanocoumarins tend to covary with both nitrogen and water (Berenbaum 1981b), two factors with profound effects on insect growth and feeding behavior (Slansky and Rodriguez 1987). Therefore, we examined the effects of variation in both nutritional and furanocoumarin content of flowering parts on webworm growth and development as well as the effects of one of these compounds, bergapten, on webworms when incorporated into artificial diet (Berenbaum et al. 1989). Nitrogen content, one aspect of the nutritional value of flowers, accounted for a significant amount of variation in approximate digestibility, as did xanthotoxin, despite the fact that nitrogen and xanthotoxin concentrations were themselves negatively correlated. Moreover, bergapten content was negatively associated with approximate digestibility. When administered in artificial diet, bergapten at comparable concentrations significantly

TABLE 4.3
Correlative associations between chemical phenotypes and insect herbivory in plant populations

Plant	Insect	Chemical
Cardamine cordifolia	"insects"	Glucosinolates[a]
Larrea divaricata	"canopy arthropods"	Resin[b]
Lupinus spp.	*Glaucopsyche lygdamus*	Alkaloids[c]
Pastinaca sativa	*Depressaria pastinacella*	Furanocoumarins[d]
Pinus ponderosa	*Dendroctonus brevicomis*	Terpenes[e]
Pseudotsuga menziesii	*Choristoneura occidentalis*	Terpenes[f]

[a]Louda and Rodman 1983. [b]Lightfoot and Whitford 1989. [c]Dolinger et al. 1973. [d]Berenbaum et al. 1986. [e]Sturgeon 1979. [f]Cates et al. 1988.

lowered both growth rate and approximate digestibility. In contrast, xanthotoxin, an isomer of bergapten, was positively associated with relative growth
rate and approximate digestibility at equal concentrations in artificial diet. Subsequent work revealed that the differential effects of the two isomers may be
attributable to pronounced differences in the efficiency with which these compounds are metabolized (Zangerl and Berenbaum, in preparation). It is therefore not at all surprising, given the completely opposite effects of just two of
the five or more furanocoumarins present in *P. sativa,* that a measure of "total
furanocoumarins" did not account for a significant amount of variation in parsnip resistance to webworm infestation (Berenbaum et al. 1986).

Quantification: How Much of Each?

Rarely is one so fortunate in a study of plant-insect interactions to come across
a system in which critical chemicals are either present or absent. Even in species in which control of expression is monogenic, mutants almost invariably
produce low (rather than no) amounts of a chemical. The lupine alkaloids are a
case in point. Of the 18 mutants at the locus controlling alkaloid synthesis, all
produce low (less than 1%) but measurable quantities of alkaloids (Waller and
Nowacki 1978). Quantitative traits—those displaying continuous phenotypic
variation—must therefore be measured in some manner. There are two standard
approaches for quantifying plant chemistry: expressing amount relativized with
respect to either weight or area, or expressing total amount in absolute terms
with respect to particular plant parts (e.g., total furanocoumarins per seed; Berenbaum et al. 1984).

A factor complicating the accurate estimation of concentration is that insects and humans do not sample plant tissue in similar ways. The effective concentration of a chemical experienced by an insect can depend to a great extent
on its behavior. The feeding mode, for example, can greatly reduce exposure of
the insect to plant allelochemicals. Furanocoumarins in plants in the Apiaceae
are localized in seeds in storage structures called vittae (Zangerl et al. 1989)
and in vegetative parts in "companion canals" (Bicchi et al. 1990). Thus, *Aphis
heraclella* encounters virtually no furanocoumarins as it feeds on phloem of
Heracleum sphondylium (Camm et al. 1976) and *Lygus lineolaris,* the tarnished
plant bug, sucks fluids from seeds without contacting the vittae at all (Flemion
and McNear 1951). Even foliage-feeding insects can avoid much of the allelochemical content of their host plants. *Trichoplusia ni,* the cabbage looper, in its
early instars skeletonizes foliage, avoiding major veins. By feeding in this manner, it can presumably avoid encountering the furanocoumarins localized in
companion canals adjacent to the veins (Zangerl 1990). The lack of a correlation between chemical variation (as measured in bulk plant extracts) and insect
herbivory may not necessarily imply a lack of importance of plant chemistry in
determining patterns of insect distribution and abundance; it may mean that the

chemical content of the plant tissue was not measured in an entomologically relevant manner.

Induced versus Constitutive Chemical Content

The fact that damage can change the chemical composition of plant tissues presents technical challenges to the investigator. Damage can produce two types of chemical changes: it can result in the de novo biosynthesis of compounds absent from intact leaves (or present in undetectable levels), and it can bring about enhanced biosynthesis and elevated concentrations of compounds already present in plant tissues (Bailey and Mansfield 1982). Again, it is virtually impossible to predict a priori whether constitutive or induced levels of compounds are more important as selective agents (Zangerl and Berenbaum 1990). In a system in which defensive chemicals are induced, measurement of chemical resistance traits is wholly dependent upon the action of the herbivore—an additional complication in studies purporting to measure selective impact of the herbivore by comparing damaged and undamaged populations. With such an approach, it is particularly important to ensure that undamaged populations remain undamaged, since even low levels of infestation may suffice to induce resistance factors, thereby obliterating differences between purported treatments.

Artifacts of Sampling

Yet another complicating factor in measuring chemical variation in plants is that different methods of extraction and quantification can yield wildly different estimates of concentration. Martin and Martin (1982), for example, estimated phenolic content of red oak foliage in six different ways and ended up with essentially six different rankings of their foliage samples. Assays based on biological properties of tannins (e.g., protein precipitation) tended to yield results that were more similar to each other than they were to assays relying on molecular structure (e.g., Folin-Denis or proanthocyanidin assays).

Even the same assay can produce different results, depending upon sample preparation. Again, in quantifying tannins, Hagerman (1988) found that the amount of tannin extracted was influenced by the way in which leaf samples were processed and the type of solvent used for extraction. Lindroth and Pajutee (1987) compared several different methods of sample preparation in estimating phenolic glycoside content of foliage from *Populus* and *Salix* species and found that leaf treatment as well as solvent extraction conditions affected the apparent phenolic glycoside content, even to the extent of producing artifactual glycosides not present in intact foliage.

These problems are particularly acute for those compounds that are destructively sampled. The concentration of cyanogenic glycosides, for example, can be estimated by the amount of cyanide they release when they are hydrolyzed (Conn 1979). The amount of cyanide released, however, does not give a clear indication of the number and relative concentrations of different cyanogenic glycosides present in the plant tissue prior to hydrolysis. The number and type of cyanogenic glycosides may be critical in determining the outcome of an encounter between a cyanogenic plant and an herbivore. For example, *Heliconius* caterpillars appear to possess substrate-specific beta-glucosidases that can inactivate plant glucosidases and prevent hydrolysis and subsequent release of cyanide (Spencer 1988a). The amount of cyanide actually released during herbivory may depend on the structure of individual glycosides rather than on the total amount of glycosides present.

Even for nondestructive sampling procedures, artifacts are facts of life. Chemicals other than the ones of immediate concern present in plant tissues can interfere with extraction efficiency (e.g., tannins interfere with the extraction and purification of just about everything else; Swain 1979), or they can simply mask the presence of the target compounds (e.g., by absorbing at the same wavelengths or by reacting with a diagnostic reagent. Thus, nonalkaloids such as conjugated carbonyl compounds can give a false Dragendorf's test by reacting with Dragendorf's reagent, and plant pigments can give a false test result by masking the color reaction; Robinson 1979).

Measuring Plant Resistance

Estimates of plant chemical variation are not necessarily estimates of plant resistance. Ascertaining the relationship between plant chemistry and resistance necessitates an independent measure of resistance. Painter (1951) defined resistance of plants to insect attack as the "relative amount of heritable qualities possessed by the plant which influence the ultimate degree of damage done by the insect." If this definition is acceptable, then measuring resistance means— at some point along the way—measuring insect damage. Unfortunately, Painter (1951) did not provide a definition for "damage."

Damage can be measured in any number of ways, which can be classified as either herbivore-free (that is, measuring the effects or consequences of herbivory) or herbivore-dependent (that is, tallying the number of herbivores). Herbivore-free estimates include area of plant tissue removed, weight of plant tissue removed, number of feeding or oviposition punctures, reduction in plant growth rate, reduction in plant photosynthetic rate, and the ultimately indisputable evidence of damage, mortality. Herbivore-dependent estimates include herbivore load (defined by Root [1973] as biomass of herbivores/100g plant), number of herbivores per plant or plant part, herbivore biomass per plant or plant part, and number or biomass of herbivores per plant biomass.

Of these alternative approaches, measuring damage independent of herbivores is in all probability a more accurate approach to obtaining an estimate of plant resistance. The actual number of herbivores on a plant, for example, may be inconsequential to the plant if none of the herbivores are feeding or ovipositing on it. However, for those herbivores whose damage does not consist of removing easily quantifiable amounts of plant tissue (e.g., sap-sucking homopterans), estimating the number of presumably feeding individuals may be the only practical route to take.

Measuring the amount of plant damage sustained may convey a confusing or misleading impression of the amount of resistance, depending on the mechanism for resistance in the plant in question. Tolerance, for example, is defined by Painter (1951) as "a basis of resistance in which the plant shows an ability to grow and reproduce itself or to repair injury to a marked degree in spite of supporting a population approximately equal to that damaging a susceptible host." A tolerant genotype may actually sustain even greater amounts of damage than would a susceptible genotype yet fail to suffer a fitness loss; wound repair, compensatory regrowth, or tissue replacement may not be measured in a timely fashion (or even be observed) by an investigator. In this form of resistance, which can be genetically based, damage may be entirely inadequate as an indicator of genotype.

One final complication in using damage as an index of resistance is that local variation in herbivore population densities may generate apparent differences in damage levels; these damage levels, however, are more estimates of variation in herbivore population densities than of variation in plant resistance. Even the most susceptible genotype experiences no damage in the absence of herbivores. This notion is far from abstruse (and in fact is the basis for a joke, popular with third graders, about the Illinois farmer and the elephant repellent) yet is extremely difficult to control for in ecological studies of plant resistance. Common garden experiments can control for local variation in herbivore population densities only if they are situated in an area known to be populated by the desired damaging species or if they are caged or enclosed in an area free from herbivores and artificially infested by herbivores provided by the experimenter (an approach fraught with perils of its own).

Measuring Plant Fitness

In order to estimate selection differentials, the covariance between fitness and the trait in question must be measured (Lande and Arnold 1983). It is therefore necessary to devise a method for measuring fitness. Fitness is yet another property of plants (as well as of insects) that is easier to define than it is to measure. Since it is generally defined in the context of reproductive success, one logical way to measure plant fitness is to measure reproductive output. This task, however, is deceptively simple. Fitness has been measured both directly, as weight or number of seeds, and indirectly, with traits that are correlated with seed production, such as total plant biomass, stem diameter, or flower number. For all but monocarpic plants, these estimates do not measure lifetime fitness unless they are measured over the (often extended) lifetime of the plant. Furthermore, counting or weighing seeds does not take into account differential viability of seeds (Hendrix and Trapp 1989).

Additional complications arise when the herbivore under study actually consumes seeds. A measure of resistance is thus at the same time also a measure of fitness. These measures are by definition not independent. Rausher and Simms (1989) in a study of resistance to herbivores in morning glory *Ipomoea purpurea* reported that *Heliothis zea* (Lepidoptera: Noctuidae) had a significant impact on plant fitness. Since this insect consumes flowers and seeds, the measure of fitness (number of seeds) and the measure of resistance (number of pods damaged) may not have been independent. If pod number is correlated with seed number, then a significant directional selection gradient may simply reflect the underlying correlation between pod number and seed number, rather than any effect of the insect per se.

Yet another complication arises when the mechanism of resistance to herbivory involves tolerance—ability to withstand damage without a loss of fitness. Damage estimates do not under these circumstances provide an accurate estimate of resistance. According to Smith (1989), in a discussion of practical methods for evaluating crop plants for resistance, "entirely different techniques are employed to assess tolerance [as opposed to antixenosis or antibiosis], because it does not involve a plant interaction with insect behavior or physiology." It is entirely possible that many studies failing to document an association between insect damage and fitness reduction, often cited as evidence that insects do not have a selective impact on plants (e.g., Jermy 1984), may actually have involved genotypes that have evolved resistance to insect damage through tolerance mechanisms (Zangerl et al., 1991).

Painter (1951) established three criteria for establishing the cause of resistance in plants to insect herbivory: first, experimental demonstration of "an intimate relation" between the putative resistance factor and insect behavior or physiology; second, evidence for the genetic control of the resistance factor, particularly in contrasting resistant with susceptible genotypes; and third, statistical correlation between the putative resistance factor and measures of resistance in the field. In virtually no study in the ecological literature have these three criteria been met. In fact, the prevailing opinion among many is that it is not necessary to identify resistance factors per se in order to discuss such arcana as costs of resistance or selective impacts of insects on resistance. Indeed, agriculturalists interested only in increasing yields have no need for identifying underlying mechanisms of resistance to accomplish their short-term goals. However, in order to address ecological and evolutionary questions, explanations must be more powerful than, effectively, "Resistant plants owe their resistance to the fact that insects don't eat them as much." Although quantitative genetic techniques are powerful, they are essentially statistical tests; significant correlations may not represent causation.

One study in which correlations led to an incorrect assumption about the mechanism of resistance involves plants in the Cucurbitaceae and their arthropod enemies. Cucurbitaceous plants are distinctive in that many species in the family produce cucurbitacins, triterpenoid compounds that are extremely bitter. Bitterness in *Cucumis sativus* is regulated by one major gene, Bi, whose expression is influenced by additively inherited intensifier genes. Selection of lines resistant to *Tetranychus urticae* spider mites led to an increase in bitterness, leading to the conclusion that the bitter principle was in fact the factor conferring resistance against spider mites (DaCosta and Jones 1971; Kooistra 1971; and Gould 1978). However, dePonti and Garretsen (1980) demonstrated that there is no relationship between the gene Bi and acceptance and oviposition by the mites; the association between Bi and degree of damage can be attributed to linkage between the gene Bi and the gene (or genes) for resistance.

Measuring Insect Responses to Plants

Reciprocal selection between plants and insects implies that a change in the distribution of plant genetic traits can cause changes in the distribution of genetically based traits in insect populations. Although the existence of insect biotypes (genetically differentiated insect populations associated with genetically differentiated crop cultivars) has been well documented and widely recognized in agriculture for over fifty years (Painter 1930), attempts to document such changes in natural situations have been written off as largely unsuccessful (Bernays and Graham 1988; Futuyma and Moreno 1988). Most of these attempts, however, have involved putative trade-offs in performance between populations of polyphagous species (Via 1984; Futuyma and Philippi 1987; Pashley 1988—but see Rausher 1984b). Limiting the search for trade-offs (or negative correlations) in performance to two host-plant environments is not necessarily justifiable for those species with potentially dozens or even hundreds of host plants. In trade-offs involving whole suites of traits, positive covariances may occur between any two traits selected at random (Pease and Bull 1988). Generally, these studies have examined populations of polyphagous species utilizing several host-plant species within a relatively narrow geographic range; insect genotypes from such populations are far more likely to be able to utilize any two host plants from within their range than would be genotypes that do not normally encounter all possible hosts within their range. In agricultural situations, biotypes are far more likely to arise under circumstances in which only a single host species or cultivar is utilized by an oligophagous species over an extensive geographic area (e.g., wheat, corn, or rice) (table 4.4).

Ecological studies of trade-offs associated with host shifts also have generally involved measurement of physiological parameters (the genetic bases for which are not known) in laboratory environments; as is true for most quantitative genetic measures, estimates of genetic correlations apply only to the situations in which they are measured, and it may well be true that in laboratory

TABLE 4.4
Ecological attributes of insect species forming biotypes

Insect species	Life history[a]	Diet breadth[b]	Host family	Growth habit[c]
Acyrtosiphon pisum	Parth	Oligo	Fabaceae	Per
Amphorophora rubi	Parth	Oligo	Rubiaceae	Per
Brevicoryne brassicae	Parth	Oligo	Brassicaceae	Ann, Per
Eriosoma lanigerum	Parth	Oligo	Rosaceae	per
Mayetiola destructor	Sex	Oligo	Poaceae	Ann, Per
Nephotettix spp.	Sex	Oligo	Poaceae	Ann, Per
Nilaparvata lugens	Sex	Oligo	Poaceae	Ann, Per
Orseolia oryzae	Sex	Oligo	Poaceae	Ann, Per
Phylloxera vitifoliae	Parth	Oligo	Vitaceae	Per
Rhopalosiphum maidis	Parth	Oligo	Poaceae	Ann, Per
Schizaphis graminum	Parth	Oligo	Poaceae	Ann, Per
Therioaphis maculata	Parth	Oligo	Fabaceae	Ann, Per

Sources: Kennedy 1978; Gallun and Khush 1980; Smith 1989.
[a]Parth = parthenogenetic; Sex = sexual.
[b]Oligo = oligophagous.
[c]Ann = annual; Per = perennial.

colonies, where insect whims and desires are met, trade-offs in performance on different hosts are not detectable (Rausher 1988b). Moreover, preference and performance are both attributes upon which selection can act (Singer et al. 1988; Gould 1984; Lockwood et al. 1984), but there is no a priori way to predict whether selection acts first or most frequently upon preference or performance (Berenbaum 1990). Exposure to synthetic organic insecticides as selective agents has resulted in populational differences in behavior—exophily in anpheline mosquitoes, for example (Matsumura 1975)—and there is no compelling reason to assume that changes in plant chemistry could not do the same. Behavioral change is particularly important in situations in which antixenosis (nonpreference) is the mode of resistance. Genetic changes in preference may entail changes in both feeding and oviposition behavior.

Studies purporting to examine (and failing to document) evolutionary trade-offs associated with host shifts should therefore take into account the inherent limitations of such studies. Futuyma and Philippi (1987), for example, examined performance of parthenogenetic genotypes of *Alsophila pometaria,* a polyphagous geometrid caterpillar, on four different plant species; after correcting for differences among genotypes in general vigor, they found, of nine possible correlations, two that were significantly positive and one that was significantly negative (for performance on white oak and chestnut). On the basis of their findings, they concluded that "genetic factors enhancing performance on one host do not generally have strong antagonistic pleiotropic effects on performance on the other hosts in this population." Considering the odds against detecting any sort of trade-offs, finding almost as many negative correlations as positive correlations is remarkable in itself.

Measuring Behavioral Traits

Behavioral resistance to host-plant chemicals can be defined as the failure of an insect to experience a lethal dose. Quantifying differences in behavior has proved difficult, at least in part because in laboratory situations behavioral responses are more context-specific than are physiological responses, and in field situations, maintaining proper controls offers insuperable challenges. A long-standing controversy exists regarding the relative merits of choice versus no-choice experimental designs. Relative rankings of preference differ accordingly. As to which design is the most appropriate for any given system, the choice should be dictated by the life history and ecology of the insect under investigation. In most holometabolous species, host selection is largely the responsibility of the more mobile ovipositing female. Indeed, butterflies are frequently observed visiting, evaluating, and rejecting individuals within a plant population (Rausher 1980; Papaj and Rausher 1983; Singer et al. 1988). In contrast, caterpillars, particularly in early instars, have little capacity to switch hosts and thus rarely have an opportunity to choose among different genotypes within a population. However, generalizations about insects, particularly about holometabolous insects, are invariably unreliable. A substantial number of caterpillars are sufficiently mobile and hardy that they can move considerable distances and routinely choose among different host plants (*Battus philenor,* e.g., Rausher 1980; *Lymantria dispar,* Barbosa et al. 1981). In such species, choice tests for taste preferences are entirely appropriate.

The technical challenges of designing taste and oviposition preference tests, which take in peculiarities of the different species, have been discussed in detail elsewhere (Miller and Miller 1986). Statistical artifacts are also problematical in the design and analysis of laboratory bioassays (Bernays and Graham 1988). The manner of presentation of plant material can greatly affect outcome of laboratory bioassays (Marquis and Braker 1987; Risch 1985; Jones and Coleman 1988), as can the number of trials (Bernays and Wege 1987).

Measuring Physiological Traits

Phenotypic distributions of physiological traits have certainly been altered in insects by exposure to synthetic organic insecticides; among these traits are those conferring physiological resistance, target site resistance, and metabolic resistance (Brattsten and Ahmad 1986). Measurement of physiological resistance can be direct or indirect. Direct measurement necessitates knowledge of the mechanism of resistance—quantifying, for example, activity levels of cytochrome P450 monooxygenases involved in metabolizing toxins (Hodgson 1985). Since this information is rarely available for natural systems (but see Lindroth 1988b; Nitao 1989; Cohen et al. 1989), indirect measurements of performance, which presumably are correlated with physiological resistance, can be made.

By far the most popular measurements of insect physiological performance are those of digestive efficiency (reviewed by Waldbauer 1968; Kogan 1986). These include ECI (efficiency of conversion of ingested food), ECD (efficiency of conversion of digested food), AD (approximate digestibility), and RGR (relative growth rate). These tend to be measured over the short term—from one day (Berenbaum 1984) to one instar (Scriber and Feeny 1978)—but rarely over the life of the insect (but see Neal 1987). Long-term measures of performance generally include development time and pupal weight.

The digestive efficiency measures are all prone not only to errors in measurement (see Schmidt and Reese 1986) but also to statistical artifacts due to autocorrelations. Farrar et al. (1989) pointed out that measures of consumption rate, relativized for average weight over the duration of a test period, presumably take into account growth and differences in feeding rates associated with growth. However, because growth is a function of digestive efficiency, relative consumption rate (corrected for average weight over the test period) is not an independent measure of behavior; relative consumption rate also encompasses physiological performance. Personal observations show that estimating genetic correlations among these variables is a statistical nightmare; since so many of the parameters involve the same weight measurements, they are of necessity autocorrelated and thus subject to bias.

Genetically Determined versus Environmentally Plastic Traits

Both behavior and physiology are subject to change phenotypically over the course of development of the insect; needless to say, these sorts of changes are not likely to manifest themselves in progeny (although the extent to which phenotypic changes occur may be under genetic control). Behaviorally, exposure to chemicals in early instars may produce learned aversions (Dethier 1980), resulting in nonpreference or reduced consumption rates in later instars or behavioral preferences that increase the likelihood of accepting a marginal host. Habituation to a deterrent can occur across instars or even within an instar (Berenbaum 1986), and motivational factors such as hunger can greatly alter the rank order of behavioral preferences for hosts. Physiologically, exposure to chemicals in early instars can lead to enhanced metabolic rates (Brattsten and Ahmad 1986) or improved performance in later instars (a form of "induced resistance") (Scriber 1981). Even within an instar, prior exposure to an allelochemical can enhance detoxification capacity later in the same instar (Nitao 1989).

Measuring Insect Fitness

Many of the problems complicating the measurement of plant fitness also complicate the measurement of insect fitness. One of the most direct measures is fecundity—offspring production—which can be measured in a variety of ways, including egg number, egg mass, or, in the case of aphids, embryo number (Llewellyn and Brown 1985). Like estimates of seed production, estimates of egg production do not take into consideration egg viability or offspring success. These shortcomings are particularly important in that effects of chemicals on fitness may extend to subsequent generations. Measuring fitness entirely by egg production also does not take into account male contributions to subsequent generations.

Use of traits correlated with egg production as estimates of fitness creates its own set of problems. These traits (such as pupal weight) are often themselves used as estimates of resistance or are correlated with such estimates (e.g., larval weight, a possible resistance measure, which is correlated with pupal weight). Notwithstanding, pupal weight or adult female weight is probably the most widely used index of insect fitness (see review in Nitao 1989). Such measures may be prone to biologically irrelevant associations. For example, Rossiter et al. (1988) found that gypsy moth (*Lymantria dispar*) growth rate was independent of foliar condensed tannin levels, that pupal mass was negatively correlated with constitutive phenolic content, and that fecundity (or egg mass and egg number) was positively correlated with condensed tannins. Reliance upon pupal mass as a measure of insect fitness would have created an entirely different impression of the impact of plant phenolics on gypsy moth reproduction and population growth.

Finding a Candidate Interaction

The importance of chemicals as mediators of insect-plant interactions in ecological time is difficult enough to assess; evaluating the importance of chemicals as mediators of insect-plant coevolution presents almost insuperable challenges. In order to maximize the likelihood of obtaining robust estimates of selection and drawing valid conclusions from such estimates, an investigator must be circumspect in his or her selection of a study system. Such a system has the following characteristics.

1. It has a well-defined natural history, such that the range and host associations of the insects are known, the range and ecological associations of the host plants are known, and the life histories of both plant and insect are such that they are conducive to study. (For instance, studying a host-alternating aphid the primary host of which is unknown presents problems not only in rearing the species in the laboratory but also in determining the nature of evolutionary changes in the aphid populations.)

2. It is well-defined systematically. Carefully constructed scenarios of host race formation can be easily scuttled by taxonomists who demonstrate that "host-specific genotypes" may in fact be different (and not necessarily closely related) species altogether.

3. It has well-known and accurately quantifiable chemistry.

4. It is suited for quantitative genetic studies, that is, the organisms have appropriate breeding biology to permit necessary crosses and have sufficient reproductive potential to generate sufficient numbers of offspring to allow reliable genetic estimates (Mitchell-Olds and Rutledge 1986).

5. It is composed of short-lived species, to facilitate actual selection studies. While selection responses in redwood trees or periodical cicadas are certainly worthy of lifelong study, these sorts of pursuit are unlikely to generate enthusiasm or support after the first few decades without results.

6. It is relatively restricted in dimensions. The use of specialist insects restricted to a narrow range of host plants greatly reduces the numbers of possible combinations and permutations in estimating genetic covariances. While these types of associations are not representative of all, or even most, insect-plant interactions, they at least have the virtue of being relatively easy to recognize and define. Whether polyphagous species are equally polyphagous throughout their range or whether they consist of more specialized subpopulations is still a matter for debate (Fox and Morrow 1981).

Final Words

In his 1990 paper entitled "When is it coevolution?" Janzen made things at the same time elegantly simple and dauntingly difficult. By clearly defining coevolution as a reciprocal evolutionary change in traits in individuals in two interacting populations, Janzen set up unambiguous criteria for recognizing a coevolutionary interaction; he did little however, to assist the next generation of ecologists and evolutionary biologists in finding and documenting those reciprocal evolutionary changes. In the context of chemically mediated plant-herbivore coevolution, such documentation is a multilevel process. In order to state unequivocally that an evolutionary change in a trait in a population of herbivores is a response to an evolutionary change in the chemistry of a plant host, it is necessary to demonstrate (1) that there is measurable variation in the chemical trait, (2) that the variation in the chemical trait is to some degree under genetic control, (3) that the variation in the chemical trait is directly responsible for plant resistance to herbivory, (4) that variation in resistance is responsible in turn for variation in plant fitness, (5) that herbivores respond differentially to plant chemical variation, (6) that the differential herbivore response has a genetic basis, and finally (7) that the genetically based differential herbivore response results in differential perpetuation of herbivore genotypes.

In view of the requirements of "proof," it is hardly surprising that there are to date no definitive studies of a chemically mediated coevolutionary interaction, even in a highly simplified system. There may never be, given the operational difficulties of even defining the requisite parameters, much less measuring them. Quantitative genetics methodologies, however, have contributed enormously to establishing the feasibility of such interactions in allowing documentation of the genetic basis for plant resistance, plant chemistry, and insect responses.

Recent innovations in genetic technologies offer great promise for alleviating other operational difficulties associated with studying phenotypic selection and evolutionary responses to selection in insect-plant interactions. Genes for resistance (that is, in this context, coding for proteins involved in the production of chemicals associated with resistance) not only can be identified, they can be isolated, sequenced, and experimentally altered. Site-specific mutagenesis can be employed to neutralize a purported resistance gene in a plant, and consequent changes in fitness in the presence of herbivores can be measured directly; by the same token, fitness effects of site-specific mutagenesis of a gene associated with insect adaptation to plant defenses (as, for example, a gene coding for an enzyme that detoxifies a particular chemical) can be measured directly. Purported resistance genes (or sets of genes) can potentially be transferred into susceptible species or cultivars and into transgenic plants used both as subjects of study, to confirm the identity of the resistance genes, or as sources of experimental variation, to determine the selective impact of plant resistance on insect populations. Most of these methods (indeed, molecular approaches in general) have yet to be utilized by evolutionary ecologists; it is fervently to be hoped that, as these technologies develop, evolutionary ecologists will exploit them (and in less than the eighty years that passed between the publication of Pearson's 1903 paper on the use of multivariate analyses to measure direct and indirect selection and the publication of Lande and Arnold's 1983 paper applying these techniques to studies of natural systems in an evolutionary context).

Of course, demonstrating the importance of chemistry in a particular plant-insect interaction is not equivalent to demonstrating that genes governing plant chemistry control the nature and extent of plant-insect reciprocal evolution; nor does demonstrating the lack of importance of chemistry in a particular insect-plant interaction indicate that plant chemistry does not ever play a role in determining patterns of insect-plant evolution. The current state of the art, however, is such that data of any sort would be helpful in constructing an overarching hypothesis concerning the evolution of plant-insect interactions. The theoretical possibilities for these interactions are numerous; there is no reason to suppose that nature has not exploited any number of them in actuality.

Acknowledgments

We thank the editors of this volume for their inexhaustible patience and enthusiasm and we thank the National Science Foundation for its generous support of research on plant-insect interactions (including NSF BSR88–18205 to the authors).

PART TWO

Evolutionary Responses to Plant Resistance by Herbivores and Pathogens

5

Evolution of Herbivore Virulence to Plant Resistance: Influence of Variety Mixtures   *Lawrence R. Wilhoit*

Understanding the adaptation of insect herbivores to resistant plant varieties is important to both basic and applied science (Gould 1983). It is important to basic science in providing ideal systems for testing evolutionary theories or, at least, for testing population genetic models of microevolutionary changes. It is important to applied science because of the economic costs of loss of effective resistance in crop varieties as pests evolve adaptions to them.

The study of natural selection in nature is fraught with difficulties (Endler 1986) yet the wealth of opportunities in agricultural systems to understand evolution has been underutilized. Some of the clearest examples of natural selection are provided by adaptations of pests to resistant varieties and to insecticides, but textbooks in evolution usually do not mention these, possibly because agricultural systems are not considered natural. The evolutionary processes that occur in agriculture, however, are natural and may be easier to study than those in more complex natural ecosystems.

The use of resistant crops has been a major means of controlling plant diseases (Van der Plank 1963, 1968) and has been important in managing insect damage (Painter 1951; Maxwell and Jennings 1980; Smith 1989). Resistant varieties offer many advantages, such as ease of use and low cost for the farmer, effectiveness, and relative safety to humans and the environment. Disadvantages are the time and expense needed in breeding resistance genes into crop plants without losing other desirable agronomic traits. These disadvantages are exacerbated by the evolution within a few years of pest genotypes adapted to the resistant crops (Gallun and Khush 1980). These problems have motivated considerable research on genetic and physiological traits contributing to the durability of resistance (Lamberti et al. 1983; Kennedy, Gould et al. 1987). Some plant resistances have proven stable for decades (Painter 1951; Russell 1978; Gallun and Khush 1980; Johnson 1983), but durability is usually difficult to predict, and for many plant-pest systems finding stable resistant factors has proven difficult.

91

Durable resistance can also be achieved by better methods of deploying resistance genes. Possible strategies involve, for example, particular spatial and temporal patterns of resistance deployment (Gould 1983). Spatial patterns include variety mixtures or multilines using different types of resistant varieties or resistant and susceptible varieties, strip cropping different varieties, or planting separate fields with different varieties. Temporal patterns include variety rotations from season to season or variety substitution only when virulent genotypes start to arise. Factors that could affect the rate of herbivore adaptation include the type of resistance (for example, whether it is a polygenic or monogenic trait and whether it achieves resistance through antibiosis, xenosis, or tolerance), the extent of genetic variation and nature of inheritance in the herbivores, the life history traits of the herbivores, and the ecological setting. Population genetic and ecological models will be essential in making these predictions. Few have examined spatial and temporal patterns of resistance deployment, especially for insects (Gould 1986a, 1986b), but these strategies offer many potential advantages. For example, variety mixtures may provide some of the pest-reducing functions of polyculture without the cost of increased farm management complexity. That is, they will increase diversity from the pests' point of view, but maintain uniformity from the farmers' point of view.

Though it is possible to test evolutionary theories and compare deployment strategies through field experimentation, these tests will be limited in many ways. It is impractical to follow the evolution of organisms over large geographic regions, or to isolate a system completely from immigration from surrounding regions, or to follow changes for decades. Thus, most of the work has utilized mathematical models based on studies limited in space and time.

This chapter will focus on variety mixtures as a strategy for deploying plant resistance. First, the concepts of biotype, virulence, and variety mixtures will be discussed. Then the nature and extent of genetic variation in pathogen and insect virulence will be discussed, since this knowledge is basic for understanding evolution. Of particular importance is the cost to the herbivore of having virulent genes when it is on a susceptible host. General theoretical studies from population genetics and ecology that relate to the question of herbivore evolution will then be reviewed, and models for predicting evolution of pathogens and insects in variety mixtures will be discussed. Finally, a simulation model will be presented predicting the evolution of virulence of aphids in fields with a mixture of both resistant and susceptible varieties. Though developed to predict the usefulness of variety mixtures for agriculture, the model can also be applied to evolution of aphids in a more natural setting, where plants typically vary in the levels of resistance to aphids and the aphids vary in levels of virulence. The model calculates the population growth of aphids and predators on individual plants and their movement between plants, all based on physiological processes. The model also includes growth of plants, damage caused by the aphids, and various assumptions regarding the genetic control of virulence.

These studies demonstrate the importance of ecological interactions in understanding evolution, especially in heterogeneous environments.

Definitions

Biotypes

In describing the evolution of pests on resistant varieties, the vague and confusing terms "biotype" (with insects) and "race" (with pathogens) have been used. The typical procedure is to claim that whenever a resistant variety is no longer effective, a new pest biotype has arisen. At present, for example, many plant pathogens have 20 or 30 named races (Robinson 1976); the greenbug, *Schizaphis graminum* (Rondani), has eight biotypes (Puterka et al. 1988); and the Hessian fly, *Mayetiola destructor* (Say), has twelve biotypes (Gallun 1977; Sosa 1981). New biotypes are constantly being added.

The biotype concept, however, is fraught with difficulties (Claridge and Den Hollander 1983; Diehl and Bush 1984; Gallagher 1988) primarily because the genetic structure of biotype groups is unclear. Eastop (1973) suggested that the term "biotype," when used with aphids, meant a clone, that is, a group of genetically identical individuals, but this would produce an unworkable number of biotypes.

A biotype is more widely defined as a population or an individual distinguished from other populations or individuals of its species by some nonmorphological trait, usually host-plant adaptation. However, the genetic makeup of such a biotype is unclear. It is commonly assumed that a biotype is defined by the presence or absence of a particular gene bestowing adaptation to particular resistance genes. This definition assumes a gene-for-gene relationship between virulence in a pest with resistance in a host (Flor 1955). However, in practice, a biotype label is given to a population that possesses a particular set of virulence genes or that responds similarly to a set of resistant varieties without determining whether a gene-for-gene relationship exists. As Van der Plank (1983) points out, however, such a scheme implies millions of different biotypes even when few genes are involved.

In this chapter, the term "biotype" will be used to signify a group of insects that respond similarly in a particular trait to a particular set of plant varieties (a defining set). A biotype group may well consist of genetically different individuals and individuals that respond differently on varieties not in the defining set. No attempt will be made to try to name all possible groups that can be defined this way.

Virulence

Virulence means the ability to overcome the defenses of a host, and usually applies to microorganisms. There are two aspects of this definition: (1) the ability of the organism to grow and reproduce on the host, and (2) the amount of harm caused to the host. While these two aspects may be related, they are not the same thing. This fact may explain an apparent contradiction between two experiments. Eisenbach and Mittler (1987a) crossed two greenbug clones, each of a different biotype, and concluded that the trait distinguishing the biotypes was inherited maternally. Puterka and Peters (1989) performed a similar cross between the same two biotypes and concluded that the trait was inherited in a normal Mendelian fashion. Both conclusions appear valid, but they are not in conflict because the traits measured in the two experiments were different— Eisenbach and Mittler measured total number of progeny produced and adult weight, while Puterka and Peters measured damage to the host plant. It is important to be clear which aspect of virulence is meant. In this paper, "virulence" will usually refer to the ability of insects to grow and reproduce, not to the level of damage to a plant.

Variety Mixtures

Variety mixture refers to a homogeneous, spatial mixture of different genotypes of one plant species in a field. Wolfe (1985) distinguishes these terms as follows: "multiline" refers to mixtures in which the component lines differ only by identified resistance genes; "line mixtures" are developed from lines selected from the hybridization of common parents; and "variety mixtures" consist of varieties differing in many characters, including pest resistance. Marshall (1977) also discusses bulk hybrid varieties that are similar to variety mixtures except that the mixture of lines is mass-propagated for many generations under standard agricultural conditions, allowing for natural outcrossing between plants and natural selection to sort out the superior locally adapted genotypes. In this paper, "variety mixture" will sometimes be used as a general term referring to any of these situations, though it will usually refer to Wolfe's (1985) more specific definition.

Genetic Variation in Pathogens and Insects

How quickly an herbivore adapts to a resistant plant depends in part on the nature and extent of genetic variation in the herbivore population. Numerous studies have found considerable genetic variation in resource use among pathogens (McDonald et al. 1989) and insects (Futuyma and Peterson 1985), which would explain the frequent breakdown in resistance. Much remains to be learned about the characteristics of this genetic variation: how much variation exists within local populations; which traits vary in association with resource use; what the modes of inheritance are; and what correlation exists between traits. To manage resistance deployment, we need to understand herbivore fitness with respect to environmental factors, such as the spatial distribution of hosts with varying levels of resistance. An especially important aspect for evaluating variety mixture strategies is the fitness of different genotypes on different host plants or varieties, that is, whether there is a cost to herbivore virulence. These issues will be discussed for pathogens and aphids.

Genetic Variation in Pathogen Virulence

Many pathogens adapt to resistant varieties within a few years, probably because of the extensive variation in the pathogen population (McDonald et al. 1989; Nelson 1973). Numerous surveys have demonstrated that as many as 40 races of several pathogens exist in the field (Luig and Watson 1977; Groth and Roelfs 1982; Barrett 1987; Burdon 1987a; Groth and Christ, this volume). However, these surveys are difficult to interpret, since the number of races depends on the number of samples and plants used to determine races (Wolfe and Knott 1982). Further, most surveys are done over large areas and include samples from fields with different plant varieties. It is not clear how much variation exists in one field subject to the same selection pressures.

Genetic Variation in Aphid Virulence

As with pathogens, aphids often become adapted to resistant plant varieties within a few years (Webster and Inayatullah 1985). Parthenogenetic reproduction facilitates the spread of favorable mutations, and sexual reproduction gives rise to new sources of variation through recombination. Because high fecundity and short generation times further enhance the opportunities for new genotypes, aphid populations should be genetically diverse. Studies using electrophoretic techniques indicate, however, that aphids may be less variable than other organisms (May and Holbrook 1978; Wool et al. 1978; Baker 1979; Mack and Smilowitz 1980; Tomiuk 1987). Nevertheless, most studies of biological measurements, such as rate of growth on different plant genotypes, photoperiodic response, and thermal requirements for development (Dunn and Kempton 1972; Campbell et al. 1974; Moran 1981; Service 1984b; Wöhrmann 1984; Weber 1985a, 1985b; Via 1989; MacKay 1989), have found considerable variation within local populations and between geographic regions (Pilson, this volume).

Development of Greenbug Biotypes

The potential for rapid evolution of virulence in aphids can create problems for crop breeders in developing resistant varieties. The greenbug, *Schizaphis graminum*, provides an ideal example of the problem of resistance breakdown. In the 1940s, various wheat varieties were tested for greenbug resistance in the United States (Atkins and Dahms 1945). The first report of differences between greenbug populations in adaptation to different varieties was by Dahms (1948). However, the different populations were not labeled as biotypes. In 1955, Dahms et al. (1955) developed a wheat variety, Dickinson Selection 28A, that was resistant to greenbugs. By 1961, however, greenbugs had adapted to this variety and these virulent genotypes quickly became prevalent in the Midwest. These greenbugs were labeled "biotype B" in contrast to the avirulent "biotype A" (Wood 1961). In 1968, greenbugs were found damaging sorghum, which previously had been a poor host (Harvey and Hackerott 1969a). These greenbugs were labeled "biotype C" and, in contrast with most biotypes, could be distinguished from earlier biotypes by some morphological characters. Biotype C was still virulent on Dickinson Selection 28A. Biotype D is defined by its insecticide resistance (Teetes et al. 1975). After greenbug-resistant varieties, such as Amigo wheat, were released in the 1970s, another biotype, E, able to attack those varieties, emerged in 1980 (Porter et al. 1982). Amigo was resistant to all three previous biotypes (A, B, and C). Soon, biotype E–resistant varieties were found, such as Largo wheat. Largo was not, however, resistant to biotype B (Webster et al. 1986). A few years later, yet another biotype, designated F, was discovered (Kindler and Spomer 1986). This was most similar to biotype A except that it could attack Amigo wheat. Most recently, two other biotypes have been named, G and H (Puterka et al. 1988).

As frustration mounts over the varieties of greenbug genotypes adapted to previously resistant crop varieties, scientists have started examining more closely the variation in greenbug populations, primarily through biotype surveys in the Midwestern United States (Puterka et al. 1982; Moffatt et al. 1983; Kindler et al. 1984; Dumas and Mueller 1986; Kindler and Spomer 1986; Bush et al. 1987; Puterka and Peters 1988; Beregovoy et al. 1988). Many genotypes coexist in a region and it is not easy to classify all greenbugs into clear-cut biotype designations (Salto 1976; Niemezyk 1980; Ratcliffe and Murray 1983; Kindler and Spomer 1986; Kerns et al. 1987; Bush et al. 1987; Michels et al. 1987; Puterka et al. 1988). Table 5.1 summarizes some of these findings. This table, of course, simplifies the interaction, since in many cases there is a continuous variation from highly susceptible to highly resistant and different studies measure different characteristics, including both damage to plant and components of fitness of the aphid. The table reveals complex interactions between genotypes and hosts, with no plant resistant to all genotypes (except possibly P8493 sorghum and Will barley, which were not tested on all biotypes) and no genotype adapted to all plants.

Variation in Greenbugs within a Small Region

In another study examining variation within greenbug populations (Wilhoit and Mittler 1991), virginoparae were collected from four fields near Davis, California, and clones were started from single individuals. Twelve clones were set up on three sorghum varieties (G1711, TX2783, and R980) to test their relative adaptation, measured by the weights of one-day-old adults. Seven clones were collected from one field, three clones from three other nearby fields, and the other two clones from the Midwest. The results are summarized in figure 5.1, in which each dot represents the mean weight on each variety. Standard errors

TABLE 5.1

Resistance ($-$) and susceptibility ($+$) of greenbug biotypes on a range of host plants and varieties

Plant varieties	Greenbug Biotypes						
	A	B	C	E	F	G	H
Wheat							
TR64							
TAM105		+[o]	+[h]	+[h]	+[p]	+[o]	+[q]
DS28A							
CI9058	−[a]	+[h]	+[h]	+[h]	−[l]	+[o]	+[n]
Amigo	−[f]	−[f]	−[f]	+[h]	+[l]	+[o]	+[n]
Largo		+[m]	−[h]	−[h]	+[p]	+[o]	−[n]
TXGH							
10563B		−[q]	−[q]	−[q]	+[q]	+[q]	−[q]
Sorghum							
wheatland			+[d]	+[q]	+[q]	−[q]	−[q]
KS30		−[c]	−[c]	+[h]	−[l]	−[q]	−[q]
P8515			−[q]	+[q]	+[q]	−[q]	−[q]
P8493			−[i]	−[i]	−[q]	−[q]	−[q]
Piper		−[b]	+[b]	+[l]	−[l]	−[o]	−[n]
TX2783		+[l]	−[j], +[l]	−[j]	−[l], +[l]		
Oats							
Nora		+[g]	+[g]	+[q]	+[q]	+[q]	+[q]
CI1580		+[i]	−[i]	−[h]	+[q]	+[o]	+[n]
Barley							
Will	−[c]	−[c]	−[c]	−[h]			
Wintermalt			+[k]	+[k]	+[q]	−[q]	+[q]
Post			−[k]	−[k]	−[q]	−[q]	+[q]
Rye							
Elbon			+[i]	+[i]	+[q]	+[q]	+[q]
Insave	−[f]	−[f]	−[f]	−[h]	+[q]	−[q]	+[q]

Sources: (a) Wood 1961; (b) Harvey and Hackerott 1969a; (c) Hackerott 1972; (d) Starks and Schuster 1976; (e) Starks and Burton 1977; (f) Sebasta and Wood 1978; (g) Wilson et al. 1978; (h) Porter et al. 1982; (i) Starks et al. 1983; (j) Peterson et al. 1984; (k) Webster and Starks 1984; (l) Kindler and Spomer 1986; (m) Webster et al. 1986; (n) Bush et al. 1987; (o) Kerns et al. 1987; (p) Puterka and Peters 1988; (q) Puterka et al. 1988.

range from about 8μg to 50 μg with larger standard errors associated with larger aphids. On sorghum variety G1711, the aphids fell into two clearly distinct groups those doing well symbolized by circles and those doing poorly by triangles. On TX2783, all clones did relatively poorly, but, again, there were two distinct groups, those doing well shown with open symbols and those poorly with solid symbols.

Sorghum G1711 is similar to sorghum KS30, which from table 5.1 is seen to be susceptible only to biotype E. Thus one might conclude that the group from figure 5.1 better adapted to G1711 (circles) should be classified as biotype E. However, not all clones of this group respond similarly to either of the other two varieties. For example, some do well on TX2783 (open circles) and some do poorly (solid circles). If biotype labels are to be made there should be two

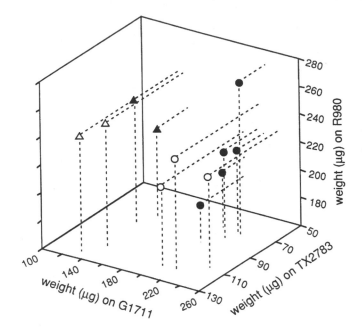

Fig. 5.1. Weights of one-day-old adult greenbugs from different clones collected in California in 1987 and from the midwestern United States. All clones were reared on three different sorghum varieties. Circles represent clones that did well on G1711; triangles, clones that did poorly on G1711; open symbols, clones that did well on TX2783; and solid symbols, clones that did poorly on TX2783.

biotypes here. A similar pattern holds for the clones poorly adapted to G1711 (the triangles): some do relatively well on TX2783 (open triangles) and some do poorly (solid triangles). Therefore, four biotypes can be classified. Including results from R980 further complicates the picture. Aphids generally do better on this variety than the other varieties, but analysis of variance reveals highly significant differences among clones on this variety and significant interactions between weights on this variety and weights on each of the other two varieties. For example, the solid circle group is clearly not uniform, since one clone does very well on R980 and one quite poorly. This continuous variation makes it even more difficult to classify these clones into distinct biotypes. Most of these clones were chosen randomly from one small field and were tested on only three varieties, and yet they fall into more than four classes. One could expect even more variation if more clones and more plant varieties were examined.

Is There a Cost of Virulence?

Do some herbivore genotypes specialize on certain host-plant varieties while doing poorly on others, or do there exist superior (or generalist) genotypes able to feed on all or most host genotypes? This question has proven difficult to answer. A cost of virulence implies a "crossing" type of herbivore genotype $\times$ host interaction in fitness on different varieties and many studies have found such interactions (Blakley 1982; Futuyma 1983b), including the greenbug experiments discussed above. However, as Rausher (1983) points out, existence of a herbivore $\times$ host interaction does not necessarily mean that there is a cost of being a generalist. If adaptation to different hosts is a result of different genetic loci, interaction will occur in a certain proportion of insect-plant systems because of association of the different alleles at the different loci (Via 1984; Futuyma and Philippi 1987; Simms and Rausher, this volume). For example, the greenbug $\times$ host interaction on G1711 and TX2783 could be explained if the greenbugs had two distinct virulence loci each adapted to one or the other variety. If there was complete dominance or recessiveness, then there should be four phenotypes with a crossing type interaction even if there was no cost associated with each adaptation.

Starting with Van der Plank (1963), plant pathologists have also discussed the importance of genotype $\times$ host interactions. Van der Plank assumed that there must be selection against unnecessary virulence on a susceptible host; that is, a cost associated with overcoming host defenses results in reduced fitness of virulent genotypes on hosts susceptible to avirulent genotypes. He called such an effect "stabilizing selection," which is a different use of the term than in population genetics (Antonovics and Alexander 1989). Crill (1977) and Nelson (1972) claimed that most studies do not confirm a cost of virulence, but Leonard and Czochor (1980) pointed out problems in these studies and discussed the difficulties in demonstrating this cost. They concluded that more appropriate studies (e.g., Leonard 1977) suggested a cost to virulence. The controversy continues as more recent studies (Grant and Archer 1983; Bronson and Ellingboe 1986) found lower costs of virulence than Leonard (1977).

Insect-plant genotype interactions have been discussed for many years, but much is still unknown (Gould 1983; Futuyma and Peterson 1985). The ability of an herbivore to feed effectively on as many different plants as possible should be beneficial by providing a wider range of food and a greater likelihood of finding food. However, since most insects feed on only a few types of plants, there must be some cost to being a generalist. Dethier (1954) was one of the first to postulate that this cost was due to the energy required to overcome host defenses, particularly toxic compounds. The greater the diversity of plants upon which an insect feeds, the more kinds of physiological mechanisms an insect must have and thus the greater the energy requirements. For each insect-plant system, there must be some balance between the costs and benefits of feeding on different plants.

However, most studies have revealed little or no evidence for a significant physiological cost of being a generalist (Smiley 1978; Futuyma and Wasserman 1981; Scriber 1983; Futuyma 1983a; Rausher 1983, 1984b; Futuyma et al.

1984). Of course, the costs may be too small to be detected experimentally, but still sufficient to be crucial over evolutionary time. There may also be other kinds of cost, such as greater exposure to predators (Bernays and Graham 1988).

Though there is little experimental evidence for a cost of virulence, one can make theoretical arguments that there must be some kind of cost even if it does not reduce physiological efficiency. One argument was mentioned previously: since there are apparent advantages to being a generalist and most insect species are not generalists, there must be some disadvantage associated with it. Even if there are genotypes that produce more progeny or are larger on all hosts (the mostly commonly measured components of fitness), there are many other important factors that determine fitness, such as overwintering ability, mating success, and predator avoidance. Thus, the cost of adaptation may not be evident from reduced reproduction or weight on other varieties. If a population has much genetic variation, then the probability that there will be a genotype superior in all fitness components should be low because such a superior genotype would quickly replace the others (the rate depending on selection intensity and the frequency of the favored genotype), resulting in little genetic variation. It is possible that such a superior genotype could arise by mutation, but it must be a very rare occurrence. The lack of information regarding cost of virulence is unfortunate, since theoretical arguments suggest that this information is critical in understanding evolution in heterogeneous environments.

Effects of Heterogeneous Environments in Evolution and Ecology

Though this chapter mostly discusses agricultural variety mixtures, a similar but more general problem is that of evolution in heterogeneous environments, which has been reviewed in other papers (Hedrick et al. 1976; Felsenstein 1976; Mitter and Futuyma 1983; Hedrick 1986). The primary question addressed has been whether genetic polymorphisms in populations can be maintained in variable environments. Mathematical models assumed that genotypes vary in fitness ranking in different habitats. The general conclusion from these models is that if the mean fitness of the homozygotes over all patches is less than that of the heterozygote, then a stable equilibrium polymorphism can result, at least, for certain values of fitness, abundance, movement between patches, and habitat selection (Maynard Smith and Hoekstra 1980; Hoekstra et al. 1985). However, these models do not predict the time needed to reach such an equilibrium, which genotypes will be predominant during the process, or how other factors in the system would effect the process. Such information is important, especially in crop protection, where a significant delay in the evolution of virulence would be a valuable contribution.

Effects of Variety Mixtures on Plant Pathogens

Considerable research has been directed toward variety mixtures as a method for controlling plant diseases and for reducing the rate of evolution of virulent pathogen genotypes. Although some sporadic work was done as early as the eighteenth century (see Barrett 1983; Wolfe 1985), modern research on this issue was initiated by Jensen (1952) and Borlaug (1953). Most research has had to depend on models because of the difficulties of experimental studies. Models of natural systems have been used to ascertain whether stable polymorphic states can exist in coevolving populations of plants and pathogens (Leonard and Czochor 1978). Models of agricultural systems have been used to determine what effect mixtures of plant genotypes would have on the ecology and evolution of pathogens (Marshall 1989). Most studies have used either ecological models to examine the effects on pathogen population size or population genetic models to examine the evolution of virulence. A few models have tried to integrate ecology and evolution (Barrett 1980; Østergaard 1983).

Plant and Pathogen Coevolution Models

The earliest models (Mode 1958, 1960, 1961; Jayakar 1970; Leonard 1969b, 1977; Sedcole 1978; Leonard and Czochor 1978) were population genetic, pairwise coevolutionary models of monogenic gene-for-gene systems. The primary question addressed was whether the plants and pathogens would form stable polymorphic equilibrium states. These studies concluded that a polymorphic equilibrium state could exist only if there was a cost of both virulence and resistance. Though this state was not stable, there was a stable cycling of plant and pathogen gene frequencies around the equilibrium state. The frequency of plant genotypes was determined by factors that affected pathogen fitness and the frequency of pathogen genotypes was determined by plant fitness. Also, at the polymorphic equilibrium points the frequency of genes for virulence was higher than the frequency of genes for resistance.

The Evolution of Pathogen Superraces in Variety Mixtures

A primary concern in agriculture is whether variety mixtures could provide more durable resistance by preventing or slowing the development of virulent pathogens. One argument (Browning and Frey 1969; Frey et al. 1977) is that the right variety mixture should result in a stable equilibrium of one or more simple pathogen genotypes with few genes for virulence, halting the evolution of a superrace (or a generalist pathogen) able to attack all varieties. This question has been examined in a series of studies using population genetic models assuming gene-for-gene systems (Groth 1976; Groth and Person 1977; Barrett and Wolfe 1978; Marshall and Burdon 1981; Marshall et al. 1986; Marshall 1989). These studies found that with either a sufficient number of plant genotypes in a mixture or a high enough cost of virulence, superraces could not evolve. It remains an open question whether the cost of virulence is generally high enough to meet this condition, given the number of resistance genes available.

Length of Resistance Effectiveness under Different Resistance Deployment Strategies.

If the cost of virulence is not large enough to prevent the evolution of a super-race, variety mixtures could still be useful by slowing the evolution of virulence, allowing sufficient time to substitute new varieties that are effective against the new pathogen races (Wolfe and Barrett 1979; Wolfe et al. 1981). Key questions are how long resistance will remain effective in mixtures and whether other deployment strategies might not be better. Kiyosawa (1972) used a model to compare the duration of effectiveness between a mixture and the sequential release of varieties in pure stands. Simple exponential growth of pathogen populations both within a season and between seasons was assumed and the between-season growth rate was determined from the population at the beginning of each season. Breakdown of resistance was assumed to occur when the population at the beginning of a season reached some given level. With sequential release each variety was used until populations reached the breakdown population level and then was immediately replaced by another variety. With these assumptions, Kiyosawa found that either strategy could be superior, depending on the number of varieties used and the between-season population growth rate in mixtures and in pure stands.

Trenbath (1977, 1984) also used models to examine the durability of different strategies. His models are among the few that include the effects of mixtures and disease interactions on plant yield. Consequently he was able to use the more realistic criteria for time of resistance effectiveness as yield loss of 20% or greater. The models included growth of both plants and disease, but the only genetic assumption was a constant mutation rate from one gene to another. Four strategies for managing a gene-for-gene system with three resistance genes and three virulence genes were compared: (1) a mixture of three varieties each containing one resistance gene; (2) one variety containing all three resistance genes (pyramiding genes); (3) a single gene variety planted for several years, to be replaced by another single gene variety for several more years; and (4) yearly rotations of different single-gene varieties. In all cases he assumed that 10% of the area was planted to completely susceptible varieties. The effects both of cost of virulence and of induced resistance (in which avirulent pathogens landing on a plant normally susceptible to a virulent strain induce resistance in the plant to the virulent strain) were examined. If there was both a cost of virulence and induced resistance then there was permanent protection from disease in all four strategies. If there was induced resistance but no cost of virulence then there was resistance breakdown in 13–15 years in all four strategies, but if neither factor was present breakdown occurred in 12–14 years. It was concluded that duration of effectiveness was nearly the same in all strategies.

Pathogen Epidemiology in Variety Mixtures

Even if variety mixtures prevent the evolution of virulent pathogens, mixtures will not be agriculturally useful if disease damage is high, as might happen if susceptible varieties are a large percentage of the mixture. Many models have been developed to predict the pathogen population size in mixtures, and experiments have been carried out to test them (Browning and Frey 1969; Trenbath 1975; Jeger et al. 1981; Wolfe et al. 1984; Burdon 1987a; Mundt 1989). Nearly all models and experiments found that pathogen population densities were less in variety mixtures than the mean pathogen density in pure stands of the component varieties. Several factors have been proposed to explain these results: (1) the presence of fewer susceptible plants reduces the amount of inoculum production; (2) increased distance between susceptible plants results in greater loss of inoculum during dispersal; (3) the presence of resistant plants interferes with the dispersal of inoculum; and (4) the presence of avirulent pathogens on plants induces resistance to the virulent pathogens. Two studies (Burdon and Chilvers 1977; Chin and Wolfe 1984) indicated that the first two factors were more important, at least in the early stages of the disease.

Leonard (1969a) developed one of the first models of pathogen populations in mixtures using a simple formula for disease growth rate assuming that spores landing on resistant plants died. Kiyosawa and Shiyomi (1972) expanded on this by assuming a distribution function for spore dispersal. They demonstrated that the effectiveness of mixtures to control diseases increased as the dispersal gradient became more shallow. This idea was later explored (Mundt et al. 1986; Mundt and Leonard 1986; Mundt and Brophy 1988) using a modified version of a more realistic, spatial simulation model (Kampmeijer and Zadoks 1977). They examined in both experiments and models the effect of genotype unit area (the area of host that is genetically homogeneous), spatial distribution of spores, dispersal gradients, and number of plants on disease progress in mixtures. They found that smaller genotype unit area and shallower dispersal gradients increased the effectiveness of mixtures. This effect resulted from a larger ratio of alloinfection (infections resulting from spores produced on other plants) to autoinfection (infections from spores produced on the same plant). At higher alloinfection rates, fewer spores will fall on the plant that produced them, which must have been susceptible, and therefore more are likely to fall on resistant plants. This increase in effectiveness was enhanced if spores were distributed uniformly over the field rather than distributed around point foci. Also, disease progress was less in fields with more plants.

A few studies have tried to integrate ecology and evolution (Barrett 1978; Barrett 1980; Østergaard 1983). Barrett added a simple dispersal function to a previous population genetic model of superrace evolution and found that high autoinfection tended to prevent the development of superraces. The magnitude of this effect was greatly affected by reproduction rates and selection coefficients. Østergaard (1983) also developed a model combining epidemiology and genetics. She assumed simple exponential growth of pathogen populations and asexual reproduction and varied parameters for the percentage of autoinfection, proportion of different varieties in mixtures, and cost of virulence. The results were similar to what others had found. Higher autoinfection resulted in more disease damage in mixtures but prevented the development of a superrace, especially at intermediate proportions of the component varieties.

Results of Studies of Pathogens in Variety Mixtures

After 30 years of studies on variety mixtures some general patterns are clear, but the effects of mixtures on plant pathogens are still mostly unknown. It is clear that variety mixtures can reduce the evolution of virulent pathogens, but the rate of reduction depends on many poorly studied factors, such as the variation in pathogens and the cost of virulence. Too much emphasis has been placed on finding conditions of stable polymorphic equilibria, which depend on the cost of virulence. Whether cost of virulence is in general great enough to satisfy conditions of polymorphic equilibria will be irrelevant if enough virulence genes have sufficiently low cost that they could lead to superraces, nullifying all the arguments and models. The alternative approach of looking for strategies that slow the rate of evolution of virulence regardless of existence of stable equilibria may be more enlightening and useful. The few models that exist are quite simplistic, and, since most strategies to reduce selection pressure for virulence will probably involve higher damage, one must examine and weigh effects of both epidemiology and genetics. To give another example of conflicting criteria, epidemiological studies found that higher alloinfection-to-autoinfection ratios provide better disease control, while genetic studies found that lower ratios slow the development of superraces. Even fewer studies have examined the effects on yields or the effects of stochastic events. Models have been useful in these studies mostly by revealing what factors need more study. As more data is gathered and models of particular systems become more realistic, it may be possible to provide clearer guidelines on when variety mixtures can provide more stable resistance.

Effects of Variety Mixtures on Insects

The Paucity of Studies

Given the extensive work with pathogens in variety mixtures it is startling to find so few studies that examine insects in variety mixtures. A few studies have examined insect population sizes in mixtures (Altieri and Schmidt 1987; Power 1988; Wilhoit 1988), and other studies have modelled the genetic consequences of mixtures on insects (Gould 1986a, 1986b; Cox and Hatchett 1986). Many of the pathogen models apply to insects, since they make fairly general assumptions, but insects have characteristics that distinguish them from pathogens and these need to be included in the models. One distinguishing characteristic is insects' ability to move actively. They can reject a host they happen to land on and use host-seeking mechanisms to find another, while pathogens are stuck with the host they happen to land on. Spore dispersal gradients and the degree of autoinfection can have big impacts on both severity of disease and evolution of virulence. These factors are likely to be quite different in insects and pathogens. Nearly all pathogen studies assume gene-for-gene relationships between pathogen and plant. Whether this type of interaction is common to insects is mostly unknown, though many authors suggest that it is not (Robinson 1976; Gould 1983; Claridge and Den Hollander 1983). There is, however, evidence for gene-for-gene relationships between the Hessian fly and wheat (Gallun 1977), the insect-plant system in which the genetics are best known.

Gould's Genetic Models of Insect Evolution to Plant Resistance

Gould (1986a, 1986b) developed two simulation models to predict the durability of resistant varieties against insects. The first (Gould 1986a) was for an insect with a typical sexual life cycle, while the second (Gould 1986b) addressed the specific case of the Hessian fly. He used these models to compare the effects of deploying resistance genes sequentially, in a variety mixture, and combined in a single pyramided variety. For each strategy susceptible varieties were either mixed in or absent. He assumed two resistance factors (these could be either monogenic or polygenic factors) in the plant and two corresponding virulence genes in the insect. He also explored the effects of different levels of dominance, epistasis, gene linkage, and initial allele frequencies for the virulence genes.

Results of his first model (Gould 1986a) are summarized in table 5.2. Fitness of all insect genotypes was assumed equal to 1.0 on susceptible plants (no cost of virulence) and was decreased by a factor of 0.24, 0.48, or 0.96 (determined by what is considered to be of economic value) for homozygous unadapted genotypes on plants with either resistance factor. Resistance was considered effective as long as the mean fitness of the insect population was below 0.8. Runs were made with the virulent allele codominant to the avirulent allele (degree of dominance = 0.5) or with the virulent allele completely recessive (degree of dominance = 0.0). Epistasis between the two insect virulence loci was dictated by interactions between resistance factors in the plant. Positive interaction occurred when two resistance factors together caused greater fitness reduction to the insects than the sum of each separately. If their presence together reduced fitness less than the sum of each separately, then negative interaction resulted. The model was run assuming either no interaction (degree of epistasis = 0.0) or complete negative interaction, in which each resistance fac-

TABLE 5.2
Summary of results from simulation model in Gould 1986a

Duration of effective resistance (in number of generations)				
Sequential	30	40	>>50	—	8
Mixed	27	40	>>50	—	8
Pyramided	7	25	15	—	>>50
Sequential + susceptible	40	>>50	>>50	40	—
Mixed + susceptible	40	>>50	>>50	40	—
Pyramided + susceptible	40	>>50	>>50	40	>>50
Genetic parameters of virulence genes				
Degree of dominance	0.50	0.50	0.00	0.00	0.00
Degree of epistasis	0.00	0.00	0.00	0.00	-1.00
Initial allele frequency	0.01	0.01	0.01	0.10	0.01
Reduction of fitness	0.48	0.24	0.48	0.48	0.96

Note: The upper part of this table shows time until mean fitness of insect population is 0.8 under different strategies for deploying two resistance factors: sequential release of each resistance factor; different resistance factors spatially mixed in field; pyramiding the resistance genes within one variety; and mixtures of a susceptible variety with each of these strategies (ratio of resistant to susceptible is 8 : 2 in sequential and mixed cases and 4 : 6 in pyramided case). Each column gives results from different simulations. The lower part of the table gives the different genetic characteristics used in the different simulations. The degree of dominance of virulent allele is 0.5 (codominant with avirulent allele) or 0.0 (virulence recessive); degree of epistasis between resistant loci is 0.0 (no interaction) or -1.0 (presence of both resistance factors has same effect on insect fitness as either alone); initial frequency of the virulent allele in the population; and decrease in fitness of homozygous unadapted insects caused by a resistance factor.

tor is redundant to the other (degree of epistasis $= -1.0$). The model was also run by adding susceptible varieties in a ratio that insured the same initial mean fitness of insects in all three strategies.

If virulence in the insect is assumed to exhibit an additive linear relationship between loci (no epistasis) and no allelic dominance, and if the initial virulent allele frequency is 0.01, then the duration of effective resistance under the sequential and mixed releases is about four times that under a pyramided release. Recessive virulent alleles greatly increase durability under sequential and mixed strategies and double it for the pyramided strategy. Negative epistatic interactions between alleles reduce durability for sequential and mixed deployment and greatly increase durability for pyramided deployment. Thus, just as was found in pathogens, the evolutionary trajectory of virulence in insects depends a great deal on which particular genetic assumptions are incorporated into the model. Generally, sequential and mixed deployment give similar results. If virulence is recessive, durability is increased. If there is no epistasis, whatever the level of dominance, sequential and mixed strategies are more durable than pyramided release. With both recessive virulence and negative epistasis, pyramided release gives longer durability than sequential and mixed release, but in these cases the pyramided release could cause a greater decrease in the local population size of the insect, at last initially.

Gould's model could apply to pathogens (at least, sexually reproducing ones), since it makes such general or simple assumptions that they are not restricted to the peculiarities of pathogen life cycles. The fact that many pathogens are haploid may not qualitatively affect the conclusions, if the results of Jayakar (1970) and Leonard (1977), who compared coevolution of haploid and diploid organisms, are generally valid. It is of interest to compare Gould's result with that of Trenbath (1984), who provided one of the few pathogen studies comparing resistance deploying strategies. Trenbath saw little difference between the strategies and found that a cost of virulence and induced resistance were needed to provide long-term durability. Gould found considerable differences between strategies under certain assumptions, and though he assumed neither cost of virulence nor induced resistance, he found in some cases very long durability. But these two studies were very different: Gould's was a population genetic model with few ecological relationships, in which time to resistance breakdown was set when the insect fitness reached 0.8; and Trenbath's was primarily an ecological model in which time to breakdown was set when yield loss exceeded 20%. It would be instructive to study the interactions between the factors present in both studies.

Effects of Variety Mixtures on Insect Population Sizes

As with pathogens, in order to understand the possible trade-offs in using variety mixtures, one must understand the ecological effect of mixtures on pest populations. The few studies in this area (Altieri and Schmidt 1987; Power 1988; Wilhoit 1988) have all found reduced populations in mixtures compared to the mean of populations in pure stands of component varieties. These studies are an insufficient basis for generalization. There have been, however, many studies on insects in polyculture (Perrin 1977; Perrin and Phillips 1978; Cromartie 1981; Altieri and Liebman 1986). Most of these studies report that polycultures reduce insect populations (Risch et al. 1983), and two hypotheses have been proposed to explain this result: the natural enemies hypothesis and the resource concentration hypothesis (Root 1973).

The natural enemies hypothesis (Russell 1989) states that a diversity of plant species may provide important resources for natural enemies such as alternative prey, nectar and pollen, or breeding sites. This mechanism may not be important for variety mixtures, since from the natural enemies' points of view, different varieties may often be nearly identical. However, van Emden (1986) suggests that in the presence of natural enemies, herbivore population may be lower on moderately resistant plants than would be predicted from the population on susceptible and more resistant plants. He hypothesizes (and provides some evidence) that increasing levels of resistance up to a point sometimes do not reduce the absolute number of herbivores attacked, possibly because in these cases natural enemies are attracted to the plants themselves rather than to the number of prey. However, plants with high levels of resistance may no longer attract natural enemies and do not provide sufficient prey to sustain a natural enemy population. These arguments could apply to variety mixtures, which can be considered a method of creating moderate levels of resistance.

The resource concentration hypothesis states that specialist herbivore abundance should be lower in diverse communities than in simpler communities because herbivores are less likely to locate hosts, their survival and fecundity will be lower, and they will remain in the diverse environments for shorter time periods (Root 1973). Many studies have supported this hypothesis for polycultures (Kareiva 1983), but none have examined variety mixtures. Whether or not the hypothesis holds for variety mixtures depends on the mechanism involved. The few studies that have tried to uncover the mechanisms (Bach 1980; Risch 1981) found that the most important mechanism was the tenure time of the herbivores in the habitat, though other mechanisms could be important in other systems. More specifically, Risch (1981) found that herbivores emigrated from diverse habitats because corn, the nonhost plant, shaded the host plants and because the corn stalks disrupted movement. This type of mechanism would not operate in variety mixtures where the varieties were physically similar. However, effects of host-plant quality on emigration, as found by Bach (1980), could play a role in variety mixtures.

Ecological Effects of Heterogeneity and Its Relevance to Evolution

Ecological studies are necessary not only for predicting insect population size and damage to plants, but also for understanding fitness and the consequences for evolution, which is the long-term result of short-term ecological interactions. The problem of determining whether variety mixtures can affect the development of resistance-breaking genotypes can be restated as determining whether host-plant variation can relieve the competitive exclusion of an avirulent genotype by a virulent genotype. The extensive work on interspecific competition and predation in heterogeneous environments can be used to provide some answers to this question, particularly for organisms with parthenogenetic life histories, since the different genotypes in those organisms are distinct, noninterbreeding lineages and can be treated as different competing species.

Several factors have been proposed that relieve the competitive exclusion of an inferior competitor by a superior one. Hutchinson (1951) was one of the first to suggest that in an environment with patchy and ephemeral resources, a species with inferior competitive abilities could nevertheless persist if it had greater dispersal abilities and thus could more readily colonize newly opened habitats, free of the superior competitors. The persistence of a competitively inferior species thus depends on its colonization rate exceeding local patch extinction rates. These ideas were later quantified in models by Levins and Culver (1971), Horn and MacArthur (1972), Slatkin (1974), S. A. Levin (1976), Hanski (1983), and Ives and May (1985). A similar idea was developed by Den Boer (1968), which he called "spreading the risk." Models by Atkinson and

Shorrocks (1981), Green (1986), Shorrocks and Rosewell (1988), and Ives (1988) further demonstrated that competitive exclusion could be reduced even without a greater dispersal ability in the inferior competitor if distributions of two species were independently clumped. Competitive exclusion was decreased as clumping in the superior competitor increased and as the number of resource patches increased.

Predators may enhance coexistence of competitors (Caswell 1978; Hastings 1978; Crowley 1979; Hanski 1981; Comins and Hassell 1987; and Worthen 1989). As predation reduces population size, competitive interactions are relaxed. Sometimes competition is lessened when predators feed on the more numerous competitor, which is often the superior competitor. In addition, if predators consume all their prey in a patch, freeing it for recolonization, coexistence is possible for dispersing fugitive species. Edson (1985) found that predation probably facilitated coexistence of two competing aphid species on goldenrod by freeing space on plants and allowing the inferior competitor a chance to become reestablished. Caswell's (1978) model demonstrated that in open, nonequilibrium systems predation could greatly extend the time to competitive displacement. His model assumes the existence of a number of discrete cells (or patches), two competitors, one of which always outcompetes the other in each cell, a predator, and random migration between cells. The model follows only presence or absence of organisms in each cell and calculates the probability of colonization of each cell as proportional to the number of cells occupied by that species. If the superior competitor completely eliminated the inferior competitor in 5 generations within a cell, then in a system with many cells without predation the two competitors coexisted a median time of about 30 generations. In the same system but with predation the competitors coexisted about 50 generations, sometimes extending coexistence for over 1,000 generations. Greater dispersal rates of the inferior competitor and the predator increased time to extinction of the inferior competitor. Not only did time to extinction increase, but the frequency of the inferior competitor was often greater than the superior competitor. Increased dispersal rate of the superior competitor decreased time to extinction.

Though Caswell's model can predict the number of patches occupied by the different species, it cannot predict abundances. The superior competitor may be far more abundant, even if it occupies fewer patches. This information is important, especially in the agricultural situation where one wants to maintain the virulent genotype at much lower abundance than the less virulent genotype. Caswell contends that ignoring abundances will not qualitatively affect the results, but this claim is questionable (Hastings and Wolin 1989). Greater abundance of either competitor will increase the probability of dispersal to other patches, and as he shows, dispersal rate significantly affects the time to extinction. For example, as the superior competitor increases in abundance, dispersal rates will increase, which will decrease the time to extinction of the inferior competitor. Caswell's model cannot determine the relative importance of this effect. Similarly, the abundances of prey will affect the abundances of their predators, which influences the dispersal rates of the predators.

A third factor that could be important in the persistence of inferior competitors in situations where organisms randomly colonize ephemeral resource patches is late-arrival disadvantage. Organisms that arrive later to a resource patch will be at a disadvantage relative to earlier arrivers if there exist mortality factors that increase with population density. If superior competitors are initially rare relative to inferior competitors, and therefore more likely to reach a patch later, the superior competitors will less likely displace the inferior competitors. When an organism is a superior competitor because it is more virulent, late-arrival disadvantage could be considered a cost—an ecological cost—associated with virulence. But the cost has nothing to do with virulence, only with the fact that the virulent organism is initially rare.

Effect of Variety Mixtures on Aphids

Rather than deal with pathogens and insects in general, the rest of this chapter will focus on aphids. It is important to analyze the evolution of virulence in aphids to plant resistance because aphids differ in important ways from pathogens, Hessian flies, and insects with typical sexual cycles; because aphids have often adapted to resistance varieties; and because aphids are common crop pests. Aphids differ from Hessian flies in having a cyclic parthenogenetic life cycle, and they differ from plant pathogens in their ability to move actively from plant to plant and in the role predators play in controlling their population size. These factors may have important effects on the outcome of the evolution of virulence. For example, because each genotype is represented by many individuals over several generations, interactions such as intergenotypic competition and predation will affect the relative fitness of the different genotypes. Thus the fitness of each genotype cannot be considered constant over time as is often assumed in many population genetic models. Even models that incorporate frequency-dependent and density-dependent selection ignore many factors such as predation that can be essential in determining fitness values.

A Stochastic, Spatially Structured Model of Two Aphid Phenotypes and a Predator

Theoretical and empirical studies of pathogens suggest that host heterogeneity will have a significant effect on preventing adaptation of herbivores to resistant hosts only if there is some cost to virulence on susceptible hosts. However, many factors important in insect systems, such as movement, host finding, natural enemies, and stochastic effects, have not been considered. To determine the importance of these factors the model discussed here assumes no cost of virulence, except the ecological cost involved in late-arrival disadvantage. The model predicts the rate at which a virulent aphid genotype will replace an avirulent genotype in a variety mixture with different proportions of resistant and susceptible plants, under various assumptions of aphid and predator movement, and with different population growth rates, predator efficiencies, initial proportion of the virulent genotype, and genetic transmission of virulence.

Full details of this model are given in Wilhoit (1991) and the most important equations are given in the Appendix to this chapter. The model assumes two plant varieties, two aphid phenotypes, and an aphid predator. The two plant varieties are homogeneously mixed in a field and the model treats plant individually arranged in space. One aphid phenotype (the virulent) does equally well on both plant types, and the other phenotype (the avirulent) does poorly on the resistant plant. Virulence is assumed to be governed by two alleles at one locus and may be inherited either dominantly or recessively. There are thus three genotypes, but two phenotypes. Both aphids do equally well on the susceptible variety and in all other ways are identical except that the avirulent aphid predominates initially. A reservoir of aphids and predators surrounds the field, each individual aphid having an equal probability of landing in the field. Once aphids and predators land in the field they randomly land on plants within the field. If an avirulent aphid lands on a resistant plant, then it is more likely to leave. Predators are more likely to land on plants with aphids. As the aphids grow and reproduce, the ratio of the plant food supply to their maximum food demand rate decreases, which causes the aphids to move off the plant at a rate determined by the parameter, ϕ. Also, the presence of predators increases the rate aphids leave the plant. When aphids leave a plant, some die trying to find another plant while the rest randomly settle on one of the neighboring plants.

In some ways this model is similar to the model of Caswell (1978), with plants as cells, two phenotypes (rather than two competing species), predators that can decimate a population on a plant, and migration of all organisms between plants. Differences are that here there are two types of patches rather than one, and the competition between the aphids on the susceptible plants is minimal. Also, dispersal of aphids is only to neighboring plants, not randomly to any plant in the field. The biggest difference is that the population dynamics on each plant is modeled.

Growth of organisms is modeled by following the physiological processes of food search, ingestion, excretion, respiration, assimilation, and reproduction. The growth of each plant and its associated population of aphids and predators is determined by an energy allocation model derived from that of Gutierrez et al. (1984) which uses realistic equations with biologically meaningful parameters that can be experimentally determined. Solar energy is captured by plants and transformed to either respiration or plant mass. Aphids consume the plants' energy (or biomass), diverting the energy to excretion, respiration, growth, or reproduction. Similarly, predators feed on aphids and the energy captured goes to excretion, respiration, growth, and reproduction. The rate of food capture for both aphids and predators is given by the Frazer-Gilbert functional response (Frazer and Gilbert 1976). Most parameter values were taken from previous experiments on aphids (Randolph et al. 1975; Frazer and Gilbert 1976; Gutierrez et al. 1984). To deal with unknown parameters, simulations were run with a full range of values to determine their effect.

Populations are followed through a season, during which the aphids reproduce parthenogenetically. At the end of the season, the plants die or are harvested and the aphids present at that time either all leave to some overwintering site where they remain until the start of the next season (the pure parthenogenetic case) or else they form sexuals, mate randomly, and lay eggs that overwinter (the sexual case). The equations for the change in the virulent allele frequency are given in the Appendix.

The primary variable of interest is the proportion of virulent aphids, whose values can be followed through time from the model. The difference between the initial and final proportion of virulent genotypes constitutes one component of the relative fitness of the virulent to the avirulent genotype. This fitness value incorporates all the ecological relationships discussed above, such as aphid-plant and predator-aphid relationships, making the fitness values much more realistic and meaningful than in the usual population genetic treatment.

Because of the stochastic factors, different computer runs made with exactly the same parameters gave different results. The simulation was run 50 times for each set of parameter values and the end-of-season proportion of virulence was compared for runs that differed in only one parameter value. A problem with varying only one parameter at a time is that interactions between factors cannot be investigated. The primary parameter of interest is the proportion of resistant plants in the field, but it is also interesting to see how variation in other parameters affects the results. The mean and standard error of the end-of-season proportion of virulence was then plotted against the parameter value.

Results within a Season

The graphs in figure 5.2 show the effects of six different parameter values on the change in proportion of virulence during a season. The mean end-of-season proportion of virulence quickly increased as the initial proportion of virulence increased for both pure stands of resistant varieties and variety mixtures (fig. 5.2a). The end-of-season proportion of virulence for a given initial proportion was considerably less in variety mixtures than that in pure resistant plots (fig. 5.2a). As the proportion of resistant varieties decreased from 100% to 80%, the end-of-season proportion of virulence decreased from 30% to 5% (fig. 5.2b).

Most ecological studies, such as Caswell's (1978), found that as the number of patches increased, the inferior competitor was less likely to be displaced. In the aphid model, the end-of-season proportion of virulence decreased as the number of plants increased to about 50, but with more plants there was little change (fig. 5.2c). Caswell also predicted that dispersal rates of competitors affected the outcome of competition. However, varying the parameter for dis-

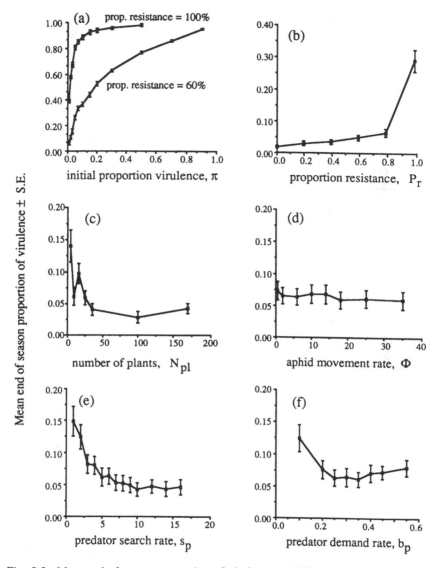

Fig. 5.2. Mean end-of-season proportion of virulence ± S.E. as a function of various simulation model parameters. In each case, all parameters, except the one varied, are held to constant values as follows: initial proportion of the virulent phenotype, $\pi = 0.01$; proportion of resistant plants in the field, $P_r = 60\%$; number of plants in the field, $N_{pl} = 9$; aphid movement rate, $\phi = 25$; predator search rate, $s_p = 5.0$; and predator demand rate for prey, $b_p = 0.3$.

persal rates of avirulent and virulent aphids showed no such effect (fig. 5.2d). As predators became more efficient searchers, up to a point, they were more effective in slowing the development of the virulent genotype, with the end-of-season proportion of virulence decreasing from 15% to 4% (fig. 5.2e). A similar decrease in the end-of-season proportion of virulence occurred as the predator demand rate (the maximum rate at which predators consume prey) increased (fig. 5.2f). Both predator search rates and demand rates were parameters in the predator functional response (see Appendix). Miscellaneous runs varying other parameters had little effect on proportion of virulence.

Results between Seasons

Three genetic assumptions were examined to determine between-season results: (1) there was no sexual phase; (2) virulence allele was dominant; (3) virulence allele was recessive. Virulence arose most quickly in the parthenogenetic case and least quickly in the recessive case (fig. 5.3). Virulence alleles

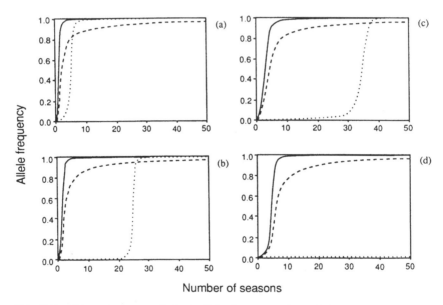

Fig. 5.3. Change in mean virulent allele frequency from the stochastic simulation model through several seasons in fields with all resistant plants or in fields with 60% resistant and 40% susceptible plants, with an initial virulent frequency of either 1% or 0.1%, and for three genetic assumptions: virulent allele dominant ----- ; virulent allele recessive ········ ; and no sexual phase ———. All other parameters were the same as in figure 5.2. (*a*) 100% resistance and 1% initial virulent allele frequency; (*b*) 60% resistance, 1% initial allele frequency; (*c*) 100% resistance, 0.1% initial allele frequency; (*d*) 60% resistance, 0.1% initial allele frequency.

dominated within 5 years for all three reproductive assumptions for 100% resistance fields with 1% initial proportion of virulence (fig. 5.3a). Adding susceptible varieties increased the time for virulence to arise in all cases, but most noticeably in the recessive case—from 5 years to 35 years (figures 5.3a and b) or from 35 years to hundreds of years (figures 5.3c and d).

Discussion of the Aphid Model

The simulated growth of predators on susceptible plants in variety mixtures slowed the development of virulent aphid genotypes. The avirulent genotypes needed no selective advantages relative to the virulent genotype, except that provided by their ecological situation. This ecological advantage was associated with avirulence only if the initial population size of the avirulent aphids was greater than that of the virulent aphids and is only a probabilistic advantage. Owing to random immigration into a field, the virulent aphids could reach the field first and the advantage would then fall to them.

Sexual reproduction slowed the development of virulence. Gould's (1986a) genetic model of sexual organisms with no ecological interactions demonstrated that the presence of susceptible plants slowed the development of virulence. Both models also demonstrated that recessive virulent alleles resulting in greater durability of plant resistance.

A rather surprising result from the aphid simulation studies was the robustness of the outcome over a wide range of values for most of the parameters. The proportion of resistant plants had a large effect, but the greatest change occurred between 80% and 100% resistance. The action of predators was also important; more effective predation (as a result of higher search rates or greater demand rates) slowed the development of virulence. Spatial structure and dispersal rates did have an effect, but it was minor. Other parameters, such as probability of landing on plants and aphid movement rates, had little effect. However, the effects of the parameters that were important depended critically on an initially low population of virulent genotypes. As soon as the virulent genotype exceeded a proportion of 1–5%, it quickly achieved predominance.

Though this aphid model included various ecological and physiological processes, it was still simplistic in many ways. For example, it included only three aphid genotypes and only two plant genotypes that differed by one resistant factor. Plant pathologists have found that variety mixtures should be most useful if they contain a large number of different resistant factors. The genetic assumptions of the aphid model were also very simple. It ignored interactions between loci, which were important in Gould's models. Thus, still more work is necessary to integrate ecology and genetics and, in particular, to explore possible synergistic or antagonistic interactions among the various factors.

This model assumed no physiological cost of virulence, which, as discussed above, probably has a great effect on the evolutionary trajectory of virulence. Cost of virulence was left out to make the model conservative. Populations would lose their genetic variation if a genotype superior in all fitness traits existed, since it would displace all other genotypes. Thus, in a population with much genetic variation, the appearance of such a superior genotype should be a rare occurrence, and the model demonstrated that if a superior genotype was rare then the presence of diverse hosts would substantially delay its spread in the population. This situation, then, is a kind of disadvantage to the genotype, a disadvantage not due to some physiological or individual characteristic but due to the type of ecological framework in which it lives, characterized by density-dependent mortality factors, and continual and random immigration into ephemeral habitats.

Late-arrival disadvantage will be of greatest significance in a patchy environment, and thus possibly most significant in natural settings rather than in monocultural agriculture. Imagine a region with several plant patches, newly emerging in the spring. Some of the plants in these patches are good hosts for all aphid genotypes (susceptible plants). Other plants (resistant plants) are poor hosts for inferior genotypes but good hosts for superior genotypes. Overwintering aphids from the surrounding area disperse and randomly land on a patch. If the inferior genotypes greatly outnumber the superior genotypes, they are more likely to reach a patch first, and if they reach susceptible plants, their population can grow rapidly. Any effects of density-dependent mortality factors will increase with population size so that by the time the superior genotype arrives, its population growth will be suppressed. In the meantime, as the population size of the inferior genotype increases, it disperses and colonizes other patches, starting the process anew. Thus, so long as there are empty patches for the aphids to colonize, the process will continue. The population of inferior genotypes will continue to increase and prevent the superior genotype from getting established. This effect depends, however, on the chance event of the superior genotype arriving later. The superior genotype may find an empty patch, but on average over a region of many patches the inferior genotype will outnumber the superior.

Conclusion

This chapter has touched on only a few of the issues involved in the evolution of pathogens and insects to resistant hosts. Most currently available information and theory deals with evolution in homogeneous situations. Although some theory has treated heterogeneous environments, it has been based on ecologically simplified systems. In contrast, in much ecological literature dealing with interactions in heterogeneous environments, the genetics has been mostly ignored. Many studies have shown the importance of various individual factors such as genetic dominance and epistasis, cost of virulence, predator mediated coexistence, and late-arrival disadvantage, but nothing is known about the interaction of these factors. There is a great need for models that further integrate population genetic and ecological factors. One reason for the lack of such studies is that systems become very complex. Probably the only way to untangle the interweaving of these processes is through computer simulations. These simulations must be based on realistic processes and knowledge of individual behavior. As computers become more powerful and more readily available, this work should become more feasible. These models will add immensely to the study, which has only just begun, of the evolution of herbivores in the heterogeneous environments created by variable plant resistance.

Appendix

Equations for the Population Growth of Plants, Aphids, and Predators

The stochastic, spatial model of virulent and avirulent phenotypes and a predator discussed in the text was based on the energy allocation model of Gutierrez et al. (1984). Growth of organisms was modeled by calculating the change in biomass of each plant and the biomass of the aphid and predator populations on each plant for each time unit:

$$\Delta M_{pl} = M_{pl}^* - Z_{pl}M_{pl} - M_a^* - M_v^* - T(M_a + M_v)$$

$$\Delta M_a = M_a^*(1 - \beta_a) - Z_a M_a - M_{pr}^* \left[\frac{M_a}{M_a + M_v} \right]$$

$$\Delta M_v = M_v^*(1 - \beta_v) - Z_v M_v - M_{pr}^* \left[\frac{M_v}{M_a + M_v} \right]$$

$$\Delta M_{pr} = M_{pr}^*(1 - \beta_{pr}) - Z_{pr}M_{pr},$$

where:

M_i = mass in mg of plant (pl), avirulent aphid (a), virulent aphid (v) or predator (pr);

M_i^* = food supply rate (amount of food consumed);

β_i = the fraction of food consumed which is egested;

Z_i = respiration rate; and

T = toxemia of plant owing to aphids.

Each equation states that the daily increase in mass of each organism or population equals the mass of food consumed times a fraction that is not egested minus the amount that goes to respiration and minus an amount that is consumed by the next higher trophic level. Plants feed on solar energy, aphids consume plant supply, and predators eat aphids.

The rate of food capture for both aphids and predators is given by the Frazer-Gilbert functional response (Frazer and Gilbert 1976):

$$M_{pl}^* = M_s\left[1 - exp(-b_{pl}M_{pl}G_{pl}/M_s)\right]$$

$$M_a^* = M_{pl}^*\left[1 - exp\left\{\left[-b_a M_a/M_{pl}^*\right]\left[1 - exp(s_a M_{pl}^*/b_a)\right]\right\}\right]$$

$$M_v^* = M_{pl}^*\left[1 - exp\left\{\left[-b_v M_v/M_{pl}^*\right]\left[1 - exp(s_v M_{pl}^*/b_v)\right]\right\}\right]$$

$$M_{pr}^* = (M_a + M_v)\left[1 - exp\left\{\left[-b_{pr}M_{pr}/(M_a + M_v)\right]\left[1 - exp(s_{pr}(M_a + M_v)/b_{pr})\right]\right\}\right],$$

where

M_s = solar energy available for plants;
b_i = maximum food demand rate; and
s_i = search rate.

These equations give the amount of food consumed as a function of the amount of food that is available for each organism, the maximum rate at which the organism can eat, and the rate at which it searches for food.

The equations for virulent aphids are the same on both resistant and susceptible plants. The equations for avirulent aphids apply when they are on resistant plants. When avirulent aphids are on susceptible plants the equations for the virulent aphids are used, that is, no cost of virulence is assumed. The difference in growth rates of the avirulent aphids on resistant and susceptible plants results from different aphid search rates (s_i) and different demand rates (b_i), though in most cases, only demand rates were varied, leaving the search rate equal to 1.0.

Equations for the Change in Virulent Allele Frequency

Let *a* designate the allele bestowing adaptation to a particular plant resistance factor and *A* the corresponding allele without the adaptation. Denote the virulent allele frequency at the beginning of the season by *q* and at the end of the season by *q'*. Denote the size of the aphid population at the beginning of a season as *N* and that of each genotype as N_{aa}, N_{Aa}, and N_{AA}, and denote the population sizes at the end of the season as N', N'_{aa}, N'_{Aa}, and N'_{AA}. Denote the proportion of the virulent phenotype at the beginning of the season as π and at the end of the season, π'. Only in the fall do the aphids form sexuals, mate randomly, and lay eggs that overwinter. All genotypes have equal overwintering survival so that the frequency of alleles after mating, denoted by q_1, remains the same to the start of the following season. The fitness of each genotype is given by:

$$W_{aa} = \frac{N'_{aa}}{N_{aa}}, \ W_{Aa} = \frac{N'_{Aa}}{N_{Aa}}, \ W_{AA} = \frac{N'_{AA}}{N_{AA}}.$$

Using the standard equation for allele frequency change,

$$q_1 = \frac{q^2 W_{aa} + q(1-q) W_{Aa}}{q^2 W_{aa} + 2q(1-q) W_{Aa} + (1-q)^2 W_{aa}},$$

one gets, assuming virulence is recessive,

$$q_1 = \frac{\pi' + q}{1 + q}$$

and assuming virulence is dominant,

$$q_1 = \frac{\pi'}{2 - q} .$$

These equations give the virulent allele frequency in one year as a function of the previous year's allele frequency and the end-of-season proportion of virulent phenotype which comes directly from the ecological model.

6

Insect Distribution Patterns and the Evolution of Host Use

Diana Pilson

Host-use patterns in herbivorous insects are the result of both evolutionary and ecological processes. Past natural selection has limited the number of plant species fed on by any individual insect species to a relative few; current insect distribution patterns thus have a historical component. Contemporary ecological processes further limit insect distribution to a subset of the potentially suitable hosts. In this chapter I will argue that genetic determination of insect distribution patterns on an ecological (within generation) time scale is a prerequisite for changes in the pattern of host utilization on an evolutionary (between generation) time scale. This argument is not new; all models of the evolution of host use in insects make this assumption. However, relatively little empirical work has examined genetic determinants of insect distribution patterns.

In the following sections I discuss evidence for genetic variation in insect host-use characters and the effect such variation can have on insect distribution patterns. If genetic variation is a determinant of insect distribution patterns, then (except in rare equilibrium situations) natural selection will occur, changing future patterns of host use. Finally, then, I discuss models of the evolution of host use. These models make assumptions about the types of variation found in insect populations and the effect such variation has on distribution patterns. Model assumptions are discussed relative to the empirical data presented earlier.

Genetic Variation in Insect Host-Use Characters

An extensive literature documents genetic variation in insect characters that can affect host use by herbivorous insects (see reviews in Gould 1983; Futuyma and Peterson 1985; Thompson 1988b; Jaenike 1990; and Via 1990). Instead of reviewing that literature again, I will examine genetic variation in aphids (Homoptera: Aphididae) as a representative example of the amount and type of genetic variation present in herbivorous insects in general. Aphids have been well studied because they are serious crop pests and because they reproduce parthenogenetically (which allows testing genetically identical individuals on different plants). In addition, many aphids have undergone host shifts between agricultural crop species or cultivars (see, e.g., Harvey and Hackerott 1969b; Nielson and Don 1974a; Porter et al. 1982), and for this reason at least some aphid populations must harbor the type of variation necessary for these shifts.

Host-use characters are usually divided into those affecting preference for host plants and those that are components of performance on particular plants. Preference characters determine where insects will attempt to oviposit or feed, while performance characters determine the fitness of an individual on a particular plant. The preference-performance distinction is important when considering insect distribution patterns and in models of host-use evolution (see below).

It is striking how many different characters vary among aphid clones or populations. Physiological and morphological characters related to performance (survival, growth rate, adult weight, feeding rate, and leg and/or rostrum length), preference characters (plant choice, alate [winged dispersing morph] production, and the tendency to drop from host plants when disturbed), and other characters whose relationship to host-plant use is less clear (temperature tolerance and the ability to transmit virus) all have been shown to vary in some aphid species (table 6.1).

From the published evidence it is impossible to determine whether some characters typically vary more frequently than others, or even whether preference or performance characters are more often variable. However, the evidence does suggest two generalizations. First, aphid populations, and by extension other herbivorous insect populations as well, contain considerable genetic variation in host-use characters, and such variation may be nearly ubiquitous (Futuyma and Peterson 1985; Jaenike 1990; Via 1990). If variation in host-use characters affects insect distribution across available hosts, then new patterns of host use may evolve. The diversity of life history patterns exhibited by aphids, including specialization and generalization, the use of herbaceous and woody host plants, feeding on aboveground and belowground plant parts, and the presence and absence of host alternation, presumably resulted from natural selection acting on genetic variation in host-use characters in ancestral populations. New patterns of host use will presumably result from selection acting on currently available variation.

The second generalization that appears upon examination of genetic variation in aphid host-use characters is that many different characters contribute to both preference and performance, and many can simultaneously vary in populations. Aphids can exercise preference when choosing a host to settle on and when dispersing away from a host (Kennedy 1950; Lowe and Taylor 1964), and probably in other ways as well. Variation in components of preference has also

TABLE 6.1
Evidence for genetic variation in aphid host-use characters

Character	Species	References*
Performance characters		
Survival	*Acyrthosiphon kondoi*	7, 16, 29
	Acyrthosiphon pisum	16
	Uroleucon tissoti	23
Colony growth	*Sitobion avenae*	36
rate, fecundity,	*Acyrthosiphon pisum*	7, 15, 16, 17, 34
development time	*Rhopalosiphon maidis*	4, 27
	Acyrthosiphon kondoi	16
	Schizaphis graminum	6, 11, 37
	Therioaphis maculata	20
	Uroleucon rudbeckiae	26
	Uroleucon tissoti	23
Adult weight	*Acyrthosiphon pisum*	3
	Rhopalosiphon maidis	4
Leg and/or	*Uroleucon* sp. (varies	19
rostrum length	between species	
	and is correlated	
	with host-plant	
	trichome length)	
	Myzus persicae	2
Feeding rate,	*Rhopalosiphon maidis*	22
probing behavior	*Therioaphis maculata*	21
Preference characters		
Plant choice	*Aphis fabae*	10
	Acyrthosiphon pisum	15
Alate	*Sitobion avenae*	14, 35
production	*Acyrthosiphon pisum*	13, 15, 16
	Rhopalosiphon maidis	4
	Schizaphis graminum	12
	Acyrthosiphon kondoi	16
	Urolcueon tissoti	23
Tendency to drop	*Acyrthosiphon pisum*	15, 24, 24
from host plants		
Neither preference nor performance		
Photoperiod	*Myzus persicae*	1
response	*Schizaphis graminum*	5
Ability to	*Myzus persicae*	31
transmit virus	*Acyrthosiphon pisum*	32
Temperature	*Acyrthosiphon pisum*	25
tolerance	*Myzus persicae*	30, 31, 33
	Schizaphus graminum	28
	Rhopalosiphon maidis	27
	Sitobion avenae	8
Resistance to	*Acyrthosiphon pisum*	9, 18
pathogenic fungus		

*(1) Blackman 1971; (2) Blackman 1972; (3) Cartier 1963; (4) Cartier and Painter 1956; (5) Eisenback and Mittler 1987a; (6) Eisenbach and Mittler 1987b; (7) Frazer 1972; (8) Griffiths and Wratten 1979; (9) Hughes and Bryce 1984; (10) Kennedy 1950; (11) Kerns et al. 1989; (12) Kvenberg and Jones 1974; (13) Lamb and MacKay 1979; (14) Lowe 1984; (15) Lowe and Taylor 1964; (16) MacKay and Lamb 1988; (17) Markkula 1963; (18) Milner 1985; (19) Moran 1986; (20) Nielson and Don 1974a; (21) Nielson and Don 1974b (22) Pathak and Painter 1958; (23) Pilson and Rausher 1990; (24) Roitberg and Myers 1978; (25) Roitberg and Myers 1979; (26) Service 1984a; (27) Singh and Painter 1964; (28) Starks et al. 1973; (29) Summers 1988; (30)

Table 6.1
(continued)

Takada 1979; (31) Tamaki et al. 1982; (32) Thottappilly et al. 1972; (33) Ueda and Takada 1977; (34) Via 1989; (35) Watt and Dixon 1981; (36) Weber 1985b; (37) Wood 1971.
Notes: Differences between aphid clones and biotypes are included.

Despite the considerable evidence of genetic variation in aphid host-use characters presented here, there are two related considerations that from an evolutionary point of view potentially cloud interpretation of most published aphid data. First, most of the agricultural literature discusses differences between aphid biotypes rather than aphid clones drawn from the same population. Most commonly insect biotypes are distinguished by survival and development differences on a particular host species or by differences in host preference (Diehl and Bush 1984; and see Wilhoit, this volume). Often biotypes have not been drawn from the same interbreeding population, and may even be drawn from geographically widely separated populations. Genetic differences between populations are not evidence of response to selection, but could instead be the result of other processes, such as genetic drift following a population bottleneck. In addition, not all individuals of a given biotype are necessarily genetically identical; they merely respond similarly to a subset of plant strains (Diehl and Bush 1984).

A second consideration in interpreting reported genetic variation in aphids is that even work done within a population measures genetic differences only between clones. Clonal differences are broad-sense genetic differences and include dominance and epistatic genetic variance and maternal effects along with any additive genetic variance (Falconer 1981). Although selection acts on phenotypic variation, any response to selection following a sexual generation is possible only if some of the phenotypic variation is due to underlying additive genetic variation. However in clonal selection (which occurs during the parthenogenetic generations in aphid populations; for example, Rhomberg et al. 1985) the response to selection depends on both additive and nonadditive genetic variation. Recent theoretical work by Gabriel (1988) suggests that even when parthenogenesis is interrupted as rarely as once every 100 generations by a sexual generation, the rate of response to selection in a parthenogenetic species can be as rapid as in a species that is always sexual. While the genetic basis of response to selection may be different in sexual and asexual aspecies, if some additive genetic variation is present in both, then evolution of new host-use patterns is possible in both.

Even with these two caveats, because differences are so ubiquitously found between biotypes and between clones, additive genetic variance in aphid populations is probably common.

been reported by Jaenike (1986a) for *Drosophila tripunctata,* and such variation is probably common in other herbivorous insects as well. Further, Jaenike (1987) found that settling and oviposition behavior, both components of preference, were each determined by several unlinked loci. Performance is also a compound character that includes at least survival and fecundity. For example, both colony growth rate and adult weight (which is probably related to fecundity) vary among biotypes of *Rhopalosiphon maidis* (Homoptera: Aphididae) (Cartier and Painter 1956). Survival and fecundity vary among biotypes of *Acyrthosiphon pisum* (Homoptera: Aphididae) (MacKay and Lamb 1988) and among clones of *Uroleucon tissoti* (Pilson 1990a; Pilson and Rausher 1990).

The Effect of Genetically Variable Plants and Insects on Insect Distribution Patterns

The environment imposes selection on populations, and spatially varying environments impose spatially varying selection on the populations that inhabit them. For an herbivorous insect species, an important component of the environment is the host-plant species on which it is found. If an insect species uses more than one host species, or if the host-plant population is genetically variable for characters that affect its use by insects, then insects on different host-plant individuals will experience different environments, and thus different selection pressures. Yet in spite of the fact that the presence of genetic variation has been demonstrated repeatedly in both insect (see above) and plant populations (see below), the importance of this genetic variation as a determinant of insect distribution patterns has largely been ignored. Characters that are important determinants of insect distribution patterns must also be the characters affected most strongly by selection on patterns of host use. Here I discuss the effect that genetically variable plants and insects may have on insect distribution patterns, natural selection imposed on insect populations as a result of those distribution patterns, and the expected response to selection by the insect population given the genetic variation that influenced the original distribution pattern.

The Effect of Variable Host Plants on Insect Distribution Patterns

Host-plant populations that vary in resistance to insects necessarily cause insects to be unevenly distributed across plants within those populations. Less resistant plants support larger numbers of insects than more resistant plants. This effect has been documented by many workers, both directly by censusing insects on plants (Fritz et al. 1986; Karban 1987a; Fritz and Price 1988; Maddox and Root 1987, 1990; Pilson 1990a, 1990b) and indirectly by censusing insect damage (Marquis 1984, 1990; Berenbaum et al. 1986; Simms and Rausher 1987; Rausher and Simms 1989).

Plant variation affects not only insect densities on individual plants, but also which insect species are commonly found together on plants. In two common garden experiments Maddox and Root (1987, 1990) censused *Solidago altissima* (Asteraceae) (tall goldenrod) for seventeen abundant herbivorous insects and found considerable genetic variation in resistance. Significant genetic correlations between resistances to the different herbivores were generally positive, although a few were negative. In addition, Maddox and Root performed a principle components analysis (PCA) on the genetic correlation matrix of insect resistances followed by cluster analysis of the principle component loadings on each of the 17 abundant insects and found that these insects fell into four clusters. Insect species within each of these clusters, which Maddox and Root call herbivore suites, respond similarly to plant resistance, while species in different suites respond differently to plant variation. Contrary to expectation, insects belonging to the same feeding guild or taxonomic group did not tend to occur in the same herbivore suite. In fact, there is no obvious pattern to the division of insects among herbivore suites. Regardless of the mechanism, however, members of the same herbivore suite will tend to be found together on plants more often than members of different suites.

There are two general mechanisms that could lead to herbivore suites such as those found by Maddox and Root. First, insects could respond directly to intrinsic properties of the plant. Characters that are known to affect the suitability of a plant include leaf toughness and the presence of trichomes (Hoffman and McEvoy 1986) and leaf chemical content (Louda and Rodman 1983b; Berenbaum et al. 1986; Bowers 1988a), and many insect species can discriminate among plants based on characters such as these. If plant properties are not modified by the presence or absence of other herbivores, then plant resistance is the direct result of plant characters alone. Insect species will fall into the same herbivore suite if they respond either in a similar way to the same character or in a similar way to different characters that are positively genetically correlated in the plant.

The second general mechanism that could result in resistance groupings such as those found by Maddox and Root results from indirect responses of insects to their host plants that are mediated by other herbivores. For example, suppose a plant population contains genetic variation in resistance to herbivore species B which leads to a nonrandom distribution of herbivore B among individual plants. Herbivore B might affect the distribution of herbivore A in one of two ways: herbivore A could respond directly to the presence or absence of herbivore B, or herbivore A could respond indirectly to herbivore B by response to an induced change in plant chemistry or morphology. Evidence of direct responses of herbivores to each other is rare (Strong et al. 1984; but see Finch and Jones 1989) and many observed patterns of positive and negative associations of herbivore species probably result from similar or opposite responses to intrinsic differences among plants (Fritz and Price 1988; Fritz, this volume) and from indirect responses of herbivores to induced changes in plant chemistry or morphology (see below).

Indirect responses of one herbivore species to another are probably common and result from insect-induced alterations in a host plant's phenotype. Interactions of this type could be either positive (facilitation), in which case early- and late-colonizing insects would be in the same herbivore suite, as defined by Maddox and Root, or negative (avoidance) in which case the insects would be in different herbivore suites. In addition, there is no a priori reason to assume that insects responding to each other in this way should be in the same feeding guild or be taxonomically related. An indirect response of one insect species to the presence of a second insect species that results in a significant genetic correlation between resistance to the two species requires: genetic variation in resistance to the early-colonizing insect, a phenotypic change in the plant caused by the early-colonizing insect, and a response by the late-colonizing insect to the phenotypic change caused by the early-colonizing insect. Both positive (Williams and Myers 1984; Damman 1989; Gange and Brown 1989; Tscharntke 1989) and negative (Harrison and Karban 1986; Faeth 1986; Moran and Whitham 1990) interactions between insects that are mediated by the host plant have been documented, but genetic variation in resistance to the insects has not been measured in these cases.

Such an indirect interaction between insect species has been documented by Pilson (1990b), who found that distribution of the aphid *U. tissoti* across genotypes of its host plant *S. altissima* depends on the assemblage of herbivore species present. *S. altissima* contains genetic variation for resistance to a group of herbivores that damage the apical meristem and cause plants to branch. In the presence of these branch-causing herbivores aphids are more abundant on branched plants than on unbranched plants, while if branch-causing herbivores are experimentally excluded, aphid distribution is random with respect to plant genotype. In an analysis similar to that performed by Maddox and Root, aphids and branch-causing herbivores would probably be found in the same herbivore suite. Despite the presence of genetic variation for resistance to aphids (measured as aphid preference and performance on different plant genotypes), aphid distribution patterns predicted by host-plant variation do not match natural aphid distribution patterns (Pilson 1990a). Thus, resistance to branch-causing herbivores, rather than resistance to aphids, controls aphid distribution. Clearly, models which assume that variation among plant genotypes (or host species) directly affects insect distribution patterns may not accurately predict the evolution of host use.

Competition between herbivorous insects could also affect insect distribution patterns. For example, Fritz (1990b) has found that the intensity of competition between sawfly species depends on host-plant genotype. Similarly, Moran and Whitham (1990) found that interspecific competition between *Pemiphigus betae* (Homoptera: Aphididae) and *Hayhurstia atriplicis* (Homoptera: Aphididae) eliminated *P. betae* (the inferior competitor) only on plant genotypes which were susceptible to *H. artiplicis* (the superior competitor). Likewise, Mopper et al. (1990) found that competition occurred only on plant phenotypes which were susceptible to both insect competitors. Results such as these suggest that, at least for insect species adversely affected by competition, the assemblage of insect species present, as well as the available array of plant genotypes, will affect the outcome of selection for preference or increased performance.

Parasitism or predation could also affect the evolution of host use in herbivorous insects (Bernays and Graham 1988). Shoot-galling sawflies are more vulnerable to predation on some willow genotypes than others (Craig et al. 1990), suggesting that preference might evolve in response to predation risk rather than in response to the effect of host-plant genotype on performance. Likewise, Denno et al. (1990) argue that female *Phratora vitellinae* (Coleoptera: Chrysomelidae) preferentially oviposit on salicylate-rich willow species, not because larval performance is high on these species, but because larvae are able to sequester salicins and incorporate them into defensive secretions.

Environmental conditions (such as temperature, light levels, and soil moisture and nutrient levels) can also have a strong effect on insect distribution patterns (Tingey and Singh 1980; Faeth et al. 1981; Coley 1983; Bach 1984; Service 1984b; Price and Clancy 1986b; Mattson and Haack 1987a; Louda et al. 1987; Bultman and Faeth 1988) but the relative importance of genetic resistance in the plant and other environmental factors as determinants of insect distribution patterns has rarely been evaluated. In three recent studies host-plant clone was found to be more important than climate (Karban 1987a) and competition (Fritz et al. 1986; Fritz 1990a) in determining insect population growth or distribution. At least in these systems, selection imposed by variable host plants is probably stronger than selection imposed by the other measured environmental conditions. In addition, environmental conditions can modify the expression of genetically determined plant resistance (Wood and Starks 1972; Service and Lenski 1982; Maddox and Cappuccino 1986; Hughes and Hughes 1988; Pilson 1992). The relative magnitude of genetic and environmental determinants of insect distribution patterns and any interactions between them will be an important factor affecting the evolution of host use in insects (Weis, this volume; Weis and Gorman 1990).

Although plant genotype can affect insect distribution patterns in a variety of ways, when considering selection imposed on insects by their host plants most workers have assumed that plant variation (either within or between species) directly affects insects. As more studies document the mechanisms by which genetic variation in plant resistance affects insect distribution patterns, the relative importance of each can be determined. Models currently available to describe the evolution of insects in response to plant variation all assume direct effects of plants on insects (see below). Incorporation of indirect effects of plant genotype and plant genotype-by-environment interactions on insect distribution patterns will add another layer of complexity to these models.

The Effect of Variable Insects on Insect Distribution Patterns

Studies of the evolution of host use in insects have generally assumed that preference and performance directly affect insect distribution. However, there is little empirical evidence available to evaluate this assumption. Thus, while current distribution must be both the consequence of past selection and the cause of current selection on preference and performance, it is usually unknown whether extant connections between insect characters and insect distribution are direct, indirect, or absent.

Preference Characters and Insect Distribution Patterns

If a plant is always rejected as a host, then there will never be selection to improve performance on that host. For this reason preference characters are thought to play at least as important a role in the evolution of host use as do performance characters (Futuyma 1983b; Jermy 1984; Thomas et al. 1987; Futuyma and Moreno 1988; and see the models of the evolution of host use discussed below). However, selection acts differently on preference and performance characters: performance characters always experience direct selection, whereas preference characters experience selection both indirectly through performance on the preferred host and directly via differential rates of host finding.

Documented genetic variation in preference characters has been of two types. In some cases different genotypes exhibit different preferences. A population of *Euphydryas editha* (Lepidoptera: Nymphalidae) butterflies studied by Singer et al. (1988) showed heritable variation in oviposition preference, and because females were generally able to locate their preferred plants, they usually laid eggs on the host species they preferred (Singer et al. 1989). Genetic variation thus results in differential distribution of *Euphydryas* genotypes across host plants. Codling moths (*Laspeyresia pomonella*, Diptera: Olethreutidae) (Phillips and Barnes 1975) and *Drosophila melanogaster* (Diptera: Drosophilidae) (Hoffmann et al. 1984) appear to prefer the host species from which either they or their ancestors were collected, indicating that natural distribution patterns are at least in part determined by preference variation in these two species as well.

In the second type of preference variation some individuals in a population exhibit preferences while others do not. In another population of *E. editha*, Ng (1988) has demonstrated that only some individuals discriminate among host plants. In this case all discriminating females chose the same plants. If females are able to locate the plants they prefer, and thus are able to oviposit on those plants, then offspring of discriminating females will be found more commonly on preferred plants. In contrast, offspring of nondiscriminating females will be found on plants in the same proportions in which they are encountered by searching females.

However, demonstrations of genetic variation in host preference in the laboratory do not imply that natural distribution patterns will result from that variation. Environmental factors that modify preference or an inability to locate preferred hosts (Singer et al. 1988) may result in distribution patterns that appear random with respect to host-plant genotype. Selection will act on genetic variation for preference only when such variation affects natural distribution patterns and insect fitness.

Variable Performance Characters and Insect Distribution Patterns

Performance characters have a more subtle effect on distribution patterns than do preference characters. In the absence of preference, individuals that vary in performance will initially be randomly distributed across host plants. However differences in survival, growth rate, or other performance characters among insect genotypes will eventually lead to more individuals showing high performance remaining on plants. For example, seasonal changes in aphid clone frequency have been attributed to natural selection (Rhomberg et al. 1985), perhaps acting on variation in performance among clones.

Because aphids reproduce parthenogenetically during the summer months, genotypes are passed intact from one generation to the next. Thus, identical genotypes reproduce and distribute themselves across host plants in the field. Because of their unusual reproductive system, aphid distribution patterns may be quantified as the number of individuals of a particular clone on a particular plant genotype or species. Aphid clones can realize different distribution patterns through differential colonization of or dispersal from host plants (preference), or through differential survival or colony growth rate (performance) once a colony has been established.

Hypothetical aphid distribution patterns that result from performance variation are shown in figure 6.1. In the example shown in figure 6.1A aphid distri-

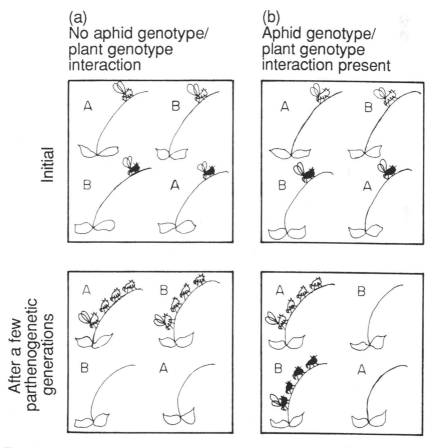

Fig. 6.1. Predicted aphid distribution patterns when no preference is present. Initial distribution is random with respect to aphid and plant genotype. If there is variation for aphid performance, but no aphid genotype/plant genotype interaction, then after a few parthenogenetic generations aphids that perform poorly will be lost from the population. If there is an aphid genotype/plant genotype interaction, then after a few parthenogenetic generations each aphid genotype will be common on the plant genotype it performs well on and absent from the plant genotype it performs poorly on.

bution is initially random with respect to plant and aphid genotype, but variable performance will eventually lead to some aphid clones being more common than others. Selection will lead to uniformly high performance on all plant genotypes. However, if the relative performance ranking of insect genotypes changes depending on the plant genotype they are tested on, then a plant genotype/insect genotype interaction is present for the tested performance character. If such an interaction is present, then after initial random distribution across plant genotypes different insect genotypes will be most common on different plant genotypes (fig. 6.1B).

Plant genotype/insect performance genotype interactions have been less commonly sought than straight genetic variation in performance, but have been documented in some (Moran 1981; Service 1984a), but not all (Pilson and Rausher 1990), aphid species (and see Blakely 1982 and Rausher 1983 for review). Unfortunately, there are presently no data concerning the natural distribution patterns that result from plant genotype/insect performance genotype interactions. Nevertheless, if these interactions affect distribution patterns, then selection for preference or improved performance on the initially less satisfactory host will result. The response to such selection will depend on the form of the plant genotype/insect genotype interaction and on the presence or absence of variation in preference.

Plant genotype/insect performance genotype interactions can take on one of the three general forms shown in figure 6.2. First consider interactions in the absence of variation in preference. In figure 6.2A half of the original individuals of each genotype perform well and half perform poorly, but neither insect genotype has a performance advantage over the other. Assuming no density dependence, only in the unusual case where the relative performances of each insect genotype on each plant genotype are exactly reciprocal will both insect

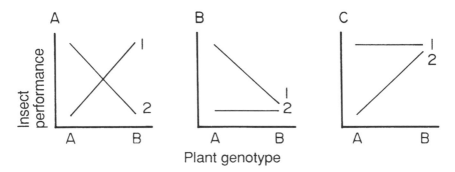

Fig. 6.2. Possible forms of an insect performance genotype/plant genotype interaction. Each line connects the performance of a single insect genotype on two different plant genotypes.

genotypes be maintained in the population. In figure 6.2B, although there is a genotype-by-environment interaction, one insect genotype is best on both plant genotypes. Over evolutionary time the inferior insect genotype will be eliminated from the population and the population will become genetically uniform (for this performance character). The situation in figure 6.2C is similar to that in figure 6.2B. Again, selection will cause the loss of one insect genotype from the population. However, in one case (fig. 6.2C) performance of the remaining insect genotype is high on both plant genotypes, while the other case (fig. 6.2B) insect performance varies among plant genotypes.

In all three of these cases, however, because insects show no preference, distribution will initially be random with respect to plant genotype. Thus, there will be continued selection for increased performance on all plant genotypes. For this reason, if there is no variation for preference and if each insect genotype is equally likely to colonize each plant genotype, it is unlikely that variation for performance will be maintained in the population. Thus, it is unlikely that specialized populations (those that use only plant A or plant B) will remain specialized.

Insect Distribution When Both Preference and Performance Vary

The expected evolutionary trajectory of the insect populations diagrammed in figure 6.2 changes if insect preference as well as insect performance varies. If there is variation for performance (as in fig. 6.2B), those individuals that both prefer and perform best on plant genotype A will be most fit. In the absence of any genetic variation in performance on plant B preference for plant B will be lost from the population. Eventually the insect population will be made up only of individuals that prefer and perform well on plant genotype A; the population will be specialized.

Compare selection for specialization with the situation in figure 6.2C. Individuals of insect genotype 2 that prefer plant genotype B will realize high fitness even though they would perform poorly on plant genotype A if they ended up there. There is thus continued selection for the ability to discriminate between plant genotypes. In contrast, because individuals of insect genotype 1 perform well no matter which plant genotype they end up on, there is no selection for the ability to discriminate between plant genotypes. In this case it is possible that both specialist and generalist genotypes will be present in the population. The scenario depicted in figure 6.2C is similar to that reported by Ng (1988) for a population of *E. editha*. The offspring of females that did not discriminate between hosts performed well on both hosts, while the offspring of females that did discriminate between hosts performed best on the preferred host.

If performance varies as depicted in figure 6.2A and preference varies as well, then individuals who prefer the plant genotype on which they perform best will be most fit. Thus, when there is a plant genotype/insect performance genotype interaction present, a correlation between insect preference and performance may build up through linkage disequilibrium. Such a correlation could be maintained in a sexual species, or in aphids at the end of the parthenogenetic season, if there is positive assortative mating by preference genotype. Mechanisms that could lead to assortative mating include preference for the larval host plant in both males and females followed by mating on the host plant, and phenological differences between host plants that affect the timing of mating in individuals with different preferences. Theoretical models suggest that sympatric host races may evolve as a result of the association between preference and performance and the resulting specialized patterns of host use (Bush 1975a, 1975b; Rausher 1984a).

A crossing plant genotype/insect genotype interaction, such as that shown in figure 6.2A, is also thought to result from many generations of selection taking place on the same long-lived host-plant genotype. Karban (1989b) documented such fine-scale adaptation of a thrips (*Apterothrips seticornis*, Thysanoptera: Thripidae) on its long-lived clonal host *Erigeron glaucus* (Asteraceae). Adaptation to host-plant clones and the maintenance of an insect genotype/ plant genotype interaction for insect performance can occur when there is little to no dispersal and thus little selection for increased performance on clones other than the natal clone. No dispersal between generations is equivalent to perfect preference for the plant on which an insect developed.

In general then, it appears that if only performance varies in an insect population, generalization will evolve, but when both preference and performance vary, generalization, specialization, or sympatric host races may evolve. The above discussion and the models of the evolution of host use described below assume that genetic variation in herbivorous insects directly affects insect distribution patterns. Unfortunately, there is little empirical evidence available to evaluate this assumption. Our understanding of the evolution of host use would be greatly improved by data detailing the effect that variable insect characters have on insect distribution patterns.

Modeling the Evolution of Host Use

A large majority of herbivorous insect species are specialized to some degree (Futuyma and Gould 1979; Chapman 1982; Price 1983b; Strong et al. 1984; Scott 1986; Bernays and Graham 1988) and the causes and consequences of specialization have received much theoretical and empirical attention. Two parallel approaches have been taken by workers attempting to elucidate the ecological and evolutionary conditions under which specialization will evolve. The first approach originally sought to describe the benefits of host specificity (implying some cost of generalization), and more recently has tried to enumerate the ecological conditions under which specialization and generalization will evolve (Thompson 1988a, 1988b; Courtney 1988; Barbosa 1988b). In these models properties of the plant (such as secondary chemistry and morphology) and insect are assumed to determine whether the insect will become specialized or generalized. Specialization may also evolve as a mechanism of enemy avoidance; some evidence suggests that some insects feed on particular plant species not because they are nutritionally superior but because they are less frequently attacked by natural enemies on those plants (Atsatt 1981a, 1981b; Jeffries and Lawton 1984; Jaenike 1986b; Sato and Ohsaki 1987; Bernays and Graham 1988; Denno et al. 1990).

The second approach to the evolution of host use has considered the conditions under which sympatric host races will be formed. This approach differs from the first in that host shifts and the addition of a new host or loss of an old host from the diet are not considered. Instead, conditions under which members of the same interbreeding population will become reproductively isolated from one another as a result of specialization on different host-plant species are sought.

Here I discuss formal models of host-plant specialization and sympatric speciation. Predictions of these models are compared to the predictions developed in the previous section from a consideration of insect distribution patterns.

Specialization versus Generalization

Four recent models have examined the conditions under which host specialization will evolve (Gould 1984; Futuyma 1986; Castillo-Chavez et al. 1988; Rausher 1988a). All four are two-locus models where one locus determines host preference and the other locus determines performance on each of two hosts. A result common to all the models is that initial gene frequencies at the two loci determine whether the population becomes specialized (via fixation of behavioral avoidance of the detrimental host) or generalized (via evolution of adaptation to the initially detrimental host), suggesting that there may not always be an ecological "reason" for specialization or generalization. Instead, the pattern of host use found in some populations may be the result of chance genetic events.

Another result common to these models is that stable polymorphisms, where some individuals are specialized and others are generalized, appear to be rare. These models would suggest that the specialist/generalist polymorphism found in *E. editha* (Ng 1988) is ephemeral and that the population is on its way either to specialization or generalization. However the genetic bases of host preference and performance are not known in *Euphydryas*. It is possible that more than two loci are involved, or that preference and performance include many characters and that the two-character, two-locus models discussed here are not relevant.

Rausher (1988a) also examined several ecological factors affecting the evolution of specialization versus generalization and found that the mechanism of population regulation affected the probability of the evolution of specialization. In his soft selection model, where each host contributed a fixed proportion of the individuals in the next generation, specialization could not evolve. In contrast, the evolution of specialization and that of generalization were about equally likely when population regulation was independent of host plant. As Rausher points out, whether population regulation in herbivorous insects is density-independent or density-dependent is an open question (see Strong et al. 1984; but see Turchin 1990), but if this prediction of his model is correct, then density-independent population regulation must be common, at least for specialist insects.

In Rausher's model increased search cost increased the probability of generalization evolving, while increased physiological cost of adaptation to both hosts increased the probability of the evolution of specialization. A physiological cost of adaptation to two hosts would manifest itself as a negative genetic correlation between performance on the two hosts. Although several investigators have attempted to detect such trade-offs (Gould 1979; Rausher 1984b; Via 1984; Weber 1985a; Hare and Kennedy 1986; Futuyma and Phillipi 1987; Fry 1990), few have found any. Interestingly, the two studies that did find some evidence of trade-offs, those of Gould (1979) and Fry (1990), both used selection experiments. In contrast, the other studies measured correlations between relatives. Bell and Koufopanou (1986) found the same pattern in a survey of studies measuring costs of reproduction: the predicted negative correlations between reproduction and later survival were found more commonly in selection experiments than in breeding experiments. Fry (1990) argues that fitness trade-offs between performance on different host plants could be common, but that laboratory experiments employing breeding designs may often be unable to detect them. If the laboratory environment is novel there may be considerable genetic variation for fitness itself that results in positive genetic correlations between components of fitness in the laboratory, and that obscures any trade-offs that might exist under natural conditions (Rausher 1988b; Fry 1990). Along similar lines, in a study of genetic correlations between life history characters, Service and Rose (1985) found that a predicted negative genetic correlation was greater in a natural than in a novel environment. Rausher (1988b) discusses other reasons why extant negative genetic correlations between performance on two hosts might be difficult to detect. Although there is currently little evidence of trade-offs in performance on two host plants, there are several plausible reasons why detecting such trade-offs might be difficult (see also Jaenike 1990). For this reason, trade-offs cannot be dismissed as a possible explanation for insect specialization.

Low search costs could also explain the evolution of specialization (Rausher 1988a). In Rausher's model search cost is the decrement in fecundity experienced by individuals who accept only one of the two plant species for oviposition. Unfortunately, there are even fewer empirical data on search costs than on physiological costs. The existing data suggest that search costs can be present or absent depending on the species which is examined (Courtney 1982, 1984, 1986). Low search costs thus may account for some specialization in some insect species. Finally, Rausher (1988a) also finds that the relative abundance of physiologically suitable and unsuitable hosts also affects the probability of specialization. The higher the frequency of initially suitable hosts the greater the chance of specialization on that host. In contrast, if the frequency of suitable hosts is low, then generalization (the ability to feed on both suitable and initially unsuitable hosts) is more likely to evolve. Rausher's model suggests that specialization evolves when trade-offs exist in performance on two hosts, when suitable hosts are abundant, and when search costs are low.

In all of these models (Gould 1984; Futuyma 1986; Castillo-Chavez et al. 1988; Rausher 1988a) the maintenance of preference or performance polymorphisms is much rarer than empirical evidence concerning the presence of genetic variation in host-use characters would suggest. Some of the genetic variation probably can be explained by a balance between selection and mutation (Lande 1976) or migration, which were not considered in these equilibrium models. In addition, the evolution of specialization appears to be more common in nature than in the models. There are both genetic and ecological explanations for these discrepancies. First, the genetic basis of host-use characters is not as simple in nature as it is in the models. All four of these models assume that preference and performance are each single characters and that each is controlled by one locus with two alleles. As discussed above, many characters are involved in both preference and performance. For example, host preference in *D. tripunctata* involves both host settling behavior and oviposition behavior once a potential host has been chosen (Jaenike 1986a). Additional characters are probably involved as well. Performance on host plants must also be a complicated combination of characters including survival, growth rate, and eventual fecundity. Second, the assumption of single-locus control of each character is often incorrect. For example, the ability to feed on different raspberry strains in the raspberry aphid is probably controlled by at least two loci, each with two alleles (Briggs 1965; and see Blackman 1977). Modifying the models to incorporate more characters and more loci, while difficult, might result in qualitatively different predictions about the frequency of the evolution of specialization and the maintenance of genetic variation in host-use characters. And finally, these models assume that an insect's genotype directly affects what plant it will feed on. However, as was discussed above, interactions with other herbivores, environmental effects, and an inability to find preferred plants can all affect insect distribution patterns in ways that are not directly related to insect genotype. Depending on the assemblage of insect species present or prevailing environmental conditions, indirect relationships between insect genotype and distribution patterns may change the trajectory of natural selection on alleles affecting host use.

Sympatric Host-Race Formation

The evolution of host use in insects is not limited to host shifts and host gains or losses within lineages; different patterns of host use within a lineage can also lead to speciation. Allopatric speciation occurs as a by-product of geographic isolation of once interbreeding populations and is thought to be common. The prevalence of sympatric speciation, however, has been more controversial. Verbal and mathematical genetic models suggest that sympatric host races can evolve (Levene 1953; Maynard Smith 1966; Bush 1975a, 1975b; Futuyma and Mayer 1980; Rausher 1984a), but until recently there has been little empirical evidence to suggest that they actually do evolve.

In general, models of sympatric speciation assume that different loci control preference for and performance on two hosts. If there is some mechanism that ensures mating among individuals that share preference genotype, these models predict that host races can develop. Because many herbivorous insects use the host plant as a mating rendezvous, it seems plausible that males and females that prefer the same host more commonly mate than males and females that prefer different hosts. Feeding primarily on one host will lead to selection for increased performance on that host. As a result, positive assortative mating based on host preference will result in linkage disequilibrium and the development of a positive genetic correlation between preference and performance. Continued mating isolation of insects from different host races is postulated to lead eventually to speciation in much the same way that allopatric speciation occurs: the reproductively isolated populations face different selection pressures, genetic drift may also occur, and both these processes serve to further differentiate the populations.

In order to determine if sympatric host races are common or can evolve in nature several studies have attempted to measure the genetic correlation between preference and performance. The predicted genetic correlation between female preference and larval performance has rarely been found (see review by Futuyma and Peterson 1985; Rausher 1984a; Hare and Kennedy 1986; Futuyma and Phillipi 1987). Via (1986) found a weak but significant genetic correlation between preference and performance in the leaf-mining fly *Liriomyza sativae* (Diptera: Agromyzidae) feeding on cowpea and tomato. Unfortunately Via's results are difficult to interpret because the flies used in her experiments were derived from flies collected in fields several kilometers apart and thus may not be members of the same interbreeding population. In addition, because *L. sativae* eggs are impossible to count directly, Via quantified female preference as the number of mines initiated. As Singer et al. (1988) point out, if failure to hatch is correlated with later larval performance, then Via's measure of female preference is confounded with larval survival, a component of performance.

More recently, Singer et al. (1988) found a significant correlation between female preference and larval performance within a population of *E. editha.* Although this correlation is probably genetic, Singer et al. (1988) point out two weaknesses in their experimental design. First, two separate experiments were performed, using different butterflies. In one the heritability of oviposition preference was demonstrated and in the other the correlation between phenotypic values of female preference and genotypic values of larval performance was measured. Second, their design was unable to remove possible maternal effects on larval performance. To the extent that maternal effects increase the genetic correlation between female preference and larval performance the reported correlation is overestimated. Nonetheless, Singer et al. have provided the most convincing demonstration to date of the preference-performance correlation predicted by models of sympatric speciation.

It is not completely clear why the predicted genetic correlations have been so difficult to document. One possibility is a result of our lack of knowledge about the genetic basis of preference and performance characters. As Futuyma and Peterson (1985) and Jaenike (1987) point out, the more loci that are involved in determining preference and performance characters, the less likely it is that linkage disequilibrium will be able to build up. It is possible that few loci determine preference and performance in the *Euphydryas* population studied by Singer et al. (1988) and that as a result it was possible for linkage disequilibrium to develop. In contrast, perhaps the existence of more loci controls preference and performance in other species in which the genetic correlation between preference and performance has been measured, and this more complex genetic determination has effectively precluded the buildup of linkage disequilibrium. If this argument is correct, then species thought to have undergone sympatric host-race formation (such as *Rhagoletis pomonella,* Diptera: Tephritidae; see below) may have a relatively simple genetic determination of preference and performance characters.

A second possible reason that significant genetic correlations between preference and performance have rarely been documented is that they typically have large standard errors (Via 1984) and thus very large sample sizes can be required to detect them. Perhaps sample sizes have often been too small to detect weak genetic correlations. If this is the case, then other approaches to sympatric genetic divergence may prove more fruitful.

Instead of measuring genetic correlations using a breeding experiment, Feder et al. (1988) and McPheron et al. (1988) measured allozyme differences between host-associated populations of *R. pomonella,* a true fruit fly. Sympatric hawthorn-associated and apple-associated populations have diverged significantly at several allozyme loci even though flies collected from the different tree species hybridize readily in the laboratory. Allozyme differences and the level of linkage disequilibrium between allozyme alleles in the two host races indicate gene flow of about 20% per generation (Barton et al. 1988). Gene flow of that magnitude should lead to the breakdown of linkage disequilibrium relatively rapidly (Barton et al. 1988), suggesting that differing selection pressures on the two hosts are also contributing to the maintenance of linkage disequilibrium in these populations.

A third possibility is that preference-performance correlations are not necessary for the formation of host races. Because the timing of emergence of apple- and hawthorn-derived flies differs such that time of emergence matches the time of either apple or hawthorn fruit ripening, and because this difference between the populations is genetically based (Smith 1988), sympatric host races could have formed even in the absence of a preference-performance correlation. In another population of *R. pomonella,* Prokopy et al. (1988) showed that although all flies preferred hawthorn, apple-derived flies were more likely to oviposit on apple than were hawthorn-derived flies. However, there were no differences in larval survival between the two host races on the two fruits. If this population shows genetic divergence similar to that present in the populations studied by Feder et al. (1988) and McPheron et al. (1988), then sympatric genetic divergence has occurred in the absence of a positive genetic correlation between female preference and larval performance.

Finally, it is possible that sympatric host-race formation via the buildup of a genetic correlation between preference and performance due to assortative mating may really be rare. Other mechanisms that ensure host-based assortative mating, such as emergence timing (Wood and Guttman 1982; Smith 1988), may be more important in sympatric host-race formation.

Conclusions, and Where to Go from Here

Theoretical models of the evolution of host-plant use by herbivorous insects make accurate predictions only if model assumptions are true. The major assumptions of all the models discussed here are single-locus control of both preference and performance, and insect genotype as an accurate predictor of insect distribution. Future research efforts should be directed toward testing each of these assumptions.

In this chapter I have stressed the relationship between genetic variation in host-use characters and insect distribution patterns. If genetic variation does not affect insect distribution patterns on an ecological time scale, then there will be no change in insect distribution patterns between generations, on an evolutionary time scale. It must follow that those host-use characters affecting ecological distribution patterns will be the characters acted on by natural selection, and therefore also the characters that will affect future patterns of host use. Research programs that carefully monitor determinants of insect distribution patterns on an ecological time scale will thus contribute greatly to our understanding of evolutionary patterns of host use. Such research programs need to take two points of view. First, which insect characters are important determinants of insect distribution patterns? Are preference characters or performance characters typically more important? Whether insects become host-specialized depends strongly on whether insect preference affects insect distribution patterns. And second, which plant characters are important determinants of insect distribution patterns? How do plant characters, other insect species, and environmental conditions interact with regard to insect distribution patterns? Do interactions between herbivorous insect species and interactions between insects and their predators affect insect distribution patterns? If so, the direct connection between insect preference and performance and insect distribution patterns may be lost. Host-plant specialization in insects is less likely if variable factors interact with plant genotype and cause variable insect distribution patterns.

The second major assumption of models of host-plant use by herbivorous insects is that insect preference and performance are each controlled by a single locus with two alleles. This assumption is often violated. Modifications of existing models to incorporate more characters or more loci controlling each character might result in different predictions about the evolution of specialization and generalization, and about the maintenance of genetic variation in host-use characters. If modified models make different predictions than currently available models, existing patterns of host use and genetic determination of host-use characters can be compared as a test of model predictions.

In the last ten years our understanding of the evolution of host-use patterns in herbivorous insects has been greatly improved, largely through the development of genetic models and empirical tests of model predictions. What is now needed is a combination of ecological and genetic approaches that will enable further refinement of models of host use.

Acknowledgements

I thank M. D. Rausher, R. S. Fritz, E. L. Simms, and an anonymous reviewer for valuable comments on earlier versions of the manuscript.

7

Plant Variation and the Evolution of Phenotypic Plasticity in Herbivore Performance   *Arthur E. Weis*

Variation in plant resistance and quality produces a heterogeneous environment to which insect herbivores must adapt. Here I consider the influence of host-plant variability on the evolution of insect growth by casting the problem in terms of phenotypic plasticity. Performance traits can be defined as those characteristics of the phenotype that are responsible for converting resources into biomass, and ultimately, into offspring. Because plasticity in physiological characters can allow performance to adjust to suit the plant occupied, phenotypic plasticity represents an additional mechanism to host-plant choice behavior for dealing with host-plant variability. Some excellent literature exists on the genetics and evolution of host-plant choice (e.g., Jaenike 1985; Rausher 1984a; Singer et al. 1988; Thompson 1988c; see reviews by Futuyma 1987; Singer 1988; Via 1990). However, the selective consequences of choosing a particular host depend on the subsequent performance on that host. For this reason a full understanding of plant choice evolution requires that performance evolution also be understood.

Adaptive evolution of host-specific performance requires within-population genetic variation in performance differences. Genetic variation of this type has been investigated in a few cases (e.g., Moran 1981; Service and Lenski 1982; Rausher 1984b; Via 1984; Ng 1988; Weis and Gorman 1990), but we are far from being able to make generalizations on levels of genetic variation and on intensity of the selection regimes that real-world ecological situations impose on host-specific performance. Further investigation of plasticity in herbivore performance is thus needed to understand questions on the evolution of host range, counterdefenses, and life history characters that have occupied students of plant-insect interactions. This chapter is intended as a guide for those who seek a better understanding of the evolution of phenotypic plasticity in general, and its importance in the evolution of plant-insect interactions particularly. It presents a selective, and perhaps idiosyncratic, review of the structure of natural selection as it acts on plasticity, of population genetic theory on the evolution of phenotypic plasticity, and of case studies which illustrate the adaptive consequences of herbivore plasticity in the face of host-plant variation.

Phenotypic plasticity can be defined as the sensitivity of genetic expression to environment. As a simple example of plasticity, an insect of a given genotype may grow to a larger size on one host plant than another. A conspecific of an alternate genotype may show the reverse relationship of size to host plant. In this example, size is influenced by an interaction between genes and host plant. As Schmaulhausen (1949) pointed out, genotypes do not code for development of a given phenotype; instead they code for development of a range of phenotypes given a range of environments. Since the range of phenotypes produced, i.e., the plasticity, is influenced by genes, it may evolve. It should be of keen interest to evolutionary ecologists to know how natural selection can influence patterns of plasticity, including its loss. There is a venerable tradition in theoretical population genetics that relates polymorphism to variable environments, and occasionally it is supposed that plasticity and polymorphism are alternate strategies for dealing with environmental uncertainty; this is a false dichotomy (Scheiner and Goodnight 1984), since populations could be polymorphic for plastic responses. As a caution, some restrict the definition of "phenotypic" to cases where change in phenotype with change in environment is clearly adaptive; but the broader view, that plasticity is any environment-dependent change in phenotype (Schlichting 1986; Scheiner and Goodnight 1984; Stearns 1989), is employed here. The focus of the chapter will be on the evolution of plastic response from nonadaptive to adaptive states, which can include the adaptive loss of plasticity.

The analysis of herbivore performance in terms of plasticity allows new views on two themes often encountered in the study of plant-herbivore interactions. These are the notion that a "jack-of-all-trades is a master of none" with reference to the relative efficiency of generalists and specialists, and the notion that "specialization is an evolutionary dead end." This chapter will show that by thinking in terms of plasticity, it can be argued that these constraints on evolution may not be so formidable as is frequently supposed.

Environment, Phenotypic Plasticity, and Natural Selection

In the simplest genetic models of natural selection, the environment, if defined at all, is treated as an external entity that imposes a selection pressure. Although ecologists have long been aware that environmental factors can influence growth and development, efforts to reconcile the role of environment as a template (Southwood 1977) defining the fittest phenotype and the environment as a factor in phenotypic development have not occupied the mainstream of evolutionary ecology. Although viewing the environment as an external entity that imposes selection has allowed an approximation of the adaptive process of great heuristic value, in reality the multiple effects of the environment on development, survival, and fertility need not be totally external, or independent of genotype (Schmaulhausen 1949; Waddington 1957; Bradshaw 1965; Lewontin 1983; Lynch and Gabriel 1987; Antonovics et al. 1988; Stearns 1989; de Jong, 1989).

Developmental programs of some species react to environmental factors in adaptive ways. Such cases include mandible size in late instars of the armyworm *Pseudaletia unipunctata,* which increases with the toughness of its food during growth and development (Bernays 1986). In other species, the influence of environment on development may simply reflect physiological limitations, as with the decreased growth efficiency of caterpillars fed on leaves of low water content (Scriber 1977). One would expect natural selection to either increase or decrease phenotypic plasticity in traits in whichever way enhances fitness; understanding the evolutionary dynamics of phenotypic plasticity requires an appreciation of its genetic basis and its potential response to selection.

Genotype and Environment in Development

Environmental perturbations during development can influence the phenotypic expression of a given genotype. A graphic model inspired by a similar graph by Lewontin (1974b) is presented in figure 7.1 to illustrate sensitivity of development to unspecified environmental factors. This graph depicts the phenotypes that can be expressed by a collection of genotypes drawn from a population. Each genotype can be mapped as a point in genotypic space. The expected values of quantitative traits x and y (e.g., length and width) can be taken as parameters that describe position in genotypic space. If development were completely insensitive to environmental perturbation, then the genotypic value of each individual in a population (though not necessarily its genotype) could be ascertained from the realized values in phenotypic space, since the latter would be a direct projection of the former; all variation in the trait would be heritable. However, the environment acts analogously to a diffusing optical filter that scatters the light emitted from a point source (fig. 7.1). In biological terms, environmental sensitivity of development causes each genotype to produce phenotypes that may deviate from the expected. Thus each genotype encodes a distribution of potential values in phenotypic space. This scatter of the genotype's potential phenotypes is precisely what is meant by phenotypic plasticity. When the genotype is sensitive to environmental perturbation, genotypic value becomes the average phenotype that gets expressed by a genotype in the average environment; additional parameters that describe the spread of potential phenotypes describe the degree of plasticity. The scatter in potential phenotypes caused by environmental sensitivity prevents the direct mapping of genetic value from phenotypic value; this is another way of saying that as environmental sensitivity increases, heritability decreases.

Just as different genotypes occupy different points in genotypic space, they may differ in the variances of the expected phenotypes (fig. 7.1). Conceivably,

DEVELOPMENT

Phenotypic Space

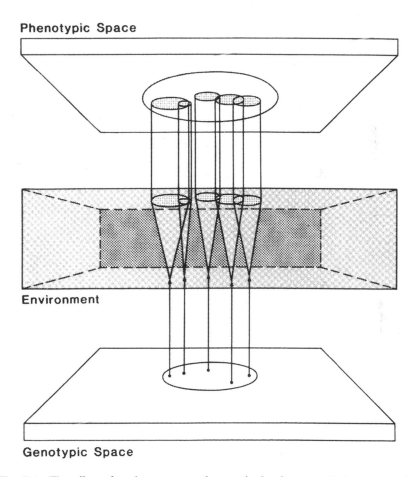

Environment

Genotypic Space

Fig. 7.1. The effect of environment on phenotypic development. Each genotype in a population can be characterized by a position in genotypic space that corresponds to its expected position in phenotypic space. During development, however, the perturbing effect of the environment can cause the realized phenotype to vary from its expected value. Thus, each genotype encodes for a distribution of phenotypes. Genotypes can differ both for the mean and for variance in the expected phenotype. The variance in the expected phenotype for a genotype is a measure of its phenotypic plasticity.

two genotypes that have the same expected phenotypic value could differ in the range of potential phenotypes. Among-genotype differences in phenotypic range indicate genetic variation in phenotypic plasticity; such variation allows plasticity to respond to selection. But how?

An episode of selection on phenotypic plasticity is illustrated in figure 7.2. Viewing phenotypic space from above, it depicts the range of phenotypes produced by three genotypes. Also indicated is a stabilizing selection threshold. All three of the genotypes depicted have average phenotypes that fall the same distance within the threshold. Despite the identical fitness for their average phenotypes, the three genotypes have different overall expectations of fitness. All individuals of the genotype with the narrowest range of phenotypes (a) will survive and reproduce. By contrast, individuals of the genotype with the broadest range (b) have a high probability of developing phenotypes that fall outside the threshold. All else equal, a population consisting of these three genotypes would show a decrease in phenotypic plasticity in the next generation, and if the selection threshold were constant, plasticity would continue to decline.

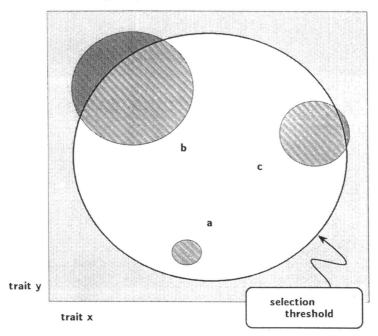

Phenotypic Space

trait y

trait x

selection threshold

Fig. 7.2. An episode of selection that acts on phenotypic plasticity. Genotype *a* has the highest expectation of fitness because its low plasticity causes all potential phenotypes to fall within the selection threshold. Genotype *b* has the lowest fitness expectation because a large proportion of its potential phenotypes fall outside the selection threshold. Intermediate fitness is expected for *c*.

This simple graphic model shows how the phenotypic plasticity of a character can be considered a character in itself (Schmaulhausen 1949). The "target" of selection may not be the genotype that gives rise to the optimal phenotype, but rather the genotype that gives rise to the array of phenotypes optimal to the normal variation in conditions. Although selection may decrease plasticity, or "canalize" development (Waddington 1957), more complex situations may favor increased plasticity. And, as I will illustrate below, within a set of developmentally linked characters, plasticity in some can lead to canalization in others (Bradshaw 1965; Weis and Gorman, 1990).

Plants as Variable Environments

From the perspective of many herbivorous insects, the host plant is a key component of the environment. By virtue of their small size, insects can experience many distinct microhabitats on the plant; the surface of a plant would constitute a "fine grained" environment (Levins 1968) for an ungulate but a "coarse grained" environment for a thrips. Further, low vagility confines many herbivorous insects to an individual plant during development. When migration to different host individuals is restricted, there will be a premium on the ability to adjust development and physiology to suit whichever plant the insect occupies.

The characteristics of host plants can influence the performance and fitness of an insect by a multitude of developmental, physiological, and behavioral mechanisms manifested through a network of cascading effects. Consider the ways lepidopteran larvae can be affected by leaf nitrogen content. When compared to performance on high-nitrogen leaves, growth and vigor in caterpillars are poor when fed the same quantity of low-nitrogen foliage. To compensate for low leaf quality, some caterpillar species increase the amount of leaf consumed (e.g., Feeny 1968; Slansky and Feeny 1977). This plastic behavioral response, in turn, may cause decreased gut retention time and increased time budgeted to feeding. Increased consumption may fully compensate for poor food quality. But if compensatory feeding is incomplete, the growth rate of the caterpillar could be slowed, which could thereby increase development time and decrease size at maturity. These changes in performance under poor conditions, in turn, may have several effects on fitness. Extending feeding and development times could increase the caterpillar's exposure to natural enemies and the elements. Additionally, increased development time could cause the individual to mature out of synchrony with potential mates or other time-critical resources. Small size could decrease the ability to attract mates and reduce fertility of those matings that are achieved. Thus an insect performance character such as ingestion rate evolves within a complex context that includes both the external environment and a chain of physiological and developmental linked characters that act as its internal environment. When considering selection on a character in that chain, it must be remembered that it develops in an environment determined by the preceding character in the chain, and that its effects on fitness are mediated by the subsequent characters in the chain.

Environmental Factors in the Evolution of Phenotypic Plasticity

In the analysis of natural selection, the environment has three types of effects in the course of a generation—first during settlement, then during ontogeny, and finally during the selection episode itself. Although these three environmental inputs can be considered separately at a conceptual level, in practice they may not be independent, especially when the relevant environmental factors can be part of an organism's internal environment. Theoretical models need not explicitly deal with each of these influences, but as will be shown below, the empirical analysis of selection depends on successfully identifying and measuring the three environmental influences. In this section I will discuss the influences of environment on selection. For clarity, they will be considered out of sequence, starting with ontogeny, followed by selection and then settlement.

Ontogeny

Factors that influence the ontogenetic development of a character can be called the "developmental environment"; phenotypic plasticity is caused by sensitivity to the developmental environment. As suggested above, the evolution of insect growth rate is influenced by the nitrogen content of the food (Scriber 1977; Scriber and Slansky 1981). Thus, when selection acts on growth rate, it in fact acts on growth rate, given the type of diet consumed.

The influence of a factor in the developmental environment can be incorporated into the analysis of selection on phenotypic plasticity when it is quantified as a reaction norm (de Jong 1989). Reaction norms are defined as the array of phenotypes produced by a specific genotype in response to variation in the developmental environment (Woltereck 1909; Schmaulhausen 1949). Although some characters change abruptly with environment, like the solitary and gregarious phases of migratory locusts, others show continuous variation. The change in growth rate with change in diet nitrogen concentration is a reaction norm that can be quantified as a mathematical function. Different genotypes in the caterpillar population could produce distinctive reaction norms which differ in slope, curvature, position, etc. (fig. 7.3). Since selection acts on a character, given its developmental environment, selection in effect acts on reaction norms (de Jong 1989, 1990).

Empirical studies of reaction norms are few (Lewontin 1983; Stearns 1989), owing in part to the difficulty in measuring them. Genotypes must be replicated and the replicates grown over an array of environments (Moran 1981; Service and Lenski 1983; Pilson, this volume). The mating systems of most insects preclude this approach, and as an alternative, the offspring of a mating can be reared under varied conditions (Via 1984; Ng 1988; Weis and Gorman

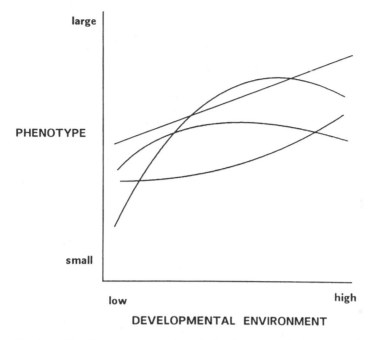

Fig. 7.3. Reaction norms can characterize the expected phenotype of a genotype as a mathematical function of the environment. Reaction norms can differ from one another in position, slope, curvature, etc.

1990); in experiments of this design, the plot of the offspring phenotype against the gradient of the developmental environment estimates the reaction norm of the parental genotypes. Regression methods can be used to find a mathematical function that describes the reaction norm. Because the same set of genes can influence many traits, and phenotypic development may respond to many environmental factors, selection may in many cases act on multidimensional reaction norms; however, to simplify the presentation, the discussion will be limited to the case of a single trait and a single factor in the developmental environment.

In summary, when selection acts on a trait, it acts on the developmental program that produces the trait, and the environment supplies input to that program.

Selection

Elements of the "selectional environment" determine the fitness of different phenotypes. The direction and form of natural selection imposed by the selectional environment on a specific trait can be quantified by the fitness function, which is a mathematical expression that relates the expected value of an absolute fitness component (e.g., survival probability, fertility) for a range of character values (e.g., from slow to rapid growth) (fig. 7.4).

Showing a correlation between variation in a trait and variation in fitness is not a sufficient condition to conclude that selection pressure acts directly on the trait (Lande and Arnold 1983; Endler 1986; Mitchell-Olds and Shaw 1987; Schluter 1988). To prove selection pressure it must be shown that there is a causal relationship between trait value and fitness dosage. Establishing the causal link entails identification of the relevant agent in the selective environment.

As a practical matter, selective agents may be simple or difficult to study experimentally, as illustrated by two contrasting examples. In some cases, the agent of selection is a component of the external environment, such as a predator. When an external entity like a predator imposes selection, the causal link between the state of a defensive character and fitness can be tested experimentally (e.g., Kettlewell 1956; Weis et al. 1985; see Endler 1986 and Manley 1985 for more examples.). In other cases, an element of the organism's internal environment mediates the relationship between the state of a phenotypic char-

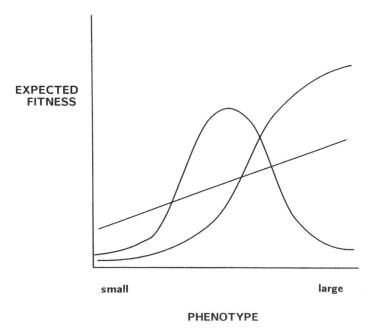

Fig. 7.4. Fitness functions depict the expected fitness across the range of phenotypes observed in a population. Likely forms for the fitness function include linear, logistic, and Gaussian.

acter and fitness. Such is the case with characters like ingestion rate, as discussed above, whereby the selective value of a particular ingestion rate depends on assimilation rate. The common view on such situations would consider leaf nitrogen as the agent of selection (e.g., Endler 1986, table 5.1), since fitness can vary with nitrogen content. I would argue, by contrast, that although ingestion-rate reaction norms evolve in response to the leaf nitrogen levels encountered, the fitness achieved by a particular ingestion-rate reaction norm depends on how well the ingested nitrogen is subsequently converted into progeny. It is the operation of this subsequent physiological apparatus that acts as the selective environment on ingestion rate. The importance of this distinction between developmental and selectional environment becomes apparent in empirical studies of selection on plasticity; paths of causality are more tractable when developmental and selectional factors are distinguished.

Inability to identify agents acting in the internal selectional environment leads to the epistemological difficulty of separating the intensity of selection on the trait of interest from selection acting on the internal selecting agent itself. If, for instance, assimilation is programmed to adjust to ingestion rate, and assimilation is also under selection (say through the agency of allocation patterns), then the correlation between ingestion rate and fitness will be confounded by any correlation between assimilation rate and fitness. In addition, if either of these two traits are correlated with others affecting fitness, additional confounding will occur. The statistical analysis proposed by Lande and Arnold (1983) can partition the influence of two or more traits on fitness into direct effects and effects through correlated traits (see below), but the accuracy of the analysis depends critically on identifying all correlated traits (Mitchell-Olds and Shaw 1987) and makes rigid assumptions on the causal relationships among the analyzed traits (Crespi and Bookstein 1989).

In summary, when phenotype has a causal influence on fitness, selection can act directly on that trait. The selectional environment, external or internal, consists of the factors that mediate the phenotype-fitness relationship. To measure successfully the intensity of selection on the reaction norm for a trait, it is critical to understand not only the influence of the developmental environment on its expression, but also the mechanisms whereby the selectional environment acts upon it.

Settlement

The importance of a population's distribution along the developmental environment in the evolution of environmental sensitivity becomes evident if one considers that selection will act most often in the segment of the environment most commonly occupied. One determinant of the population distribution across the environment is the statistical distribution of the environmental factor itself. Returning to the leaf nitrogen example, it is very unlikely that individuals in a caterpillar population would be evenly distributed among leaves of all nitrogen concentrations. More than likely, leaf nitrogen will be normally distributed, and so even if the caterpillars are distributed at random among leaves, most will occupy near-average leaves. When multiple host species are consumed, the frequency distribution of the relevant plant attribute may be skewed, platykurtic, or polymodal and so an insect population that settles randomly among hosts will have a similar distribution along the gradient of the developmental environment.

Variation in host preference behavior will have an additional influence on the distribution of the population along the gradient of the developmental environment. Preference for a particular host type will mean that genes influencing performance are most often expressed against the environmental background provided by the preferred host. If preference for a host and performance on it are positively genetically correlated, owing to linkage disequilibrium or pleiotropy, improved performance will evolve rapidly because of correlated selection responses between preference and performance. If not correlated to performance, the genetic variation in host choice can act as yet another source of variation in environmental background against which performance evolves.

Plasticity versus Canalization

One challenge to those investigating the evolution of host use is to determine the conditions that select for plasticity in performance characters and those that select against plasticity. To clarify, there are two components to plasticity that can respond to selection; these are developmental noise, and what can be called, for want of a better term, directional plasticity. Selection on directional plasticity will change the slope, elevation, etc. of the population average reaction norm, whereas selection on developmental noise will change the degree to which realized phenotypes deviate from the value expected from the reaction norm. In the parlance of regression statistics, directional plasticity is described by the regression model parameters and developmental noise by the residual variance. Directional plasticity connotes a predictable shift in phenotypic development in response to an identifiable environmental variable. A character showing developmental noise is sensitive to slight differences in the timing and rate of unspecified developmental processes; random differences between the right and left halves of bilaterally symmetrical organisms offer one example of developmental noise (Palmer and Strobeck 1986). Some cases that might appear as developmental noise may in fact be directional plasticity to unidentified environmental factors. The fitness consequences of each type of plasticity can be quite different. Selection against either type of plasticity can lead to canalization of development (Waddington 1957); canalized traits are insensitive to environmental perturbation.

Developmental Noise versus Canalization

Just as fluctuating environments can in some cases favor genetic polymorphism, fluctuating environments may also favor plasticity (Cohen 1967; Slatkin and Lande 1976; Bull 1987). If the selectional environment is unpredictable, and thus the most fit phenotypes are also unpredictable, then genotypes which give rise to a large array of phenotypes will have a good chance of producing a fit one. Similar arguments have been constructed as explanations for a short-term advantage to sex (Williams 1975). A mother that produces phenotypically variable offspring could be at an advantage in a highly variable environment, and that advantage would accrue whether the variable offspring were the result of genetically diverse sibs or phenotypic plasticity among genetically similar sibs or some combination of the two.

Slatkin and Lande (1976) used a model to assess the evolution of developmental stability in a quantitative trait when optimal phenotype varied among generations. They asked how variable the environment would have to be to favor increase in a rare allele at a modifier locus that decreased phenotypic variance among offspring, i.e., decreased developmental noise. Although a full exposition of the results would require an extensive mathematical treatment, suffice it to say that noisy developmental systems are favored only when all phenotypes have low fitness most of the time. Interestingly, that model and others' showed that these conditions also maintain genetic variation (Bull 1987).

If environments are constant, then over the long run selection will favor canalized development. Genotypes that produce phenotypes predictably at or near the optimum will have greater fitness than those that produce less predictable phenotypes (Slatkin and Lande 1976; Lynch and Gabriel 1987).

Directional Plasticity versus Canalization

On first consideration it would seem likely that selection will favor genotypes coding developmental programs that take advantage of occasional resource bonanzas. Many examples exist which show that herbivore growth performance or fitness increases with host nutrient content (Tabashnik and Slansky 1987; Mattson and Scriber 1987). Does this type of phenotypic plasticity represent an adaptation to maximize reproductive success when conditions permit or does it reflect physiological limitations imposed by low-quality resources?

Some simple genetic models can show the limitations of a bonanza exploitation strategy. Gillespie (1973, 1974) has argued that genotypes with a lower variance in reproductive success across environments would have a long-term advantage over more variable genotypes even if the mean reproductive success of the two were identical. The advantage of increased success in bonanza generations will not necessarily offset losses in poor generations. As a numerical example, take an annual, asexual population composed of two genotypes that show different fitnesses during "good" and "bad" years owing to climatic effects on host-plant quality. Suppose genotype A is variable and leaves behind 1 successful offspring in a bad year, but 3 in a good year. Genotype B, by contrast, leaves behind 1.8 successful offspring regardless of the conditions. If good and bad years occur in equal proportions, the average fitness of genotype A will be 2.0, while that of genotype B will be 1.8. Over the long run, however, genotype B will increase at a faster rate. Consider a two-year run with a good and a bad year. The expected number of descendants at the end of the two years for individuals of genotype A is 3.0 (1 offspring in the second generation from each of the 3 offspring in the first) whereas individuals of genotype B can expect to leave 3.24 (1.8 offspring in the second generation per the 1.8 offspring in the first). Gillespie and others (Dempster 1955; see also Seger and Brockmann 1987) have shown that in temporarily variable environments, the genotype with the highest geometric mean fitness is favored over the long run (but cf. Frank and Slatkin 1990). In the numerical example, the geometric mean fitness of A ($\{1 \times 3\}^{1/2} = 1.73$) is less than that of B ($\{1.8 \times 1.8\}^{1/2} = 1.80$), even though A's arithmetic mean fitness is greater. In this numerical example, a bonanza exploitation genotype is inferior even though bonanzas occur half the time. If bonanza years were rarer, the geometric mean fitness of this genotype would be even lower. As good years become more common, the so-called bonanza strategy is the favored genotype, but its fitness would still be less than a new, implastic genotype that equalled its arithmetic mean but bettered its geometric mean.

These results suggest that reaction norms showing improved performance with increase in resource quality may more often reflect physiological limitations than adaptation to capitalize on occasional bonanzas—there is a great advantage to improve under poor conditions even if it involves moderate deterioration of performance under good conditions. If selection acts to increase the rate of conversion of food into viable progeny, evolution of increased conversion at low resource levels may also lead to an evolutionary increase in conversion at high resource levels and vice versa (Weis et al. 1989; Via and Lande 1985). When phenotypic plasticity reflects physiological limitations in this way, it can be considered a nonadaptive consequence of directional selection for improved performance.

When will directional plasticity be favored? The place of the characters in the developmental hierarchy is one factor that can determine the fitness consequences of plasticity (Bradshaw 1965). When one character directly influences the expression of another, canalizing selection on the latter can favor plasticity in the former. Take for instance the size of the gall induced on goldenrod (*Solidago altissima:* Compositae) by the tephritid *Eurosta solidaginis*. Although the gall is plant tissue, gall size can be considered a phenotypic character of the insect (Weis and Abrahamson 1986) that develops in response to the insect's physiological stimulus. Weis and Gorman (1990) argue that stabilizing selec-

tion favors canalization of gall size, but that this can be achieved only if the
larva adjusts its stimulus to suit the host plant. Natural selection imposed by
natural enemies generally favors *E. solidaginis* individuals that induce galls of
intermediate size (Weis and Abrahamson 1986; Abrahamson et al. 1989). Gall
makers that induce small galls are vulnerable to the attack of a parasitoid which
must penetrate the gall's central chamber with its ovipositor to successfully
parasitize *E. solidaginis*. On the other hand, individuals that induce large galls
are obvious targets for attack by downy woodpeckers in winter when food is in
short supply (fig. 7.5). Gall size is influenced by insect genotype (Weis and
Abrahamson 1986; Weis and Gorman 1990) and so in principal can evolve in
response to these selective pressures on the insect. However, plant characteristics, particularly the time taken to respond to the gall maker, also influence gall
size. From the insect's perspective, plant response time can be considered an
environmental influence on gall size. Insect genotypes that produce an inter-
mediate-sized gall on any plant, regardless of plant response time, will have a
higher fitness than those genotypes that produce more variable gall sizes (see

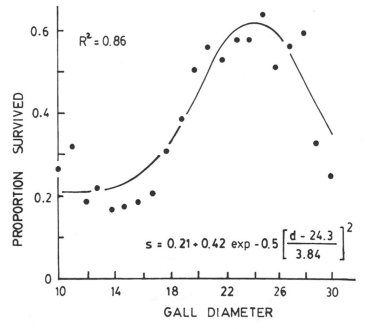

Fig. 7.5. Expected fitness of the gall-maker *Eurosta solidaginis* as a function of its gall
diameter. Gall makers in small galls are vulnerable to attack by parasitoid wasps,
whereas those in large galls are more often attacked by woodpeckers. From Weis and
Gorman (1990).

below). Thus selection on the gall maker favors the canalization of gall size against the perturbing effect of plant response time. It appears that insect genotypes which code for the ability to adjust the stimulus to suit the individual plant's reactivity will more consistently induce an intermediate gall size than those that cannot. In summary, the selective factors favoring canalization of gall size also favor directional plasticity in the intensity of the induction stimulus. Many other cases of adaptive plastic responses are likely to follow this pattern, as does the increase in consumption rate with decreases in food quality seen by Slansky and Feeny (1977), since the compensatory increase in ingestion maintains nitrogen intake rates.

Phenotypic Plasticity and Some Common Concepts of Evolutionary Trade-offs Involved with Ecological Specialization

Trade-offs are commonly invoked in evolutionary ecology (Rose et al. 1987). This logical structure is at the heart of much conventional thinking on the evolution of herbivore performance as well. One very compelling application of the trade-off paradigm is the food specialization hypothesis (Brues 1924; Deither 1954; Scriber 1983), which states that monophagous species will utilize their food more efficiently than polyphagous species. Presumably, ecological specialization on one food type would select for metabolic specializations that make the most efficient use of the single food used. Ecological generalization, on the other hand, would favor a metabolic machinery that could handle a variety of foods. However it is assumed that flexibility comes at the cost of overall performance in a way reminiscent of the proverb "A jack-of-all-trades is a master of none." A second trade-off notion related to the cost of metabolic flexibility concerns the evolutionary cost to specialization, i.e., that specialization is an evolutionary dead end. This evolutionary cost is thought to come about because selection for improved performance in a narrow niche eliminates genetic variation that would be required to adapt to new conditions. The place of these notions in the conventional wisdom of evolutionary ecology owes much to their intuitive appeal. Although there is at present no compelling reason to reject their relevance to the evolution of herbivore performance characters, there are too few data to conclude how frequently they apply. A look at the genetic mechanisms that impose these trade-offs will reveal how they can be overcome.

Can a Jack-of-All-Trades Master Them All?

The comparative method has been used to test the feeding specialization hypothesis: the results have been mixed. Take for instance work by Scriber and Feeny (1979) and Scriber (1983) which measured the growth rates, ingestion rates, and conversion efficiencies of papilionid and saturniid larvae that covered a range of host-plant specificity. Among papilionids, specialists were more efficient than generalists, but this difference could be explained by the fact that the former feed on forbs, which have higher water and nitrogen content, while the latter feed on the nutritionally inferior leaves of woody plants. All saturniids studied fed on woody plants, and in this taxon specialists were on average more efficient, but with exceptions. Futuyma and Wasserman (1981) compared the feeding efficiency of the polyphagous *Malacosoma disstria* to its monophagous congener *M. americanum* on the latter's host plant and found no difference. Thus it seems that the cost of generalization, in terms of efficiency, need not be very high.

There is a genetic assumption that underlies the feeding specialization hypothesis whose generality is yet to be established. This is the assumption that the same set of insect genes influences performance and efficiency in the same way on all plants. As a hypothetical example, imagine a generalist herbivore that uses two host plants with different toxic secondary compounds which are detoxified by enzymes coded at a single genetic locus. It is conceivable that an allele that is mediocre at detoxifying the compounds from both plants, and thus contributes to mediocre growth performance, could be favored over alleles that are superior in dealing with one plant but inferior on the other. However, the metabolic mechanisms for growth and efficiency can involve many loci, and there is the opportunity for inequalities in performance owing to the action of an allele at one locus to be compensated by the action of alleles at other loci. For instance, herbivorous insects commonly have several esterase enzymes encoded by different loci. Loci encoding efficient detoxification of one host may be different from those encoding efficient detoxification of the other. If the two detoxification systems are genetically independent, high growth performance could evolve on both plants. But when genetic control of performance on two hosts overlaps, there can be a constraint on the evolution of a "jack that masters all trades." Experimental approaches are required to determine the importance of genetic constraints in natural systems.

Using the techniques of quantitative genetics, the degree of genetic overlap in performance on two hosts can be measured as genetic covariance (Via 1984; see Simms and Rauscher, this volume). The genetic covariance between a pair of traits measures the degree to which the two are influenced by the same genes, i.e., pleiotropy, and by linkage disequilibrium. For convenience of interpretation, genetic covariances can be transformed to genetic correlations. Falconer (1952) noted that if one considers a single character expressed in two different environments as two character states (e.g., growth rate on host X and growth rate on host Y), the genetic correlation between the two character states quantifies the extent to which the variance in the two character states share a common basis. A population in which a performance trait expressed on host X shows no genetic correlation with the same trait as expressed on host Y would be composed of genotypes that had different polymorphic loci influencing the trait on the two hosts. A genetic correlation of 1.0 would indicate that the same polymorphic loci affect trait expression in the same way on the two hosts.

Via and Lande (1985) have developed a quantitative genetic model which explores the nature and consequences of genetic covariance in performance in alternate niches that has direct application to the evolution of performance on different hosts. Their model rests on frequent assumptions made in quantitative genetic theory, such as multivariate normal distribution of phenotypes and breeding values, linkage equilibrium, a lack of dominance and epistasis (all genetic variance is additive), constancy of genetic variances and covariances over time, panmixia, and a convex fitness function (i.e., selection is stabilizing when the population mean phenotype is at the optimum, but directional if it is far from the optimum) sufficiently broad to yield weak selection.

Since the evolution of phenotypic plasticity is the evolution of reaction norms, the relationship between reaction norms and genetic correlations should be made clear. This relationship was made clear by Via (1987), and is depicted in figure 7.6, which shows hypothetical reaction norms (solid lines) for three populations. They depict the average ingestion rate on two equally abundant host species as exhibited by five genotypes sampled from each of the populations. Next to each reaction norm distribution is the scatter plot illustrating its associated between-host genetic correlation. Panel 7.6A illustrates a case in which there is no genetic variation in the between-host difference in phenotype. The reaction norms for the five genotypes are parallel, which means that every genotype changes expression equally between hosts. There is genetic variation for ingestion rate within each environment, but there is no between-environment genetic variation. The corresponding genetic correlation, illustrated in panel 7.6A', is 1.0. In panels 7.6B and B', the reaction norms for the five genotypes cross one another, which indicates that the genotypes change their expression between the two hosts by different amounts. In this case there is genetic variance for ingestion rate both within and between environments; the genetic correlation between hosts is positive, but small. Finally, panels 7.6C and C' show a situation where all reaction norms cross at the same point. Here the genetic correlation between hosts is −1.0. There is genetic variation for ingestion rate within each host, but its contribution to total genetic variance is canceled out by the negative genetic covariance between hosts. What then are the consequences of these patterns of genetic variation for the evolution of ingestion-rate reaction norms?

Suppose the optimal reaction norm, the one that maximizes fitness on both plants, is as depicted by the dashed lines in figure 7.6. Imagine that host Y is less nutritious and so more of it must be eaten to reach the optimal pupation size. Will all three populations evolve to the optimal reaction norm? In population A all genotypes do poorer on host Y than on host X by the same amount. Because the optimal reaction norm has a positive slope (ingestion rate is higher

REACTION NORMS   GENETIC CORRELATIONS

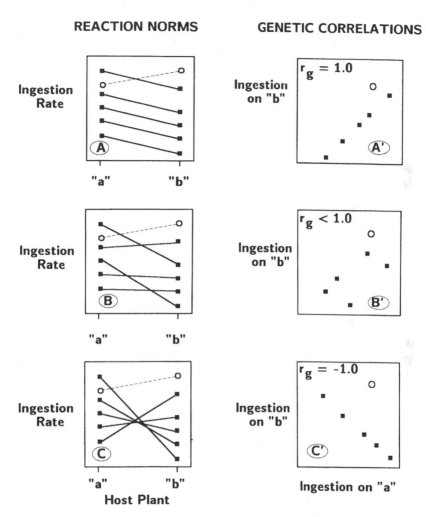

Fig. 7.6. The relationship between reaction norms for a trait expressed on alternative host plants, and the genetic correlation between the host-specific expressions of the trait. Dashed lines among the reaction norms indicate the optimal reaction norm (yields highest fitness). Circles among the correlations depict the combination of trait values corresponding to the optimal reaction norms. After Via (1987).

on *Y*) and all genotypes have the same negative slope, selection is pushing in a direction in which the population has no genetic variation. Since this example has been constructed so that population mean ingestion starts below the optimal on both hosts, the overall genetic variation in population *A* allows an increase on both hosts. However the end result of selection will be a population mean reaction norm that is a little above optimal on host X and a little below on Y. In population C, there is a genetic constraint that prevents selection progress toward the optimal reaction norm. Those genotypes that do well on plant X do poorly on plant Y and vice versa, and intermediates do well on neither; alleles do not exist that improve performance on one host without also diminishing it on the other. Only in population B is there the possibility of reaching the optimal reaction norm. Since in this population the genes that influence the trait on the two hosts are not identical, some of the loci that affect ingestion rate on host Y are not expressed when individuals develop on host X and vice versa. Allelic differences at loci affecting performance on one host are of neutral value to individuals occupying the other host. Shifts in allele frequencies for the two sets of host-limited genes are independent, and so selection progress toward the optimum on X can occur simultaneously with progress toward the optimum for Y. Although none of the genotypes initially in population B show the optimal reaction norm, when the model's assumptions hold, successive generations of selection and recombination alter allele frequencies so optimal genotypes are eventually produced.

Translating the information from reaction norms into genetic correlation allows the multivariate selection model (Falconer 1952; Via and Lande 1985; see Simms and Rausher, this volume) to be used for quantitative predictions of selection response. One can ask, how much does the population mean ingestion rate change on the two hosts after one generation of selection? The change in the two character states can be expressed in the following pair of complementary equations:

$$\Delta z_X = G_{XX}\,\beta_X\,q(\bar{W}_X/\bar{W}) + G_{XY}\,\beta_Y\,(1 - q)\,(\bar{W}_Y/\bar{W})$$
$$\Delta z_Y = G_{YY}\,\beta_Y\,(1 - q)\,(\bar{W}_Y/\bar{W}) + G_{YX}\,\beta_X\,q(\bar{W}_X/\bar{W}),$$

where Δz_X and Δz_Y are the changes in ingestion rates on the two hosts caused by a single generation of selection. The variables G_{XX} and G_{YY} are the additive genetic variances in ingestion rate on hosts X and Y respectively and G_{xy} is the additive genetic covariance in ingestion between the hosts. The strength of selection acting in environments X and Y is given by β_X and β_Y respectively; these are the partial regressions of relative fitness onto phenotype in the two environments (see Lande and Arnold 1983; Simms and Rausher, this volume). The proportion of individuals maturing on host X is given by $q(\bar{W}_X/\bar{W})$ and on host Y by $(1 - q)(\bar{W}_Y/\bar{W})$, where q is the proportion of all hosts that are X, $\bar{W}_X$ and $\bar{W}_Y$ are the mean fitnesses of individuals on the two hosts, and $\bar{W}$ is the population mean fitness. The right-hand side of the first equation can be divided into two terms; the first term denotes the change in the mean phenotype that will be expressed on host X owing to the direct effect of selection operating on the individuals occupying that host in the current generation, and the second term is the change in the phenotype on host X caused by the indirect effect of selection acting on individuals occupying host Y. The second equation shows the corresponding change in the mean phenotype as expressed on host Y. The indirect effect comes about since individuals occupying one host nonetheless carry genes which influence phenotypic expression on the other. To the extent that the genes influencing the phenotype on both hosts are the same (i.e., the genetic covariance is large), gene frequencies at loci affecting performance on one host can be changed by selection on performance on the other host.

Via and Lande (1985) examined the evolution of reaction norms by applying the multivariate selection equation iteratively to hypothetical data. If the genetic assumptions of the model hold, a panmictic population that occupies two stable habitat types will eventually evolve to a mean reaction norm that includes the optimal phenotype for each habitat—the jack-of-all-trades masters all. The only exception to this outcome occurs when the absolute value of the genetic correlation of performance between hosts is 1.0. These calculations suggest little reason to expect that generalist herbivores are by necessity constrained to poorer performance than specialists.

It is important to note that although the Via and Lande model predicts no absolute genetic constraint (except for perfect genetic correlation) on evolution of an optimal reaction norm for generalists, it also shows how realistic ecological and genetic situations can greatly slow the rate of reaction norm evolution, so that many thousands of generations are required. Thus one can speculate that the time needed to reach an optimal reaction norm will in some fraction of cases be longer than the time between major ecological perturbations that redefine the optimum. The population will move directly to the optimal reaction norm when the genetic variances and selection gradients are high and equal between hosts, the genetic covariance is zero, and the population evenly distributed between hosts. However, when there is asymmetry between hosts in the initial conditions (i.e., genetic variances, selection gradients, or host abundance) the population will evolve toward the optimal phenotype for one host more quickly than for the other. Via (1987) offered some generalities on the influence of asymmetry in starting conditions. For instance, the optimum in the performance trait will be reached more quickly on the host that is more abundant, on the host where the intensity of selection is stronger, or where the genetic variance is higher. The model showed other interesting dynamics when the sign of the genetic covariance and the direction of selection were discordant. For instance, progress toward the optimal reaction norm was slowed when selection favored changes in the same direction on both hosts but trait expression on the two hosts was negatively correlated. When combined with unequal host-plant frequencies, discordance between selection and the genetic correlation can cause performance to evolve away from the optimal on the rarer host for several hundred generations before reversing direction. The asymmetry in initial conditions can thus cause rapid adaptation to one host while adaptation to the other is so slow that long-term environmental changes may intervene before the optimum is achieved.

The application of the multivariate selection model by Via and Lande (1985) to performance in polyphagous herbivores serves as an important alternative to the food specialization hypothesis; it potentially explains why generalists are not always less efficient than closely related specialists. The model has excited much interest because it is based on parameters that population biologists can measure in natural or seminatural situations (although often with difficulty). Its careful application to real systems may allow short-term inferences on the direction and rate of selective change. However, long-term extrapolation from observed parameters would be suspect, since the model may not be robust against deviations from the underlying genetic assumptions (Gillespie and Turelli 1989; Barton and Turelli 1989).

The important genetic assumptions that affect the validity of the long-term extrapolations from initial conditions include the constancy of the genetic variance and covariances over time and the completely additive nature of the genetic variation. The first of these assumptions allows information on the current levels of genetic variability to be used as estimates of genetic variability many generations into the future. However, continued selection erodes genetic variability, as can be seen in the limiting case where the frequencies of the favored alleles go to 1.0. The generality of the constant-variance assumption has been tested in several comparative studies (Arnold 1981; Lofsvold 1986; Kohn and Atchley 1988). The results of these comparisons indicate that variances and covariances often do change, but it is unclear whether the observed changes are sufficient to completely discredit models that assume constancy. Given the ambiguous support of the constant variance–covariance assumption, long-term predictions based on it should be viewed with caution.

A second genetic assumption of the Via-Lande model which may not hold in nature is that of completely additive gene effects—that the effect of an allele m on the phenotype is independent of the allelic constitution of all loci. It is the additive effects of genes that are responsible for phenotypic resemblance among relatives, which is why the level of additive genetic variance controls the response to selection (see Crow 1986 for a readable explanation). Given the constraint of complete additivity, pleiotropy is the only way to generate phenotypic plasticity in the Via-Lande model. In many cases, however, the contribution made by some alleles can depend on which other alleles occur at the same locus (dominance) or at other loci (epistasis). Epistasis may be an important genetic mechanism that allows plastic developmental responses. For instance, one can imagine two nonoverlapping sets of loci that influence development of a trait, one of which controls the average of the phenotype and a second that controls the phenotype's sensitivity to the developmental environment. With this type of genetic control, one could divide a single trait into two separately evolving traits, i.e., the trait itself and the plasticity of the trait (Schmaulhausen 1949).

Because the "plasticity loci" would act as modifiers of the trait loci, their influence would be epistatic. Some empirical studies have shown independent genetic variation for a trait in a standard environment and that trait's environmental sensitivity (Caligari and Mather 1975; Khan et al. 1976; Scheiner and Lyman 1991). This epistatis provides an additional mechanism to pleiotropy to generate phenotypic plasticity (Scheiner and Lyman 1989).

Epistatis introduces nonlinearity into the selection response; the between-generation change in population mean of a trait caused by selection is no longer the product of the selection gradient and the additive variance. Unfortunately, current quantitative genetic theory lacks the general methods to deal with such nonlinearities (Barton and Turelli 1989). However, two- and three-locus models (see Nagylaki 1977) can be useful in exploring the influence of epistatis on the evolution of plasticity. For instance, Gonzalez-Candelas and Mueller (1989) have constructed a model with direct application to the evolution of herbivore performance on alternating host plants. In their model the herbivore's base-line performance was controlled by one locus, and the plasticity between host plants was caused by a modifier locus that expressed on only one of the plants. Their results show that under a wide variety of conditions, selection will push the allele frequencies to equilibria that do not correspond to the optimal reaction norm—epistatis can constrain the evolution of plasticity.

At present there is little empirical information that can shed light on the prevalence of additive pleiotropic effects versus those of epistatic effects in the genetic control of reaction norms. The quantitative genetic breeding experiments required to obtain this information are complex, and require mating schemes that include multiple matings for both males and females (Mather and Jinks 1982). Crosses among inbred lines offer an alternative method for distinguishing additive and epistatic influences, but inbreeding makes results of such experiments difficult to interpret if the base population is normally outcrossing (Service and Rose 1985). Unfortunately, these methodological requirements make herbivorous insects generally unsuitable for resolving the additive-versus-epistatic question in this way. As an alternative, artificial-selection experiments can be performed to determine whether selection on the phenotype in the mean environment also results in a correlated change in environmental sensitivity and vice versa (Scheiner and Lyman 1991). Again, most herbivorous insects are impractical subjects for selection experiments conducted under realistic environmental conditions, although groups such as mites (e.g., Gould 1979) and stored-grain pest species of bruchids (e.g., Wasserman and Futuyma 1981) offer exciting opportunities.

Despite its untested assumptions, the Via-Lande model for the evolution of phenotypic plasticity gives an important starting point for exploring the genetic factors that can impose a short-term cost to generality. Quantitative genetic experiments that measure the additive components of within- and between-host variance and covariance should be encouraged, especially if they can be conducted under field conditions. Only after a number of such studies accumulate will there be a general indication if the typical jack-of-all-trades is genetically constrained over the short term from mastering all. Interestingly, the studies that have been performed tend to argue against the prevalence of negative genetic covariance in performance between host plants (Via 1990). Complete interpretation of such results, however, requires quantifying the selection pressures acting on performance-trait reaction norms. The next section will show how such quantification can be achieved.

Is Specialization an Evolutionary Dead End?

When continued adaptive evolution in a stable environment leads to specialization, the macroevolutionary potential of a species may be diminished. An evolutionary decrease in niche width can shorten expected time to extinction because environmental changes of a given magnitude will more often shift key biotic or physical factors outside the narrow range required by a specialist than outside the broader range tolerated by a generalist. Some cases of extinction are reputed to be the consequence of overspecialization (e.g., Jablonski and Lutz 1983). If the assumptions of the fundamental theorem of natural selection hold (Fisher 1959), genetic variation for niche-use characters will be eroded by natural selection. If the environment changes, the low levels of genetic variation will constrain the specialist's adaptation to new conditions.

However, the impact of the overspecialization principle depends on several contingencies. These include the rapidity and magnitude of the shift relative to the level of genetic variation available for adaptive tracking of the environment. Further, as will be argued here, the nature of the shifting environmental factors is of special importance because change in the distribution across the developmental environment will affect the availability of genetic variance, while shifts in the selectional environment will not. By explicitly evaluating the way that selection acts on reaction norms, this section shows how the combined action of mutation and selection on reaction norms can allow the buildup of genetic variation that is hidden under typical environmental conditions, but expressed when the distribution along the developmental environment shifts. Because a shift in the developmental environment can increase the expression of underlying genetic variation, specialists may often be able to keep up with environmental changes (Waddington 1957).

A thought experiment on the change in phenotypic plasticity in response to selection (fig. 7.7) will reveal how hidden variation can be maintained in a population that inhabits a narrow, stable environment. Consider two populations that start with high variability in reaction norms, i.e., the elevations, slopes, and curvature of these reaction norms follow a multinormal distribution with large variance. Although no explicit assumptions are made concerning the genetic mechanism giving rise to the reaction norms in the thought experiment, de Jong (1989, 1990) presents a simple model of single-locus effects on slope

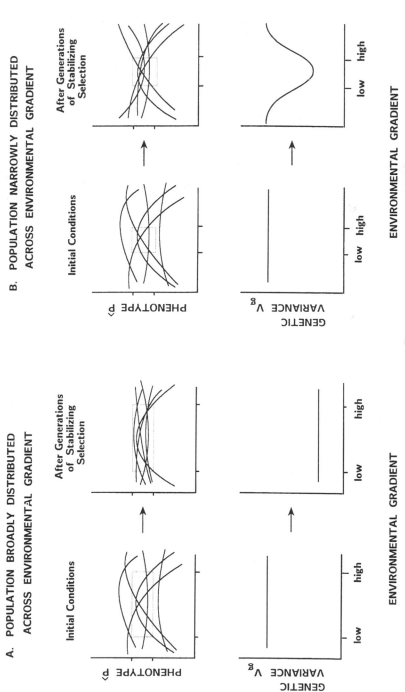

Fig. 7.7. Effect of stabilizing selection on the population distribution of reaction norms. (*A*) In a broadly distributed population, selection reduces the variation in position, slope, and curvature of reaction norms. (*B*) In a narrowly distributed population, selection reduces variation in position, but not in curvature and slope, of reaction norms.

and elevation of linear reaction norms that is a paradigm for the evolution of reaction norm parameters generally. In both populations stabilizing selection favors the same narrow range of phenotypes (P) at all points in the developmental environment (fig. 7.7). In the broad-ranging population, selection will reduce the frequency of alleles that cause deviation away from the optimal phenotype, so that development will become canalized (fig. 7.7A). After many generations of selection and recombination, the population will consist of genotypes with fairly flat reaction norms of similar elevations. By contrast, weaker canalization can be expected in the population occupying the narrow range. Genotypes which express the optimal phenotype when development occurs within the typical narrow range of environments will be favored by selection, and so allele frequencies should shift so that most genotypes will express reaction norms that include the optimal phenotype in the occupied range. But while this process is occurring, selection will not favor reaction norms that also code for the optimum over the unoccupied segments of the environment (selection cannot act in an environment that is not occupied), so that the potential deviation from the optimal phenotype under unusual conditions is selectively neutral. This genetic variation in environmental sensitivity of development is unexpressed as long as the population continues to occupy the same narrow segment of the developmental environment; genetic differences may be present, but because their expression is environment-dependent, and the environment is constant, the differences are not reflected in the phenotype. However, if the same collection of genotypes were forced to develop on some different segment of the environmental gradient, the underlying genetic differences among individuals could be expressed as phenotypic differences. This thought experiment shows when natural selection will have limited effect on existing genetic variation in environmental sensitivity, but how might such genetic variation arise?

Assuming genetic additivity and absence of genetic corrections, selection will cause the population's mean reaction norm to include the optimal phenotype within the typical developmental environment and thereby cause a loss in genetic variation in some reaction norm parameters. However, mutation can continuously reintroduce genetic variation to replace that eroded by selection so that some genetic variation can remain even when the population is at evolutionary equilibrium. The level of expressed genetic variance in a trait depends on the rate at which mutation introduces new genetic variation and the rate at which variation is removed by selection (Lande 1976; Turelli 1984). Consider then the fate of a mutation that contributes to the optimal phenotype when given the typical developmental environment but expresses a nonoptimal phenotype in an atypical environment. Such an allele will be favored, since selection acts only on the gene effects that are actually expressed. The genetic variance that could be potentially expressed in other environments is not removed by selection (except through indirect selection [Via and Lande 1985]) and so is free to accumulate. The result is a buildup of potential gene effects that can be expressed only if the developmental environment shifts.

Based on these arguments, it can be predicted that the genetic variance expressed for performance traits in insect herbivores should be greater on less frequently used hosts than on commonly used hosts. Reaction norms corresponding to this pattern were found in the gall maker *Eurosta solidaginis* on goldenrod. Weis and Gorman (1990) performed a quantitative genetic experiment to measure the variation in gall size reaction norms by growing different gall maker sibships on a genetically and phenotypically diverse array of *Solidago altissima* plants. In this case, host-plant properties are part of the gall maker's developmental environment. The gall sizes induced by some *Eurosta* sibships were sensitive to plant response time (days between egg deposition and gall initiation), while others were not (fig. 7.8A). Plant response time was unimodally distributed (fig. 7.8B) and so most individuals developed on near-average plants and few on extreme plants. The experiment uncovered a pattern of reaction norm variation consistent with the mutation-selection balance prediction—all sibships tended to produce similarly sized galls on plants that had response times near the mode (the typical environment), but gall size varied

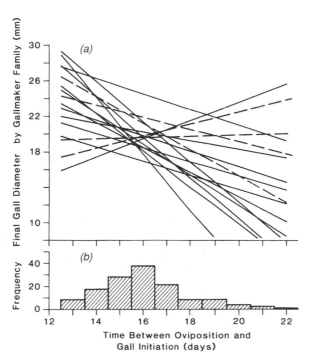

Fig. 7.8. Reaction norms of *Eurosta solidaginis* gall diameter as a function of plant response time. (*A*) Reaction norms (depicted as linear functions) calculated for 16 full-sib families of gall makers converge near the average environment. (*B*) Frequency distribution of plant response times. From Weis and Gorman (1990).

among sibships much more on plants with extreme response times (atypical environments).

The natural history of the *Eurosta-Solidago* systems allows the influence of the environment at the three salient points during the course of a generation (i.e., during settlement, ontogeny, and the selection episode) to be evaluated so that the intensity of selection on reaction norms can be estimated (Weis and Gorman 1990). The fitness function for *Eurosta* (expectation of survival versus gall size) was determined from field data. The population distribution or reaction norms (gall size versus plant response time, fig. 7.8A) was measured for the 16 sibships in the greenhouse, as was the distribution across the developmental environment (number of observed galls versus plant response time, fig. 7.8B).

The basic procedure that integrated information on the three influences of environment (developmental, selectional, and settlement) to estimate the fitness expected of each of the 16 gall size reaction norms is illustrated in figure 7.9. Two reaction norms, A and B, are depicted as linear functions relating expected phenotype to a developmental environmental gradient; thus the reaction norms can then be characterized by their elevation and slope parameters; the elevation is the expected phenotype in the average environment and also the average phenotype expressed across the gradient, while the slope denotes the environmental sensitivity of phenotypic expression to the developmental environment. Assume that absolute fitness is a quadratic function of phenotype. By substituting the reaction norm equation into the phenotypic term of the fitness function, fitness reaction norms for the two genotypes are derived (fig. 7.9a). The same intermediate phenotype is assumed optimal at all points of the environmental gradient, so that the steep slope of A yields poor expectation of fitness in extreme environments, but higher fitness at intermediate points. The zero slope of reaction norm B yields a uniform expected fitness across the gradient.

The fitness reaction norms quantify expected fitness as a function of environment, but to evaluate selection intensity, the expectation of fitness must be integrated across environments. This is equivalent to multiplying the expected fitness at each point along the environmental gradient by the frequency with which the point is occupied, and summing over all points (fig. 7.9b). Thus the area under the curve of "fitness reaction norm × frequency distribution of environment" yields the overall expected fitness. In this example, the optimal reaction norm is defined as a line with elevation equal to the optimal and a zero slope. Of the two reaction norms, A has the optimal elevation and B the optimal slope. It is interesting to note that in this case A has the greater expected fitness. This is because the A genotype produces the optimal phenotype in the most commonly encountered environment. By contrast B produces the same suboptimal phenotype at all points of the environment. Although B performs better than A in extreme environments, these are unlikely to be encountered and thus make small impact on B's total fitness.

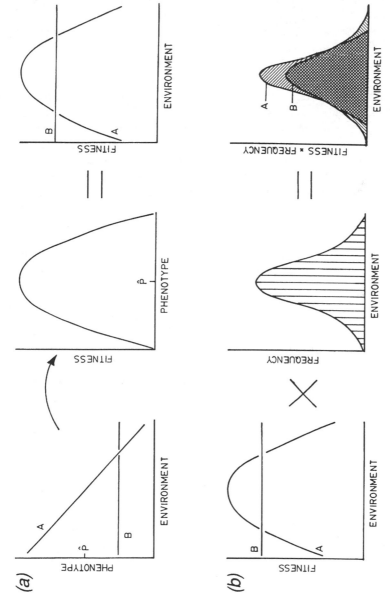

Fig. 7.9. Calculation of expected fitness of reaction norms. (a) Phenotype varies with environment, and fitness varies with phenotype, so fitness will vary with environment. (b) Environments are not all occupied with equal frequency, so overall expected fitness is the sum of the environment-specific fitnesses weighted by the frequency of each specific environment. From Weis and Gorman (1990).

With the expected fitness across environments in hand, the intensity of selection on the reaction norm parameters can be evaluated with an adaptation of the approach by Lande and Arnold (1983) whereby fitness is regressed over phenotype (see Simms and Rausher, this volume). Selection gradients on reaction norm parameters are calculated as their standardized partial regression coefficient over expected fitness. The interpretation of selection gradients on reaction norms differs from that of the Lande and Arnold approach in an important way. The regression of fitness on phenotype reveals the change in the distribution of phenotypic values caused by an episode of selection, thereby estimating the strength of a selection pressure. By contrast, the regression of fitness on reaction norm parameters reveals the change in the distribution of genetic values (Weis and Gorman 1990; Rausher and Simms 1989); if all genetic variation is additive, then this regression estimates the response to selection.

For *Eurosta,* there is stronger selection for increasing the elevation of the reaction norm than for leveling the slope (table 7.1). Since the mean gall size in the population is below the optimal size predicted from the fitness function (fig. 7.5), genotypes producing larger galls are favored, hence strong selection on elevation. Directional selection on slope is positive (the average slope evolves to be less negative), but only one fourth as strong as on elevation. Reaction norm slope may have strong influence on gall size on plants with extreme response time, and thus strong influence on *Eurosta*'s survival on extreme plants, but plants with extreme response times are rare, and so the influence of slope on overall survivorship is very small. Stabilizing selection on elevation and slope are of similar magnitude.

To evaluate the importance of the distribution of plant response times on selection intensity, several recalculations of the selection gradients were done with hypothetical response time distributions (Weis, unpublished data). When it was assumed that all observed plant response times had equal frequencies

TABLE 7.1
Selection gradients on parameters of the reaction norms of *Eurosta solidaginis* toward plant response time

Reaction norm parameter	Selection gradient	Standard error
	Directional selection	
Elevation	0.122	0.014
Slope	0.028	0.014
	Stabilizing selection	
Elevation	−0.072	0.016
Slope	−0.052	0.016
Elevation × slope	0.062	0.014

Source: Weis and Gorman 1990.
Note: Selection is quantified as the partial regression of the family-specific expected relative fitness over the reaction norm parameters.

(distribution platykurtic), selection on slope was equal to selection on elevation. But when all plants had the average response time (distribution leptokurtic), selection on slope vanished. Thus, selection on sensitivity to the developmental environment depends on how widely the population is distributed across the environmental gradient. These analyses confirm the qualitative expectation that selection will be ineffective at eliminating genetic variance in environmental sensitivity in populations that occupy a narrow range on the developmental environmental gradient.

Returning to the evolutionary dead end issue, if selection cannot counterbalance the buildup of genetic variation in environmental sensitivity caused by mutation, environmental shifts will not necessarily doom specialists to rapid extinction, since the shift may cause expression of sufficient underlying genetic variation to allow an adaptive response to the new conditions. As an extreme case, imagine a host-plant species being replaced by a related species through competitive exclusion. Further suppose that host choice behavior permits colonization of the novel host. Selection may have greatly diminished expressed genetic variance in performance on the original, which would seem to lessen chances of the necessary adaptive response for a successful host shift. But the accumulated genetic variation in environmental sensitivity might be expressed if the population comes to occupy the invading host (fig. 7.10), which would then allow the herbivore to evolve to a new optimum. Of course, the specialist's success in making a forced host shift depends on the difference between the old

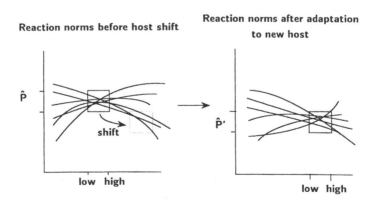

DEVELOPMENTAL ENVIRONMENT

Fig. 7.10. Although stabilizing selection may reduce the genetic variance expressed in the occupied segment of the environment, the remaining variation in environmental sensitivity can be converted into genetic variance under altered environmental conditions. This could allow a specialized population to evolve to a new phenotypic optimum when the environment shifts.

and new optima relative to the amount of genetic variance in sensitivity, but the important point is that success is not precluded by the low genetic variance expressed on the old host. The logic applied to forced host shifts also applies to host range expansions. Rather than be constrained by an absence of genetic variation, evolution of an extended host range by a specialist may more critically depend on ecological factors that bring potential host plants to high abundance, on the flexibility of the host choice behavior, and on external selective factors such as host-dependent predation rates. To summarize, the accumulation of genetic variance for environmental sensitivity in specialists may allow them to adapt to changed conditions following shifts in the developmental environment.

Despite these arguments based on the influences of selection and mutation on genetic variance, a specialist's success in adapting to an environmental shift can nevertheless be constrained by other factors. Even if most traits express increased genetic variation when placed in the new environment, lack of sufficient expressed variation in a single key trait can lead to extinction. For instance, in the process of specialization some traits useful in ancestral populations but neutral in the specialist may be lost owing to the accumulation of dysfunctional alleles through mutation and drift; such a trait could be irretrievable (Moran 1988). Antagonistic pleiotropy among selected traits is of course a possible constraint, and it has been shown that genetic correlations can change with environment (Gebhardt and Stearns 1988; Service and Rose 1985). Although theory predicts that traits closely tied to fitness will show no or slightly negative genetic correlations among one another at equilibrium (Lande 1976), this would apply only to the correlation expressed in the developmental environment where selection has occurred. Without further quantitative modeling, it is unclear how pleiotropy will affect trait correlations in atypical environments. However, it is often observed that life history traits are positively correlated in populations that are brought from the field and reared in the lab. It has been suggested that these correlations can be due to the overall poor performance of a few individuals in the new environment (Service and Rose 1985). Also unclear at this time is how genome-wide heterosis, which could favor a reduction in developmental noise (Gillespie and Turelli 1989), could act to maintain genetic variability in directional plasticity.

In summary, a population specialized to a narrow niche probably may on average have a shorter time to extinction than an otherwise similar generalist, but the reduction in expressed genetic variation that comes with specialization does not make this average outcome inevitable. Genetic variation in environmental sensitivity, which is selectively neutral in a specialist, can arise through mutation. The accumulated store of hidden genetic variance can then be expressed when conditions change, allowing adaptation to new situations.

Some Additional Thoughts

The focus of this chapter has been on the evolution of plasticity in herbivore performance in response to host-plant quality. One clear conclusion is the major influence of the population's distribution along the host-plant quality gradient. Host choice behavior is of course a major determinant of that distribution. On the other hand, performance on alternative hosts will determine the fitness expectations for choice phenotypes. In the parlance introduced here, choice behavior determines the population's distribution along the developmental environment, whereas performance traits determine the selection environment on choice. It could be beneficial to the field for theorists to construct models that examine the joint evolution of performance, of plasticity in performance, and choice as an evolving determinant of the expression of genetic variance for performance.

Models of the joint evolution of performance plasticity and choice should also be applicable in understanding the evolution of plant resistance to insect attack. Plant resistance to insect attack can be divided into components that lessen the frequency or intensity of tissue damage (defense) and components that act to compensate physiologically and developmentally for lost tissue (tolerance). These components are directly analogous to insect choice and performance, respectively. For instance, loci regulating production of a chemical deterrent can determine which point along the herbivory intensity gradient the plant will develop; and the plant phenotype produced by a given set of growth and development genes will thus depend on these deterrence loci. The plasticity in plant growth and development will then evolve against an environmental background influenced by herbivore deterrence traits. On the other hand, the performance of the plant in the face of herbivore damage will in turn determine the selection intensity acting on deterrence traits. There is a touch of irony here, since it is quite possible that by understanding the genetic basis for the joint evolution of plant defense and tolerance, an analytical framework can emerge that will allow us to better understand the joint evolution of insect herbivore host choice and growth performance.

8

Diversity in Plant Pathogenic Fungi as Influenced by Host Characteristics

James V. Groth and Barbara J. Christ

Like other organisms, plant pathogenic fungi possess genetic variation. The description of this variation in plant pathogens is incomplete, and receives much attention from researchers. In addition, a major goal of current study is to determine the forces that maintain such variation, including (for diploid or dikaryotic fungi) heterozygote advantage, patchiness of host or physical environment, and frequency or density-dependent selection. (Parker discusses some of these forces in chapter 15 of this book.) These cannot be elucidated from superficial field observation. Unusual patterns of variation such as inhomogeneous or discontinuous variation, or correlation of certain patterns with geographical or ecological attributes, can suggest how the variation is maintained. Yet establishment of the forces underlying patterns of variation constitutes an experimental challenge that has as yet to be met in most organisms.

In this chapter, we will review existing knowledge of the impact of plant-host genetic diversity on the genetic diversity of plant pathogenic microorganisms, concentrating on fungi. Because very little is known, we will also discuss the historical development of the field and speculate about future directions of research.

Detection of Diversity in Plant Pathogenic Fungi

Our concept of species in plant pathogenic microorganisms has followed the evolution of ideas regarding species in general. Plant pathologists and mycologists began with a strongly typological view just after the discovery in 1853 that plant diseases are often caused by parasitic fungi (Ainsworth 1981). The typological view precluded acknowledgment of genetically determined variation within species. Even today, tension exists between the typological treatment of species that is embodied in the dependence of systematic biologists on collections of "type" organisms, and the evolutionary biologist's view of species as an imperfectly delimited group of organisms that are more or less similar by virtue of their sharing in the same gene pool. The fields of mycology and plant pathology seem to be especially affected by this tension. The population view of species became prevalent rather late, and is not as firmly grounded in these fields as in more unified areas of biology such as genetics. This may help to explain the present state of knowledge of, and enquiry into, diversity of plant pathogenic fungi. Later, we will briefly document how these two views of species can affect particular investigations.

Variation within species of plant pathogenic fungi was first documented by Erikssen in 1894 (Ainsworth 1981), who found that *Puccinia graminis* isolates are specialized on one or more species of host plants, without being consistently morphologically distinguishable. Progress in defining rather complex and hierarchical variation in this rust fungus continued when, in 1913, Stakman (Ainsworth 1981) showed that *P. graminis* on wheat could be further subdivided into races based on differential compatibility on a series of host cultivars. The earliest documented case of variation in virulence at the intraspecific host (cultivar) level was by Barrus (1911) with the imperfect fungus that causes anthracnose of edible beans. These and similar studies on a small number of other obligate fungal parasites can be considered early pioneering studies of nonmorphological, intraspecific variation that assume an important place in the history of biology. Pathogens receiving the most attention were those for which genetically determined host resistance was short-lived owing to the rapid ascendence of "new" (actually extant but rare) phenotypes with the capacity to negate the resistance. Little attention seems to have been paid to pathogens which did not have this appalling habit, and because practical reasons for studying variation in other fungi did not exist, a typological view of all but a few plant pathogenic fungi still held firm, even as more and more evidence of variation in the rusts, smuts, and mildews was accumulating.

Problems in Studying Variation in Fungi

The species concept in fungi has never been as firm as in other organisms. There are a number of historical reasons for this, centering around a poor understanding of their ecology and population genetics as well as poor definition of the basis (genetic versus environmental) for minor differences in morphology commonly observed in the field. Some mycologists, citing the ambiguity in defining fungal species, especially using genetic versus morphological criteria, have recently questioned whether the concept of species as discrete gene pools is as appropriate or useful for fungi as it has been for higher animals and plants (Burnett 1983; Boidin 1986; Kemp 1986).

To be able to measure confidently intraspecific variation, one must first distinguish populations that have no or very restricted gene flow and are thus isolated species. This is often not possible with fungi, because morphological definitions of species do not always agree with genetic definitions (Burnett

1975, chap. 11). It may even be impossible to delimit an individual so that the unit of selection or the population size is known. Moreover, plant pathogenic fungi have often been excessively taxonomically split based solely on host range, usually without definitive cross-inoculation studies to determine the limits of host range. In some cases a new species has been named for almost every host that harbors a particular genus of fungus (Alexopolous and Mims 1979).

Diversity is a term with a number of meanings. Groth and Roelfs (1987b) have discussed these various meanings with regard to plant pathogens. Diversity often refers to a larger number of phenotypes without regard to their frequency distribution. Rapid appearance of "new" phenotypes can also lead to a view that a pathogen is diverse, as this component contributes greatly to the boom-and-bust resistance cycles of crop plants and diseases. Other elements of diversity, such as evenness of frequency and phenotypic or genetic distance, have only rarely been used for defining diversity of plant pathogenic fungi.

Variation in Plant Pathogens in Relation to Host Variation

Following Robinson (1976), we define a plant pathosystem as a subset of an ecosystem centering around parasitism of a host taxon by a pathogen taxon. It is clear that in studies of plant pathosystems emphasis has not been placed on the pathogen, but rather on the host (chapters 2, 3, 7, 9, 12, 14, 15, 16, this book). Many more people are directly interested in variation in the host plants than in variation in the pathogen. In addition to plant pathologists, plant physiologists, plant anatomists and morphologists, plant breeders, agronomists, and horticulturalists all work with variation in higher plants, whereas only a few plant pathologists and a small fraction of mycologists concern themselves with intraspecific pathogen variation. Even among less commodity-oriented disciplines such as genetics or ecology, the emphasis has been on higher plants and their biotic and abiotic environment, with pathogenic microorganisms being considered secondarily if at all.

It has not been possible to design experiments that demonstrate a clear host effect on plant pathogen diversity in the field. This requires that the pathogen be sampled in identical environments, but where subpopulations have adapted to host populations that differ in diversity. Environments are never identical when subdivided pathogens are examined, and other aspects of the biology of the pathogen have varied considerably. Several examples are presented. Stem rust of wheat, *Puccinia graminis* f. sp. *tritici,* is considerably more diverse, as measured by numbers and frequencies of phenotypes, in the Pacific northwest of the United States than in the Great Plains (Roelfs and Groth 1980). The difference in prevalence of sexual reproduction in these two locations is striking. The alternate host, barberry, which supports sexual reproduction in the stem rust fungus, has been eradicated from the Great Plains, and the pathogen survives vegetatively, cycling from south to north annually. Collections made in the Pacific northwest were all associated with the sexual stage on barberry. While the diversity of the pathogen is greater in the Pacific northwest, the diversity of the wheat host, as number of cultivars and sources of resistance, is lower than in the Great Plains. Combined diversity of both hosts in the Pacific northwest is greater than that of the single host species, but the effect of the (largely unmeasured) diversity of barberry on diversity of the fungus cannot be compared directly with the effect of wheat diversity, because even a uniform stand of barberry permits sexual reproduction in the pathogen.

Other examples where differences in pathogen diversity can be more readily ascribed to sexual reproduction than to host diversity include that of *Bremia lactucae,* which causes downy mildew of lettuce. In California, this fungus has fewer virulence phenotypes than in Britain or Europe, largely because the sexual oospores function as primary inoculum in Britain or Europe much more often than in California (Illot et al. 1987). Al Kherb et al. (1987) and Welz and Kranz (1987) show cases where sexually versus asexually derived collections from a single population and season of *Puccinia coronata* and *Erysiphe graminis* f. sp. *hordei,* respectively, resulted in different levels and patterns of diversity. Sexually derived collections gave more phenotypes with a more even distribution of frequencies.

Finally the late blight fungus, *Phytophthora infestans,* has been transported throughout the world from its center of origin in southern Mexico. Inexplicably, until recently only one of the two mating types necessary for fertilization leading to sexual reproduction could be found outside the center of origin, and the number of races is correspondingly less outside the center of origin (D. S. Shaw 1987). Unfortunately, diversity in the host (as number of phenotypes and/or genetic distance) in this case is greater in the center of origin, since several species of wild and weedy *Solanum* species harbor the fungus there, so the influence of sexual reproduction and that of host diversity on pathogen diversity are confounded.

These examples illustrate the difficulty in ascribing characteristics of the population structure of a pathogen solely to characteristics of the host. In reality, the presence or absence of sexual reproduction is but one of the more important of a number of influences on pathogen diversity that are always confounded with the influences of the host in field settings. Long-term experiments, in which all effects except host diversity are controlled, may be the only way of clearly showing the host effects on pathogen diversity.

Pathogen Diversity in Agroecosystems

In economically important pathogens, an uneven picture of pathogen diversity has emerged. In many ways, our understanding of diversity is more complete, but the urgent, applied nature of the work has not fostered balance in setting objectives.

Genetic Relationships between Host and Pathogen

Emphasis in understanding genetic interactions between pathogens and hosts has clearly been based on simple inheritance patterns, resulting in qualitative descriptions of pathogen variation (Barrett 1985).

Stepwise Gains in Virulence in Response to Incorporation of Resistance

For many agricultural crops, specific resistance genes conferring a high level of resistance have been incorporated into germplasm and released as commercial cultivars that are grown over a large acreage. As resistance is overcome by the pathogen, another resistance gene is incorporated. Each "race-specific" resistance gene corresponds to a specific virulence gene in the pathogen. The gene-for-gene hypothesis, as first described by Flor (1955), stated that for every resistance gene in the host there is a corresponding gene in the pathogen specifically governing virulence to that resistance gene. Not all gene-for-gene relations fit this exact pattern; rather there are variations on the same theme, as reviewed by Christ et al. (1987). All these variations exhibit extreme specificity between resistance and avirulence alleles. Because of the specificity of gene-for-gene relations, the standard agricultural practice of planting commercial cultivars over large acreage can produce strong directional selection on the pathogen population for virulence at loci matching the cultivar's resistance gene or genes. The stepwise incorporation of resistance into the host, over a period of time, results in stepwise accumulation of virulence in the pathogen population, which may lead to a pathogen genotype with virulence to all known resistance genes, the "superrace" of the pathogen.

In the majority of genetic studies of virulence, avirulence rather than virulence is found to be dominant. For the population of a dikaryotic or diploid pathogen to shift quickly in response to new race-specific resistance genes, ideally the pathogen population should contain individuals that are heterozygous for the corresponding locus. Therefore, the virulence allele already exists in the population and only needs recombination and selection to be expressed and eventually predominate in the population. Young and Prescott (1977) examined populations of leaf rust (*Puccinia recondita*) on wheat, and found a higher frequency of avirulence heterozygotes than homozygotes at a number of loci. Heterozygosity for virulence tends to occur at high frequency in a number of rust fungi (Groth and Alexander 1988).

Host-Parasite Coevolution and Gene-for-Gene Relations

The evolution of a gene-for-gene relationship is of interest in terms of pathogen diversity for virulence. The following is one evolutionary scheme involving

single-gene changes only, as suggested by Person (1959). This scheme can be
applied to host-pathogen interactions occurring on agricultural hosts as well as
unmanaged hosts, but was inspired by agricultural systems. If a host population
is exploited by a pathogen so that there are high levels of disease, the pathogen
selects for increased frequency of any alleles that confer resistance. In agricultural systems a resistance gene would be incorporated by the plant breeder into
common cultivars that would then be widely grown. Farmers would initially
see little or no disease. This would be called the incompatible reaction of the
gene-for-gene relationship, and it is always attributed to the interaction of a
particular resistance gene or its product in the host with the avirulence gene in
the pathogen. Resistance and avirulence are conditional and are expressed only
when both are present. The only way that the specific resistance gene was expressed in the first place was that the pathogen population was nearly fixed for
the corresponding avirulence gene. Thus, the pathogen experiences no or
greatly reduced reproduction on the resistant cultivar, which represents a new
environment for the pathogen. Because the new cultivar is widely cultivated, it
exerts strong selection for virulence to its resistance gene. If this virulence allele is present in the pathogen population, it will spread quickly, resistance will
be overcome, and the pathosystem will again exhibit a compatible reaction. The
particular resistance gene will no longer be of value to the host. The resistance
allele may be reduced or lost within the host so that there may be another resistance gene selected in place of the allele that was removed. Now the virulence
gene in the pathogen is no longer of selective advantage and the pathogen may
(through poorly understood mechanisms) revert to the avirulent form at that
locus (see Parker [chap. 14] and Simms [chap. 17], this volume, for further
discussions of this phenomenon). This change would return the cycle to the
original starting point for that particular resistance-avirulence gene. The cycle
of changes in the host-pathogen relationship may then continue for another
resistance-virulence gene combination. Both natural and artificial selection will
favor mutations in the host that place the pathogen at a disadvantage, and likewise natural selection will favor mutations in the pathogen that increase its reproductive capacities. Perhaps the most unusual feature of this system is the
extreme changes in pathogen reproduction that are characteristic of gene-for-
gene systems. In agroecosystems, humans have clearly selected for extremes
(Barrett 1985). But such variation is not the whole picture.

Quantitatively Inherited Variation in Compatibility

The variation described above is qualitative. Quantitative variation for compatibility within host-pathogen relations is also important (Hooker and Saxena
1971). Although quantitative variation is assumed to be polygenically controlled, very little is known about the underlying genetics of this kind of variation in the pathogen or the influence of host selection pressures on shaping such

variation. The effect of quantitative variation in plant resistance on specific adaptation of pathogen populations to host cultivars is likewise unknown. With at least some quantitative resistance in the host, it is expected that resistance will be more stable because the pathogen would have to gain appropriate alleles at many loci to overcome resistance (Parlevliet and Zadoks 1977). This accumulation of a number of additive genes will (based on selection theory expectations for many genes) proceed more slowly than for a single locus. Consequently, adaptation of pathogen populations to quantitative resistance in the host would most likely proceed at a slower rate than adaptation to single-gene resistance.

A major question concerning quantitatively determined variation in host-pathogen systems is whether the gene interactions are specific or nonspecific, i.e., whether gene-gene recognition exists. One argument for nonspecificity is that quantitatively determined resistance is a cumulative effect of several specific resistance genes (with gene-for-gene relations in the pathogen) that were matched by virulence genes but still provide a residual level of resistance (Nelson 1978). The level of nonspecific resistance observed would depend on the number of defeated genes combined into a single host background. A second argument is that quantitative variation is due to different genes than is qualitative variation and that these genes can nonspecifically modify the phenotype produced by the genes involved in gene-for-gene relations.

Most of the qualitative variation in pathogens has been observed in obligate parasites where there is an obvious intimate association between host and pathogen. Knowledge of quantitative and qualitative variation for compatibility to its host is provided by the obligate parasite *Ustilago hordei*, which causes the covered smut disease of barley. In this pathosystem the phenotype typically expressed by a specific virulence gene can be modified by several other genes (Christ et al. 1987). Additive and nonadditive gene action, genotype-by-environment interactions, and interaction of genes conferring small effects with specific genes for virulence all appear to be involved in the genetic determination of virulence (Pope 1982).

Quantitative variation for compatibility is also noted for nonobligate parasites or necrotrophic fungal pathogens. Two examples are *Gaeumannomyces graminis* on wheat and *Ophiostoma ulmi* on elms. There is no evidence of host-specific virulence genes or of nonadditive effects on compatibility in *G. graminis*. The estimated number of loci involved in virulence varied between five and nine, depending on the isolates used in the cross (Blanch et al. 1981). For *O. ulmi* there is preliminary evidence that genes with major effects act in concert with genes conferring minor effects, and that sometimes there is an interaction between these two types of genes (Brasier 1987b). Brasier points out that the complexity of components of pathogenesis by necrotrophic organisms, ranging from toxins and enzymes to growth parameters, add to the complexity of the genetics as well as the diversity in these fungi.

Multilines and Mixtures

In agriculture, through the use of excessive monoculture and pure lines, we have unwittingly guided the evolution of pathogens. One means of maintaining genetic diversity in pathogen populations to reduce losses to disease owing to simple changes in the pathogen is the use of multilines or cultivar mixtures. A multiline is a mixture of several lines, all of similar genetic background, but each differing at one or more resistance genes. Seed of these lines can be combined in different ratios based on prevalence of races virulent to each line of the pathogen population. As the pathogen population shifts in response to selection, component lines containing other resistance genes can be substituted. A genetically heterogenous host population presents the pathogen population with an evolutionary dilemma. Assuming some degree of directional selection against virulence, the pathogen must maintain enough virulence to reproduce as propagules move from one host genotype to another, yet not so much virulence that directional selection, assuming it exists, will overcome the advantage of attacking many host lines. The outcome of this dilemma depends on population genetics and pathogen dynamics within the host mixture. The predicted result is that host diversity will help maintain pathogen diversity with a low probability of the superrace developing (Barrett 1980). Multilines or mixtures reduce the probability of successful infections and subsequent pathogen reproduction, which results in reduced inoculum dispersal (Browning and Frey 1969). Burdon and Shattock (1980) identify several other mechanisms by which multilines reduce disease progress. Many models have been developed to predict the ability of multilines to stabilize the pathogen population. Some of the assumptions of these models are unrealistic but were made because of our lack of knowledge concerning selection coefficients of virulence genes and fitness changes conferred by these virulence genes (Groth 1976; Marshall and Pryor 1978, 1979). More accurate measurements of pathogen population dynamics and fitness are required to predict with certainty whether multilines will select for the superrace.

Regional Heterogeneity

Another means of using pathogen genetic diversity as an ally in managing plant disease is to deploy specific resistance genes over different parts of a single epidemiological unit (the geographical area of host stands which shares the same pathogen population). For example, several regions within a large geographical area could be assigned different resistance genes, or a management plan could cover a smaller geographical area with different genes assigned to different fields within that area. Major stumbling blocks to this strategy are the logistics and regulation involved in assigning genes to specific areas. To a degree this type of gene deployment existed in the wheat stem rust pathosystem in different regions within the North American *Puccinia* pathway (that area of northward movement of urediniospores in central North America from Mexico into the prairies of Canada) (Van der Plank 1968).

Host-Species Heterogeneity

Alternative host species may also influence genetic diversity within pathogen populations. In plant pathology, most research focuses on the host that is economically important and fails to consider fully the wild/weedy species that serves as the alternative host for the remainder of the life cycle of a fungal pathogen. Such hosts can be important for the survival of the pathogen as a reservoir oversummering or overwintering site. Pathogens with critically important alternative hosts must maintain competence to survive on both hosts, and this may influence the diversity of the pathogen on the economic host in ways that are not understood. A special case of alternative hosts is that of the alternate hosts of obligate rust parasites, many of which require two unrelated plants to complete the life cycle. The alternate host can obviously affect the genetic structure of the pathogen population. For example, Dwinell (1985) showed that bulk collections of teliospores of fusiform rust of southern hard pines produced significantly different levels of infection on slash versus loblolly pines when the teliospores were from different oak species. Variation in ability to infect pine species was generally less in collections made within oak species.

The View of Pathogen Diversity from Economically Important Systems

The purpose of most studies of variation in host disease resistance in economically important plants (or related species) is to find useful resistance for incorporation into cultivars of the host. To be deemed useful, resistance must be of an adequate level and durability. Durability of resistance is lost once critical virulence in the pathogen predominates. While details of this process are not understood, pathogen genetic diversity is clearly behind this change. In the past the assumption often was that a single pathogen isolate was sufficient to define host resistance, at least against some pathogens. Even today many who would give lip service to the existence of intraspecific variation in pathogens conduct research as though such variation were effectively absent. Perhaps this provides the best evidence of all that many scientists continue to down play genetic variation in plant pathogens. Examples of use of single isolates to represent whole species in studies whose objective is to identity useful, durable host resistance still can be found readily in the literature (e.g., Riemenschneider 1990), but perhaps less so in the plant pathology literature than in more general agronomic, horticultural, or forestry literature. Nevertheless, there has been some improvement. In some instances, an isolate was selected based on some knowledge of locally important variation in the pathogen, and is more or less justified as being representative. Such isolates can be used as tools for selecting resistance genes among segregating progeny. Usually the resistance is known to be race-specific in these instances (Van Ginkel and Scharen 1988; Wilson and Shaner 1989).

Other workers have used mixtures of pathogen variants to provide a potentially more representative sample of a diverse pathogen. Some of these mixtures are artificial, with all components more or less known (Burnes et al. 1988), while others are mass field collections of unknown composition (Kuhlman and Powers 1988; David et al. 1988). With some pathogens, only tests using field collections (in soil for instance) are practical, so that diversity of the pathogen is more or less sampled automatically (e.g., Cook and El-Zik 1991). Finally, several isolates are often used in separate trials, with or without knowledge of whether they are representative of important diversity in the pathogen (Killebrew et al. 1988).

Not all work with resistance is designed to provide immediately useful resistance for plant breeders. In some studies the objective of obtaining clear patterns of inheritance or of accurately measuring some component of resistance requires that a single genetically pure isolate be used. As long as it is made clear in such studies that nothing is known about the utility or durability of the subject resistance, such studies will not be misconstrued.

Overall, pathogen diversity is more commonly acknowledged in resistance work than in the past. The great increase in studies in the last several years that deal exclusively with some aspect of pathogen diversity should ensure that this trend continues.

Pathogen Diversity in Unmanaged Host-Parasite Systems

It is possible to divide pathosystems into two categories: (1) disturbed, where the host is highly modified, not freely evolving, and dependent on humans, and where the pathogen has undergone significant (often poorly understood) change from its original wild state, and (2) unmanaged, where both host and pathogen are freely evolving and likely have been long associated. Our view is simplistic in that degrees of coevolution exist, from relatively short-lived associations to very long-term associations that have resulted in cospeciation. Disturbed systems have little or no stability, being maintained only with human intervention. We believe that most pathosystems can be placed in either the disturbed or the unmanaged category.

The primary value of studying pathogen diversity in unmanaged systems is that patterns of pathogen diversity can be associated with presumably long-standing patterns of host and environmental diversity. The hope that unmanaged systems will prove less complex than agricultural systems is unfounded. Host diversity in natural systems is often more complex and less well understood than it is in agricultural systems. In cultivated plants the larger spatial and temporal scale of often remarkably closely related genotypes of the host (Barrett

1981) makes this side of the host-parasite equation well documented relative to even the simplest natural system.

Agricultural systems lack continuity, especially in intensive cropping systems of developed countries. Over a period of years, many things change, including cultivars, crop rotation, kinds and degree of use of pesticides, size of fields and of noncultivated areas, and large-scale epidemiological variables, such as changes in continental extent of a crop or of alternative hosts for a pathogen. These changes are both too rapid and too serendipitous to permit much consistent pathogen evolution, aside from the rise of virulent forms that so commonly follows the widespread use of race-specific resistance in the host. Too often the patterns of diversity of a pathogen cannot be related to any long-term trends of diversity of the system that might permit a view of long-term patterns in coevolution. Rather, these patterns are better related to short-term patterns of dispersal and other factors that are more properly in the realm of epidemiology than population genetics. The diversity of the pathogen is real, of course, and can be measured. What cannot be known is the degree of stability of diversity, and therefore the role of present host and environmental diversity on pathogen diversity. The boom-and-bust cycles of agriculture are not generally considered good examples of evolutionary trends that have any kind of stability; rather they are cases where the pathogen has responded in a way that is dramatically effective in the short run, but has unknown implications in the long run.

Unfortunately, if pathogen diversity has been neglected in economically important, managed plant populations, it has been even more neglected in wild or weedy plants. There are two major reasons that we lack information on unmanaged pathosystems. First, there are far fewer sources of funding for studies of these systems than there are for studies of economically important hosts and their pathogens. Plant pathology has a strong commodity orientation and has only rarely contributed to knowledge of pathogens on wild or weedy host species. Work on such species is usually justified by its direct benefits to agriculture, as these hosts provide potential sources either of disease resistance or of inoculum for diseases of crop or ornamental plants. For example, a study of rust resistance in several species of wild sunflowers (Zimmer and Rehder 1976) produced immediate practical value by providing new resistance sources for cultivated sunflowers. More recently, interest in pathogen diversity of unmanaged systems has been stimulated by attempts to use pathogens as agents of biocontrol of weeds. For example, Burdon et al. (1981) demonstrated the importance of understanding interactions among weed-mycoherbicide variants when they found selection for forms of the target weed *Chondrilla juncea* that are resistant to the races of the rust pathogen applied for biocontrol.

Second, other deterrents to studying natural pathosystems are problems with defining genetic systems in fungi, the difficulty of manipulating plant pathogenic fungi genetically, and problems in obtaining differentially susceptible host materials for use in defining pathogen diversity for virulence. These kinds of problems are not exclusive to unmanaged pathosystems, but in agricultural pathosystems earlier workers have largely solved them, making prospects for short-term productivity in studies of agricultural pathosystems brighter.

Phenotypic Diversity

Knowledge of diversity of pathogens in unmanaged pathosystems is essentially phenotypic. Phenotype is easier to characterize than genotype; the lack of genotypic information is more likely because we are at an early stage in studies of pathogen diversity in unmanaged pathosystems.

Attempts by Plant Pathologists to Sample Pathogen Diversity in Disease Resistance Screening

The few studies that deal with pathogen diversity in unmanaged systems simply show as a first step that there is more than one pathogen phenotype. Information on extent of diversity is scant, but includes estimated numbers or frequencies of phenotypes, and phenotypic distance based on a large number of markers or markers of more than one kind (virulence, isozymes, and morphological traits, for instance) (Burdon 1987a). Many reports of phenotypic polymorphism in pathogens result indirectly from studies of isolate or race specificity of resistance in the host, and were never primarily intended to be studies of the pathogen (e.g., Zimmer and Rehder 1976; Harry and Clarke 1986). Obviously such studies cannot give very thorough descriptions of pathogen variation. Nevertheless, these studies have greatly expanded the number of coadapted pathogenic fungus species for which there is evidence of phenotypic variation.

The Lack of Adequate Data on Pathogen Diversity in Unmanaged Systems

A recent review by Burdon (1987a) of the extent and kinds of knowledge regarding nonagricultural pathogen diversity devotes about two pages to the subject. An important conclusion made by Burdon is that adequate data to illustrate the extent or scale of diversity do not exist in any unmanaged system except where the host range of the pathogen includes agricultural species. The current state of knowledge can be summed up as follows: in a few systems that are presumably independent of agricultural influence, some degree of virulence polymorphism and host specialization has been qualitatively demonstrated (Harry and Clarke 1986; Miles and Lenné 1984; Parker 1985; Clarke et al. 1987; De Nooij and Van Damme 1988b). In a larger number of systems in which an agriculturally important host may be influencing the pathogen's diversity, somewhat more detailed knowledge of pathogen diversity is available (e.g., Zimmer and Rehder 1976; Brasier 1987a; Dinoor and Eshed 1987; Segal et al. 1980). These studies are both too few and too brief to allow generalizations about pathogen diversity.

In most of the studies cited above, phenotypic diversity in the pathogen was uncovered while studying host resistance, and only because the pathogen was sampled more than once, either through using more than one isolate in controlled inoculations or through observing differential disease levels in plantings at different sites. Needless to say, this constitutes neither extensive nor representative sampling of the pathogen, and it certainly does not permit quantitative description of the scale or pattern of diversity. One of the best surveys of a pathogen in the literature was summarized in Clarke et al. (1987), who sampled the powdery mildew fungus (*Erysiphe fischeri*) on wild groundsel (*Senecio vulgaris*) in two localities in Scotland, and could conclude from what was a small but random sample that diversity for host genotype-specific virulence is striking, even in a small area. Isolates of this pathogen are not likely to be phenotypically identical for virulence on 50 host lines.

Unmanaged systems can suffer various degrees of disturbance by humans, and while pathogen diversity is undoubtedly being affected by disturbances, details of the changes and forces that determine them are lacking. Changes in host heterogeneity and culture in an increasingly intensive regime of forest management in North America provide perhaps the best examples of transition from unmanaged to strongly human-influenced systems. Fusiform rust (*Cronartium quercuum*) of native southern pines is one of the better-documented examples. Unfortunately the picture is complex, with several host pine species, all of which have been modified with respect to genetic diversity (usually in the direction of reduced diversity) and sometimes planted in inappropriate sites. Interestingly, while slash and loblolly pines in part of their range are seriously damaged by the disease, other species, such as shortleaf pine, are considered to be still in a state of equilibrium with the pathogen. The degree of disturbance of loblolly and slash pines is correspondingly greater than for the other species. The pathogen is specialized with respect to pine species and even within species (Powers et al. 1981). Unfortunately, the applied objectives of studies of this problem have not fostered detailed description of pathogen diversity except as it influences management decisions and programs to breed resistant pines.

In summary, with the possible exception of the groundsel-mildew example, even at the more accessible phenotypic level, there exist today no published detailed or quantitative studies of diversity of any pathogen as it occurs on a wild or weedy species. The authors are aware, however, that greater interest in direct studies of economically important plant pathogens has stimulated projects that will attempt to examine aspects of diversity on a worldwide scale, to include pathogen populations collected in the center of origin on wild hosts. The fungi responsible for Dutch elm disease (*Ophiostoma ulmi*) (Brasier 1988), potato late blight (*Phytophthora infestans*) (D. S. Shaw 1987) and rice blast (*Pyricularia oryzae*) (Leung and Taga 1988) are three pathogens receiving such attention. Comprehensive studies like these will be especially important in documenting the changes that can occur in pathogen diversity as the pathogen moves from a relatively natural ecosystem into agroecosystems.

Genetic Diversity

There is virtually no information on the genetic diversity of plant pathogens in unmanaged systems. The major problem has been with the amount of groundwork that is necessary before genetic analysis can begin. Genetic markers must be found, and methods of crossing and selfing isolates of the fungus must be perfected.

With economically important pathogens it has been customary (and understandably attractive) to begin genetic analysis using host cultivar specific virulence as the first kind of genetic marker, if for no other reason than that alternative kinds of markers were unavailable (or at least unexplored) in obligately parasitic fungi that can be grown only on the living host. To use virulence as a marker, genetically uniform host lines of differential resistance are necessary. Such inbred lines are readily obtained in agriculturally important species that are grown as pure lines or as hybrids. With wild or weedy plants, however, such material is not immediately available.

More recently, isozyme markers have become available that should be equally applicable to pathogens from unmanaged or agricultural pathosystems, since these markers do not involve the host. An example of the use of isozymes is found in Linde et al. (1990). Restriction-fragment-length polymorphisms (RFLPs) as phenotypic and genetic markers hold even more promise for the future (McDonald and Martinez 1990).

Only a small number of plant pathogens, nearly all from agricultural pathosystems, have been employed as subjects for extensive genetic analysis. These are surveyed in a recent volume on the genetics of plant pathogenic fungi (Sidhu 1988). The kinds of genetic analyses that can be or have been done with different pathogens vary greatly, with many of the pathogens being amenable only in limited ways. Population genetics has not been emphasized for the majority of these organisms. To examine the genetic diversity of a pathogen species, certain methodologies are necessary. The crossing and selfing of isolates should first and foremost be possible with the majority of isolates obtained in the field. If the field population is to be adequately characterized genetically, many isolates will need to be screened for many genetic markers, preferably of more than one class (see below). Shaw (1988) reviews recent progress made in genetically defining the pathogen *Phytophthora infestans* as an example of what is being done with a few species. For many markers, once their mode of inheritance has been elucidated, the genotype of the isolate with respect to the marker can be determined directly (with no need for participation by each isolate in crosses) for isozymes and RFLPs, and through progeny testing after selfing for virulence or other markers for which dominance is nearly or quite complete (Groth 1988).

General Considerations of Pathogen Diversity

Since so few data are available regarding phenotypic or genetic diversity of pathogens in unmanaged pathosystems, it seems appropriate to discuss some of the problems and expectations that might be applicable to a variety of pathosystems.

Problems with Marker Selection

A thorough knowledge of the extent and scale of phenotypic or genetic diversity of any pathogen is going to require more than one class of marker (morphological, nutritional, virulence, isozyme, or DNA). If they exist, differences in host-parasite compatibility that can be attributed to the pathogen are likely to be selectively important, to the extent that such traits cannot serve reliably as markers representative of the genome. In fact, there are some data from agricultural pathosystems that suggest that virulence traits are usually more variable than are isozymes (Burdon, Luig et al. 1983; Linde et al. 1990) in the same populations. Comparisons of levels of diversity of this kind are dangerous, since they depend a lot on which specific markers (enzymes or host differentials) are used. We have shown how different levels of absolute phenotypic diversity can result from substitution of one differential host line for another in cereal rust fungi, with the potential that substitutions can change diversity values interactively (Groth and Roelfs 1987a). The same can be said for isozyme marker substitutions. Considerations of the possible basis for diversity (or lack of it) for virulence markers follow.

Diversity for Virulence in Pathogens in Unmanaged Pathosystems: A Neglected Topic

It is not reasonable to expect that measureable diversity for virulence (broadly defined, and including all measureable levels of difference in compatibility of the parasite to the host) will always occur. The numerous and complex ways in which plant pathogens have evolved would argue against general patterns that apply to every case. There are at least two conditions that must be met for pathogens to adapt to their hosts by becoming phenotypically and genetically diverse for virulence: (1) host genotype–specific virulence among pathogen variants must exist, and (2) there must be an ecological advantage to possessing virulence diversity in space or time.

Criterion (1) has been documented in a number of pathosystems (Burdon 1987a), but its universality is far from established. In agricultural pathosystems, the nature and even the existence of nonspecific host-parasite genetic interactions are controversial, to the extent that sets of concepts and terms have been largely abandoned as being logically or empirically unsound (Johnson 1984). From a practical standpoint, even if all compatibilities between host and pathogen genotypes have some degree of genotype-genotype specificity, it may be impossible to detect with the accuracy in measurement that we possess. This problem is especially acute when many genes with additive effects govern compatibility of host and parasite (Parlevliet and Zadoks 1977). Pathogen diversity is a concept that centers almost exclusively around qualitative characters and strongly interactive host × parasite specificity. As Caten (1987) points out, race designation (in agricultural pathosystems) based on quantitatively varying traits has led only to confusion. More subtle levels of variation may be important in many unmanaged pathosystems if for no other reason than that plant breeders have not selected and isolated genes for high-level resistance. Accumulation of small, individually immeasurable differences may result in measureable diversity without specific interactions, but such differences could be maintained in the population only if mobility and resultant direct competition among pathogen genotypes were absent. Even limited mobility would be expected to eliminate those variants of the pathogen that possessed reduced fitness on all host genotypes, as a lack of host × parasite interaction would dictate.

So perhaps the biggest distinction between managed host-parasite systems and unmanaged systems is in how the genes influencing compatibility that lie along a continuum of strength and specificity of effects are distributed, relative to one another, among the host and parasite genotypes.

The question of whether virulence diversity is ecologically advantageous to the pathogen, criterion (2), depends upon many aspects of the interaction between host and pathogen. Perhaps the most obvious advantage for possessing genetic diversity is the capacity of a parasite to change rapidly in response to host changes or movement of the pathogen to different host genotypes and different environments. Direct evidence for this advantage is difficult to obtain as it entails a holistic view of the pathosystem in all its complexity.

One other aspect of the host-parasite relationship that ecologists have been especially interested in is whether the pathogen exerts selection pressure on the host, that is, whether disease curtails host survival or reproduction. Many pathogens are associated with the host in ways that appear to minimize detrimental effects on the plant. Examples include many of the imperfect leaf spot fungi, powdery mildews, and some rust fungi of temperate dicots, which increase to noticeable levels only very late in the growing season, when leaf tissues are senescing. If indeed there is no appreciable selection pressure for the host in such cases to evolve higher levels of resistance, and there is little diversity for resistance among host genotypes, then there may be little need for variation in virulence in the pathogen. This is speculative, with no data to support it. Another perspective on pathogens that avoid damage to the host is that such associations represent a coevolved state. Looking at putatively coevolved pathosystems, it is always difficult to place a given level of compatibility along the scale of potential levels. One could argue that coevolution implies that loss of phenotypic variation has occurred in both organisms. For either organism, the expression of compatibility is conditional, being totally and reciprocally dependent on the status of the symbiont. Practically, such interdependence results in problems such as is shown in figure 8.1 from Segal et al. (1980). In expressing levels of compatibility of plants of the wild oat *Avena sterilis* to crown rust (*Puccinia coronata*), only when susceptible cultivars of cultivated oats (*A. sativa*) were included was the potential for a high level of compatibility seen. The conclusion of Segal et al. was that all plants of *H. spontaneum* possessed some resistance. In systems with wild hosts unrelated to cultivated species such high compatibility "controls" may not be available from the cultivated species. Indeed, some workers have concluded that in unmanaged pathosystems, inter-

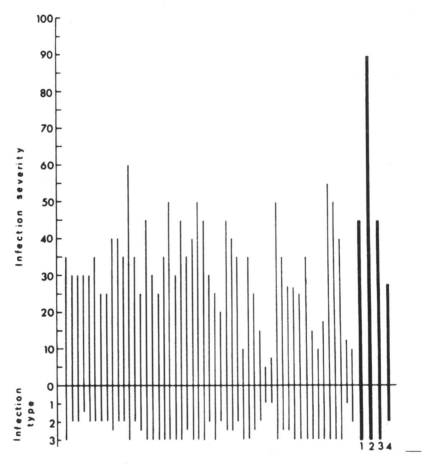

Fig. 8.1. Two measures (top = infection severity as percentage of maximum disease, bottom = infection type on a 0–4 scale) of crown rust (*Puccinia coronata*) in plants of *Avena sterilis* from a transect sample of a natural stand in Israel. The four bold, numbered lines on the right are four accessions of cultivated oats, *Avena sativa*.

mediate and less variable levels of compatibility (compared with agricultural pathosystems) are the rule (Segal et al. 1980). In the case of those fungi whose mode of parasitism results in minimal impact on host reproduction, it is possible that this represents a successful strategy for the parasite, and is a kind of accommodation of host and parasite. Such parasites would have stability in that the host is not subject to strong depression of fitness due to parasitism, yet the parasite would still be able to reproduce well at least for a brief, important period of time preceding the dormant state of the fungus. Indeed, interference with the pathosystem for foresters may indicate how one pathogen, one of the fungi that causes leaf rust of poplar species (*Melamspora medusae*), has become a problem in the north central United States on hybrid poplars containing exotic, susceptible germplasm grown in intensively managed stands (Ostry and McNabb 1985). In natural populations of poplars this pathogen generally increases rapidly about the time that the leaves are turning yellow in the fall. Observing only natural populations, it might be tempting to assume that this rust species is capable of rapid increase only on senescing tissues. The experience with hybrid poplars in which the pathogen can be epidemic on healthy green leaves indicates that this is not entirely true; the pathosystem has apparently moved toward such an association. Exactly how this comes about is not clear. Native poplars may possess a kind of resistance that breaks down with extreme leaf age. A similar case involving another species of *Melamspora* on western black cottonwood showed variation for resistance among host genotypes, but no detectable specific isolate × host genotype interactions (Hsiang and Van der Kamp 1985), suggesting that specific adaptation is not an important part of host-parasite diversity; it has been not advantageous for the pathogen to adapt strongly and specifically to a portion of the host population because of the intermediate and uniform levels of compatibility of host and pathogen.

The above interpretation is oversimplified in many respects. It depends upon the assumptions that the pathogen is highly mobile and that each pathogen individual is likely to encounter a number of host individuals during a reproductive cycle. This would select for broad compatibility. The rust pathogen is heteroecious, ensuring that the pathogen moves at least to the nearest stand or individual of the alternate host to complete its life cycle. This increases its chance of encountering different poplar genotypes each cycle. Also oversimplified are assumptions about the stability of the strategy, including its origin, maintenance, and the potential for either partner to gain an advantage. The dynamics of host and parasite spatial and temporal diversity are unknown; they may be more complex than the apparent stability in overall compatibility would indicate.

Parker (1985) has documented another pattern which we believe is likely to be common in unmanaged pathosystems where neither partner is particularly mobile. In such cases, selection favors specialization of the local pathogen population to the "home" genotype of the host at the expense of compatibility to other host genotypes. For such a strategy to be advantageous to the pathogen, temporal stability must be assumed, so that the pathogen is likely to have compatible host genotypes in the same area for many years. There is very little evidence available about diversity within and among locally adapted pathogen populations. At the two extremes, the pathogen at each site may maintain considerable genetic diversity that permits them to cope with change, or it may have lost most diversity owing to a founder effect if only a few propagules reached the host clone or deme originally. That there is a trade-off between wide host adaptation and narrow specialization is itself speculative; it has not been documented as a reliable expectation in any pathosystem. In the absence of recent coevolution and the likelihood that host breeding and pathogen range extension have muddied the waters, it remains for questions regarding scales of diversity and their genesis and maintenance to be answered in unmanaged rather than agricultural pathosystems.

Predictions

Studies of host-parasite diversity and interactions that center on the pathogen appear to be increasing in number. Aside from the realization by more and more investigators who work with disease systems that intraspecific pathogen diversity is common and can be important, new tools or new views are providing greater opportunity for understanding pathogen diversity. These include:

1. The availability of new DNA markers. RFLPs can provide an unlimited supply of genetic markers suitable for use with all kinds of pathogens. Not only are methods now available that will allow fungi to be characterized with as much detail as is necessary, but the accessibility of these methods is rapidly increasing as they become more automated, cheaper, and simpler. While setup costs will remain high for some of these markers, running costs are probably decreasing overall. For example, polymerase chain reaction methods to amplify DNA will permit RFLP markers to be used without the need to produce large amounts of either nuclear or mitochondrial DNA. This will also allow larger samples of individuals of a pathogen to be characterized with RFLP markers, and unculturable organisms to be used.

2. The advent of new diagnostic kits utilizing monoclonal antibodies or DNA probes. The utility of any kit for diagnosing the presence of a causal organisms is a function of its accuracy. That is, the kit must react positively only to the pathogen. Selection of antibodies with a high degree of accuracy will entail more detailed knowledge of the limits of variation in pathogens serologically or genomically, and is likely to provide useful basic information on how serological and DNA diversity relate to phenotypic diversity. Ultimately, such knowledge may influence our views and research goals in studies of coadapted pathogens.

3. Gradual and total replacement of a fixed, typological concept of fungal species by a gene-pool concept, along with the accumulation of data showing that morphological differentiation need not accompany genetic isolation in nature. The prevalence of intraspecific variation in fungi is a fact that impinges on an ever expanding sphere of research into their biology. This variation is no more important now than before; its importance is simply more often recognized as we gain more detailed knowledge of the molecular genetics of fungi. To curious biologists, such detailed knowledge constantly raises questions about the ecological and evolutionary forces leading to the often odd and unexpected patterns of variation that are found. Simplistic views of what constitutes the pathogen will not survive in such a research environment.

PART THREE

Population and Community Responses to Plant Resistance Variation

9

Plant Variation: Its Effects on Populations of Herbivorous Insects

Richard Karban

Plant Variation: Population Ecologists Take Note

Recently ecologists have become increasingly interested in intraspecific variation of host plants with respect to the herbivorous insects that feed on them. One consequence of such variation is that populations of herbivores may behave quite differently on different individuals of the same host species. In this chapter I review the evidence for intraspecific phenotypic host variation in natural plant populations and consider some of the consequences of such variation on insect populations.

The notion that host variation is important for insect populations involves three premises. (1) Individual plants within a population differ from one another in characteristics that affect herbivores. (2) As a result, herbivores will be better at exploiting some host-plant individuals than others. (3) These differences will produce significant differences in the population dynamics of herbivores on different plant individuals.

Plant Variation: A Truism

The first premise is essentially a truism; it depends solely upon our ability to measure differences in host-plant characteristics. It is similar to saying that no two people are identical. Once the investigator becomes sufficiently familiar with either people or host plants, numerous differences in characters of interest are certain to become apparent. Clearly, the identification of host-plant differences in traits that we know to affect herbivores tells us little about the effects of such traits on herbivore population dynamics. Yet it is surprising how often authors are willing to make the leap from identifying differences in traits presumed to be related to "plant defense" to assumptions about the effects of those traits on insect populations.

The second premise often follows from the first. In practice, workers most often raise insects for a portion of their life cycles on plants which have been shown to differ in some traits. In these experiments, the insects are used as bioassays of the effects of plant quality on insect population dynamics. Correctly interpreting the results of these experiments can be problematical and this issue will be discussed below. The third premise is the most difficult to satisfy but it provides the most convincing evidence that host variability is important to insect populations.

An Agricultural Precedent

While interest in the effects of variation of natural host plants on herbivores has sparked among ecologists within the past decade, agriculturists have appreciated for at least a century that different varieties of cultivated crops support different pest populations. Unfortunately, agricultural workers most often concerned themselves with the extent of pest infestation and damage to various crop cultivars. They tended not to measure insect populations directly. Nevertheless, there are many convincing examples of crop varieties supporting very different populations of insects. This literature will not be covered in this chapter but was reviewed by Painter (1951), Maxwell and Jennings (1980), and Gould (1983).

Host Variability: Effects on Insect Performance and Populations

Ecologists who are trying to assess the effects of host-plant quality on insect herbivores most often place eggs or young immatures (usually in cages) on the plant material which is being evaluated. They then measure the performance (growth rate, efficiency of assimilation or conversion, size at maturity, survival, etc.) of these bioassay insects. These tests are usually conducted over a very small spatial scale (i.e., the insect has no choice in selecting its own host, and processes that require larger arenas, such as predation, are not included). These tests are commonly conducted for very short periods of time, usually less than the time required for a single insect generation. They are often conducted under very restricted environmental conditions that may or may not include important sources of mortality for the insects being tested. The reasons for these shortcuts are obvious. It is often impossible to set up experiments over longer time periods and larger spatial scales or with more environmental realism. Indeed experiments conducted in natural settings may have so many different variables uncontrolled that interpretation becomes exceedingly difficult even if the experiments are possible. Large-scale field experiments are particularly poor for acquiring information about physiological mechanisms, for example. Thus, I do not wish to advocate that one experimental approach be used in preference to another, since each provides different and useful information. I do advocate caution in extrapolating results from small-scale, highly controlled experiments in order to make inferences about the behavior of insect populations in the field.

Evidence that host-plant variability affects insect performance should not be equated with a demonstration that natural insect populations will be necessarily affected. In other words, fulfilling the first two premises discussed above should not be misconstrued as evidence that variability matters to populations. Four problems often make the results of small-scale bioassays different from situations involving insect populations in nature.

First, there are numerous situations in which experiments conducted on a small spatial scale give misleading results. For example, in small-scale experiments it was found that chinch bugs laid far fewer eggs on Atlas Sorgo, a variety of sorghum, than on other varieties (Dahms et al. 1936). As a result of these small-scale tests, Atlas Sorgo was released and largely replaced other varieties in much of eastern Kansas. However, for several reasons, chinch bug populations increased over the same time period (Painter 1951, 7).

Second, experiments that are conducted over short periods of time may fail to characterize adequately the actual effects of plant traits on herbivores. *Trifolium repens* (white clover) is a plant that exhibits large differences in the frequencies of individuals that are cyanogenic versus acyanogenic over distances of a few meters in nature (Dirzo and Harper 1982a). This trait has consequences for many different herbivores, although sheep reverse their preference for cyanogenic and acyanogenic morphs as the season progresses (Dirzo and Harper 1982b).

A third potential pitfall of many bioassays is that they measure only a part of the insect life cycle. Plant characters that favor larval growth may be unsuitable for successful development of eggs. White et al. (1982) found that periodical cicada nymphs were able to feed successfully on the roots of conifer trees. However, pines do not support large populations of periodical cicadas because eggs are unable to hatch from the resinous twigs of these hosts. An investigator who is able to follow the insects for only a limited portion of the life cycle may not be in a very good position to generalize about the effects of different hosts on population dynamics. Whenever possible, bioassays should be maintained for more than one complete insect generation.

A fourth problem involves variation in plant traits that are expressed only under certain environmental conditions. Resistance in different varieties of wheat to the wheat stem sawfly was correlated to the solidness of the straw; larvae apparently could not move to obtain sufficient food if the pith was firm and compact (Painter 1951, 169–173). However, expression of stem solidness was influenced strongly by the amount of light that the stem received when it was elongating (Platt 1941). In another example, four different clones of goldenrod were found to support very different numbers of aphids (Maddox and Cappuccino 1986). However, these differences in aphid numbers on clones were observed only when the plants were well watered, and the effect was not found in dryer environments. These examples of genotype × environment interactions suggest that we must be cautious in generalizaing results found under one set of conditions.

Variation in Natural Plant Populations

This review considers phenotypic variation that affects invertebrate herbivores within natural plant populations. All of the studies that I was able to find in the literature that look for this variation are listed in table 9.1. Studies of cultivated crops are not considered here, nor are studies comparing plants of a single species that grow in distant locations and have very little chance of interbreeding. Distinguishing the variation within a plant population from that among populations depends on the plant breeding system, barriers to gene flow, etc. Since this information was rarely available, the decision to include a study was based largely on whether the plants were found within several kilometers of each other in nature.

Variation among plants that affects herbivores may be caused by genetic differences and/or by environmental differences between plant individuals. Many environmental factors have been shown to influence plant resistance to insect herbivores, directly (see review by Tingey and Singh 1980) and indirectly (see examples above). Simply noting that individual plants consistently support different numbers of herbivores does not suggest that the plants are genetically different, because factors such as microenvironment, competitors, predators, parasites, pathogens, plant size, age, and history have not been controlled. The most commonly sought evidence for a genetic basis for plant-caused differences in insect numbers involves moving the plants (cuttings, asexual propagules, sexual propagules of known parentage) to a common environment. In a common garden, greenhouse, or lab, environmental heterogeneity is reduced and differences in herbivore numbers (a phenotypic trait) are likely to reflect some underlying genetic influence. This argument is strengthened considerably if replication is possible by planting many individual propagules representing each genotype with proper interspersion (Hurlbert 1984). Studies were not included in table 9.1 if they did not demonstrate that plants were genetically different with respect to herbivores with at least the rigor of a common garden. An exception was made for several studies involving tree species where such a design would have been very difficult. Unfortunately, many of the studies of herbivore population dynamics on different plant individuals did not identify whether the cause of the variation was environmental or genetic. For example, Southwood and Reader (1976) observed large, consistent differences in whitefly populations on different viburnum bushes at Silwood Park over a 10-year period. However, there is no way to know the cause of these host-plant differences.

Even with a common garden, observed differences among individual plants are not necessarily genetic. Many of the common gardens in the studies listed in table 9.1 were made up of replicated cuttings from individual plants. Most workers assumed that growing the plants in a homogeneous environment negated the effects of heterogeneous collection sites. However, it is possible that

TABLE 9.1
Variation in natural plant populations that affects herbivores

Plant	Insect	Scale of plant variation	Evidence for genetic basis	Plant reproductive biology	Herbivore parameters affected	No. of generations of herbivore assessed	References
Trifolium repens (white clover)	weevils	within population	common garden	largely clonal	no preference	—	Dirzo and Harper 1982b
	snails/slugs	within population	common garden		preference for acyanogenic	—	Dirzo and Harper 1982b
Rudbeckia laciniata	Uroleucon rudbeckiae (aphid)	3 sites within Chapel Hill	common garden	largely parthenogenetic	fecundity, no effect on survivorship, age of reproduction	1	Service 1984b
Populus angustifelia (cottonwood)	Pemphigus betae (aphid)	within population	common garden	largely clonal	preference and survival	several	Whitham 1983
Solidago canadensis (goldenrod)	Uroleucon caligatum (aphid)	within population	common environment	largely clonal	adult weight	1	Moran 1981
Solidago altissima (goldenrod)	Eurosta solidaginis (galling Tephritid)	within population	common garden	largely clonal	pop. density (probably survival)	3	McCrea and Abrahamson 1987
Solidago altissima (goldenrod)	Lygus lineolaris (Plant bug)	within population	cmmon garden	largely clonal	numbers	4	Maddox and Root 1987
	Corythuca marmorata (Lace bug)	within population	common garden	largely clonal	numbers	4	Maddox and Root 1987
	Philaenus spumarius (Spittlebug)	within population	common garden	largely clonal	numbers	4	Maddox and Root 1987
	Uroleucon caligatum (aphid)	within population	common garden	largely clonal	numbers	4	Maddox and Root 1987
	Uroleucon nigrotuberculatum (aphid)	within population	common garden	largely clonal	numbers	4	Maddox and Root 1987

TABLE 9.1
(*continued*)

Plant	Insect	Scale of plant variation	Evidence for genetic basis	Plant reproductive biology	Herbivore parameters affected	No. of generations of herbivore assessed	References
	Exema canadensis (Chrysomelid)	within population	common garden	largely clonal	numbers	4	Maddox and Root 1987
	Microrhopala vittata (Chrysomelid)	within population	common garden	largely clonal	numbers	4	Maddox and Root 1987
	Ophraella conferta (Chrysomelid)	within population	common garden	largely clonal	numbers	4	Maddox and Root 1987
	Trirhabda spp. (Chrysomelid)	within population	common garden	largely clonal	numbers	4	Maddox and Root 1987
	Epiblema scudderiana (galling Tortricid)	within population	common garden	largely clonal	numbers	4	Maddox and Root 1987
	Epiblema spp. (boring Tortricid)	within population	common garden	largely clonal	numbers	4	Maddox and Root 1987
	Asteremyia carbonifera (gall midge)	within population	common garden	largely clonal	numbers	4	Maddox and Root 1987
	Rhopalomyia solidaginis (gall midge)	within population	common garden	largely clonal	numbers	4	Maddox and Root 1987
	Eurosta solidaginis (galling Tephritid)	within population	common garden	largely clonal	numbers	4	Maddox and Root 1987
	Ophiomyza sp. and *Phytomyza* sp. (mining flies)	within population	common garden	largely clonal	no effects on numbers	4	Maddox and Root 1987
Erigeron glaucus	*Apterothrips secticornis* (thrips)	within population	common garden	clonal, apomictic	population size	approx. 10	Karban 1987a
	Philaenus spumarius (spittlebugs)	within population	common garden	clonal, apomictic	no effect on survival, growth, adult size	2	Karban 1989a
	Platyptilia williamsii (plume moth)	within population	common garden	clonal, apomictic	no effect on survival, growth, adult size	6–8	Karban 1989a
Xanthium strumarium (cocklebar)	*Euaresta aequalis* or *Phaneta imbridana*	between populations, within populations	common environment	self-compatible annual	% damage to plant	<1	Hare 1980 and pers. comm.

		probably within populations	common garden	probably sexual	% damage to plant	several	Marquis 1990a
Piper arieianum	many species						
Ipomoea purpurea (morning glory)	Chaetocnema confinis (Chrysomelid)	within population	full-sib design common garden	self-compatible annual	% damage to plant	<1	Simms and Rausher 1987
Oenothera biennis (evening primrose)	Schinia florida (Noctuid)	within population	common garden	apomictic seed	density and preference	<1	Kinsman 1982
	Macrosiphum gaurae (aphid)	within population	common garden	apomictic seed	preference	<1	Kinsman 1982
	Popillia japonica (Japanese beetle)	within population	common garden	apomictic seed	preference	<1	Kinsman 1982
	Plagiognathus spp. and Lygus lineolaris (Mirid bugs)	within population	common garden	apomictic seed	no effect on density, preference	<1	Kinsman 1982
Betula pubescens (mountain herb)	Epirrita autumnata (Geometrid)	within population	—	largely clonal	growth rate, pupal weight	<1	Ayers et al. 1987
Salix lasiolepis (willow)	Euura lasiolepis (galling sawfly)	within population	common garden	largely clonal	larval density	1	Fritz et al. 1986
	Pontania sp. (galling sawfly)	within population	common garden	largely clonal	larval density	1	Fritz and Price 1988
	Phyllocolpa sp. (folivorous sawfly)	within population	common garden	largely clonal	larval density	1	Fritz and Price 1988
	Euura sp. (galling sawfly)	within population	common garden	largely clonal	no effect, ¾ years	1	Fritz and Price 1988
Pinus sylvestris (Scots pine)	Neodiprion sertifer (sawfly)	?	common garden	sexual	larval mortality and development rate	<1	Larsson, Bjorkman et al. 1986
Pinus ponderosa	Nuculaspis californica (scale)	?	—	sexual	survival	<1	Edmunds and Alstad 1978
Fagus sylvatica (beech)	Cryptococcus fagisuga (scale)	?	common garden	sexual	survival and fecundity	1	Wainhouse and Howell 1983
Pinus monophylla	Matsucoccus acalyptus (scale)	within population	—	sexual	survival	<1	Unruh and Luck 1987

the cuttings retain some characteristics determined by the environment from which they were removed. For example, grafted trees produced with scions (cuttings) taken from old eastern white pines were more resistant to white pine blister rust than were ungrafted seedlings (Wright 1976, 129–133). Results of common garden studies that indicate significant effects due to host-plant clones should not be taken as conclusive evidence of heritable variation. In a study of 16 species of herbivores raised in a·common garden of goldenrod cuttings, 14 of the herbivores were affected significantly by plant clone (Maddox and Root 1987). However, the effect of plant clone was found to be heritable for only 10 of these species in a parent-offspring regression and for only 9 species in a sib-correlation. These results suggest that evidence for a genetic basis of plant variability to herbivores from common garden studies may overestimate the actual variation that is available to natural selection. However, common garden experiments are much easier to conduct than breeding experiments and can provide useful information about the expression of the underlying genetics. It is this expression and its consequences that are most relevant to population ecologists interested in plant-herbivore interactions.

It may be reasonable to expect that the mode of plant reproduction will influence the likelihood that genetic variability of plants will be important to herbivores. A large number of the studies reported in table 9.1, in which plant genotype was found to be important, involve plants that spread largely asexually. There are many possible explanations for such a pattern. Plants that reproduce clonally may present a constant phenotype in the same place for many insect generations, leading to the possible evolution of demes of herbivores adapted to individual genotypes (Edmunds and Alstad 1978; Whitham 1983; Karban 1989b). An alternative explanation for the high frequency of studies in table 9.1 involving plants that spread asexually is that such plants are much easier to establish in common gardens. Because negative results are not always reported, we cannot know how many studies involving plants exhibiting various reproductive modes failed to show a significant effect of plant "genotype." At this time it is difficult to assess whether the observed correlation between clonal reproduction and important plant variability is a real pattern.

It is sometimes possible to learn about the mode of action of an ecological factor by examining the insect life history stage that it affects. Insect numbers on different plants may vary as the result of differences in their preferences for different hosts, in their growth rate and survival once on a host, and/or in their fecundity. Plant resistance that is based on differences in preference will exert less direct selective pressure on the insects than resistance based on differences in survival, particularly if alternate hosts are readily available (Gould 1983, Kennedy, Gould et al. 1987). Differences in preference or in plant tolerance may have little effect on insect fitness.

What parameters of insect performance are affected by host-plant variation? Unfortunately, different studies have measured different performance variables for herbivores. Studies that lasted for more than one generation most often measured insect numbers or density. Those lasting less than one insect generation tended to measure preference, survival, development rate, fecundity, or size at adulthood. Unfortunately, no clear pattern emerges from a review of these studies (table 9.1) that might indicate that some estimates of insect performance are more likely to be affected than others. Similarly, few of the studies in table 9.1 provided information on particular plant traits that were responsible for resistance to insects.

Consistency of Variation over Time

Even though a population of plants may be variable with regard to insect herbivores, particular individuals within the population may always be excellent hosts or poor hosts. Alternatively, both populations and individuals may vary over time. A particular clone may be an excellent host during one year and a poor host during the next. The studies of natural plant populations that considered temporal consistency of variation in plant genotypes are presented in table 9.2. If the effects of different genotypes are consistent over time, then the interaction of plant genotype × year will not be significant. Even if the interaction term is significant, the magnitude of the effect may be small relative to the effect of plant genotype. Indeed this was often the case (table 9.2). Censuses of aphids on clones of willows during consecutive years showed that aphid preference for individual trees remained relatively constant, despite large year-to-year fluctuations in aphid numbers (Whitham 1983). For three of four species of sawflies that fed on willow, genotype × year interactions were small (Fritz et al. 1986; Fritz and Price 1988; Fritz et al. 1987). McCrea and Abrahamson (1987) found that the patterns of variation among clones of goldenrod were stable over three years with regard to numbers of *Eurosta solidaginis,* a gall-forming fly in Pennsylvania. Maddox and Root (1987) studied this same system in New York and similarly found that goldenrod genotype accounted for a large proportion of variation in numbers of *E. solidaginis.* They also found that the effects of plant genotype were relatively consistent over four years. In all, they considered 15 species of herbivores that fed on goldenrod. For most of these, the interaction between genotype and year was statistically significant, but for many the interaction explained relatively little of the variance in herbivore numbers (table 9.2). One of the herbivores that was significantly affected by goldenrod clone was the meadow spittlebug, *Philaenus spumarius.* However, when this same insect was examined in California on clones of *Erigeron glaucus* (Karban 1989a), plant clone had very little effect on the spittlebugs' performance. Similarly, plant clone did not influence performance of plume moth caterpillars on *E. glaucus,* and these results varied little from year to year (Karban 1989a). The situation was very different for a third herbivore on *E. glaucus,* a thrips. Plant clone explained 63% of the variance in thrips numbers, an estimate

TABLE 9.2
Consistency of the effect of plant variation on herbivore performance

Plant	Insect	Plant genotype		Genotype × year		References
		Significant?	Variation	Significant?	Variation	
Populus angustifolia	*Pemphigus betae*	yes	?	?	small	Whitham 1983
Salix lasiolepis	*Pontania* sp.	yes (¾ yrs)	9–11%	probably	large	Fritz et al. 1986; Fritz and Price 1988; Fritz et al. 1987b; Fritz 1988, (pers. comm.)
	Phyllocolpa sp.	yes	48–51%	usually no	small	
	Euura lasiolepis	yes	21–31%	no	small	
	Euura sp.	yes (½ yrs)	9–28%	no	small	
Solidago altissima	*Eurosta solidaginis*	yes	?	?	small	McCrea and Abrahamson 1987
Solidago altissima	*Lygus lineolaris*	yes	5%	yes	5%	Maddox and Root 1987
	Corythuca marmorata	yes	11%	yes	13%	
	Philaenus spumarius	yes	8%	yes	6%	
	Exema canadensis	yes	8%	yes	6%	
	Microrhopala vittata	yes	13%	yes	6%	
	Ophraella conferta	yes	7%	yes	3%	
	Trirhabda spp.	yes	15%	yes	9%	
	Epiblema scudderiana	yes	12%	yes	11%	

Host plant	Insect species					Reference
	Epiblema spp.	yes	6%	yes	10%	
	Eurosta solidaginis	yes	24%	yes	5%	
	Uroleucon caligatum	yes	11%	no	1%	
	Uroleucon nigrotuberculatum	yes	9%	no	1%	
	Ophiomyza sp. and *Phytomyza* sp.	no	2%	yes	6%	
Erigeron glaucus	*Apterothrips secticornis*	yes	63%	yes	9%	Karban 1987a
Erigeron glaucus	*Philaenus spumarius*	no	1–4%	no	1–6%	Karban 1989a
	Platyptilia williamsii	no	0–3%	no	0–5%	
Piper arieianum	Many species	yes	6–21%	sometimes	var.	Marquis 1990a

that is probably exaggerated somewhat by autocorrelation of the data, but none-theless correctly represents a very strong effect that varies only slightly between years (Karban 1987). In summary, most of the studies reported a significant effect of plant genotype on herbivore populations. Different genotypes tended to vary in their effects on herbivores in different years, although the magnitude of this variance was often relatively small.

Plant Variation Relative to Other Factors That Affect Herbivore Populations

Demonstrating that a particular ecological factor has a significant effect on a population is more an exercise in clever design and persistence than it is a useful scientific endeavor. It is far more useful for ecologists to understand the impor-tance of plant variation to herbivores relative to other ecological factors that we have some intuition about.

Several of the studies that evaluated the effects of plant variation simulta-neously considered other ecological factors that potentially affected herbivore populations. Table 9.3 allows a comparison between the importance of plant genotype and other factors, although such comparisons are qualitative only and do not represent a demonstration that one factor or another is statistically more important.

Plant genotype explained less variation than yearly variability in the num-bers of herbivores for 13 of 15 species reported on goldenrod (Maddox and Root 1987). Similarly, plant genotype explained essentially none of the varia-tion in performance for 2 of 3 herbivores on *Erigeron glaucus* (Karban 1987, 1989a). However, the herbivores that were most strongly affected by host clone, *Eurosta solidaginis* and *Rhopalomyia solidaginis* on goldenrod and *Apteroth-rips secticornis* on *E. glaucus* were as influenced by genotype as they were by year-to-year variation (McCrea and Abrahamson 1987; Maddox and Root 1987; Karban 1987). Plant clone accounted for 35% of the variance in growth rate and 25% of the variance in pupal weight of *Epirrita autumnata* on mountain birch (Ayres et al. 1987). Other studies of this caterpillar reported changes in growth rate of 42% due to seasonal changes in birch foliage (Ayres and MacLean 1987) and 32% due to long-term induced resistance (Neuvonen and Haukioja 1984).

Plant genotype was found to be more important than interspecific compe-tition for two sawfly species on willow (Fritz et al. 1986) and for thrips on *E. glaucus* (Karban 1987). However, interspecific competition exerted a much stronger influence on performance of spittlebugs on *E. glaucus* compared with plant clone (Karban 1989a). Plant clone was a better predictor of thrips num-bers on this host plant than were temperature and precipitation (Karban 1987). Finally, in an experiment that varied both plant clone and aphid clone, both were found to have roughly equal effects on aphid performance (Service 1984).

TABLE 9.3

Importance of plant genotype relative to other ecological factors

Plant	Insect	Ecological factor compared	Plant genotype[a] More important	Equally important	Less important	References
Rudbeckia laciniata	Uroleucon rudbeckiae	aphid clone		X		Service 1984b
Betula pubescens	Epirrita autumnata	long-term induced resistance		X		Ayers et al., 1987; Neuvonen and Haukioja 1984
		seasonal changes in foliage			X	Ayers and MacLean 1987
Solidago altissima	Eurosta solidaginis	increased numbers in subsequent gen.	X			McCrea and Abrahamson 1987
Solidago altissima	Eurosta solidaginis	var. between years	X			Maddox and Root 1987
	Lygus lineolaris	var. between years			X	
	Corythuca marmorata	var. between years			X	
	Philaenus spumarius	var. between years			X	
	Uroleucon caligatum	var. between years			X	
	Uroleucon nigrotuberculatum	var. between years			X	
	Exema canadensis	var. between years			X	
	Microrhopala vittata	var. between years			X	
	Trirhabda spp.	var. between years			X	

TABLE 9.3
(continued)

Plant	Insect	Ecological factor compared	Plant genotype[a] More important	Equally important	Less important	References
	Epiblema scudderiana	var. between years			X	
	Epiblema spp.	var. between years			X	
	Asteromyia carbonifera	var. between years			X	
	Rhopalomyia solidaginis	var. between years		X		
	Ophiomyza sp. and Phytomyza sp.	var. between years			X	Fritz et al. 1986
Salix lasiolepis	Pontania sp.	interspecific competition	X			
	Phyllocolpa sp.	interspecific competition	X			
Erigeron glaucus	Apterothrips secticornis	interspecific competition	X			
		var. between years	X			
		precipitation	X			
		temperature	X			Karban 1987
Erigeron glaucus	Philaenus spumarius	interspecific competition			X	
		predation		X		
		var. between years			X	Karban 1989a
	Platyptilia williamsii	interspecific competition		X		
		predation			X	
		var. between years			X	

[a]Statistical significance is not implied in these comparisons.

Variability in Insect Populations

Much of the discussion thus far has considered variability in plant populations but has treated insect populations as if they were composed of numerous identical individuals. Obviously, this is not the case; insect populations are at least as variable as their host plants and some of this variation is genetically based. Soon after agriculturists started selecting and employing crop varieties that were "resistant" to insects, insect variability became all too obvious. New plant cultivars that were released as resistant to specific pests lost their resistance within a few years. The insect population had contained a small number of individuals that were only minimally affected by the host-plant resistance. As the selective environment changed, so did the frequency of insects with traits that allowed them to overcome the current resistance. Members of an insect population that vary in their ability to use or to damage a plant are termed biotypes, or races. They differ from other individuals of the same insect species in their ability as parasites, although they are most often indistinguishable by other characteristics used by taxonomists and morphologists. There are many well-documented examples of genetic variation within herbivore species in their ability to parasitize plant varieties. Most of these examples are from the agricultural literature, and Gould (1983) provided an excellent review.

Since individual plants or plant genets may be genetically constant, or nearly so, for many insect generations, it may be possible for sedentary herbivores to become specialized on individual hosts. Sedentary herbivores with short generation times can complete hundreds or even thousands of generations on a single host genotype. Edmunds and Alstad (1978) noted large and persistent differences in the numbers of scale insects found on neighboring individuals of ponderosa pine. Some trees were perennially infested with high densities of scales, whereas nearby trees were perennially free of these herbivores. This observation led Edmunds and Alstad to develop the hypothesis that scale populations had become differentiated into small demes, each specialized on an individual pine tree. A deme, in this context, can be thought of as a biotype adapted to an individual plant. Edmunds and Alstad conducted transfer experiments of scales within a single host (intratree transfers) and transfers between the original host and other conspecifics located several kilometers away (intertree transfers). They found higher survival for scales in intratree transfers compared to intertree transfers. Because all of the intertree transfers involved a set of trees located at a second, distant site, we cannot separate the hypothesis of adaptation to particular trees from the alternative that trees from the two sites differed for other reasons (Kareiva 1982b). In addition, it was not possible to establish that the traits that resulted in differential survival of scales on different trees were genetically based and not caused by physiological acclimation (Kareiva 1982b; Alstad and Edmunds 1983).

The hypothesis of deme formation put forward by Edmunds and Alstad predicts that an herbivore that is found on (and hence adapted to) host individual A should have a higher fitness than should insects from other unrelated conspecific host individuals when both are reared on host A. This prediction can be tested by transferring and rearing insects on the host of origin and also on other hosts. Since we most often want to establish that the differences are genetically based, rearing insects on plants in situ does not allow us to differentiate between environmental (site) effects and effects caused by plant genotype. Therefore we should rear the insects on replicated cuttings or apomictic plant propagules that are randomly interspersed in a common garden (see above). Similarly, insects should be reared on a common host clone, other than one of the plant genotypes used in the experiment, for at least one complete generation prior to the start of the experiment. This should reduce any possible effects of insects showing a preference or improved performance on plant types to which they have been previously exposed (Jermy et al. 1968; Jaenike 1982). These refinements make interpretation of the experiment more likely to reflect accurately genetic differences among insects. The experiment outlined above will produce an analysis-of-variance table with the main effects being (1) the clone that the insects were taken from (sometimes referred to as donor plant or insect line), (2) the clone that the insects were placed on (receptor plant), and (3) the interaction between these two. If insects have formed demes on specific individual hosts, we would expect each deme to do relatively well on its adapted host and relatively poorly on other nonadapted hosts. In other words, we would expect the interaction between the main effects to be large and significant. However, other deviations from the marginal probabilities will also cause the interaction term to be significant. Therefore finding a significant interaction term is a necessary condition for supporting the hypothesis, but it is not sufficient, since many other factors can give rise to such an interaction (Rausher 1983; Unruh and Luck 1987). Once a significant interaction has been identified by analysis of variance, the hypothesis that insects adapt to specific individuals further predicts that the interaction will be caused by increased survival or fecundity of insects on their plant of origin. This specific prediction can be tested by using prior contrasts between insects on their clone of origin versus insects on unrelated clones (Kirk 1982, 95–98).

Numerous studies have directly or indirectly tested the hypothesis that insect populations are structured in demes, each of which is adapted to a particular host-plant clone or genotype (table 9.4). Many of the studies that found significant interactions between the plant clone and insect line did not use replicated plant propagules in a common garden (e.g., Edmunds and Alstad 1978; Moran 1981; Wainhouse and Howell 1983). These significant interactions are therefore difficult to interpret, since they may have been caused by undetected environmental differences between plants, rather than by genetic differences. One study found that the interaction between plant clone and insect line was not

TABLE 9.4
Studies examining the deme formation hypothesis

Plant	Insect	Replicate plants interspersed in common garden?	Insects reared on common host prior to expt?	Fitness parameter	Plant clone × Insect line interaction	Contrast: Clone vs. unrelated	References
Pinus ponderosa	*Nuculaspis californica*	no	no	survival	significant	not conducted	Edmunds and Alstad 1978
Pinus lambertina	*Nuculaspis californica*	yes	no	survival	?	not signif.	Rice 1983b
Pinus monophylla	*Matsucoccus acalyptus*	no	no	survival	significant	not signif.	Unruh and Luck 1987
Fagus sylvatica	*Cryptococcus fagisuga*	no	no	survival	significant	inconsistant	Wainhouse and Howell 1983
Morus alba	*Pseudaulacaspis pentagona*	no	no	survival	not signif.	not conducted	Hanks and Denno 1989
Solidago canadensis	*Uroleucon caligatum*	no	no	size 1st instar	significant	not conducted	Moran 1981
		no		mean adult wt.	significant	not conducted	
		no		colony weight	not signif.	not conducted	
Rudbeckia lacinata	*Uroleucon rudbeckiae*	yes	yes	pop. growth rate	significant	not conducted	Service 1984b
Betula pubescens	*Epirrita autumnata*	no	yes	larval growth	not signif.	not signif.	Ayers et al. 1987
Pseudotsuga menziesii	*Choristoneura occidentalis*	yes	yes	survival	probably	inconsistant	Perry and Pitman 1983
Erigeron glaucus	*Apterothrips secticornis*	yes	yes	pop. growth	significant	significant	Karban 1989b

significant (Ayres et al. 1987) and another detected an interaction only when gene flow was artificially restricted (Hanks and Denno 1989), findings that do not support the hypothesis that insect populations have become specialized into host demes. For several studies the interaction between plant clone and insect line was significant but was probably not due to insects performing better on the plant clone of their origin than insects from other unrelated clones (e.g., Rice 1983; Wainhouse and Howell 1983; Perry and Pitman 1983; Unruh and Luck 1987). In only one study (Karban 1989b) did the insects on their home clone consistently outperform insects originally from other plant clones (table 9.5).

TABLE 9.5
Number of thrips (*Apterothrips secticornis*) after three generations per plant (*Erigeron glaucus*) in a common garden

Plant clone	Thrips line		
	V	VII	IX
5	22.9 ± 7.7	5.8 ± 2.2	7.9 ± 2.3
7	7.8 ± 1.6	17.8 ± 5.4	11.8 ± 4.0
9	8.2 ± 2.3	10.8 ± 2.6	21.9 ± 4.6

Source: Karban 1989b.

Notes: All plants received three immature thrips at the start of the experiment. Numbers shown are mean ± 1 s.e., each treatment had 10 replicates.

Thrips from line V were taken from plant clone 5, thrips from line VII were taken from plant clone 7, and thrips from line IX were taken from plant clone 9. Thrips from all lines were reared for two generations on another universal plant clone prior to the start of the experiment to minimize effects of plant conditioning on thrips.

The examples presented in table 9.4 provide rather equivocal evidence for the hypothesis that insects commonly form adapted demes. Many of the studies that report patterns consistent with the hypothesis could have been caused by mechanisms other than fine-scale adaptation to individual hosts. The hypothesis is plausible only for small insects with very short generation times relative to their hosts and with mechanisms that limit gene flow among insects on different plant clones (Edmunds and Alstad 1978, 1981; Ayres et al. 1987; Hanks and Denno 1989). These restrictions greatly limit the frequency with which we can expect to encounter herbivore adaptation to particular plant clones in nature. However, because the consequences of deme formation are so important (Edmunds and Alstad 1981), workers should certainly continue to test the deme formation hypothesis in plant-insect systems for which it seems plausible.

Experiments That Explicitly Manipulate Plant Variation

Few studies have explicitly manipulated intraspecific plant variation and observed the effects on herbivore populations. Porina caterpillars (*Wiseana* spp.)

were reared on plantings of a single variety of ryegrass (*Lolium perenne*) and on plantings that were mixtures of two different varieties. The same design was also employed for white clover (*Trifolium repens*). There were no significant differences in the number of larvae surviving or in their weights as a function of host variation in either of these cases (Harris and Brock 1972). In another experiment, Altieri and Schmidt (1987) planted plots of broccoli with either a single variety or with a mixture of four different varieties. They found that populations of the cabbage aphid, *Brevicoryne brassicae*, were no different on the two kinds on plots during the spring, but during the autumn, aphid densities were significantly greater on the plots with only one variety. In another experiment, the proportion of broccoli varieties was varied. Aphid densities were inversely proportional to the "evenness" (sensu Pielou 1966) of the plant varieties that were present. A similar experimental design was used to compare numbers of leafhoppers, *Dalbulus maidis*, found on single-variety plantings of maize compared to plantings of five different varieties (Power 1988). Leafhopper densities were lower in the genetically diverse stands than in the pure stands.

Gould (1986a, 1986b) developed a series of simulation models that considered how long it would take an herbivore to evolve a means of circumventing two different host-plant resistance factors. The models considered three options: (1) releasing the factors sequentially, i.e., using the second factor only after the first had failed, (2) releasing the two factors simultaneously, combined in a single variety, and (3) releasing two different varieties, each with one of the resistance factors to be used in a mixed planting. The first option calls for the available variability to be spread out along a time dimension and the third option calls for variability to be spread out along a spatial dimension. The best strategy depended on the details of the genetics of host resistance and on the relationship between the resistance and insect fitness. Of particular interest was the finding that resistance was made more durable (longer lasting) if susceptible genotypes were grown adjacent to the resistant ones. Gould considered the Hessian fly and its host, winter wheat, in detail (Gould 1986b). He estimated that mixing some susceptible plants in with plants containing both resistance factors should maintain the effectiveness of resistance for over 400 generations, considerably longer than past resistance factors have lasted in monoculture.

Only a single study has considered the effects of genetic variability of a natural plant population on its herbivores. Plots of evening primrose (*Oenothera biennis*) were established in a common garden which contained genotypes A and B grown separately and in mixtures (Kinsman 1982). Kinsman was most interested in measuring the rate of insect colonization and could not assess their population dynamics owing to the small scale of her experiment. However, she found that plants of the attractive genotype were equally attractive to their herbivores regardless of the genetic diversity of their neighbors.

Conclusions

Many recent studies have found that intraspecific variation in natural populations of host plants has a significant effect on the herbivores that use them. That such differences can be detected comes as no surprise, given the large agricultural literature detailing the genetic basis of host-plant resistance. For most of the examples from natural plant populations, the effect on insect success due to individual host plants was relatively consistent from year to year. The influence of host-plant phenotype was not consistently larger or smaller than other ecological factors that also influenced herbivores. Insects were also variable in their ability to exploit individual host plants, although the evidence that herbivore populations are structured in demes, each adapted to particular plant genotypes, remains equivocal. Similarly, the few experiments that have explicitly varied host-plant phenotypes and recorded effects on insect success have produced mixed results.

If we are to make efficient progress toward understanding the importance of host-plant variability on the population dynamics of herbivores, we must change what we choose to measure and what we choose to publish. As mentioned earlier, many studies, particularly those that involve testing agricultural varieties for resistance against pest insects, measure only the percentage of infestation or the level of damage received. These factors are certainly important but they provide only suggestions about the effects of the different varieties on populations of herbivores. To address this question effectively, we must measure the actual effects on herbivore populations. Such studies will be most convincing if they are properly replicated and controlled and if they are carried out for at least one complete herbivore generation on an appropriate spatial scale.

If we are ever to gain an understanding of the situations and conditions where host-plant variability is likely to affect populations of herbivores, we must report experiments that fail to show an effect as well as those that do reveal host variability to be important. Only when we are armed with both negative and positive results can we answer most of the questions considered in this review.

Since host plants are certain to vary in traits that affect the insects feeding on them, why should ecologists be interested in noting this variation in natural populations? Will characterizing this variation provide ecologists with any information that is more tangible than characterizing the electrophoretic variation in plant enzymes? I believe that host-plant variation is interesting because it can be linked to the sizes of insect populations. It can be compared directly to other ecological factors that have been shown also to affect populations of insects. Host-plant variation can also be experimentally manipulated to establish cause-and-effect relationships with natural insect populations.

If natural populations of herbivores are strongly affected by host-plant variation, be it genetic or environmental, then managed ecosystems may do well to imitate or incorporate host-plant variability. Plant pathologists have been concerned about genetic homogeneity for many decades and have incorporated intraspecific variability into the agricultural practices used for several crops (Simmonds 1962). For example, diseases caused by rusts have been managed using programs that explicitly include genetic variability. Stem rust in wheat has been managed successfully in this manner for at least 25 years in Australia and CIMMYT in Mexico uses plantings that include a mix of different resistant lines of wheat to control several rust diseases (Watson 1970; Marshall 1977). In the United States, growers found that planting single "resistant" varieties of oats repeatedly failed against crown rust (Frey et al. 1973). Instead, different resistant genes are deployed exclusively to different regions of the country with the result that large-scale epidemics have been eliminated.

Following the lead of plant pathologists, some agricultural entomologists are beginning to consider the possibility of managing crop variability. Knowledge of the effects of natural plant variability on populations of herbivores is essential to the intelligent planning of deployment of host resistance in agriculture.

Acknowledgments

My work on plant variation and herbivore performance has been supported by grants from the National Science Foundation (BSR-8614528, BSR-8818019), the Sloan Foundation, the University of California, and the Center for Population Biology.

10

The Importance of Deterrence: Responses of Grazing Animals to Plant Variation   *A. Joseph Pollard*

Progress toward a general understanding of plant-herbivore interactions will require recognition that different species of herbivores consume plants in different ways, with very different consequences at the population, community, and evolutionary levels. This chapter is about one mode of consumption, grazing, that is underrepresented in theoretical and experimental treatments of herbivory. In keeping with the theme of the book, the chapter concentrates on the interactions of grazers with intraspecific plant variation and the evolutionary outcomes of such interactions. Literature in these areas is meager compared to studies of phytophagous insects whose feeding mode is essentially parasitic (e.g., Van Emden 1972; Wallace and Mansell 1976; Denno and McClure 1983; Strong et al. 1984; Juniper and Southwood 1986; Spencer 1988b). I have attempted to pull together information from basic and agricultural sources, to point out unique features of the plant-grazer interaction, and to suggest fruitful directions for further studies.

Although "grazing" is often used loosely when referring to any herbivory, I will use the term to denote one particular form of trophic interaction (Price 1980; Thompson 1982). A taxonomy of trophic interactions is shown diagrammatically in figure 10.1, in which the axes of lethality and intimacy represent features of the interaction between the consumer and its food organism (interactions between individuals, not species-pairs). Grazing is an interaction with both low lethality and low intimacy. Only a portion of the grazed individual is removed, and death does not usually result directly. The grazer does not long remain in contact with any one food organism. Unfortunately, grazing lacks a convenient term, like host or prey, to indicate the consumed individual. Because this chapter focuses on grazing herbivores, I shall simply refer to the consumed organism as the plant. However, consumption of animal tissue can be a matter of grazing, as in the cases of bloodsucking insects (mosquitoes, fleas) or eaters of coral, bryozoans, etc. The prevalence of plants as the consumed organisms is not so much a unique feature of grazing, as a consequence

216

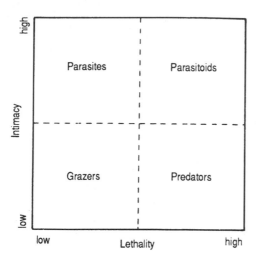

Fig. 10.1. A conceptual taxonomy of trophic interactions between individuals. Lethality represents the probability that a trophic interaction results in the death of the organism being consumed. Intimacy represents the closeness and duration of the relationship between the individual consumer and the organism it consumes (i.e., an index of symbiosis). The four quadrants of the figure represent four modes of trophic interaction; however, both axes are continua, so classification is not always unambiguous.

of the modular architecture of plants (White 1979; Dirzo 1984), which permits survival of the individual despite destruction of modules.

In contrast to grazing, predation is quickly and predictably lethal to the prey, as in the cases of most animal-animal interactions, seed-eating mammals and birds (Janzen 1971), and consumers of young seedlings. Parasitoids have a long-term relationship with their individual hosts that is predictably though not immediately lethal; in addition to the usual cases of interactions among arthropods, parasitoids could also be seen to include bruchid beetles and other insects whose larvae live in seeds (Janzen 1971; 1971a; 1975c; 1980a). Parasites have a long-term, intimate relationship with their hosts, in which the host is neither immediately nor predictably killed. Among herbivores, most phytophagous insects are best seen as parasitic (Price 1980; Thompson 1982), especially if they live for an extended period on a single plant, have poor powers of dispersal from one plant to another, or are small compared to the plant upon which they feed. The intimacy of the host-parasite interaction allows ample opportunity for specialization and pairwise coevolution, including changes in resistance and virulence (Futuyma 1983; Wilhoit in this volume).

Characteristics of Grazers and Grazing

The distinction among modes of consumption is not a trivial exercise in semantics. An understanding of lethality and intimacy is essential in making predictions regarding diet breadth, foraging strategy, defense mechanisms, and coevolution. Although grazing may be defined in principle on the basis of lethality and intimacy, other characteristics are suggested by the term. Table 10.1 lists nine generalizations about grazing.

Item 9 in the table is of course not unique to grazing; all living things are

TABLE 10.1
Generalized characteristics of grazing interactions

Characteristic	Comments/references
1. Time spent feeding on a single plant is small relative to grazer lifetime; individual grazer feeds from very large number of plants during its life.	Corollary of low intimacy
2. Time during which a plant is eaten by any one grazer is small relative to plant lifetime; individual plant may encounter many grazers.	Corollary of low intimacy and lethality
3. Grazers perceive their food environment as fine-grained because they are highly mobile and because they are often large animals eating small plants.	Price 1980; Thompson 1983
4. Although sublethal, grazing by a large animal can be very harmful to a plant; effects may be asymmetrical, with one grazer affecting a plant much more than any one plant affects the grazer.	Fitness effects on plants may be low for clonal individuals.
5. Grazers consume large quantities of poorly digestible plant material and have low assimilation efficiency compared to more specialized feeders.	Few data exist for these efficiencies (Ricklefs 1973)
6. Grazers are dietary generalists, i.e., the individual grazer consumes many different food-plant species; however, not all foods are equally preferred.	For hypotheses on generalism, see Westoby 1974; Freeland and Janzen 1974; Pulliam 1975; Bryant, Chapin et al. 1985.
7. Grazer populations may be more often food-limited than populations of specialist insects, which tend to be predator- or disease-limited.	Crawley 1983
8. Large mammal grazers are capable of a high degree of neurological and behavioral complexity.	Lindroth 1988b
9. Both grazers and the plants they eat possess intraspecific genetic variation, and can evolve in response to one another.	See comments in text.

variable and evolve. Biological variability exists at many levels, including differences between species, genetic differences between populations of a given species, genetic polymorphism among individuals within a population, developmental variability between stages or parts of a given individual, and phenotypic plasticity in response to environmental stimuli. All these are relevant to grazing. Different grazer species feed on different plant species, but as pointed out by Janzen (1979), "herbivores do not eat Latin binomials" and "Latin binomials do not eat plants." The most important level for evolutionary biology is within-population genetic variation. Study of this variation allows testing of evolutionary hypotheses such as the defensive function of plant chemicals (Jones 1971) and permits observation of changing gene frequencies over time, i.e., evolution in action (Dirzo and Harper 1982a).

Kinds of Grazers

Many of the best examples of grazers are terrestrial mammals, especially ungulates (hoofed animals), but also lagomorphs (rabbits and hares), rodents, elephants, primates, and marsupials. Grazing is of great economic importance because of domestic livestock, and has been manipulated by human activities such as introduction of species into new geographic areas, range management, and selective breeding. An inadvertant result of these manipulations may be the creation of "natural experiments" that can provide insight into the plant-grazer interface. There is also a large body of potentially relevant animal husbandry literature.

In the context of resource partitioning, it may be important to distinguish between grazing in a strict sense (eating grass and forbs) and browsing (eating leaves and young shoots of woody plants) (McNaughton and Georgiadis 1986). In a functional and evolutionary context, however, the distinction is less pronounced, and many animals employ both feeding modes. Browsing generally removes a lower proportion of total plant biomass than does grazing; however, much of the uneaten biomass consists of nonproductive woody tissues. Browsing may in fact be more destructive than grazing, owing to the different architectures of woody plants and graminoids (Bryant, Chapin et al. 1983a).

The insects that best fit the characteristics of grazers listed above are grasshoppers and locusts. For example, Ueckert and Hansen (1971) reported that species of grasshoppers surveyed in northeastern Colorado had diets ranging from as few as 6 to as many as 38 plant species. Otte (1975) found that species of *Schistocerca* would accept up to 91 different food plants in laboratory feeding trials, and noted that these grasshoppers can fly for distances of 100 m or more. Nonetheless, grasshoppers do show definite grazing preferences (Mulkern 1967; Bernays et al. 1976).

Other invertebrates that are highly polyphagous include snails and slugs (Cates and Orians 1975; Mason 1970; Pallant 1969). Grime et al. (1968, 1970) found that the snail *Cepaea nemoralis* showed some degree of acceptance of 19 plant species in feeding trials, and Dirzo (1980) found that the slug *Agriolimax carunae* would accept 19 species under similar conditions. Once again, both snails and slugs displayed hierarchies of food plant acceptability.

Herbivorous fish may act as generalist grazers on seaweeds (Hay and Fenical 1988). It is ironic, considering the entomological emphasis of terrestrial herbivory studies, that in marine systems the grazers have received most of the attention, while less mobile, more specialized invertebrates like amphipods have been little studied (Hay et al. 1987).

Effects of Grazers on Plants

This chapter deals primarily with the responses of grazing animals to intraspecific variation among plants. In this section, however, I discuss the way grazing affects plants. This topic is explored further in a later chapter (Marquis), but I have considered certain aspects of it here because much of the evidence concerns grasslands and ungulates, and because intraspecific variation among plants may itself be a consequence of grazing.

The potential harms and benefits incurred by plants as a result of grazing are the one aspect of plant-grazer interactions that has been very thoroughly reviewed (McNaughton 1979a, 1979b, 1983a, 1983b; Dyer et al. 1982; Belsky 1986); indeed, the amount of review and discussion in the literature possibly exceeds the amount of original research (Stenseth 1983). Few studies deny that grazers can have an impact on plants. Controversy surrounds the issue of whether this impact is uniformly harmful or whether it may be beneficial to the plant. It is unfortunate that so extensive a controversy has been generated primarily by ambiguity over the meanings of "harm" and "benefit." In a brief summary of the issues, I offer five generalizations.

1. Being grazed is harmful to many, probably most, plants. Studies supporting this generalization are numerous (e.g., citations in Harper 1977; Crawley 1983). The harms of grazing have been documented on many levels, including decreased survivorship, growth, and reproduction (as well as "degradation" of the community and ecosystem, Ellison 1960). If the tissue eaten is productive (e.g., leaf blades), then the harm to the plant may be more than proportional to the amount of material removed (Louda 1984).
2. Plant species differ in the extent to which grazing harms them, i.e., some species are more grazing-tolerant than others.
3. Individuals of the more grazing-tolerant species may be capable of some kind of compensatory increase in productivity or growth following defoliation (McNaughton 1983a; Wallace et al. 1984). This is essentially a physiological phenomenon (Belsky 1987). Many of the studies in this category are only partial, that is, they show compensation in the short term rather than over the plant lifetime, or they report aboveground rather than whole-plant measurements, or they report only vegetative and not reproductive output. Studies claiming to show overcompensation (improved performance following defoliation) on a wholeplant, lifetime basis have received methodological and interpretational criticism (Belsky 1986).
4. Members of more grazing-tolerant plant species become more common in communities as a result of grazing on the community. This generalization is not new (Clements 1920), but is embodied in more recently proposed terms such as competitive fitness (McNaughton 1979b) and grazophile (McNaughton 1986). It is a community-level phenomenon generated by grazers suppressing more sensitive species and releasing the more tolerant ones from competition.
5. Grazing may perhaps stimulate rates of ecosystem productivity (Chew 1974; Mattson and Addy 1975), although this too is controversial (Belsky 1987).

Even if the five generalizations above are all accepted despite their controversies, they tell us little about evolution. It seems that a tendency in this debate has been to overgeneralize items 3 through 5 above to a simple statement that "herbivory can be beneficial to plants," and then follow with corollaries that being grazed increases plant fitness, and that plants have evolved increased palatability and mutualistic relationships with grazers (Mattson and Addy 1975; Owen 1980; Owen and Wiegert 1976, 1981, 1982a, 1982b, 1984; McNaughton 1984; Georgiadis and McNaughton 1988).

To document an increase in Darwinian (relative) fitness resulting from grazing, it would be necessary to show that grazed individuals have greater reproductive success than ungrazed plants of the same species in the same population. For increased palatability and mutualism to evolve would also require that heritable variation in palatability exist in the population. In the words of Belsky (1986, 884), "The only evidence that would prove unambigously that herbivory increases plant fitness would be the documented increase of more-palatable genotypes relative to less-palatable genotypes in a grazed population." No such evidence is presently available.

McNaughton (1986, 765) has stated that he has "nowhere written that the mere act of herbivory is beneficial to any affected plant." However, he and others have generated confusion through inconsistent usage of the term fitness. "Competitive fitness" (McNaughton 1979b, 698) does not correspond to individual Darwinian fitness; it refers to the performance of one species relative to other species in a grazed environment. If the members of this species receive a benefit, it is not because they themselves have been grazed, but because the other species around them have suffered from grazing more than they have. McNaughton has stated (1983a, 334) and reiterated (1986, 768) the hypothesis that plants can "overcompensate for damage so that fitness may be increased." The statement begs the question "Increased compared to what?" If it does not refer to fitness compared to ungrazed plants of the same species, then it must refer to fitness relative to plants in the same population that are grazed but cannot overcompensate. Since intraspecific polymorphism in compensatory growth ability has never been investigated, the latter is at best a speculation as to the evolutionary origins of overcompensation. It would be far better if all parties in this debate could avoid ambiguous and undefined use of terms like "harm" and "benefit," and restrict use of the term "fitness" to its proper evolutionary context.

One recent study deserves mention. Paige and Whitham (1987) have shown that *Ipomopsis aggregata,* a monocarpic herb, produces more seeds following either grazing by elk and mule deer or experimental clipping, than do uneaten controls. If there were genetic variation in seed production following defoliation (as might conceivably arise through mutation), then genotypes that increased seed production following defoliation would become more common in the population, and if there were genotypic palatability differences, then the evolution of increased palatability could also occur. Further study is needed to confirm Paige and Whitham's findings and to address these genetic questions.

Grazing Preference and Plant Defense

The remainder of this paper assumes that grazing results in reduced fitness for the grazed plant (Dirzo 1984). As stated previously, grazers are generalists, but they are not indiscriminate; they respond to characteristics of their food plants (Ivins 1952; Garner 1963; Archer 1973). Based on the premise that grazing reduces plant fitness, numerous authors, dating back to pioneering studies by Stahl (1888) and Fraenkel (1959), have suggested that some of these plant characteristics have evolved as "defense mechanisms" against herbivory.

The Meaning of Defense

"Defense mechanisms" is another phrase that is loosely if at all defined and flirts with teleology. There appear to be at least three shades of meaning attached to the term "defense," each supported by a different kind of evidence. I will call these antibiosis, functional defense, and adaptive defense.

Studies demonstrating antibiosis are those showing that chemicals or structures possessed by a plant are harmful to an herbivore. For chemicals, the term "allelochemics" or "allelochemicals" is often used in this sense (Whittaker and Feeny 1971). While interesting, this phenomenon alone is not really evidence of defense as such, because it does not show how the plant benefits from having the harmful features.

What I have designated a functional defense is a feature of a plant that causes herbivores to eat less of this plant than of others, either because they do not initiate feeding (deterrence), or because they feed at a slower rate, or because their population levels are suppressed. Most case studies of functional defense involve comparisons between different plant species, perhaps backed up by demonstration of reduced feeding on some normally palatable foodstuff if it is adulterated with the supposed defensive substance. In other cases, functional defense may be proposed by showing that one group of plants suffers lower rates of damage than others, without knowing the cause of differential eating. This is usually termed "resistance," and has mostly been explored in interactions of plants with parasitic insects (e.g., Marquis 1984; Simms and Rausher 1987).

Adaptive defenses are those that have evolved in response to selective pressures imposed on plants by herbivores. It is often assumed that a defense demonstrated at the antibiosis and/or the functional levels does in fact constitute an adaptive defense, but there is an inherent danger of "adaptive storytelling" (Gould and Lewontin 1979). Chemicals and structures of plants may have evolved for many reasons (Siegler and Price 1976). If they are harmful to herbivores, it remains to be shown how the plant benefits. If the plant does benefit, this still does not prove that herbivory was the underlying selective force that produced the feature. Comparisons between monomorphic species are particularly susceptible to this limitation. A functional defense demonstrated through interspecific comparisons may easily be an "abaptation" (Harper 1982) or "exaptation" (Gould and Vrba 1982), i.e., a feature evolved under selection for something other than defense (or under no selection at all), with defensive function merely coincidental.

The phrase "defense mechanism" has an underlying connotation of adaptation. Defensive adaptation is not impossible to document, but it must be done based on intraspecific genetic variation (Jones 1971; Dirzo and Harper 1982a). If such variation exists, then three testable predictions become available for experimental study. First, plants with the putative defense should suffer lower rates of herbivory than those without; this is functional defense at the intraspecific level. Second, the benefits accrued by the plant should exceed any costs associated with the defensive system; consequently, in a grazed environment the defended plants should have a higher fitness than conspecific but undefended plants. Third, defended plants should become more common in grazed populations than undefended plants. If variation in defense is continuous rather than discrete, then these predictions may be modified to compare varying levels of defense. Although it would be desirable to assemble all three lines of evidence for a given defense mechanism, this has rarely been accomplished (see Simms and Rausher 1989 for studies of fitness costs and benefits associated with varying resistance to insect herbivores).

In the sections that follow, I have not attempted to summarize every plant defense mechanism ever suggested, especially those based purely on antibiosis or functional defense evidence. Because of the focus on grazing, I have omitted references to most of the entomological literature (reviewed in Beck 1965; Van Emden 1972; Wallace and Mansell 1976; Rosenthal and Janzen 1979; Futuyma 1983a). I have concentrated on grazers and plant variation. Genetic variation is emphasized, but some studies of nongenetic (environmental, ontogenetic) variation have been included in which the results are particularly revealing. Most of the literature involves constitutive defenses; inducible defenses are discussed toward the end of the chapter.

Intraspecific Variation in Plant Chemistry

Grazers are known to choose food-plant species based on differences in positive nutritional content of the plants (Blaxter et al. 1961; Raymond 1969; Westoby 1974), differences in toxic secondary chemicals (Arnold and Hill 1972; Arnold et al. 1980; Freeland and Janzen 1974; D. A. Levin 1976b), or a combination of the two. Within species, differences in nutritional content (gross energy, protein, sugars, minerals) are less pronounced. Sheep and cattle select the most nutritious parts and growth stages of grasses when grazing (Arnold 1964; Raymond 1969); similar processes occur among wild ungulates in the Serengeti (Gwynne and Bell 1968) and among folivorous primates (Hladik 1978; Milton 1979). Genetic polymorphism in nutritional content is not reported in the literature for native forage plants (although it has been the subject of breeding programs in domesticated pasture species; Raymond 1969). Genetic variation in nutrients could be constrained by architectural or competitive factors unrelated to grazing. Furthermore, reduction in nutrients as a defense mechanism might be counterproductive, by stimulating herbivores to consume more plant materials to meet their nutritional needs (Moran and Hamilton 1980).

Digestibility is known to vary between and within plant species (for methods of determination see Van Dyne 1968). Digestibility is often considered a nutritional measure, but it is affected by secondary chemicals, particularly tannins and lignins (Swain 1979), and by structural attributes, such as fibers and silica bodies. In ruminants the main effect of tannins, specifically condensed tannins, is reported to be increased resistance of ingested food to the activity of rumen bacteria (Zucker 1983; Cooper and Owen-Smith 1985); by contrast, hydrolyzable tannins appear to be more important to insects and act by interfering with insect digestive enzymes. Tannins may also have deterrent properties unrelated to digestibility per se (Bernays 1981; Zucker 1983; Mole et al. 1990). Browsing ruminants in South African savannas vary their feeding preferences in response to seasonal changes in protein and condensed tannins (Cooper, Owen-Smith et al. 1988), and howler monkeys choose young leaves based on protein, fiber, and tannin contents (Milton 1979). In marine systems, snails preferentially graze low-phlorotannin plants of the brown alga *Fucus distichus*. Elevated tannin levels in this species are induced by grazing or clipping; it is not known whether additional genetic variation exists (Van Alstyne 1988). In one of the few cases clearly involving genetic variation in digestibility, cattle and sheep were shown to prefer strains of *Lespedeza cuneata* with low levels of stem fiber and tannin (Donnelly and Anthony 1970, 1973, 1983). This genetic variation has been used in breeding improved forage varieties.

Intraspecific genetic variation is better documented among other distasteful or toxic secondary chemicals. Heritable genetic polymorphism in indole alkaloids of reed canarygrass (*Phalaris arundinacea*) correlates with feeding preferences of cattle, sheep, rabbits, and voles (Roe and Mottershead 1962; Asay et al. 1968; Simons and Marten 1971; Marten et al. 1973; Kendall and Sherwood 1975). Alkaloid polymorphisms in *Lupinus* spp. have permitted breeding of forage and fodder cultivars that are acceptable and nontoxic to livestock (Gladstones 1970). Heritable variation in monoterpenes of *Thymus vulgaris* affects slug grazing preferences (Gouyon et al. 1983), and between-tree differences in the monoterpenes of *Pinus ponderosa* predict squirrel browsing intensity in field and laboratory studies (Farentinos et al. 1981). Green algae in the genus *Halimeda* produce diterpenoids that deter feeding by fish (Paul and Van Alstyne 1988). More potent chemical forms are found in higher proportions in some heavily grazed populations; however, some heavily grazed populations do not show elevated chemical defense, and some grazing preferences exist in the absence of differences in diterpenoids.

Grazing preferences may be affected by many plant characteristics acting simultaneously (Berenbaum 1985). Genetic variation has only rarely been studied in such complex systems. Douglas fir (*Pseudotsuga menziesii*) has been shown to vary in digestibility, essential oils, resins, fats, phenolics, flavonols, terpenoids, leucoanthocyanins, and chlorogenic acid. Between-plant variation is genetically determined. Grazing preferences of snowshoe hares and black-tailed deer are positively associated with high digestibility and chlorogenic acid, and low levels of the other chemicals (Radwan 1972; Dimock et al. 1976). Intraspecific differences in tannin and alkaloid levels of the food plants of howling monkeys are reported to affect feeding preferences (Glander 1981), but the genetic system underlying the polymorphism is not known.

Undoubtedly the best-studied defensive polymorphism involves cyanogenesis, the ability of plant tissues to release HCN (Jones 1972; Conn 1979). Approximately 1,000 plant species are known to be cyanogenic; the most common sources of HCN are the cyanogenic glycosides linamarin and lotaustralin (Gibbs 1974). Genetic polymorphism for cyanogenesis is known in a number of plants, and has been very well studied in two species in the Fabaceae, *Trifolium repens* and *Lotus corniculatus*. In these species, variation in cyanogenesis results from polymorphism at two unlinked loci, one controlling cyanogenic glycoside synthesis, and one controlling the production of linamarase, a β glucosidase that hydrolyzes the glycoside. Cyanogenic individuals contain the dominant allele at both loci, produce both the glycoside and the enzyme, and liberate cyanide if the leaves are damaged. Homozygous recessives at either or both loci are acyanogenic. However, if plants containing glycoside but no linamarase are consumed by animals that produce β glucosidases among their normal gut enzymes, there is still a possibility for HCN release during digestion (Dirzo and Harper 1982a).

The potential toxicity of cyanide is well documented (Conn 1979). Avoidance of grazing on cyanogenic morphs of variable species has been shown for a number of grazing animals under both laboratory and field conditions. Cyanogenic individuals of *L. corniculatus* are avoided by snails and slugs (Jones 1966; Crawford-Sidebotham 1972; Ellis et al. 1977a; Compton and Jones 1985), by several generalist insects (Compton and Jones 1985), and by lemmings and voles (Compton, Newsome et al. 1983; Jones 1966). Cyanogenic *T. repens* is avoided by snails and slugs (Whitman 1973; Angseeing and Angseeing 1973; Angseeing 1974; Dirzo and Harper 1982a, 1982b; Horrill and Richards 1986; Kakes 1989) and rabbits (Corkill 1952). Among other plant species, *Pteridium aquilinum* (bracken fern) is polymorphic, and cyanogenic morphs are avoided by sheep and deer under field conditions and by the locust *Schistocerca gregaria* in laboratory experiments (Cooper-Driver and Swain 1976). Developmental variation in cyanogenesis by *Sorghum bicolor* affects it acceptability to *Locusta migratoria* (Woodhead and Bernays 1978).

The results above convincingly demonstrate that cyanogenic glycosides act as functional defenses at the intraspecific level, despite some flaws in the studies (Hruska 1988). The adaptive nature of cyanogenesis is further supported by studies showing that populations with high frequencies of cyanogenic morphs occur most commonly in areas heavily grazed by mollusks (Ellis et al. 1977a; Dirzo and Harper 1982b; Compton, Beesley et al. 1983). Costs and benefits of cyanogenesis have been studied (Kakes 1989) but not in terms of plant fitness. Fitness effects of grazing can be highly dependent on the plant phenostage in which it occurs. Horrill and Richards (1986) report that for the first 5 days after germination, slugs do not differentiate between cyanogenesis morphs, but that from 5 to 36 days most acyanogenic seedlings are killed while most cyanogenic ones survive. It may be that the strongest selective pressures are thus exerted at the seedling stage, in an interaction more similar to predation than is the nonlethal grazing on mature plants.

An exemplary study of selection in action, based on plant demography (Dirzo and Harper 1982b), was complicated by the fact that cyanogenesis has pleiotropic effects on fitness. Cyanogenic morphs may be more susceptible to rust infections, frost damage, drought, salinity, and trampling (Dirzo and Harper 1982b; Daday 1965; Foulds 1977; Keymer and Ellis 1978), but may be superior competitors in closed swards (Compton et al. 1986). The importance of grazing as a selective force influencing the characteristics of plant populations must be weighed against the degree to which these other factors act on any given population.

Intraspecific Variation in Plant Structures

In contrast to the chemical features described above, little is known about the responses of grazers to variation in physical structures of plants. Potentially defensive structural attributes include "toughness," cuticular features, and the possession of thorns, spines, prickles, or trichomes.

Leaf toughness and hardness have been implicated in defense against insect herbivores (reviewed in Grubb 1986), but rarely against grazing mammals, presumably because their greater size and muscularity allow them to chew a wider range of materials. Toughness may be correlated with digestibility as discussed above, since both are influenced by content of hard materials like fibers and silica bodies. *Deschampsia caespitosa* varies in its silica content, and cattle reportedly avoid the rough, high-silica variants found in lowlands but eat low-silica variants from higher altitudes readily (Davy 1980). However, there is no evidence that the variation is genetic, and grazing preferences in this case may be confounded by differences in alternative foods available in upland and lowland sites. McNaughton and Tarrants (1983) found evidence of elevated silica levels in plants from heavily grazed sites; the differences were constant in cultivation, but comparison is difficult because different species were included from the two areas. For molluscan grazers, leaf texture is an important feature affecting the acceptability of plant species (Dirzo 1980; Grime et al. 1968; Jennings and Barkham 1975), but no studies have examined intraspecific variation.

The response of insects to the plant cuticle and epicuticular wax have received some attention (Woodhead and Chapman 1986), but no studies have considered either genetic polymorphism or vertebrate grazing. Removal of surface waxes from a number of plants stimulated feeding by *Locusta migratoria* (Chapman 1977; Woodhead 1983). Artificial removal of the cuticle of the red alga *Iridaea cordata* resulted in greater grazing damage in the field, and in laboratory studies stimulated feeding by an isopod, though not by three snail species and a sea urchin (Gaines 1985). These patterns may explain differential herbivory on vegetative and fertile blades in this species.

It is very commonly asserted that thorns, spines, and prickles of plants deter mammal grazing, despite a dearth of experimental evidence. Reports of postgrazing increases in the armament of prickles in *Rubus trivialis* (Abrahamson 1975) and of thorns in *Acacia senegal* (Seif el Din and Obeid 1971), although often quoted, are in fact quite anecdotal. Likewise, the conventional wisdom that leaf prickles of hollies (*Ilex* spp.) deter herbivory was unsubstantiated in experimental studies in which prickles were artificially removed, and in comparisons of natural variation in prickles (Potter and Kimmerer 1988). Thorns and spines have been found to interfere with browsing by ungulates in African savannas, and to influence species acceptability (Cooper and Owen-Smith 1986). Many plants show intraspecific variation in thorns, spines, and prickles (e.g., Bieniek and Millington 1968; Spangelo et al. 1970). The consequences of genetic polymorphism in large structures such as these remain virtually unexplored and are worthy of further study.

The role of trichomes as plant defenses has received considerable attention in the entomological literature (reviewed in Levin 1973; Johnson 1975; Juniper and Southwood 1986). Much of this information is of great relevance in applied fields involving pest resistance by crop plants, but none is related to grazers as here defined. As grazers that come into intimate contact with plant surfaces, mollusks might be expected to be highly sensitive to plant pubescence; however, leaf hairs do not appear to affect plant species preferences of either snails (Grime et al. 1968) or slugs (Dirzo 1980).

Stinging trichomes are the one type of specialized plant structure that has been studied with regard to both mammalian grazers and intraspecific variation. These structures occur in four plant families, the Urticaceae (*Urtica, Laportea* and several other genera), the Euphorbiaceae (*Cnidoscolus, Tragia, Dalechampia*), the Loasaceae (*Loasa, Cevallia*), and the Hydrophyllaceae (*Wigandia*) (Thurston and Lersten 1969). The structure of the trichomes is remarkably similar in the diverse families that possess them (with the exception of *Tragia* and *Dalechampia;* see Thurston 1976). A generalized stinging hair consists of a single, elongate cell, from 1 to 8 mm long, with distally brittle walls, atop a multicellular pedestal. Brushing contact with a hair causes its brittle tip to break off, leaving a beveled point that can easily enter the skin, injecting an irritating fluid contained in the hair. Plants with stinging trichomes are commonly called nettles.

The pain-producing properties of nettle stings are known to anyone who has contacted the plant. The chemical basis of the sting has been examined among members of the Urticaceae. Bioassays of the effects of nettle extracts on excised muscle have indicated that the plants contain histamine, acetylcholine, and 5-hydroxytryptamine (serotonin) (Emmelin and Feldberg 1947; Collier and Chesher 1956; Barlow and Dixon 1973); however, chromatographic and spectroscopic techniques have usually been unable to corroborate these identifications (Thurston 1974; Willis 1969; Oelrichs and Robertson 1970). Recent studies using such methods have shown evidence of serotonin, but not histamine or acetylcholine, in a member of the Euphorbiaceae, *Cnidoscolus texanus* (Muell. Arg.) Small (Lookadoo and Pollard 1991). Knowledge of nettle chemistry will allow assessment of whether the convergent evolution of stinging in these four families extends to the chemical level as well as to the morphological level.

Urtica dioica L. is known to be genetically variable in number of stinging hairs per unit area, as well as other characters, with significant differences between populations and among plants within populations (Pollard and Briggs 1982). This variation in trichome density has permitted testing of the long-held assumption that stinging hairs are an adaptive defense against mammalian grazing (e.g., Salisbury 1961; Uphof 1962). In laboratory tests, sheep and rabbits preferred plants with lower numbers of stinging hairs; when plants with different trichome densities were transplanted into the field, those with fewer hairs were more heavily grazed by rabbits (fig. 10.2; Pollard and Briggs 1984). The patchy nature of livestock grazing, owing to differences in land use, allows testing of the evolutionary responses of plants to localized grazing intensity. In both *U. dioica* and *C. texanus,* populations from heavily grazed areas have higher trichome densities than those from lightly grazed areas (Pullin and Gilbert 1989; Pollard 1986).

Do stinging trichomes deter nonmammalian herbivores? *U. dioica* is a common food source for snails and slugs (Grime et al. 1968, 1970; Pallant 1969; Mason 1970; Wolda et al. 1971; Cates and Orians 1975). The species is also highly attractive to many insects (Davis 1973, 1983). Fiercely stinging Australian Urticaceae in the genus *Dendrocnide* (= *Laportea*) are subject to extremely high levels of insect herbivory (Lowman and Box 1983; Lowman 1985; Southwood 1986). My observations in Oklahoma indicate that *C. texanus* is used as a resting and perhaps feeding plant by grasshoppers, and is a common larval food of *Cycnia tenera* Hbn. (Leipidoptera: Arctiidae). (This moth is usually described as a specialist on Asclepiadaceae and Apocynaceae; all these plants have milky latex, though *Cnidoscolus* in the Euphorbiaceae is

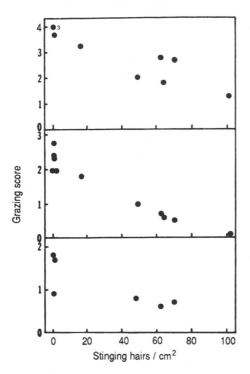

Fig. 10.2. Relationship between number of stinging hairs and grazing intensity. *Top*, grazing by sheep in pens. *Middle*, grazing by rabbits in individual cages. *Bottom*, grazing damage observed on plants transplanted into a field site with high rabbit populations. High grazing scores represent larger amounts of plant material removed. Each point represents the mean of 2–14 replicated trials. For further details of experimental design and scoring, see Pollard and Briggs (1984).

not closely related to the others. *C. tenera* reared on milkweed are reported not to sequester cardenolides [Rothschild et al. 1970].) Invertebrates seem able to circumvent stinging hairs, either by clipping them off at the base (Davis 1983; Dillon et al. 1983; Pollard and Briggs 1984) or by passing them intact through the gut (Grime et al. 1970). Presumably the kind of brushing contact needed to cause stinging hair penetration is simply not made by such animals: they may maneuver around the relatively large hair, or they may appress it to the plant surface (mollusks), or their exoskeleton may provide sufficient protection, given their low mass (insects). However, no studies have examined this issue by comparing herbivory on intraspecific variants.

Symbiotic "Third Parties"

Herbivore choice may also be influenced by symbiotic organisms that live in or on plants. Such systems are fascinating in their intricacy; however, this same intricacy may make them more difficult to study.

Undoubtedly the best-known "defensive mutualisms" are those that involve ants that are attracted to extrafloral nectaries of plants, and subsequently attack any herbivores attempting to eat the plants. The subject has been extensively reviewed elsewhere (Janzen 1966; Bentley 1977; Huxley 1986). Herbivores reported to be deterred by ants are mostly insects, but some observations suggest that mammals may also avoid ant-bearing plants (Brown 1960; Hocking 1970; Janzen 1972b; McKey 1974b). It is possible to compare plants with and without symbiotic ants in a manner similar to studies of intraspecific variation in chemicals or structures mentioned above. Differences in ant populations sometimes occur naturally, or they may be generated by herbicide treatments (Janzen 1966) or by application of artificial nectar (Bentley 1976). Intraspecific functional defense and increased phenotypic fitness for ant-bearing plants have been shown in many of these studies (Huxley 1986). However, there seem to be no reports of heritable intraspecific variation in the likelihood that a plant will attract ants in nature. This is especially unfortunate given the doubts that have been expressed elsewhere as to whether ant-plant systems actually represent coevolution: facultative (i.e., noncoevolved) mutualism has been invoked to explain defensive interactions between ant and plant species recently brought into contact with one another (Koptur 1979), and there is little evidence for the evolution of specializations in the ants in any case (Huxley 1986).

Recently, attention has been focused on a potentially defensive symbiosis between plants and endophytic fungi. A number of Ascomycete species form systemic infections inside host plants. Many of the best-documented cases involve fungi in the family Clavicipitaceae, tribe Balansiae, and host plants in the Poaceae, Cyperaceae, and Juncaceae (Clay 1988). However, a diversity of other fungal endophytes occur in a wide range of host groups, including conifers and multicellular algae (Carroll 1988). In some cases they have overt effects on plant development, including suppression of sexual reproduction, but in other cases they are asymptomatic.

Endophytes may have a number of effects on herbivores. The best documented of these is the economically important toxicity toward livestock of grasses such as *Festuca, Lolium, Stipa,* and *Melica,* with symptoms of toxicity generally termed "staggers" (White 1987; Clay 1988). These effects may be considered antibiosis; reports of functional defense such as avoidance of infected plants are much more anecdotal (Bailey 1903). For insect herbivores there is evidence for deterrence as well as antibiosis (Barker et al. 1984; Hardy et al. 1985). Among endophytes of plants other than graminoids, most of the currently available data again show antibiosis toward insect herbivores (Carroll 1988). Reports of reduced grazing by marine mollusks on green algae (Cubit 1974) do appear to constitute functional defense against a grazing animal.

Endophyte-infected individuals probably have higher fitness than plants that lack endophytes. Infected grass clones have higher survival, growth, and reproduction under grazed conditions (Clay 1990), and older pastures typically have higher frequencies of endophyte infection than do young pastures or commercial cultivars (Lewis and Clements 1986; Latch et al. 1987). However, endophyte-infected grasses may also perform better when not grazed (Clay 1987b), making it less clear whether increased fitness is related to defense.

Standard concepts of fitness are especially hard to apply to the endophyte case, both because of difficulties in relating fitness to asexual reproduction, and because of a lack of data on how plant genetics influence endophyte infections. Many grass endophytes cause host sterility; however, vegetative vigor and clonal expansion may be stimulated (Coughenour 1985; Clay 1990). The issue is even more cloudy for endophytes that do not sterilize their hosts. Outside the Poaceae, most endophyte infections are transmitted horizontally, i.e., by spores from host to host (Carroll 1988). Without evidence of genetic differences in plant susceptibility to endophyte infection, it is speculative and perhaps premature to discuss plant-endophyte interactions as coevolved defensive mutualisms. In grasses, many endophytes are transmitted vertically, in seed. *Acremonium,* an "imperfect" fungus, produces hyphae that infect developing seeds without impairing seed function. The fungus is dispersed with the seed, and has no other route to move from plant to plant. Thus, although there is no evidence for genetic differences in plant susceptibility, the fungus itself is analogous to a maternally inherited, extrachromosomal trait (Clay 1990). Examination of these issues from the standpoint of plant evolutionary biology is a promising area for further studies.

Evolution and Coevolution in Plant-Grazer Interactions

"Which came first: the chicken or the egg?" This dilemma is inherent in studies of the evolutionary biology of plant-herbivore interactions. Do grazers select for plant defense mechanisms, or do the characteristics of plants, including defenses, select for the herbivore behavior defined here as grazing? The answer, of course, can be both. In a coevolutionary process, selection works in both directions. Pairwise (stepwise) coevolution, in which one species applies selection pressure upon a second, and subsequently the second species applies reciprocal selection on the first, is limited in plant-grazer interactions, because a single grazer feeds on many plant species and a single plant is grazed by many animals. Coevolution in such systems has been termed "diffuse coevolution" (Janzen 1980b; Fox 1981). In most studies which have invoked the concept of diffuse coevolution, there appears to be an assumption that the coevolving entities are groups of populations (e.g., grasses and ungulates), or at least that if there are any species-specific relationships, they cannot readily be observed because of the multitude of other such relationships coevolving simultaneously.

Gould (1988), who prefers the term "multispecies coevolution," argues that pairwise adaptations may still be discernible and important in a multispecies system, depending on factors such as the extent of polyphagy, the intensity and additivity of selection pressures, and the existence of genetic and ecological trade-offs. As Gould points out, pertinent data are minimal; moreover, data that are available are mostly entomological. Much less is known about the highly polyphagous grazers, and it seems that grazing would commonly result in coevolution that is diffuse, in the sense of non–species-specific (though not necessarily confused or vague; see Gould 1988). However, within this matrix of generalized coevolution of the plant and grazer communities, specific evolutionary and coevolutionary relationships can still arise and be important. These relationships will be structured by aspects of the plant-grazer interaction such as potential selective asymmetries (table 10.1, item 4).

As previously stated, intraspecific plant variability provides an avenue for direct study of evolutionary and coevolutionary processes. Especially for coevolution, there has historically been a tendency to focus on spectacular coadaptations that are presumably the end products of evolution over very long time scales (Ehrlich and Raven 1964). The within-population selection that leads to such relationships is often the subject of hypotheses and models, but relatively seldom the subject of direct empirical studies of evolution in action. The sections that follow attempt to generalize from the existing studies of grazing in relation to intraspecific plant variation, and to make predictions and suggestions regarding future research directions.

The Modus Operandi of Antigrazer Defenses

Effective plant defense against grazing must go beyond antibiosis, because of the fundamental nature of grazing. Parasitic herbivores are relatively sedentary, spending much of their lives, or even several generations, on a single plant. In such cases, reduction of herbivore vigor, feeding rate, or population growth translates into reduced consumption of plant material. The same properties of antibiosis are less useful against grazers, because a grazer interacts with a particular plant only briefly, but can inflict considerable damage during the interaction. More effective defense against grazers involves deterrence, i.e., curtailing grazing before significant quantities of plant material are consumed. A similar point was made by Thompson (1982), who pointed out that interactions of short duration favor plant defenses that reduce the amount of plant material consumed and thus shift the outcome of the interaction away from a lethal, predation-type pattern, toward nonlethal grazing.

Several examples of deterrence of grazing are available. Slugs consume small quantities of cyanogenic *Trifolium repens* leaves, and then quickly cease eating (Dirzo and Harper 1982a, 1982b). Grasshoppers eating cyanogenic *Lotus corniculatus* take a few initial nibbles, then jolt backward, shake their heads, and move away from the plants (Compton and Jones 1985). When rabbits or sheep encountered *Urtica dioica* with many stinging trichomes, they nibbled or nuzzled the foliage, then almost immediately jumped back, shook their heads, licked their lips, and lost interest in the plant (Pollard and Briggs 1984). Thorns and spines may deny access to leaves, or limit bite sizes (Cooper and Owen-Smith 1986). Observations of feeding on willows by mountain hares have suggested deterrence by olfaction (Tahvanainen et al. 1985). On the other hand, many studies provide little information on the proximate behavioral responses of herbivores to plants, choosing instead to examine amount of leaf area removed, bioassays among chemically adulterated foods, and herbivore performance on various foodstuffs. Such approaches are very important, but additional observations of herbivore behavior are very much needed in order to gain further understanding of plant defensive function.

Grazing Choices as Selective Forces on Variable Plants

Plant properties affect grazing preferences, but the remaining question is: *why* do herbivores reject some plants; that is, what are the ultimate causes underlying the proximate response of deterrence (Rhoades 1979)? Two alternative explanations are usually advanced, both based on a premise that animals use sensory cues to avoid plants that are, ultimately, bad for them (toxic and/or of low nutrient quality). The first involves learned aversion (Westoby 1974; Zahorik and Houpt 1981; Lindroth 1988a). Researchers do not agree on how common learned aversion is, but concur that it is least likely if diets are diverse and food intake is a continuous stream rather than discrete meals. These are often properties of grazing, especially by large mammals. Alternatively, animals may evolve innate grazing preferences. Theoretical treatments of evolved grazing preferences are generally based on optimal foraging theory extended to take into account variable food quality (Westoby 1974; Freeland and Janzen 1974; Pulliam 1975; Belovsky 1981).

Both the learning and innate models have been developed almost exclusively with regard to between-species differences in grazing preference. However, in order to understand the kind of selection that might lead to the evolution of defenses, they need to be extended to food choice at the level of intraspecific plant variation. The food environment of grazers is fine-grained in terms of plant species (Price 1980; Thompson 1983); it is even more fine-grained in terms of intraspecific variants, and the differences are generally more subtle. Although experimental evidence is lacking, it seems ludicrous that either the memory or the genetic code of a grazer could accommodate information about all the individual plant morphs found in its environment, especially when this food environment is itself dynamic at both the genetic and community levels.

It is more probable that grazers do possess learned and evolved responses in which common classes of plant chemicals act as positive or negative feeding stimuli based on their proximate quantities of taste or smell. Taste and smell are difficult to describe except in human terms of reference, but an interesting speculation (Bate-Smith 1972; Rhoades 1979) is that diverse chemicals may taste similar because of Mullerian mimicry (convergence); for example, all alkaloids and cyanogenic glycosides taste bitter, and all tannins taste astringent, despite a diversity of physiological effects that each may produce (Robinson 1979; Conn 1979; Zucker 1983). Grazers typically move through their environments sampling small quantities of many different plants, rejecting some, and eating more extensively from others. Sampling behavior has been observed in herbivores as diverse as snails (Dirzo and Harper 1982a), ungulates (Freeland and

Janzen 1974), and primates (Glander 1981), and may itself represent an important adaptation to allow detection of feeding cues in a fine-grained, temporally variable environment.

How accurately do grazers optimize their food choices? Domestic livestock sometimes consume toxic plants, much to their detriment (Keeler et al. 1978); much less information is available regarding poisoning of native grazers. More interesting is the question of whether plants that are avoided are always those that are potentially harmful. Slugs and grasshoppers reared exclusively on cyanogenic *Lotus corniculatus* were not found to suffer higher mortality than those eating acyanogenic plants (Compton and Jones 1985). Despite its stinging hairs, *Urtica dioica* is a nutritious plant, low in secondary chemicals, and capable of supporting vigorous growth of the generalist invertebrates that do eat it. The only documented cases of animals being harmed by stinging hairs relate to humans and dogs and not to feeding herbivores (Burrows et al. 1985; Pollard and Briggs 1984), yet mammals are deterred by the stings.

Avoidance of nutritious and nontoxic species is explicable by, and perhaps even to be expected from, the sampling behavior of grazers and the difficulty of optimal foraging in fine-grained habitats. In a variable plant population, plants that evoke a negative feeding response in herbivores will be selected for, whether or not they actually harm the herbivore. Herbivores are, of course, evolving as well; however, their perception and behavior will change in order to use unexploited food sources only if the plants constitute such an important resource that ignoring them is significantly harmful to the herbivore. Given the potential asymmetry between the effects of a single plant on a large grazer versus the effects of a single large grazer on a plant, and the diffuse or multispecies nature of plant-grazer coevolution, there is ample scope for evolution of defenses unrelated to plant quality. This argument can also be phrased in terms of Batesian mimicry (Rhoades 1979; Pollard and Briggs 1984), in which harmless plants mimic the stimuli that evoke negative feeding responses evolved in response to harmful plants (or in the case of nettles, mimic pain produced by injury). As with any mimicry, effective deterrence will be greatest for mimics that are relatively seldom encountered by the consumer.

Population and Evolutionary Responses of Grazers

In a seminal but controversial paper, Hairston et al. (1960) postulated that the herbivore trophic level is rarely limited by its food supply. Even if resource limitation of herbivores as a trophic level is rare, plants and herbivores as individuals and species can still apply strong selective and coevolutionary forces on each other; however, if grazers are resource-limited, as has been suggested most forcefully for mammals (Sinclair 1975; Crawley 1983), these evolutionary interactions might be even stronger and more common. Microtine rodent population cycles have been related to their interactions with secondary chemicals in their food plants (Freeland 1974; Bryant and Kuropat 1980; Batzli 1983). It is worth noting that recent articles (Rhoades 1983, 1985 and references therein) have proposed that cyclic population fluctuations in a number of species may be related to inducible variation in plant defensive chemistry, which is discussed further below.

A full understanding of the evolutionary responses of grazers to plant defenses will require examination of intraspecific variation in the behavioral responses of grazers. This is yet another area that is much better documented for parasitic phytophagous insects than for grazers (see Pilson, this volume; Mitter and Futuyma 1983). Livestock are known to have individualized grazing preferences (Arnold and Hill 1972). Dirzo (1980) and Bernays et al. (1976) have shown within-population variation in the acceptability of cyanogenic and acyanogenic plants to slugs and locusts, respectively. Variant individuals appeared consistent in their behavior, but the genetic basis of the differences was not investigated. Burgess and Ennos (1987) have demonstrated between-population differences, with slugs from habitats with mostly acyanogenic clover plants showing a greater selectivity than those from sites with a high frequency of cyanogenic plants. This may represent ecotypic variation, but once again the genetic basis of this behavior remains unknown.

Grazing in Relation to Defense Theory

Localization of Defenses

A corollary of the deterrent nature of defenses against grazers is an expectation that antigrazer defenses will be localized on the external surfaces of plants or in such a way as to cause a rapid cessation of grazing. The presentation of spines and stinging hairs on the external surfaces of plants is well suited to their role as deterrents, and some chemical defenses may be deposited on the exterior of plants (Bryant, Chapin et al. 1985). There are only a few known examples of olfactory aposematicism ("advertisement") in grazing animals (Arnold et al. 1980; Tahvanainen et al. 1985; Camazine 1985), while the nibbling of small quantities of plant material before rejection is frequently reported. This may stem from insufficient observation of grazers (nibbling can be seen post facto on damaged plants). However, it may perhaps be that chemoreception through taste is more universal, and that the diffuse nature of grazing thus might not select for aposematic cues of more restricted usefulness, especially if outwardly expressed chemical properties can also be used as feeding stimulants for specialist herbivores.

Defenses are expected to be concentrated in plant structures and growth stages that are most vulnerable to herbivory, and in those that are most valuable to the plant in terms of reduced fitness if they are eaten (McKey 1974a, 1979). Chemicals that deter snowshoe hare browsing on paper birch and green alder are most concentrated in juvenile stages and in buds and catkins (Bryant,

Chapin et al. 1985). Defensive investment in *Urtica dioica* also corresponds with potential grazing damage, as shown by experiments described in the next paragraph.

In a study of ontogenetic changes in stinging hair numbers, stems were harvested on four occasions during a growing season, from plants representing five populations, clonally replicated and grown in a common garden. The earliest harvest sampled dwarf overwintering leaves on this herbaceous perennial; at the latest harvest leaves were beginning to senesce. Most plants maintained high numbers of stinging hairs throughout the growing season, until leaf senescence began in the autumn (figure 10.3). In some populations there was even an increase in mean trichome density between the first and second harvests. However, a different pattern was seen in plants from a population in which mature plants have very few trichomes (Pollard and Briggs 1982). In the early spring, immature leaves and stems were moderately clothed with stinging hairs, and there was a large decrease in number of hairs between the first and second harvests. Differences in the patterns of defensive ontogeny were substantiated by a significant ANOVA interaction of harvest × population. A possible explanation lies in the fact that the site of this population does not support a significant community of large herbivores, but does have voles and other rodents.

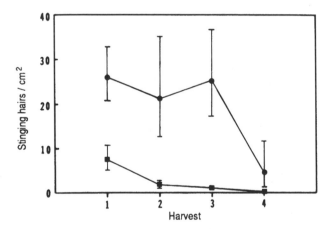

Fig. 10.3. Ontogenetic changes in stinging trichome density of *Urtica dioica* grown in a common environment. Trichome densities were measured on lower surfaces of five leaves from each of two ramets per genotype. Harvests 1 through 4 took place in early June, mid-July, late August, and mid-September, respectively. Bars represent 95% confidence limits of mean.

● Geometric mean of four individuals, each from a different population, of the common "weedy" variant of *U. dioica*.

■ Geometric mean of four individuals from the Wicken Fen "stingless" variant of *U. dioica* (Pollard and Briggs 1982).

Overwintering meristems may thus be protected at the stage when they are most vulnerable. This explanation is also consistent with the observation that even on mature plants of this "stingless" variant, the bottom few centimeters of the stem may be densely armed.

Induction of Defense

Until now, this chapter has concentrated on preformed constitutive defenses against grazing. Recently, much attention has been focused on defenses that are mobilized in response to herbivore attack, so-called inducible defenses (reviewed in Rhoades 1983, 1985; Havel 1986). Most of the evidence for inducible defense relates to insects and pathogens, but a few studies have implicated grazers. Following mechanical clipping, grasses from the Serengeti increase their silica content (McNaughton and Tarrants 1983); *Carex aquatilis* Wahl., a food plant of lemmings, increases its phenolic content (Rhoades 1983); *Urtica dioica* increases its stinging trichome density (Pullin and Gilbert 1989); and *Fucus distichus* increases its phlorotannin levels (Van Alstyne 1988). Defoliation by insects induces cyanogenesis in *Cynodon plectostachyus* at levels that may be toxic to cattle (Georgiadis and McNaughton 1988). More studies are needed, however, investigating induction of defense through before-and-after comparisons of actual herbivore damage.

In one of the few attempts to generalize the conditions that favor constitutive versus inducible defenses, Harvell (1986) has pointed out that inducibility will be favored when initial encounters with consumers are nonlethal, spatially and temporally coarse-grained, and reliably predictive of future encounters. In addition, the benefits of such defense must exceed the costs, including the loss of tissue in the initial, undefended encounter; such costs may be a function of growth rates and nutrient availability for plants (Bryant, Chapin et al. 1985). Plant-grazer interactions are often transient, with one grazing encounter not necessarily presaging further attack. It is therefore not surprising that inducible defenses are much better documented for sedentary or specialized herbivores. A potentially testable prediction that arises from these considerations is that inducible defenses will be most common against grazers that are gregarious, territorial, or low in the mobility spectrum, and least common against those that are more evenly dispersed and highly mobile.

Grazing and General Defense Theory

If citation in undergraduate ecology textbooks is used as a criterion, theories of plant defense based on plant apparency (Feeny 1976; Rhoades and Cates 1976; Rhoades 1979) have attained the status of paradigms. Briefly, plants that are apparent, predictable, and available are expected to have defenses that are quantitative (dose-dependent), digestibility-reducing, difficult to circumvent, and expensive, and are expected to be eaten mostly by generalist herbivores. Conversely, plants that are unapparent, unpredictable, or unavailable are expected to have toxic, qualitative, cheap defenses and to be eaten mostly by specialists coevolved to avoid or tolerate the defenses.

The theories are not unchallenged. Some authors have questioned whether chemicals like tannins, commonly assumed to be quantitative defenses, do in fact act in this manner (Bernays 1981; Zucker 1983). Others have challenged the evolutionary mechanisms proposed in the earlier theories. Fox (1981) has emphasized the importance of animal community diversity and dynamics. Perhaps a more fundamental challenge (Coley et al. 1985; Bryant, Chapin et al. 1985) asserts that apparency is less important than the availability of resources to the plant. Slow-growing species of nutrient-poor environments, especially evergreens, are predicted to use large quantities of immobile carbon-based defensive chemicals, while fast-growing plants of nutrient-rich environments are predicted to be under less selection for defense, and to use nitrogenous toxins with high turnover rates. This argument relies less on coevolution and more on characteristics of the physical environment in determining the range of defensive strategies available to a given plant species.

Grazing, especially by large mammals, is difficult to reconcile with apparency theory. It is hard to define what constitutes apparency to very large, very mobile generalists that employ sampling behavior in foraging. The theory predicts that defenses against generalists will be quantitative digestibility-reducers; however, grazers appear equally likely to be deterred by qualitative toxins (see review above), as well as by some deterrents that truly neither reduce digestibility nor cause toxicity. Resource availability theory, by emphasizing variation in the physical environment, goes farther toward explaining differences in defensive strategy by plants experiencing similar grazing or browsing pressures in different habitats.

Apparency theory also attempts to predict specialization and generalism of herbivores entirely on the basis of plant characteristics. In other words, in the "chicken and egg" dilemma posed above, it assumes that the plants come first and everything else follows through adaptive processes. The only independent variables influencing plant-herbivore coevolution in the scheme of Rhoades (1979) are the effects of the physical environment on plants and plant-plant interactions. The effects of predators, parasites, and pathogens on herbivores are recognized but explicitly ignored, and the constraints of the physical environment on animals are not mentioned. Furthermore, there appears to be an implicit assumption that all currently associated plants and herbivores must have coevolved with each other, which ignores biogeographic and phylogenetic history.

The viewpoint just described is largely based on an "adaptationist program," which has been criticized elsewhere (Gould and Lewontin 1979). Furthermore, while it is useful to consider independent ecological variables, it must be remembered that the evolution of a species (or the coevolution of a group of species) is constrained by its genetic makeup as determined through its phylogenetic past. A cow cannot readily evolve into a caterpillar, nor can it feed in a substantially similar manner, no matter what the advantages of doing so. Natural selection works on the available genetic variability. For plant defense against grazers, variation in the proximate characteristic of deterrence may be of equal or greater importance than the ultimate characteristics of antibiosis, including digestibility and toxicity. For our understanding of plant-herbivore interactions in general, further studies of genetic variability and its effects on fitness in both plant and herbivore populations are sorely needed, and are an exciting aspect of the papers in this volume.

Summary Statements

1. Interactions between plants and grazing herbivores are fundamentally different from interactions with more specialized insects. Although neither is necessarily lethal to the plant, the size, mobility, and generalism of grazers contrast strongly with more intimate relationships many insects have with their food plants. The distinction is basically one of grazing versus parasitism.

2. Grazers and plants probably have coevolved: they apply selection pressures to each other, and each has evolved in response to the other. However, this multispecies coevolution is "diffuse" from the standpoints of both plant and grazer; strong pairwise coadaptations between plants and grazers are unlikely.

3. There is no evidence that grazing results in increased fitness, in the Darwinian sense, for the individual plant that is eaten. Therefore, hypotheses of coevolved mutualism between plants and grazers are unsubstantiated.

4. Studies of intraspecific variation in plant defense mechanisms are necessary. They provide stronger evidence for defense as an adaptation than simple demonstration of antibiosis or functional defense, and they potentially allow examination of plant-herbivore coevolution as a dynamic, ongoing process.

5. There is also a need for studies of the behavioral and evolutionary responses of herbivores to intraspecific plant variation. Only when we understand why, in proximate and ultimate terms, a grazer eats more of one morph than another will we understand how defenses evolve.

6. The generalism and mobility of grazers, the diffuse nature of plant-grazer coevolution, and the asymmetry in the effects of grazers and plants on each other combine to predict that plants may be able to repel grazers through proximate deterrent properties that are not necessarily correlated with actual nutritional quality.

11

Community Structure and Species Interactions of Phytophagous Insects on Resistant and Susceptible Host Plants   *Robert S. Fritz*

Phytophagous insects that coexist on a plant species with their natural enemies comprise a natural and convenient community, which Root (1973) called a component community. The processes that structure these communities and their spatial and temporal predictability are central issues in insect community ecology (Strong et al. 1984). Lawton and Strong (1982) state that "the primary null hypothesis for community ecology is that species coexist independently, without effective interaction." They argue that horizontal (within trophic level) interactions, such as competition, are generally unimportant, but that vertical (between trophic levels) interactions are the principal ones that structure insect communities (but see Karban 1986 and Faeth 1987 for alternative views). Among the vertical interactions, natural enemies and host-plant factors (seasonal and environmental variation in defenses, phenology, leaf morphology, and abscission) have received the most attention. Absent from theory of insect community structure until recently (Fritz et al. 1987b; Maddox and Root 1987; Fritz and Price 1988) has been a consideration of the role of genetic variation in resistance in explaining community variation among conspecific host plants. Genetic variation among conspecific host plants could be one of the principal vertical interactions structuring phytophagous insect communities.

There is ample precedent for the viewpoint that variation in plant resistance is important in structuring communities of phytophagous insects. Interspecific variation in host-plant chemistry is correlated with community patterns of insect herbivores (Berenbaum 1981a; Rowell-Rahier 1984). Specialist insects tend to feed on plant species with more specialized secondary chemistry (e.g., angular furanocoumarins in umbellifers [Berenbaum 1981], and phenolglycosides in willows [*Salix*] [Rowell-Rahier 1984]). Plants that have evolved in rich and poor habitats may have different levels of plant defenses, which may in turn influence species number, herbivore specialization, and total herbivory (Coley et al. 1985).

Attention is beginning to focus on the role of intraspecific variation in plant resistance in structuring insect communities for several reasons. Phytophagous insects can vary greatly in population sizes among conspecific host plants in the field (White 1971; Clark and Dallwitz 1974; Southwood and Reader 1976; Edmunds and Alstad 1978; Journet 1980; Faeth et al. 1981; Coley 1983; Wainhouse and Howell 1983; Whitham 1983; Lowman 1985; Wool and Manheim 1986; Fritz et al. 1987b; Shelly et al. 1987; Simberloff and Stiling 1987; Unruh and Luck 1987; Crawley and Akhteruzzaman 1988; Anderson et al. 1989). Positive year-to-year correlations of herbivore abundances on the same plants suggest that variation in insect populations can be consistent among plants over time (Southwood and Reader 1976; Edmunds and Alstad 1978; Fritz et al. 1987b; Unruh and Luck 1987). Determining the causes of such striking patterns of herbivore population variation is a major issue in plant-herbivore ecology. While there are many possible explanations for these patterns, genetic variation in resistance is an important one.

Genetic variation in resistance in natural plant populations to herbivores is well documented. Genetic variation in herbivore abundance, population growth, fitness, and damage caused to host plants among clones or half-sib families has been documented (Hare and Futuyma 1978; Hare 1980; Moran 1981; Kinsman 1982; Wainhouse and Howell 1983; Whitham 1983; Service 1984b; Berenbaum et al. 1986; Fritz et al. 1986; Maddox and Cappuccino 1986; Karban 1987a; Maddox and Root 1987; McCrea and Abrahamson 1987; Simms and Rausher 1987, 1989; Fritz and Price 1988; Fritz and Nobel 1989; Marquis 1990a). Genetic variation in plant resistance is often correlated with phenotypic variation in population sizes of insects in the field (Karban 1987a; McCrea and Abrahamson 1987; Fritz and Nobel 1989; Whitham 1986, personal communication), indicating that genetic factors can be important in phenotypic variation in herbivory among plants.

Single-factor studies may be of limited value in explaining community patterns that must certainly be caused by multiple factors that act and interact simultaneously. An important question is whether structure of communities can be understood by analyzing component parts separately (Fowler and Rausher 1985). There are several possibilities. (1) Single dominant forces (e.g., competition or natural enemies) structure communities. (2) Several factors may act additively to structure communities (Fowler and Rausher 1985). (3) If several factors are important, species may respond individualistically to different factors, so that community patterns cannot be understood by studying factors in isolation (Lawton and Strong 1981; Strong et al. 1984; Karban 1989a). Predation, interspecific competition, and plant resistance were independently important factors for different herbivores that coexisted on *Erigeron glaucus* (Karban

1989a), but community patterns as a whole could not be understood by considering only a single factor. (4) Several factors may have higher order interactions that determine herbivore abundance and structure communities (Roughgarden and Diamond 1986). Variation in plant resistance may interact with natural enemy impact and/or competition (Fritz et al. 1986; Karban 1987a, 1989a) to produce complex interspecific interactions that contribute to community patterns.

Why is analysis of insect communities among genetically variable host plants important? First, the component community (sensu Root 1973) is a basic unit of community organization for phytophagous insects, but little is known about the role of intraspecific host-plant variation in structuring insect communities at this scale. Herbivore abundances and cooccurrence vary among leaves, branches, and positions within leaves as well as among plants that vary in quality owing to environmental causes. Much attention has been given to other plant and patch attributes in community structure of component communities (Kareiva 1983, 1986; Faeth 1987), but genetic variation in resistance has received very little attention. Genetic variation in resistance is probably ubiquitous in natural plant populations and may cause predictable patterns of insect community structure. We need to know: (1) what general patterns in insect communities are caused by genetic variation among plants, (2) if plant resistance variation can generate predictable patterns of insect species associations, and (3) the importance of plant variation, relative to other factors that structure component communities.

Species interactions, which are frequently considered to be of prime importance in structuring communities, are influenced by resistance variation of host plants (Bergman and Tingey 1979; Price et al. 1980; Hare, this volume). Thus, plant variation may affect the importance of species interactions in structuring herbivore communities. Alternatively, understanding the dynamics and outcome of species interactions may require knowing how plant variation affects densities and species interactions (e.g., competitive ability) Thompson 1988d).

Finally, current theory of insect-plant coevolution rests on the assumption of heritable genetic variation in plant resistance to herbivores. Coevolution is often viewed as a one-on-one reciprocal interaction between an herbivore and its host plant, but multispecies coevolution may be a more general model of coevolution (Janzen 1980b; Futuyma and Slatkin 1983; Gould 1988). Plants support many herbivore species that together may select for evolution of resistance (Gould 1988; Marquis, this volume). To understand evolution of plant resistance in multispecies communities the impact of each herbivore on plant fitness, heritability of plant resistance to each herbivore, and the additive genetic correlations between plant resistances to different herbivores must be known. Therefore, for the model of multispecies coevolution the relative abundances of insect species is a central component determining the trajectory of evolution of resistance.

This chapter considers the role of genetic variation in resistance of host plants *within populations* on the community structure of herbivorous insects, explores some expected patterns of herbivore association on genetically variable plants, discusses implications for understanding multispecies coevolution, and discusses the potential consequences of genetic variation among plants and variation in community structure on herbivore interactions with competitors and natural enemies. I will focus on genetic variation in resistance because several recent reviews have thoroughly discussed other factors important to community structure of phytophagous insects (Price 1983b; Jefferies and Lawton 1984; Kareiva 1983, 1986; Faeth 1987). The main questions addressed in this chapter are:

1. What are the patterns of variation in herbivore communities on genetically variable plants?
2. What are the genetic correlations in plant resistances to different herbivores?
3. Can species associations be predicted from individual herbivore response to resistance variation?
4. How important is genetic variation in resistance relative to other sources of variation in community structure?
5. What are the potential consequences of host-plant resistance on interactions of phytophagous insects?

Causes of Variation in Plant Quality

Plants do not live in homogeneous environments and are not intrinsically equal in quality for insect herbivores. Phenotypic variation in host-plant quality results from intrinsic factors (genotype, ontogeny), extrinsic factors (collectively referred to as environmental variation), and their interaction. Agricultural studies of plant resistance document extensive genetic variation in plant defense to herbivores within cultivated species and among their wild progenitors (Gallun and Khush 1980; Harris and Frederiksen 1984). Chemical and physical defenses are important resistance mechanisms that reduce herbivore damage to plants (Maxwell and Jennings 1980; Boethel and Eikenbary 1986). For natural systems, with a few exceptions (Hare and Futuyma 1978; Berenbaum et al. 1986), the inheritance of specific resistance traits is unknown. Berenbaum et al. (1986) and Berenbaum and Zangerl (1988) showed that amounts and proportions of several furanocoumarins are heritable in *Pastinaca sativa,* and that levels of some furanocoumarins are negatively genetically correlated, indicating that evolution of more resistant plants may be constrained. Genetic variation in resistance is usually demonstrated by a bioassay, where herbivore oviposition, density, growth rate, reproductive rate, or survival is measured among half sibships or clones (Kennedy and Barbour, this volume; Simms and Rausher, this volume) rather than by quantifying the expression of traits responsible for resistance.

Ontogeny generates temporal and spatial heterogeneity in plant resistance to herbivores (Zagory and Libby 1985; Kearsley and Whitham 1989). Plants may increase or decrease in resistance to different herbivores as they grow and age (Kearsley and Whitham 1989). Resistance of narrowleaf cottonwood (*Populus angustifolia*) to leaf beetles (*Chrysomela confluens*) increased with ramet age, but resistance to the galling aphid *Pemphigus betae* decreased on older ramets. Plant parts of different ages also differ in susceptibility to herbivores and pathogens (e.g., leaves: Raupp and Denno 1983; Whitham 1983; Coleman 1986; Coleman et al. 1987; Coley 1980; Gall 1987; Quiring and McNeil 1987; Larsson and Ohmart 1988). Ontogenetic changes in resistance may be heritable and subject to selection by herbivores. For example, different plant genotypes could differ in their timing and extent of change in resistance as they grow and age. Demonstrating ontogenetic variation in resistance does not mean that genetic differences among plants are not contributing to these differences or that genetic variation in resistance is not also important.

Variation in resistance that is related to plant age may be due to induced resistance that persists in mature plants (Karban 1987b). Karban reviews a number of studies that show age-related resistance variation in plants but hypothesizes that this could be due to the acquisition of induced resistance.

Numerous environmental factors (nutrients, light, water, temperature, and previous herbivory) affect plant quality and consequently herbivore damage (Tingey and Singh 1979; Bach 1984; Price and Clancy 1986b; Louda et al. 1987; Mattson and Haack 1987b; Bultman and Faeth 1988; Strauss 1987; Preszler and Price 1988). The proportion of phenotypic variation in resistance explained by variation in environmental factors rather than genetic differences is unknown for virtually all plant-herbivore systems. This is an essential issue if we are to evaluate the importance of factors influencing herbivore population dynamics and community structure.

Interaction between genotype and environmental causes of plant variation have not been well studied in most plant-herbivore systems (Maddox and Cappuccino 1986). Often ecologists study environmental factors alone, but studies of both genetic and environmental causes of variation in herbivory are important in understanding how the importance of genetic factors changes among different habitats and in heterogeneous environments. Genotype-environment interaction measures phenotypic plasticity of resistance among genotypes. Plant plasticity in herbivore resistance is measured as the proportion of resistance variance that is due to genotype-by-environment interaction. Phenotypic plasticity can be heritable (Scheiner and Lyman 1989), and for plants that live in variable environments, where they encounter different levels of herbivory, plasticity in resistance could be adaptive, especially if costs of resistance differ between habitats (Simms, this volume).

A variety of other factors, in addition to those mentioned above, contribute to herbivore community patterns. Patchy distributions of plants (resource concentration, patch sizes, isolation), vegetational diversity, and successional stage influence herbivore composition and relative abundances (reviewed recently by Kareiva 1983, 1986; Brown 1985; Brown and Hyman 1986). Therefore, the role of genetic resistance variation in determining herbivore population dynamics and community structure needs to be understood along with these other mechanisms. In this chapter I will focus on genetic variation in plant resistance; it is understood, however, that environmental variation, ontogenetic variation, and a range of other factors are important for herbivore communities and may interact with genetic variation to explain patterns in herbivore communities. My focus should not be misconstrued to suggest that genetic variation in resistance is the primary or only plant factor structuring herbivore communities, but rather that it can be important and has received little attention.

Herbivore Community Variation among Host Plants

Theory and Predictions

Theory of insect community structure has virtually ignored genetic variation among conspecific host plants, while exploring many other potential causes of community structure (e.g., Faeth 1987). Strong et al. (1984) discussed some aspects of genetic variation in plants, but at present there are few predictions of how herbivore communities could be structured on genetically variable host plants. Price (1983b) proposed the "genetic-heterogeneity hypothesis," which stated that "as the genetic diversity in a plant population increases the diversity of insect herbivores increases." Price hypothesized that herbivores use only a subset of plant genotypes and that herbivore species are added to communities as the diversity of plant genotypes increases, thus proposing that insect response to plant variation is largely a qualitative phenomenon. This hypothesis describes how herbivore communities could vary among genetically diverse or uniform plant populations, but it fails to explain the considerable variation in herbivore abundances among host plants within populations (see below). For most plants the important question is whether populations show quantitative variation in plant resistance and whether this results in variation in the relative abundance of herbivore species. Without a conceptual model of how genetic variation among plants can structure phytophagous insect communities it will be difficult to incorporate genetic variation in plants into the framework of herbivore community theory.

Fritz et al. (1987b) proposed a model that described possible patterns of co-occurrence of herbivores among conspecific host plants that vary in resistance. That model sought to explain the patterns of species associations contained in the phenotypic variance-covariance matrix of herbivore abundances among plants. The matrix contains: (1) the variance in species densities among individual host plants, and (2) the covariance of pairs of species among host plants in a population. The extent to which the patterns contained in this matrix can be explained by genetic differences among plants will indicate the importance of genetic factors in structuring herbivore communities. Here I present a graphical form of the model (Host Plant Resistance Model) of Fritz et al. (1987b) and discuss predicted patterns of species co-occurrence.

Consider two herbivore species that coexist on a host plant; the density of each species of herbivore is assumed to be inversely related to variation in plant resistance traits (fig. 11.1). Resistance is specific to the herbivore and to the range of plant genotypes and environments being considered, and is not a fixed trait of the plant to all herbivores. Variation in herbivore density among plants is established through differential colonization, survival, or population growth on host-plant genotypes. The relative abundances of the herbivores among plants will be determined by the similarity of their responses to plant resistance traits.

In the simplest case, let two herbivore species respond to variation in a single resistance trait (e.g., leaf toughness or a particular component of leaf chemistry). The response of each herbivore to the resistance trait may differ so that one herbivore may be more strongly affected by the resistance trait than the other. If the two species respond similarly to the resistance trait (or to different traits that covary positively in the plant) they will be positively correlated in their density (fig. 11.1A). Susceptible plants will have higher populations of both herbivores, while resistant plants will have lower populations of both herbivores. The strength of the correlation among the herbivore densities will depend on the similarity of their responses to plant variation. When resistances are positively genetically correlated, species should show little variation in their relative abundances (fig. 11.1C).

Two herbivores are predicted to be negatively correlated in density if they respond inversely to the same resistance trait (fig. 11.1B) (or to different traits that are negatively genetically correlated; Hare and Futuyma 1978; Juvik and Stevens 1982b). A chemical trait that confers resistance to one herbivore species can function as an attractant to another herbivore (DaCosta and Jones 1971). The stronger the negative correlation in resistance to the two species, the greater the differences in their relative abundance (fig. 11.1D). Densities of two species are predicted to be uncorrelated among host plants when plant resistances are uncorrelated, even though each herbivore may be strongly affected by plant resistance. Thus, variation in relative abundance of herbivores among host plants is predicted to be inversely related to the similarity of herbivore responses to plant resistance.

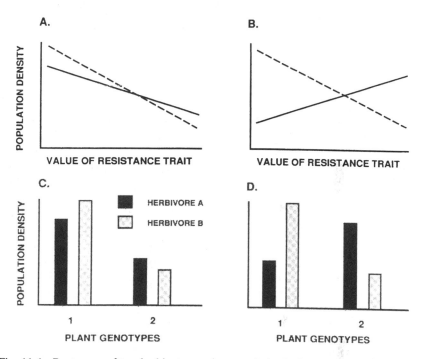

Fig. 11.1. Responses of two herbivore species to variation in the value of a single plant-resistance trait. (A) Species responses to plant resistance are positively genetically correlated so that population densities of species 1 (solid line) and species 2 (dashed line) both decrease as the value of the resistance trait increases. (B) Species are negatively correlated in their responses to plant resistance so that population density of species 1 increases with increasing plant resistance while population density of species 2 decreases. Positive correlation in density of the two species in (A) may result in different relative abundances of the two species among genotypes (C) if they differ in response to the resistance trait. In (B) since the two species are negatively correlated, large differences in relative abundance (D) among genotypes are predicted. Similar patterns can result with two herbivores responding to two different resistance traits that are positively or negatively genetically correlated.

Relative abundances of herbivores should commonly vary among conspecific host plants that vary in resistance. If phenotypes of plants do not vary because they have similar genotypes or genetic differences among plants are not expressed in a certain environment, then differences in the relative abundances of two herbivores may be small and relatively unimportant. Communities dominated by one herbivore would be unlikely to exhibit large variation in relative abundance among plants.

Patterns of herbivore cooccurrence that can result from plant variation have often been attributed to other causes. Negative correlations in herbivore densities among plants due to negative correlation of resistances (fig. 11.1B) could be interpreted as evidence of interspecific competition. This I call "plant-mediated apparent competition," adapting Holt's (1977) definition (except that changes in herbivore density are not expected in response to increases or decreases in density of the other species). Furthermore, lower fitness of one herbivore species in the presence of high densities of a second herbivore could be interpreted as a competitive effect when it could also result from a negative correlation between herbivore performance and herbivore density among plant genotypes. Positive correlation of two herbivores among variable plants could be interpreted as evidence for mutualism or facilitation, but again could be due to plant genotype effects on herbivore abundance or fitness. Plant resistance must be considered a likely cause of herbivore association patterns on plants.

The hypothesis that plant resistance traits determine patterns of species associations and relative abundances can be tested by independently determining for two herbivores how each responds to variation in host-plant resistance. Their relative abundances when cooccurring on plants should be predictable from their independent responses to plant variation when occurring alone on the same genotypes. Patterns of relative abundance when occurring together substantially similar to those predicted would support this hypothesis. Deviations from expected patterns could suggest important interactions.

Evidence and Patterns

Significant variation in relative abundances of herbivorous insects among plant genotypes must necessarily result from herbivores that vary in abundance among host-plant genotypes and whose response to host-plant variation differs from that of other herbivores. In this section I review evidence of variation in community structure among genetically variable hosts and present results from my work on willows and their herbivores in Arizona and New York.

Table 11.1 lists the plants and herbivores from studies that provide direct or indirect evidence of variation in aspects of herbivore community structure among genotypes of host plants. The types of variation in communities considered are variation in: (1) relative abundances of herbivores, (2) guild structure, and (3) the preference of specialist or generalist herbivores. Evidence of genetic differences among plants is derived from potted or common garden experiments on cloned plants, sibships, or comparison of cultivars.

Relative Abundance of Herbivores

Variation in the relative abundance of one or more herbivores among genotypes is suggested in a number of studies. Marquis (1990) found significant variation among clones of *Piper arieianum* (Piperaceae) in amount of damage caused by six herbivore species or feeding groups over a two-year period. Clones with the greatest damage by one herbivore frequently had lower amounts of damage caused by other herbivores, suggesting differences in the relative abundance of the herbivores. Differences in relative abundance of herbivores on *Ipomoea purpurea* and *I. batatus* are suggested by genetic variation in damage by several herbivores and by reversals in the importance of certain herbivores among genotypes (Simms and Rausher 1989; Cuthbert and Davis 1970; Jones et al. 1979). Karban (1987a, 1989a) showed significant variation in density of folivorous thrips (*Apterothrips secticornis*) (Thysanoptera: Thripidae) on three *Erigeron glaucus* (Asteraceae) clones in the field and in a common garden. Although densities of two other herbivores did not vary significantly among clones, the relative abundances of these species would differ among the clones because of variation in thrips density. Hare and Futuyma (1978) and Hare (1980) documented inter- and intrapopulation variation in resistance of cockleburs (*Xanthium strumarium*) to two seed predators and different responses of the seed predators to resistance traits, which also demonstrated variation in the relative abundances of these herbivores among plants. Fritz and Price (1988) demonstrated that densities and relative abundances of four sawfly species varied significantly among willow clones.

Guild Structure

If plant resistance affects members of feeding guilds similarly but affects species in other guilds differently (i.e., positive genetic correlations of resistance among guild members and negative genetic correlations among species in different guilds), then abundances of insects in some feeding guilds may be greater on some plant genotypes than on others. This is an appealing hypothesis, since it can be predicted that herbivores feeding on the same plant parts in a similar way would be affected similarly by resistance mechanisms. Maddox and Root (1990) found only limited support for this hypothesis. Patterns of genetic correlations among herbivores on goldenrod did not generally conform to functional feeding groups (guilds). Often the strongest positive genetic correlations were between species in different feeding guilds. This suggested that resistance traits affect species within guilds idiosyncratically.

Other research supports a portion of this hypothesis, that there should be positive correlations among species that are normally grouped into guilds. Jones et al. (1979) and Cuthbert and Davis (1970) found significant positive genetic correlations between resistances of sweet potato cultivars to several species of root-boring insects. Resistance to seed predation by two insects on different populations of cocklebur was most affected by burr size, which affected both herbivores similarly (Hare and Futuyma 1978). Other plant traits that conferred susceptibility to each insect were independent of burr size, and some traits that were positively correlated with damage by one species were negatively correlated with damage by the other seed predator, suggesting negative genetic correlations. Hypotheses that need to be tested are: (1) that plant resist-

TABLE 11.1

Summary of studies that show variation in structure of herbivore communities among plants in the field (phenotypic variation) and among plants growing in common environments (genotypic variation)

Plant	Insect	Phenotypic variation in community?	Plant genotype significant?	Type of variation	References
Salix lasiolepis	*Euura lasiolepis* *Euura* sp. *Phyllocolpa* sp. *Pontania* sp.	yes	yes	relative abundance	Fritz et al. 1986 Fritz et al. 1987b Fritz and Price 1988
Salix sericea	*Pontania* sp. *Phyllocolpa* spp. *Phyllonorycter salicifoliella* *Rabdophaga* sp.	yes	yes	guild	Fritz unpublished data
Erigeron glaucus	*Apterothrips secticornis* *Pilaenus spumarius* *Platypitilia williamsii*	yes	yes	relative abundance	Karban 1987a Karban 1989a Maddox and Root 1987; 1990
Solidago altissima	*Lygus lineolaris* *Corythuca marmorata* *Philaenus spumarius* *Exema canadensis* *Microrhopala vittata* *Ophraella conferta* *Trirhabda* spp. *Epiblema scudderiana* *Epiblema* spp. *Asteromyia carbonifera* *Rhopalomyia solidaginis* *Eurosta solidaginis* *Uroleucon caligatum* *Uroleucon* *nigrotuberculatum* *Ophiomyza* sp. *Phytomyza* sp.	yes	yes	guild	

Plant	Herbivores	Response measured			Reference
Populus angustifolia	Pemphigus betae, wood boring beetles	relative abundance	yes		Whitham (unpublished data)
Oenothera biennis	Plagiognathus cuneatus, Plagiognathus politus, Adelphocoris rapidus, Adelphocorus lineolatus, Lygus lineolaris, Macrosiphum gaurae, Schinia florida, Popillia japonica	guild	yes		Kinsman 1982
Piper arieianum	Ambates sp., Anacrucis piriforana, Anacrucis stapiana, Atta cephalotes, Eois sp., Homeomastox robertsi, Peridinetus spp., Quadros evans, Dipteran leaf miners, Phasmidae, Tettigoniidae, Tortricidae	relative damage	yes	?	Marquis 1990
Ipomoea batatus	Diabrotica balteata, Diabrotica undecimpunctata, Systena elongata, Systena frontalis, Plectris aliena, Euphoria sepulcharalis, Conoderus falli, Chaetocnema confinis	relative damage	yes*	?	Cuthbert and Davis 1970, Jones et al. 1979
Xanthium strumarium	Phaneta imbridana, Euaresta aequalis	relative abundance	yes†	yes	Hare and Futuyma 1978, Hare 1980

† Variation among populations.
* Variation among cultivars or breeding lines.

ance affects guild members similarly, and (2) that resistance between species in different guilds should be negatively correlated more often than the correlations of resistances between species in the same guild.

Specialists and Generalists

Specialist and generalist herbivores may be affected differently by plant resistance; the presence and proportion of specialist and generalist insects in communities could vary among plant genotypes. If plant resistance evolves in response to selection by generalist herbivores, and herbivores that adapt to plant resistance become specialized and respond positively to the resistance factor, then the negative correlations of resistances to specialists and generalists would be predicted to be widespread. Kinsman (1982) showed that two specialist insects preferred one genotype of evening primrose, *Oenothera biennis* (Onagraceae), while several generalist herbivores preferred another genotype in a common garden. The genotype resistant to specialists generally had higher fitness, except where populations of the primary generalist herbivore, the Japanese beetle (*Popillia japonica*) (Coleoptera: Scarabaeidae), were high. Beck and Schoonhoven (1980) suggest that allelochemicals that are effective repellants against generalist insects could serve as attractants to specialist insects. Cucumber beetles (*Diabrotica* spp.) and two-spotted mites (*Tetranychus urticae*) are negatively correlated in resistance to different genotypes of the cucumber (*Cucumis sativus*) (Cucurbitaceae) (DaCosta and Jones 1971). The bitter gene product (cucurbitacin Z) repels the generalist mites but attracts cucumber beetles, which are specialists, and the relative abundance of these two pests should vary between these genotypes. More studies of the responses of specialists and generalists to plant resistance would help to establish whether complementarity between specialists and generalists is a common pattern of community organization on variable plants.

Genetic Correlations of Plant Resistances to Herbivores

The patterns of herbivore community structure among plants will be determined in part by genetic correlations of plant resistances to herbivores in the community. Genetic correlations (see Simms and Rausher, this volume) between resistances measure the extent to which resistance to one herbivore species is associated with resistance to another herbivore species. Positive genetic correlations indicate that at least some resistance genes confer resistance to both herbivores (positive pleiotropy). Negative genetic correlations indicate that some genes confer resistance to one species and susceptibility to another herbivore species (negative pleiotropy). Genetic correlation may also be caused by linkage disequilibrium between loci conferring resistance (Falconer 1981). The absence of genetic correlation indicates the absence of common genetic control of resistances to two herbivore species. The sign and strength of the genetic correlation matrix of resistances of plant genotypes to herbivores provide part of the information necessary to determine whether evolution of resistance of plant populations is independent for different herbivore species, whether there are trade-offs in resistance to different species, or whether there is the potential for multispecies selection on resistance (Gould 1983, 1988).

Positive genetic correlations among resistances have been found frequently. In an extensive analysis of resistance of goldenrod sibships to 17 species of herbivores, genetic correlations were significant for 25.7% (35 of 136) species pairs, and positive correlations occurred in 27 of the 35 cases (77.1%) (Maddox and Root 1990). Whitham (1989, personal communication) found that wood boring beetles had higher densities (4.85 larvae per stem versus 2.31 larvae per stem, $P < 0.019$) on poplar trees that were susceptible to the galling aphid *Pemphigus betae,* than on resistant poplars, also suggesting a positive genetic correlation. Positive genetic correlations have also been found by Cuthbert and Davis (1970), Hare and Futuyma (1978), Jones et al. (1979), Kinsman (1982), Gould (1983), Lambert and Klein (1984), Gill et al. (1986), Fritz and Price (1988), and Fritz (1990a). Positive genetic correlations suggest that plant resistances may often affect different herbivores similarly and suggest the potential for multispecies selection for resistance (Gould 1988).

Negative genetic correlations have been demonstrated less often (von Schonborn 1966; DaCosta and Jones 1971; Hare and Futuyma 1978; Kinsman 1982; Maddox and Root 1990). In Maddox and Root's (1990) study two genera, *Ophraella* and *Trirhabda* (Coleoptera: Chrysomelidae), both specialists on *Solidago,* frequently had significant negative correlations with other goldenrod herbivores. Negative correlations of resistances between generalist and specialist insects might be expected if these herbivores have opposing behavioral responses to plant chemicals or morphological traits (Beck and Schoonhoven 1980). Rowell-Rahier and Pasteels (1986) suggest that chrysomelid beetles are attracted to plants with high salicin content for antipredator defense and that this allows them to utilize a set of plants that are protected from generalist herbivores by the phenolglycosides.

Plant resistances to herbivores may be uncorrelated. Resistance is considered to be species specific for cultivars of crop plants (Ortman and Peters 1980). This could be an artifact of analysis of crops that have been selected for specific single-gene resistance to one herbivore, where chances of pleiotropy might be less, or due to analysis of resistance of crops to only single-herbivore species. There are few studies of multiple or cross resistance to herbivores on crop plants relative to the number of resistance studies overall. In natural systems there are too few studies to determine whether resistances are generally uncorrelated. Nonsignificant genetic correlations of resistances were common in Maddox and Root's (1990) study, and Simms and Rausher (1989) found no significant genetic correlations between species or feeding groups of insect herbivores of *Ipomoea purpurea.* Fritz and Price (1988) and Schowalter and Haverty (1989) also found nonsignificant correlations between resistances of insect species. Generally, uncorrelated resistances imply separate genetic control of resistance and lack of potential for multispecies selection on the same resistance trait.

Community Structure of Herbivores on Willow

Arroyo willow (*Salix lasiolepis* Bentham [Salicaceae]) is host to four closely related species of gall-forming sawflies (Hymenoptera: Tenthredinidae). This shrubby willow grows abundantly along streams in northern Arizona. The stem-galling sawfly (*Euura lasiolepis*) forms a large gall on shoots (Price and Craig 1984). The petiole galler (*Euura* sp.) forms a gall in the petiole of leaves, which becomes greatly elongated and swollen, and the larva mines through the solid tissue of the gall. The leaf-galling sawfly (*Pontania* sp.) (Clancy et al. 1986) forms a gall along the midvein on the underside of the leaf. The gall develops gradually as the larva grows inside the round, hollow gall. The fourth species is the leaf-folding sawfly (*Phyllocolpa* sp.), which forms longitudinal folds of the leaf margin as a result of repeated stinging by the female sawfly prior to oviposition.

Silky willow (*Salix sericea* Marshall) is host to a diverse community of insect herbivores in New York state. On silky willow four guilds are prominent: shoot gallers, leaf gallers, leaf miners, and leaf folders. The beaked willow galler (*Rabdophaga rigidae*) (Diptera: Cecidomyiidae) and two other species of *Rabdophaga* form stem and shoot tip galls. *Pontania* sp. and *Phyllocolpa* spp. gall the leaves, along with a rare, unidentified species of Cecidomyiidae. The leaf miner *Phyllonorycter* sp. (Lepidoptera: Gracillariidae) is common at some sites. There are also three unidentified species of leaf folders, which have distinct folds. Leaf-chewing insects and phloem feeders are rare and were not recorded in this study.

Herbivore densities on arroyo willow varied significantly among conspecific host plants in the field at four sites over three years in Arizona (Fritz et al. 1987b). Species were highly correlated in their densities on the same clones between years when data from all sites were combined and frequently when each site was considered separately. Pairs of species generally varied independently among field plants, but all the significant correlations were positive, suggesting a trend for positive association between species. On field plants, relative abundances of the four sawfly species varied significantly among individual willow clones at each site (fig. 11.2). At different sites the dominant sawfly species differed among clones, although on many clones only small changes in the relative abundances occurred.

To test the hypothesis that field differences in herbivore densities and relative abundance among plants are due to genetic variation in resistance, I performed experiments on replicate cuttings of field plants (Fritz and Price 1988). Densities and relative abundances of sawflies differed significantly among pot-

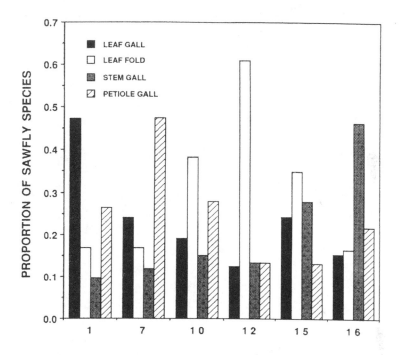

Fig. 11.2. Relative abundances of four sawfly species on ten field plants of arroyo willow at Northland Press site in northern Arizona. There are significant differences among clones in the proportions of sawfly species. (Data are from Fritz et al. 1987b).

ted clones in two years (table 11.2). These data firmly established that variation in resistance can directly cause variation in herbivore community structure among plants.

Clonal heritabilities (estimates of total genetic variation) for plant resistance ranged from 0.086 to 0.507, with the highest heritabilities being for stem-galler resistance and leaf-folder resistance. Phenotypic correlations of densities were positive and mostly significant, and genetic correlations (correlations of clone mean densities, i.e., broad-sense genetic correlations) were significant and positive between the leaf folder, stem galler, and petiole galler in 1985 (table 11.3). In 1986, correlations were generally positive but were significant only between the leaf folder and petiole galler (table 11.3). Because these experiments were performed on cloned shoots, the results indicate the presence only of total genetic variation in resistance, not the presence of additive genetic variation, which could respond to selection by herbivores. Although it seems likely that there is heritable variation in resistance of willows to these herbivores, confirmation needs to come from sibship analysis or parent-offspring

TABLE 11.2
Summary of analysis of clone effects on species density and species proportion among willow clones in experiments conducted in two years

Sawfly Species	1985		
	Density $F_{(11,77)}$	h_c^2	Proportion $F_{(11,77)}$
Leaf galler	2.00*	0.111	12.45***
Leaf folder	9.22***	0.507	9.02***
Stem galler	3.08**	0.207	2.64**
Petiole galler	1.75	0.086	1.44

Sawfly Species	1986		
	Density $F_{(5,122)}$	h_c^2	Proportion $F_{(5,122)}$
Leaf galler	3.17**	0.086	23.42***
Leaf folder	21.84***	0.476	22.48***
Stem galler	11.11***	0.305	7.01***
Petiole galler	9.86***	0.278	3.63**

Source: Fritz and Price 1988.
Note: Clonal heritability (h_c^2) estimates for species density are presented.
$* - P < 0.05.$ $** - P < 0.01.$ $*** - P < 0.001.$

TABLE 11.3
Phenotypic (upper right quadrant) and genetic correlations (correlation of clone mean density) (lower left quadrant) of resistance of arroyo willow to four species of gall-forming sawflies in northern Arizona

1985	Leaf galler	Leaf folder	Stem galler	Petiole galler
Leaf galler		0.481***	0.310**	0.572***
Leaf folder	0.406		0.408***	0.605***
Stem galler	0.441	0.896***		0.213*
Petiole galler	0.616*	0.895***	0.832***	
1986				
Leaf galler		0.466***	0.106	0.567***
Leaf folder	0.171		0.411***	0.713***
Stem galler	−0.510	0.529		0.323***
Petiole galler	−0.115	0.958***	0.697	

Source: Fritz and Price 1988.
$* - P < 0.05.$ $** - P < 0.01.$ $***P < 0.001.$

regression. Other pitfalls and caveats of analysis of cloned plant material are discussed in Fritz and Price (1988), Fritz (1990a), and Marquis (1990a).

Several host-plant traits were correlated with resistances to these herbivores. Shoot length, node number, and leaf length differ significantly among willow clones and were inversely genetically correlated (broad-sense correlations) with plant resistance. Thus, plants with long shoots are most susceptible to sawfly galling. On field plants shoot length is also correlated with herbivore density for each sawfly species (Fritz et al. 1987a). These data suggest that genetic variation in these plant traits may be expressed among field plants and affect the densities of the gall-forming sawflies. Other sawfly species have also been shown to be positively correlated with shoot length (Price et al. 1987; Price 1988) on field plants. The similarity of herbivore responses to shoot length among plants suggests a basis for the positive correlations of plant resistances to these herbivores.

Density of many herbivores of *S. sericea* in New York differed significantly among field plants over two years (table 11.3). The proportion of all insects in different guilds varied significantly among field plants (fig. 11.3). Herbivore species were usually not significantly correlated in density on field plants, results similar to those for *S. lasiolepis*. Significant variation in densities of the leaf-galler and leaf-folder sawflies, and an unidentified leaf-folding moth among replicated two-year-old potted cuttings of *S. sericea* clones was found in 1988 (table 11.4). Densities of the other species were probably too low to detect differences among clones. Three species showed both positive and negative genetic correlations (table 11.5). The leaf-folding sawfly (*Phyllocolpa* sp.) was positively correlated with the leaf miner (*Phyllonorycter salicifoliella*), but was negatively correlated with *Pontania* sp.

Proportions of the leaf-galling and shoot-galling guilds differed significantly among clones (fig. 11.3B), again demonstrating significant differences in community structure among plants that are most likely due to genetic differences among plants. Results of the studies in Arizona and New York both demonstrate a direct effect of genetic variation on community structure.

Importance of Other Factors

Undeniably, a variety of factors influence phytophagous insect communities (Strong et al. 1984; Kareiva 1986). An important question is: how important is plant resistance relative to other factors in affecting herbivore community structure? The answer to this question will require empirical study and will depend to some extent on the scale of analysis of communities. Analysis of herbivores on plants at a particular place and time may show that plant variation in resistance is quite important. Comparison of communities among sites or over time may show that abiotic factors at particular sites or that affect herbivore populations differently between years could be important. An important aspect of studying genetic and environmental effects on herbivore resistances and community structure is to discover whether different genotypes are favored in different environments. If different habitats favor genotypes with different resistances to herbivores, resulting community structures of herbivores could be quite different.

The effects of clone (11), site (2), and year (2) variation on densities of the four gall-forming sawflies were studied on *S. lasiolepis* in northern Arizona (Fritz 1990a). Clone was a significant effect for three of the four species, but

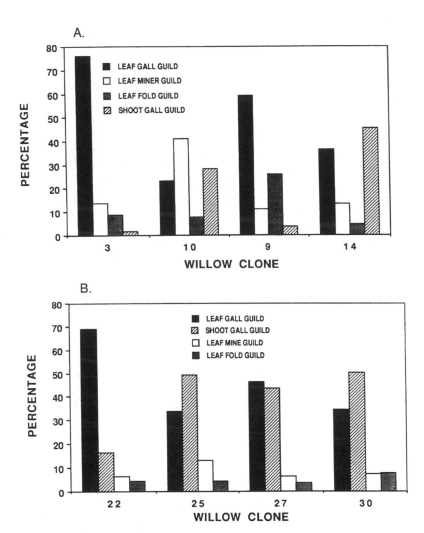

Fig. 11.3. (A) Proportions of insects in four guilds (shoot gallers, leaf gallers, leaf miners, and leaf folders) among four field clones of S. sericea in New York. Clones were chosen from among 30 that were censused to emphasize differences in guild composition. Differences in guild proportion among clones were significant ($X^2 = 359$, $P < 0.001$). (B) Proportions of insects in four guilds (summed over replicates) on four clones (genotypes) growing in pots, which were placed in a randomized array in the field. Differences in guild proportion among clones were significant ($X^2 = 89.663$, $P < 0.001$).

TABLE 11.4

Summary of variation in densities of herbivores of *Salix sericea* among field plants and among cloned plants in pots in 1987 and 1988

Species	Field 1987 $F_{(24,172)}$	Field 1988 $F_{(31,160)}$	Clones 1988 $F_{(8,103)}$
Leaf-gall guild			
Pontania sp.	2.021**	4.410***	2.234*
Phyllocolpa sp.	2.554***	5.941**	6.892***
Mite gall	4.946***	16.736***	0.890
Leaf-mine guild			
Phyllonorycter salicifoliella	5.345***	2.886***	0.636
Leaf-fold guild			
Sewn-up leaf fold	1.221	1.232	2.182*
Shoot-gall guild			
Rabdophaga sp.	3.141***	2.151**	0.734
Rabdophaga rigidae	2.151**	1.232	1.862

* $-P < 0.05$.   ** $-P < 0.01$.   *** $-P < 0.001$.

TABLE 11.5

Genetic correlations of densities of three herbivores of *S. sericea*

	Pontania	*Phyllocolpa*	*Phyllonorycter*
Pontania	—		
Phyllocolpa	− 0.648	—	—
Phyllonorycter	− 0.828**	0.699*	

Note: Correlations were calculated using clone mean densities of each species.
   * $-P < 0.05$.   ** $-P < 0.01$.

year and site considered separately were not significant. However, year and site interacted significantly to affect densities of three of the sawfly species and may have been as important as clonal variation overall. Clonal variation among willows was very important at one site, but at the other site it had little effect on herbivore densities. Longer-term studies of this sort are needed to determine how much variation in herbivore density is explained by various factors and which community properties are affected most (e.g., dominance rank, community similarity).

Fritz et al. (1987b) found that relative abundances of the willow herbivores varied among field clones within four sites over a three-year period. The dominant herbivore varied among clones, but only at one site did it change over time. At another site there was a marked shift in the relative abundance of the species owing to a nearly twofold increase in density of one species followed by a sharp decline of that species the following year. Despite the community shifts at some sites and over time, the variation in relative abundance among clones was stable and was of large magnitude in some cases. Fritz and Price (1988) found for potted clones that community similarity was often lower between years for four of six clones than among clones within years (table 11.6).

TABLE 11.6
Table of proportional similarity values for sawfly communities on MNA clones used in experiments in 1985 (upper quadrant) and 1986 (lower quadrant)

Clone	MNA 31	MNA 36	MNA 37	MNA 38	MNA 40	MNA 42
MNA 31	*.835*	.518	.536	.532	.567	.947
MNA 36	.732	*.522*	.895	.979	.951	.528
MNA 37	.661	.898	*.530*	.898	.891	.589
MNA 38	.752	.844	.855	*.669*	.965	.585
MNA 40	.772	.856	.867	.930	*.637*	.617
MNA 42	.897	.770	.684	.761	.743	*.787*

Source: Fritz and Price 1988.
Note: Proportional similarity (PS $= 1 - 0.5 \Sigma |p_{ij} - p_{hj}|$ of each clone with itself; between years is shown on the main diagonal.

Maddox and Root (1987) found significant year-to-year variation in density of 15 herbivores on genotypes in a common garden. Genotype effects were significant and accounted for 2–24% of the variation in herbivore density. Although genotype-by-year interactions were also significant for about half of the species, significant genetic correlations between years suggested that ranking of clone resistances generally remained the same. Thus, even with substantial year-to-year variation in abundances of these goldenrod herbivores, genetic effects of host plants were important for this community. Karban (this volume) reviews the relative importance of genetic and year-to-year environmental variation for herbivore population dynamics and concludes that genetic variation is often as important as or more important than year-to-year variation. Even though it is obvious that environmental variation has important effects on herbivore population dynamics and communities, it would be wrong to conclude that genetic differences between plants are therefore unimportant or cannot cause patterns of associations among herbivores in communities.

Implications for Multispecies Coevolution

Variation in herbivore community structure among conspecific host plants is central to understanding multispecies coevolution. Multispecies coevolution consists of two parts (Gould 1988): selection by two or more host-plant species on traits of an herbivore and selection by several herbivores on the resistance traits of a host-plant species. Selection imposed by one plant species on herbivore performance may change its performance on another plant species. Alternatively, evolution of plant resistance may be caused by selection by two or more herbivores. If multispecies selection for the same resistance traits is to occur, there must be positive genetic correlations between plant resistances to the herbivores. Furthermore, the contributions of each herbivore species to decreases in plant fitness must be significant, either each alone or in combination, and must cause decreases in plant fitness (Marquis 1990a; Marquis, this volume). Gould (1988) points out that even though one herbivore species alone may not select for resistance if the damage it causes is below a level that reduces plant fitness, two herbivore species in combination may cause a substantial reduction in fitness. If damage by two or more herbivores is correlated among plant genotypes, multispecies selection can occur. The common existence of positive genetic correlations in resistance may suggest that multispecies selection of plant resistance is likely in natural populations.

Evolution of plant resistance can also change community patterns of herbivores. The structure of genetic correlations in resistance will predict how other herbivore species will respond to an evolutionary change in plant resistance to one species (fig. 11.4). If herbivore resistances are not genetically correlated, abundances of other species may not change. Positive and negative correlations in resistance will result in decreases and increases of other species, respectively, even if they did not originally select for the evolution of the resistance trait. Changes in the proportion of guilds and specialists and generalists could occur because of similar correlations in resistances.

Genetic correlation matrices may have quite different forms in different plant populations (Antonovics 1976; Wolff and van Delden 1987). Gould (1988) reviewed evidence of between-population variation in defensive chemicals in the mint family (Lamiaceae). If the genetic mechanisms that confer resistance to herbivores differ between plant populations, the structure of herbivore communities could differ between populations as a result of genetic differentiation, as predicted for within-population differences. Parker (1985) found population differentiation in plant resistance to fungal attack at a small spatial scale. Hare and Futuyma (1978) found genetic differences among populations in resistance traits associated with seed predation. The consequences of between-population differences in genetic correlations of resistances can result in: (1) different responses of plant populations to multispecies selection by herbivores, and (2) consequently different responses of herbivore communities to evolution of resistance (fig. 11.4).

Variable Species Interactions on Variable Host Plants

Far from being a homogeneous, neutral substrate with respect to interactions of herbivores with other herbivores or with natural enemies, host plants vary in traits that influence the outcome of interactions and conversely can be influenced by interactions between herbivores. The host plant is the environment in which interspecific interactions take place and can therefore influence the outcome of these interactions. The frequency and distribution of outcomes of interactions can be important information in understanding the evolution of interactions (Thompson 1988d). Because interspecific interactions such as competition and natural enemy impact may be important in structuring herbi-

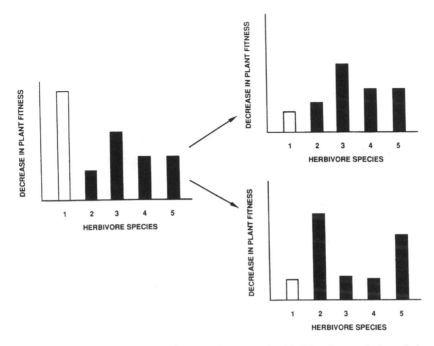

Fig. 11.4. Potential responses of four herbivore species (dark bars) to evolution of plant resistance to a fifth herbivore species (open bar). Decrease in plant fitness caused by herbivore 1 is lower after evolution of resistance. In the upper right graph the amount of damage caused by the other herbivores does not change after evolution of resistance to species 1, indicating lack of genetic correlation of resistances between the species. The lower right graph shows that damage caused by the other herbivores changed in response to evolution of resistance. Damage caused by herbivores 2 and 5 increased and damage caused by herbivores 3 and 4 decreased after evolution of resistance, indicating negative and positive genetic correlations of resistances, respectively. Thus, evolution of resistance can cause evolutionary changes in community structure of herbivores depending on the genetic correlation of resistances between them.

vore communities (Schoener 1983; Strong et al. 1984; Diamond and Case 1986; Faeth 1987), the influence of genetic variation in resistance upon the outcome of species interactions needs empirical and theoretical study.

In this section I review evidence of how plant resistance affects interactions of herbivores with other herbivores and natural enemies, and I suggest other possible effects of resistance on the outcomes of interactions that have not been studied. I divide the interactions roughly between horizontal interactions, that is, within–trophic level interactions, and vertical interactions, between–trophic level interactions. As will be seen, this distinction is blurred when indirect interactions between herbivores are mediated by natural enemies. Figure 11.5

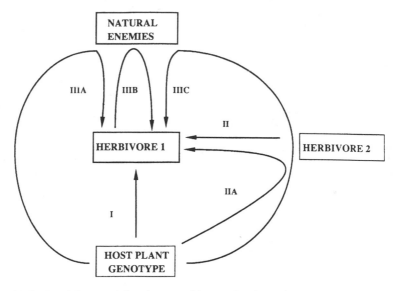

Fig. 11.5. Several potential pathways of interaction between plant resistance (geno-type), a competitor (herbivore 2), and natural enemies in affecting herbivore 1. Table 11.7 contains explanations of each pathway. Effects of interactions may be positive, negative, or neutral. Arrows passing by a box indicate that the species is affected by the previous species or factor, which then affects the next species until the herbivore 1 is affected.

TABLE 11.7
Explanation of the pathways shown in figure 11.5

Pathway	Interpretation
I	Direct effect of plant genotype on resistance (reciprocal of herbivore density)
II	Direct competition of herbivore 2 with herbivore 1
IIA	Direct competition of herbivore 2 with herbivore 1 that varies among plant genotypes (i.e., plant-mediated interspecific competition)
IIIA	Effect of plant genotype in attracting or repelling natural enemies, independent of the presence of herbivores (e.g., chemical cues, trichomes, stature)
IIIB	Characteristics of herbivores (e.g., gall size) or herbivore density varies among genotypes owing to pathway I, and enemy impact varies among host plants because of differential susceptibility to enemies or density-dependent attack.
IIIC	This pathway has two courses. Percentage parasitism of herbivore 1 varies among plants because of presence or abundance of herbivore 2. The presence and abundance of herbivore 2 may or may not be due to genetic variation among host plants in resistance. Thus, the path could start from herbivore 2 rather than from host-plant genotype. This pathway is a form of indirect competition (Faeth 1988). Enemy impact may vary among plants but may also vary among leaves or shoots within plants depending on cooccurrence (see text).

and table 11.7 illustrate several pathways of interaction between plant genetic variation, competition, and natural enemy impact.

Horizontal Interactions: Competition and Plant Resistance

The role of interspecific competition in structuring communities of phytophagous insects is an important issue in community ecology. Interspecific competition among herbivorous insects may occur infrequently because of regulation of herbivore population sizes, below levels where competition can be important, by abiotic factors and natural enemies (Lawton and Strong 1981; Price 1983b; Strong et al. 1984). In two studies, where competition was expected, only one or a few pairs of species were found to compete (Rathcke 1976; Gibson and Visser 1982), while other studies have found no evidence of interspecific competition (Wise 1981; Strong 1982a, 1982b; Howard and Harrison 1984; Lawton and Hassell 1984; Karban 1987a). However, other experimental studies demonstrate interspecific competition (McClure and Price 1975; McClure 1980a; Peckarsky and Dodson 1980; Stiling 1980; Gibson and Visser 1982; Kareiva 1982a; Stiling and Strong 1984; Karban 1986; Fritz et al. 1986; Crawley and Pattrasudhi 1988). These mixed results suggest that questions about the role of competition in community theory must change from "Is competition important?" to "When is competition important?" "What are the factors that influence the outcome of competition?," and "How much variation is community patterns does competition account for relative to other processes?"

Resource competition between herbivorous insects is manifested through effects on the quantity and quality of host plants (Kareiva 1982a, 1986; Crawley and Pattrasudhi 1988). Effects of herbivory on plant quality may be immediate or may affect plant nutritional quality for herbivores that colonize weeks or months in the future (Roland and Myers 1987; Faeth 1986, 1988). Plant resistance can limit the effects of herbivores on plant nutritional quality and quantity and therefore may be integral to understanding dynamics of competition between herbivores. For this reason it is important to consider how interspecific competition may vary among host plants that vary in resistance.

Theory and Predictions

Competition theory suggests that habitat heterogeneity will promote species coexistence if the competitive superiority of species is reversed in different habitat types or if heterogeneity provides spatial and temporal refuges in changing patches (Shorrocks et al. 1979; Atkinson and Shorrocks 1981; Ives and May 1985; Chesson 1986). Coexistence is predicted to be more likely if patches are utilized by species that tend to aggregate among patches independently of other species (Shorrocks et al. 1979; Atkinson and Shorrocks 1981; Ives and May 1985). Inferior competitors can then exist in patches that are not occupied by the superior competitor.

Ives (1988) has recently shown that the intraspecific aggregation of eggs and patterns of covariance of oviposition of two species among patches influences their probability of coexistence. His model assumes that patches are of equal quality for herbivore survival but differ in the probability of being discovered for oviposition. Variation in density of herbivores among patches can be due to variation in apparency, variation among patches in oviposition cues (i.e., preference, so long as preference is not completely correlated with performance), or a tendency for conspecifics to oviposit in patches already containing eggs. Positive covariance of species densities among patches is predicted to result in greater competition and lower probability of coexistence, whereas negative covariance of species densities among patches is predicted to result in higher probability of coexistence. Variation in plant resistance to two species due primarily to nonpreference and the genetic correlations of nonpreference to the two herbivores therefore are predicted to affect the dynamics of species coexistence.

The theory described above considers whether species coexist or not under patchy distributions. If species do coexist, then it is also of interest to consider what spatial patterns of competition might result from the effects of covariances of plant resistances and how this might affect communities of herbivores. Consider a plant species where resistance affects densities of two herbivores via nonpreference, and both herbivores are abundant enough, at least on some plants, so that resources are limited. (If both species are rare, even on the most favorable plants, then competition will not occur regardless of their patterns of cooccurrence, because resources are not limited.) Competition is predicted to occur on genotypes that are susceptible to both herbivores (positive genetic correlations and higher densities), and competition could decrease the abundance of an herbivore on susceptible plants more than on resistant plants (lower densities) (fig. 11.6A). Competition would be found unevenly across host plants regardless of which species was a superior competitor, but if competition were asymmetrical (Lawton and Hassell 1984), then it could result in a change in community structure among conspecific plants.

If the proportion of susceptible plants in a habitat is high, a larger fraction of an herbivore population should experience competition, and the opposite should be true if most plants are resistant by virtue of nonpreference. However, even if susceptible plants are rare, but a large fraction of the herbivore population occurs on these plants, then competition could be important ecologically, and perhaps evolutionarily as well, but would apparently be rare because most plants would have low densities of both herbivores. Incorporating plant variation as an effect in competition experiments could reveal significant interspecific competition that is unevenly distributed among individual plants.

When plant resistance to two herbivores is negatively correlated, herbivore densities will be negatively correlated even in the absence of competition (fig. 11.6B). The pattern of negative associations alone cannot be taken as evidence of competition, but the pattern caused by plant variation may influence the com-

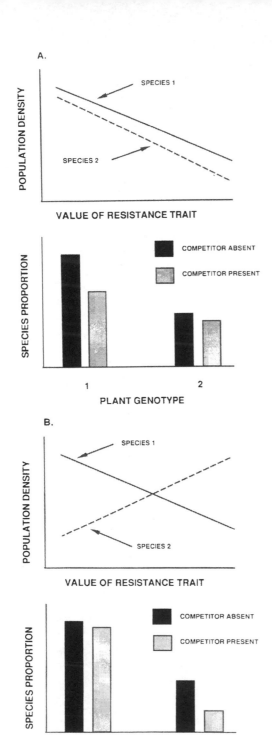

Fig. 11.6. (A) Effects of competition between herbivore species 1 (solid line) and species 2 (dashed line) when they compete on susceptible plants at high density (genotype 1) but do not compete at low density on resistant plants (genotype 2). If competition is asymmetrical (2 > 1) then the relative abundance (or relative survival) will be lower for species 1 when it cooccurs with species 2 on susceptible plants (hatched bars) compared to when it is alone (solid bars). Relative abundance would be only slightly affected or not at all on resistant plants (genotype 2). (B) When two species are negatively correlated in density among genotypes, species 1 will not be strongly influenced by species 2 even if 2 is a superior competitor, because its density is low (genotype 1; hatched versus solid bars). On plants resistant to species 1, but susceptible to species 2 (genotype 2) species 2 may have a strong competitive effect on species 1 (hatched versus solid bars).

petitive interactions between the species. On plants that are highly susceptible to one herbivore and resistant to a second herbivore, the rare species is predicted to be more affected by interspecific competition, which could lower its density and relative abundance (fig. 11.6B). The outcome of competition between these species may be reversed when the second species is on susceptible plants, which are resistant to the first species. Negative covariance of species among patches (i.e., plant genotypes) is predicted to cause asymmetries among conspecific plants in the competitive interactions between herbivores, even as it increases the likelihood of coexistence (Ives 1988).

In addition to variable competition between herbivores, due to density effects, variation and covariation in resistances (antibiosis) could affect competitive abilities of herbivores on different plant genotypes. Variable competitive abilities between species in different habitats have been demonstrated for a variety of organisms, including insects (Park 1954; Edson 1985). Resistance may decrease competitive abilities of herbivores because of the negative effects of resistance traits on feeding efficiency, and on access and utilization of plant resources compared to when herbivores are on susceptible plants (Beck and Schoonhoven 1980; Norris and Kogan 1980). For example, some biochemical resistance factors reduce feeding, growth, and development (Norris and Kogan 1980), which could create asymmetries in herbivore sizes and consumption rates between two species that in turn could affect competitive outcomes. These potential but largely unknown effects of plant resistance on competition need to be considered as a possibly important mode of horizontal interaction between herbivores.

Manipulating Competitors on Willow

Fritz et al. (1986) found significant competitive effects of the stem-galling sawfly on three other gall-forming sawfly species on willow. Competition was probably caused by reduced production of new leaves on plants with stem galls. The large galls formed by this species apparently usurped resources being directed to shoot growth and leaf production. There were also significant differences in new leaf production among clones and a significant clone-by-competition interaction. Stem gallers reduced new leaf production more on some clones than on others, perhaps because of different densities or sizes of stem galls (Fritz, unpublished data). Densities of two sawfly species were positively correlated with new leaf production, which explained the significant reduction in density of these species. The clone effect explained more variation in sawfly species density in these experiments than did competition, suggesting that it was more important in determining community structure. Because stem-galler densities are usually lower in the field, the suggestion that plant variation is likely to be more important than interspecific competition for community structure is further reinforced (Fritz et al. 1986; Fritz and Price 1990).

Fritz et al. (1986) did not find a significant interaction between competition and plant clone, but that analysis did not examine the reduction of sawfly densities after adjusting for the number of stem galls on plants. Fritz (1990b) calculated competition coefficients for each plant with stem gallers present and showed that there were significant differences among clones in the competition coefficients (table 11.8) for four species and year combinations. Competition coefficients varied from 2.35 to nearly zero; thus competitive effects were relatively strong on some clones and nonexistent on others. These data demonstrated clone-by-competition interaction, supporting the hypothesis that competition between herbivores will vary among plant genotypes, but the causes of variable competition effects were not clear.

In another study of competition and plant resistance, Karban (1987a) found no effect of interspecific competition on thrips (*Apterothrips secticornis*, Thysanoptera: Thripidae) on three clones of seaside daisy, *Erigeron glaucus* (Asteraceae). Clone was an important factor affecting thrips abundance, but clone did not interact with competition treatments.

Recently, Moran and Whitham (1990) reported results of field observations and experiments on competition between a root-feeding aphid (*Pemphigus betae*) and a leaf-galling aphid (*Hayhurstia atriplicis*) on *Chenopodium album*. The host plant varied in resistance to the leaf-galling aphid; resistant plants do not form galls but permit small leaf colonies only. Plants did not vary in resistance to the root-feeding aphid. On plants susceptible to the leaf-galling aphid,

TABLE 11.8
Results of analysis of covariance on competition coefficients among six willow clones for the leaf-galler, leaf-folder, and petiole-galling sawfly species in 1983 and 1984 experiments

Effect	DF	F-Ratio	Probability
Leaf galler 1983			
Leaf-galler density	1	18.47	.0011
Stem-galler density	1	12.14	.0045
Clone	4	5.14	.0120
Error	12		
Leaf folder 1983			
Leaf-folder density	1	15.85	.0018
Stem-galler density	1	30.25	.0001
Clone	4	6.28	.0058
Error	12		
Leaf galler 1984			
Leaf-galler density	1	10.55	.0025
Stem-galler density	1	1.67	.2036
Clone	5	4.99	.0014
Error	37		
Petiole galler 1984			
Petiole-galler density	1	11.36	.0018
Stem-galler density	1	5.35	.0263
Clone	5	2.53	.0459
Error	37		

Source: Fritz 1990b.

populations of the root-feeding aphid were suppressed by an average of 91%. Resistant plants were a refuge for the root-feeding aphids, since populations of the leaf-galling species were so low. This study provides a clear demonstration of the relationship between plant resistance and interspecific competition. Mopper et al. (1990) found that although sawfly survival was higher on susceptible phenotypes of pinyon pine (*Pinus edulis*), sawfly fecundity was lower because of competition with the stem moth *Dioryctria albovitella*, which was also more abundant on susceptible phenotypes. Both of these studies suggest that interaction between competition and plant resistance may be common.

Other Mechanisms

Classical resource competition may be rare in phytophagous insect communities, but interactions between herbivores mediated by changes in host-plant quality may be common and have important effects on herbivore communities (Faeth 1987). Change in host-plant quality may occur as a result of induction of defenses, changes in phenology, and patterns of leaf abscission. Plants probably vary genetically in their induction of chemical, physical, nutritional, and abscission responses to herbivores; indeed, evolution of these traits depends on there being genetic variation in their expression and a correlation between these traits and plant fitness. Because some of these mechanisms can be subtle and effective at low levels of herbivory, they may have been overlooked by ecologists. Temporal separation between herbivores that interact via changes in plant quality may also make these interactions difficult to observe and measure. How these mechanisms interact with variation in plant resistance is the subject of this section. Since these mechanisms are only recently getting attention and no one has examined how they vary among plant genotypes, this section is largely speculative.

Induced defenses of plants can affect herbivores intra- and interspecifically (Schultz and Baldwin 1982; Tuomi et al. 1984; Haukioja and Neuvonen 1985; Faeth 1986; Hunter 1987; Karban et al. 1987; Leather et al. 1987; Neuvonen and Danell 1987; Baldwin 1988). Conspecific plants showing strong induction responses to herbivore attack are predicted to suffer less herbivory (Haukioja, Soumela et al. 1985; Harrison and Karban 1986) than plants showing weaker induction responses, assuming constitutive plant defenses affect herbivore species similarly. Variation in plant induction could lead to positive correlations among densities of herbivores that respond similarly to induced defenses. Some herbivores, however, can circumvent induced defenses (Carroll and Hoffman 1980; Tallamy 1985) or could tolerate the induced defense. Herbivores unable to circumvent induced defenses should have reduced densities and/or survival on plants that have strong induction responses but might not be inhibited on plants that lack an induction response, show weaker induction, or sustain induction for shorter periods.

There is currently little evidence of genetic variation in induction of plants.

Karban (1990, personal communication) has found a great amount of variability among cotton varieties in induced resistance to spider mites. Cotton varieties with the highest levels of constitutive resistance were the ones that showed the greatest level of induced resistance. Zangerl and Berenbaum (1990) recently showed significant heritabilities in wild parsnip (*Pastinaca sativa*) for the extent of constitutive and induced production of several furanocoumarins. Genotypes with strong induction responses could show strong negative associations between temporally separated herbivores, whereas plants with weak induction might show few negative interactions between herbivores mediated by induction. Studies are needed that test for variable induction responses among host-plant genotypes and examine their effects on communities of herbivores.

Plant variation in phenology and abscission could also influence the structure of herbivore communities. Plants that break bud later in the spring could exclude or reduce the abundance of early-season herbivores relative to plants that break bud earlier. If plants vary genetically in abscission responses, then these abscission responses to one herbivore could have detrimental effects on species that exist on the same leaves. On the other hand loss of herbivores on abscised leaves could reduce competition between herbivore species, although the effects of a lost leaf on other herbivore species need to be considered. There may be genetic variation in plants for response to herbivores by refoliation or leaf abscission, but there is little evidence to support this. Fritz (personal observation) found clonal variation in silky willow (*S. sericea*) in the number of leaves abscised during the summer. Most of the abscised leaves did not have herbivores, however. If plants varied in extent of refoliation after herbivore attack, this could affect colonization by later attacking herbivore species (Damman 1989). If plants vary in constitutive defenses and consequently herbivory, then variation in refoliation or leaf abscission could vary phenotypically among conspecific plants even if there is not genetic variation in these traits directly. This phenotypic variation could then affect other herbivores (Pilson, 1989, personal communication).

Horizontal Interactions: Facilitation and Plant Resistance

Herbivores can benefit from the presence or feeding of other herbivore species (Danell and Huss-Danell 1985; Faeth 1986; Hunter 1987, Neuvonen and Danell 1987; Damman 1989). The effects of facilitation may vary among plant genotypes. If the benefits of prior feeding to other herbivores is dependent on the extent of previous attack, then susceptible plants would be predicted to become more suitable than resistant plants for later attacking herbivores. Alternatively, resistant plants could become more susceptible if prior feeding by other species substantially reduced resistance.

Williams and Myers (1984) noted increased pupal weight of fall webworm (*Hyphantria cunea*) that fed on plants defoliated the previous season by western tent caterpillar (*Malacasoma californicum pluviale*). Damman (1989) showed that feeding by a pyralid moth on pawpaw (*Asimina* sp.) resulted in new leaf flushes and greater population sizes of *Eurytides marcellus* (Lepidoptera: Papilionidae), an obligate young-leaf feeder. Faeth (1986) found a decrease in mortality of leaf miners due to "other" causes (including fungal attack and disease) on leaves that had been previously fed on by leaf-chewers compared to leaves that had not been fed on, but this was not enough to offset the increased effects of parasitoids on the leaf miners. Prior feeding by biotype C of the greenbug *Schizaphis graminum* (Rondani) (Homoptera: Aphididae) on genotypes of winter wheat (*Triticum aestivum*) resistant and susceptible to greenbug biotype E made the resistant genotypes more susceptible to biotype E but did not affect greenbug success on susceptible genotypes (Dorschner et al. 1987). Only in this last example of facilitation is there evidence that genetic differences among plants is important. More studies of facilitation are needed, and studies that consider genetic differences among plants in resistance to one or more of the herbivores will be most valuable in determining whether there are effects of facilitation on herbivore community structure.

Vertical Interactions: Natural Enemies and Plant Resistance

Natural enemies regulate sizes of some herbivore populations and are believed to be of major importance in structuring herbivore communities (Hairston et al. 1960; Root 1973; Lawton and McNeill 1979; Lawton and Strong 1981; Price 1983b, 1987; Jefferies and Lawton 1984; Strong et al. 1984; Faeth 1987). Patterns of herbivore species diversity on plants (Price 1983b), host use (Sato and Ohsaki 1987; Denno et al. 1990), and niche differentiation (Lawton and Strong 1981; Lawton 1986) are hypothesized to result from selection by natural enemies. Many hypotheses of enemy impact on herbivores consider differences *between* host-plant species, but these hypotheses apply equally well to variation of enemy impact on herbivores *within* host-plant species. Two concepts of variable enemy impact on herbivores relate directly or indirectly to variation in host-plant resistance: (1) the three trophic level concept; and (2) the enemy impact concept (including the concept of enemy-free space).

Three Trophic Level Concept

The three trophic level concept predicts that plant traits affect the interactions between herbivores and their natural enemies (Feeny 1976; Bergman and Tingey 1979; Price et al. 1980; Schultz 1983b; Price 1986). Empirical evidence linking genetic variation in resistance and impact of natural enemies on herbivores is abundant for agricultural systems (Bergman and Tingey 1979; Price et al. 1980; Boethel and Eikenbary 1986). Plant resistance in some crops may be enhanced by the action of natural enemies (e.g., Starks et al. 1972), and be compatible with biocontrol efforts (Boethel and Eikenbary 1986). Interaction of plant resistance with natural enemies is poorly understood in natural systems (but see Price and Clancy 1986a; Fritz and Nobel 1990; and Craig et al. 1990), and little is known about how quantitative variation in plant resistance affects natural enemy impact. Advances in understanding variation of herbivore population dynamics among plants will require considering interactions of genetic variation in plant resistance and natural enemy impact.

Plant resistance may affect enemy impact through effects on herbivore susceptibility or effects on parasitoid or predator search (fig. 11.5 and table 11.7). Plant resistance traits may increase or decrease susceptibility of herbivores to natural enemies (Boethel and Eikenbary 1986). Higher herbivore densities on susceptible plants may attract predators and parasitoids (fig. 11.5, IIIB), which may have greater impact on herbivore population growth through numerical and functional responses (Hassell and Waage 1984; Lessells 1985; Stiling 1987). Low nutrient availability or digestibility reducers can prolong development and thereby increase risk of enemy attack on resistant compared to susceptible plants (IIIB) (Feeny 1976; but see Damman 1987). Plant variation may affect sizes of insect galls (I) that in turn affect susceptibility to enemy attack (IIIB) (Price and Clancy 1986a; Price 1988; Craig et al. 1990; also see Jones 1983; Weis and Abrahamson 1986; Weis et al. 1985). Plant chemical defenses may be sequestered by herbivores to defend against enemy attack (Campbell and Duffey 1979; Smiley et al. 1985) (IIIB). Plant defenses may also affect the natural enemies directly (IIIA). Defensive chemicals may be used as cues by natural enemies to find herbivores (Vinson 1976, 1981; Price 1981; Weseloh 1981; Schultz 1983b; Van Alphen and Vet 1986), or physical defenses, such as trichomes, can also inhibit parasitoid searching (Obrycki 1986). Thus plant resistance variation can affect natural enemy impact in a variety of ways (Price 1986).

Enemy Impact Concept

The enemy impact concept asserts that natural enemies have been of major importance in determining host-plant use, densities, species diversity, niche use, and relative abundances of herbivores on plants (Holt 1977, 1984, 1987; Lawton and McNeill 1979; Lawton and Strong 1981; Jefferies and Lawton 1984; Lawton 1986). Enemy impact could favor interspecific avoidance of herbivores among host plants, or shoots or leaves within a plant (Holt 1977; Jefferies and Lawton 1984; Lawton 1986). A major mechanism proposed by the enemy impact concept is that natural enemy impact varies with the presence and/or abundance of coexisting herbivore species (Holt 1977; Jefferies and Lawton 1984; Lawton 1986). Natural enemies shared by two herbivores may have a greater impact on survival of one or both herbivore species when herbivores cooccur versus when they do not cooccur, and the effect of natural enemies may depend on relative abundances, "apparent competition" (Holt 1977, 1984, 1987; Lawton 1986). Thus natural enemies may impose selection on herbivores that choose plants that minimize enemy impact (Jefferies and Lawton 1984; Sato and Ohsaki 1987). Settle and Wilson (1990) identified differential impact by a shared parasitoid as the major cause of the increase in abundance of the introduced variegated leafhopper (*Erythroneura variabilis*) (Homoptera: Cicadellidae) and the concomitant decline in abundance of the native grape leafhopper (*E. elegantula*).

An alternative mechanism of variable enemy impact argues that rates of predation or parasitism may be greater when two species cooccur on the same plant, shoot, or leaf, even if enemies are not shared (e.g., Faeth 1986, 1988). Feeding damage by other herbivore species could enhance visual or chemical cues used by predators or parasitoids to find their hosts (Vinson 1976, 1981; Weseloh 1981; Schultz 1983a, Niemelä 1987). Faeth (1986, 1988) demonstrated increased parasitism of leaf miners on leaves of oak that had previous herbivory, and normally leaf miners avoid ovipositing on damaged leaves (but see Bergelson and Lawton [1988] and Hawkins [1988], who found no effect of foliage damage on herbivore predation or parasitism). Parasitism was higher owing to physical cues from prior feeding and changes in leaf chemistry (Faeth 1988). While variable enemy attack in response to damage by other species has been demonstrated, there are several points that need to be tested. (1) Does damage by different herbivore species affect enemy attack differently, i.e., are some combinations of herbivores less favorable for enemy avoidance? (2) Does

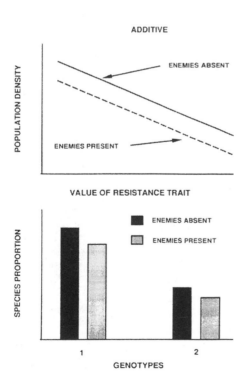

Fig. 11.7. If natural enemies reduce density or survival rates of herbivores similarly across susceptible and resistant genotypes (hatched line) compared to when enemies are absent (solid line), then enemy impact is additive. Relative abundance of an herbivore may decrease but the effect will be symmetrical across genotypes.

the amount of damage or number of herbivores affect the level of enemy attack? (3) Does plant variation, which is predicted to affect the relative abundances of herbivores among conspecific host plants (Fritz et al. 1987b; Fritz and Price 1988), cause herbivore association patterns that affect enemy impact (fig. 11.5, IIIC)? Higher levels of herbivory by other species on susceptible compared to resistant plants may cause greater enemy impact on herbivores, providing a basis for indirect interactions between them. Studies so far have not distinguished between susceptible and resistant plants and have not demonstrated responses of enemies to different levels of herbivory by other species.

A graphical model of the effects of interactions of natural enemies and plant resistance on relative abundances of phytophagous insects is presented in figures 11.7, 11.8, and 11.9 (Hare, this volume). Assume that herbivore density, survival, or population growth varies among plant genotypes. If natural enemy attack is independent of plant genotype and herbivore density, then the effect of enemy attack on herbivore population size among plants is additive (fig. 11.7), that is, there is an equal proportional reduction of herbivore population sizes on all plant genotypes. Enemy impact may reduce the relative abundance of the herbivore but will do so equally across the range of resistant and

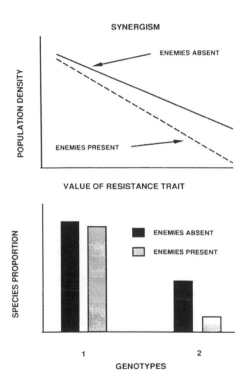

Fig. 11.8. Enemy impact on herbivores may be greatest on resistant plants; thus there is a positive correlation between direct plant resistance to herbivores and resistance mediated by natural enemies (synergism). This could result in a reduction in the relative abundance of an herbivore on resistant plants (genotype 2) but would result in little effect on relative abundance of the herbivore on susceptible plants (genotype 1).

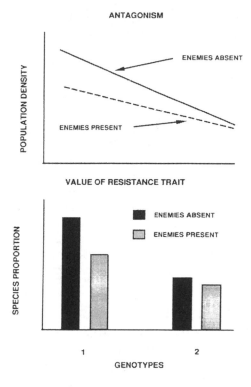

Fig. 11.9. If enemy impact is greatest on susceptible plants, then direct resistance and indirect resistance may be negatively correlated (antagonism). This could result in lower relative abundances of the affected herbivore in the presence of natural enemies on susceptible plants (genotype 1; hatched versus solid bars) but would cause little effect on resistant plants (genotype 2).

susceptible plants. If predation or parasitism is greatest on resistant plants, then enemy impact is synergistic with plant resistance (fig. 11.8). Resistant plants are predicted to have lower relative abundance of the herbivore than predicted from plant resistance effects alone. Antagonism (compensation) between enemy impact and plant resistance occurs if predation or parasitism of herbivores is higher on susceptible plants (fig. 11.9). This could be caused by enemy response to higher herbivore densities (Lessells 1985; Stiling 1987), but greater attractiveness of susceptible plants to parasitoids and predators and reduced defenses of herbivores on susceptible plants are also possible mechanisms. Antagonism is predicted to result in reduced relative abundance of an herbivore on susceptible plants compared to resistant plants (fig. 11.9).

Starks et al. (1972) found that aphid population growth was reduced more on resistant than on susceptible plants in the presence of natural enemies compared to when enemies were absent, supporting the synergism model. Van Emden (1978) found that parasitism greatly reduced aphid population sizes on both susceptible and resistant plants but the effect was much greater on susceptible plants supporting the antagonism model. In the presence of parasitoids resistant plants supported more aphids than susceptible plants, the reverse of the situation when no parasitoids were present. These studies show that natural enemy impact may be an important factor influencing the expression of plant resistance, and could alter patterns of herbivore associations among plants. Variable natural enemy impact among resistant and susceptible plants can change the correlations of herbivore densities among plants; increasing or decreasing the strength of positive or negative correlations. Variability in the host-parasitoid interaction is predicted to stabilize their interactions over time (Hassell and May 1988).

Future Directions

One of the goals of community ecology is to estimate the relative contribution of various processes in structuring herbivore communities. Experimental studies that manipulate several factors will make the most progress toward this goal (for example, manipulation of specific environmental factors, plant dispersion, and genetic variation). We are beginning to understand that two of the important factors in community structure of herbivores are genetic variation among plants (Fritz and Price 1988) and subtle indirect interactions between herbivores and plant quality (Faeth 1988).

Tests of predictions of the genetic variation hypothesis will determine whether plant resistance to herbivores can lead to predictable effects on community structure of phytophagous insects. It will be important to know the general patterns of phenotypic and genetic correlations between herbivores, since they will influence community patterns. More studies of clones and half-sib families in common gardens are needed (Maddox and Root 1987; Fritz 1990a; Karban 1989a; Marquis 1990a). With some limitations on the interpretation of results (Fritz and Price 1988; Marquis 1990a; and Fritz 1990a), studies of clones of plants can also be useful for studying genetic effects on community patterns. Half-sib analysis (Maddox and Root 1987; Simms and Rausher 1987, 1989) will permit estimation of additive heritabilities and genetic correlations. These studies along with estimates of damage caused by each herbivore and the effects of each herbivore on plant fitness will be essential to understanding the potential and trajectory of multispecies coevolution of plant resistance and maintenance of resistance variation in plant populations.

Herbivore interactions with plants, competitors, and natural enemies are complex and frequently indirect. Multifactor experimental studies are needed to dissect the important interactions that affect herbivore population dynamics and community structure. Genetic variation in plant resistance directly affects herbivore numbers and life table parameters, and may indirectly affect herbivores through predation, parasitism, and competition. To understand the ecology and evolution of herbivore interactions with competitors and natural enemies the role of plant variation in structuring the outcome of these interactions is important. Differential effects of competition among genetically variable plants may be as important in determining herbivore fitness and plant choice as plant resistance directly. Likewise, natural enemies contribute to the expression of plant resistance, but we do not know whether synergism, antagonism, additivity, or a more complex relationship (Hare, this volume) between these forces affects herbivore communities. Understanding the genetic correlations between direct resistance of plants to herbivores and indirect resistance that is mediated by natural enemies may suggest whether natural enemies facilitate or constrain the evolution of resistance.

Competitive outcomes between herbivores on plants theoretically can vary, owing to the effects of resistance on density and competitive abilities. Empirical and theoretical studies are needed that incorporate plant resistance variation with competition studies to understand: (1) the distribution of competitive outcomes between herbivores across plant genotypes, (2) the effects of competition on community structure, (3) coexistence of herbivores, and (4) evolution of competitive interactions.

Thompson (1988d) argues that the distribution of outcomes of interspecific interactions is important in understanding the evolutionary trajectory of interactions. This is necessary information in studying the evolution of interactions, but understanding the inheritance of traits of the interacting species is equally necessary information. Genetic aspects of host plants are as important for understanding competitive interactions and predator/parasitoid interactions of herbivores as are the genetics of the herbivores and natural enemies themselves.

Acknowledgements

I gratefully acknowledge funding from National Science Foundation grant DEB 82–05904 to P. W. Price, the Research Corporation Cottrell College Science Fund, and the Sloan Foundation. I thank Mark Rausher, Diana Pilson, and Rick Karban for critical comments on the chapter. Mark Rausher and Janis Antonovics generously provided a place to work and stimulating discussions during a sabbatical at Duke University. Numerous students from Vassar College, Northern Arizona University, and Duke University have assisted me in fieldwork on willows in Arizona and New York over the years and their help has been invaluable. J. R. Miller has generously assisted in the fieldwork. For several years Len and Ellie Sosnowski have permitted me to conduct research on their property near Milford, New York. Their generosity and enthusiastic support are gratefully appreciated.

12

Effects of Plant Variation on Herbivore–Natural Enemy Interactions

J. Daniel Hare

This chapter reviews how genetic variation within plant species affects the natural enemies of herbivores in ecological time and explores the consequences of such variation on plant–herbivore–natural enemy interactions in evolutionary time. The primary genetic question underlying this chapter is whether the impact of natural enemies on herbivores is sufficiently strong and systematic to cause changes in gene frequencies in plant traits affecting the impact of natural enemies on those herbivores. Most examples of plant–herbivore–natural enemy (i.e., tritrophic) interactions are derived primarily from comparisons of herbivore–natural enemy interactions on different plant species and the role that interspecific phytochemical variation plays in those interactions. Interspecific variation in plant morphological characters (e.g., trichome length and density, leaf wax thickness, fruit wall thickness) also may introduce variation in the ability of natural enemies to discover and utilize their prey on different host-plant species. Interspecific plant variation in tritrophic interactions illustrates some of the ways in which similar intraspecific variation may introduce variation in tritrophic interactions among populations of particular plant species.

In some cases, specific phytochemicals mediate the interaction between herbivores and their natural enemies. This has been especially well studied for parasitic Hymenoptera. Particular natural enemies exploit some host-plant chemicals as cues to locate the habitats of their host or host individuals. Also, some herbivores sequester host-plant chemicals either directly or after modification for use as chemical defenses against natural enemies. These phenomena have been reviewed several times recently (Duffey 1980; Price et al. 1980; Price 1981; Vinson 1981; Pasteels et al. 1983; Huheey 1984; Vinson 1984; Barbosa and Saunders 1985; Duffey et al. 1986; Price 1986; Vinson and Barbosa 1987; Barbosa 1988a; Bowers 1988b; Nordlund et al. 1988; Pasteels et al. 1988). Clearly, particular herbivorous insects and parasitoids benefit by utilizing host-plant chemicals to mediate their interactions, but the relative benefits to particular herbivores and natural enemies obviously varies substantially among systems.

Although particular host-plant chemicals may mediate herbivore-parasitoid interactions, it is far more difficult to demonstrate intraspecific variation in plant fitness within populations caused by differential production of those phytochemicals. Price et al. (1980) were among the first to point out that a direct plant–natural enemy association deleterious to herbivores does not always benefit plants. An important example of such a paradoxical situation involves sublethal plant defenses such as digestibility reducers. The mode of action of these compounds might actually lead to increased foliage consumption and reduced plant fitness if herbivores could compensate fully for reduced nutritional quality through increased feeding rates. Such potential increases in foliage loss could be offset only if the increased time spent feeding resulted in increased exposure and mortality of relatively young stages of the herbivore population to natural enemies (Price et al. 1980). Plant factors that increase the duration of older stages would be counterproductive, since the majority of foliage consumed by phytophagous insects occurs in the later instars. Although widely accepted, there is relatively little empirical verification that increases in time spent in early instars increases vulnerability of herbivores to natural enemies (Clancy and Price 1987; Damman 1987).

The considerable number of examples of variation in tritrophic interactions due to interspecific host-plant variation contrasts with our knowledge of the importance of intraspecific host-plant variation in the evolution, variation, and maintenance of herbivore–natural enemy interactions on different plant populations. Opportunities to investigate the importance of intraspecific host-plant variation in herbivore–natural enemy interactions in managed systems are available in conjunction with research aimed toward integrating host-plant resistance with biological control in pest management programs.

Historically, pest managers assumed that host-plant resistance and biological control were compatible and largely independent pest management strategies (Adkisson and Dyck 1980; Kogan 1982). One early theoretical approach reflecting this view was derived from deterministic mathematical models of host-parasitoid interactions, which show that effective control by natural enemies is enhanced when the rate of increase (r_m) of the host population is reduced (Van Emden 1966; Beddington et al. 1978; Hassell 1978; Lawton and McNeil 1979; Hassell and Anderson 1984). Such a prediction assumes that the host plant affects only the growth rate of the prey population and not the attractiveness or quality of prey individuals for discovery and utilization by natural enemies. A substantial number of cases are now known where the assumption of uniformity of the prey population on different host plants either does not hold or has limited applicability (for reviews see Bergman and Tingey 1979; Hare 1983; Duffey and Bloem 1986; Vinson and Barbosa 1987; Barbosa 1988a).

Interactions between Host-Plant Resistance and Biological Control

Mechanisms

Variation in genetically determined host-plant characteristics may alter the effectiveness of the natural enemies of herbivorous insects in a number of ways, which can be grouped broadly into categories relating to changes in the risk of discovery of herbivores by their natural enemies (mostly by hymenopterous parasitoids; there is comparatively little information available on predators) and changes in their suitability for utilization after discovery. Host plants may vary qualitatively or quantitatively in the production of allelochemicals. This variation may affect host location by parasitoids (synomones), herbivore defense (when toxins are sequestered), and nutritional suitability of herbivores as hosts. Also, plant surface or other physical characteristics may influence the searching behavior of parasitoids and predators for their prey. Examples within each of these four areas will be discussed in the following sections.

Synomones

Among the best documented of the direct effects of host-plant characteristics on natural enemies is the importance of plant-produced synomones to attract parasitic Hymenoptera. (Synomones are semiochemicals released by one organism that evoke a reaction in an individual of another species that is beneficial to both. Whitman [1988] provides a recent review of semiochemical terminology.) Research in this area has been reviewed several times recently (e.g., Vinson and Iwantsch 1980; Nordlund et al. 1981; Price 1981; Price 1986; Nordlund et al. 1988); however, in most cases, experiments were designed mainly to demonstrate if parasitoids were indeed attracted to the particular host-plant species of their prey when compared to non–host plants or neutral substrates. While this information is interesting and of potential use in pest management (e.g., Nordlund and Sauls 1981), it does not directly address the question of the evolution of plant defense against herbivorous insects via the attraction of natural enemies. This question cannot be addressed without a better understanding of the role of *intraspecific* variation in synomone production as a causal agent for intraspecific host-plant variation in herbivore mortality caused by natural enemies.

One system that illustrates, at least in part, the potential effects of within-species variation in synomone production involves naturally and artificially selected lines of cotton, *Gossypium hirsutum* L., the phytophagous insects *Heliothis zea* (Boddie) (Lepidoptera: Noctuidae) and *Heliothis virescens* (F.), both of which feed upon cotton as well as other plant species, and the parasitoid *Campoletis sonorensis* (Cameron) (Hymenoptera: Ichneumonidae), which attacks several noctuid species on a number of their host plants. The parasitoid is attracted to the odors of leaves and flowers of several plant species, simple chemical extracts of cotton and sorghum (*Sorghum bicolor* [L.]) (Elzen et al. 1983), and at least a few particular sesquiterpenes isolated from cotton (Elzen et al. 1984).

These sesquiterpenes, as well as a number of other important allelochemicals including gossypol, occur almost entirely in pigmented glands on the cotton plant (Bell et al. 1987). Glandless cultivars, developed through plant breeding, do not produce these synomones and are less attractive to *C. sonorensis* (Elzen et al. 1986). Thus it would seem that the glanded condition in undomesticated cotton varieties would be favored by natural selection through the influence of synomones attracting *C. sonorensis* to glanded plants. However gossypol, and perhaps other compounds in these glands, also directly reduces *Heliothis* spp. feeding behavior, survival, and growth (Lukefahr and Houghtaling 1969; Zummo et al. 1983; Bell et al. 1987). Thus, in an evolutionary context, it may be difficult to separate the direct benefits of the glanded condition on plant fitness via reducing *Heliothis* growth and survival from the indirect benefits of the plant being more attractive to *Heliothis*'s natural enemies.

Moreover, the phytochemicals attractive to *C. sonorensis* produced by cotton apparently are not produced by sorghum (Elzen et al. 1984). This suggests that *C. sonorensis* exploits different phytochemicals to locate *Heliothis* spp. on different host plants. Additionally, the fact that *C. sonorensis* also has alternate prey species may make it difficult to defend a simple evolutionary scenario emphasizing the role of *C. sonorensis* as a direct agent of selection for terpenoid production by cotton. While there is no doubt that *C. sonorensis* exploits cotton phytochemicals as synomones, the mechanisms underlying the evolutionary development of such behavior are unknown.

Sequestration

In addition to the direct effects of plant characteristics on the behavior of natural enemies outlined above, intraspecific plant genetic variation may affect the palatability of herbivores to predators via sequestration of plant allelochemicals. This general topic has also been reviewed several times recently (e.g., Duffey 1980; Duffey et al. 1986; Pasteels et al. 1988; Bowers 1988b), and again my purpose is not to review this vast field in total, but to concentrate on those aspects relating to the role of intraspecific plant genetic variation and the natural enemies of herbivores.

Sequestration is not a passive process, and qualitative and quantitative patterns of chemical composition of herbivores need not mirror their particular host plant (Brower et al. 1982, 1984a, 1984b; Bowers and Puttick 1986; Lynch and Martin 1987; Martin and Lynch 1988). Insects that feed on milkweed (*Asclepias* sp.) regulate uptake of total and individual cardenolides, selectively sequester particular compounds of those available, and convert plant-produced precursors to forms more effective or more easily stored. Thus, genetic variation for allelochemical production within plant species may make a relatively minor contribution to overall variation in the palatability of individuals within an insect population as compared to interspecific differences in plant chemical composition.

Nevertheless, there is at least one case known where herbivorous insects feeding on plant genotypes producing relatively low levels of an allelochemical were themselves more extensively preyed upon to the extent that their host plants suffered less defoliation. The beetle *Chrysomela aenicollis* (Shaeffer) (Coleoptera: Chrysomelidae), like many other willow-feeding chrysomelids (see Pasteels et al. 1988 for a recent review), utilizes salicin and perhaps other phenolglucosides of its host species, *Salix orestera* Schneider and *Salix lasiolepis* Bentham, as precursors for the synthesis of salicylaldehyde, the active component of its defensive secretion. Willow clones vary over a fivefold range in phenolglucoside concentration, and defoliation by *C. aenicollis* increases with increasing phenolglucoside content. When larvae were experimentally placed on adjacent high- and low-salicin clones after having their defensive secretions depleted, larvae on the high-salicin clone survived better, grew faster, and more rapidly filled their defensive glands than did beetles reared on the low-salicin clone (Smiley et al. 1985). Although predation was not measured directly, the effect of host-plant clone on salicylaldehyde regeneration suggested that the reduced observed survival of larvae on the low-salicin clone was the result of higher predation. From these results, the authors suggested that herbivory by *C. aenicollis* constitutes a selection pressure favoring willow clones that have deleted salicin from their defensive chemical repertoire, at least in areas where beetles were not restricted by cold temperatures (Smiley et al. 1985).

Herbivore Quality

Plant factors can also alter the suitability of herbivores for survival, growth, and reproduction, not only of their parasitoids and predators, but also of their pathogens. Modification of the physiological suitability of herbivorous insects for utilization by their natural enemies is probably both the most widespread and least understood manifestation of host-plant variation on herbivore–natural enemy interactions.

In only a few cases do we know how particular plant-produced allelochemicals alter herbivore suitability for natural enemies. To date, the best-studied case of intraspecific phytochemical modification of herbivore–natural enemy relationships involves lines of cultivated tobacco (*Nicotiana tabacum* L.), some specialized and generalized herbivorous insects associated with tobacco, and similarly specialized and generalized parasitoids of those herbivores.

Differential parasitization of the tobacco hornworm, *Manduca sexta* (Johannson) (Lepidoptera: Sphingidae), on tobacco genotypes selected for differential nicotine production was among the first observations suggesting a role of genetic variation among plants on relationships between herbivores and their natural enemies (Morgan 1910). Only recently has the role of variation in dietary nicotine content on herbivores and their parasitoids been examined in detail, however. In the field, Thorpe and Barbosa (1986) found no significant difference in survival of *M. sexta* on high- and low-nicotine tobacco genotypes. Rates of parasitization by an endemic population of the parasitic wasp *Cotesia congregata* (Say) (Hymenoptera: Braconidae) also did not differ between plant genotypes; however, the number of parasitoids surviving to adulthood was significantly greater on *Manduca* reared on low-nicotine tobacco, and development time of surviving female parasitoids was greatest on the high-nicotine *Manduca* (Thorpe and Barbosa 1986). In related laboratory studies, Barbosa et al. (1986) found that the more polyphagous parasitoid *Hyposoter annulipes* (Cresson) (Hymenoptera: Ichneumonidae) successfully parasitized a smaller proportion of its host, the fall armyworm, *Spodoptera frugiperda* (Smith) (Lepidoptera: Noctuidae), when caterpillars were reared on an artificial diet containing nicotine compared to caterpillars reared on a nicotine-free diet. Parasitoid developmental time was lengthened and size was reduced when parasitoids were reared on nicotine-fed hosts. The more specialized *C. congregata* can both tolerate higher quantities of nicotine in its tissues and more efficiently eliminate nicotine acquired from its host's hemolymph than can the more polyphagous *H. annulipes* (Barbosa et al. 1986). However, because *S. frugiperda* developmental time and pupal weight were both adversely affected by dietary nicotine, it is not clear whether nicotine ingested by *S. frugiperda* directly affected *H. annulipes* survival and growth, or whether the observed reductions in parasitoid life history parameters were an indirect consequence of the effects of nicotine on *S. frugiperda*'s metabolism and development (El-Heineidy et al. 1988). The problem of isolating the direct effects of a plant-produced allelochemical through a food chain on a natural enemy from its indirect effects, acting wholly on the herbivore, unfortunately, remains to be directly addressed for most tritrophic associations studied so far.

Pubescence

Foliar pubescence is an effective plant defense against a number of herbivorous insect species (Levin 1973); however, the mode of action of such a defense is relatively nonspecific. Pubescence often inhibits phytophage and entomophage alike. Dense glanded or nonglanded trichomes on foliage may be detrimental to relatively small predators, such as early-instar chrysopids (Treacy et al. 1985), coccinellids (Putman 1955; Banks 1957; Plaut 1965), and egg parasitoids (Obrycki 1986). Often, nonglandular trichomes inhibit movement of natural enemies, thus decreasing their efficiency by increasing search time, while the sticky and/or toxic constituents of glandular trichomes may be as detrimental to natural enemies as to several herbivorous species (Obrycki and Tauber 1984).

Under certain conditions, moderate levels of pubescence may actually increase the efficiency of some natural enemies by increasing their turning rate while walking, thus increasing their probability of encounter with potential prey. In contrast, smooth, waxy leaf surfaces may be too slippery for natural enemies to conduct a thorough search (Lauenstein 1980; Shah 1982). The potential conflict between the positive value of dense leaf pubescence in plant resistance and negative value in biological control suggests that where pest suppression by predators and parasitoids is to be integrated with host-plant resistance, plant genotypes having either moderate levels of pubescence (Obrycki 1986) or only particular kinds of trichomes (Ruberson et al. 1989) may offer the best compromise between either glabrous or highly pubescent forms.

Exceptions to this generalization are known, and no doubt more will be found. For example, while leaf pubescence is an effective resistance mechanism for the cereal leaf beetle *Oulema melanopus* (L.) (Coleoptera: Chrysomelidae) in that beetles oviposit less on pubescent genotypes of wheat (*Triticum aestivum* L.), pubescence apparently has no adverse effect on parasitization by one egg and three larval parasites (Casagrande and Haynes 1976; Lampert et al. 1983). In contrast, *Heliothis* spp. prefer to oviposit on pubescent cotton leaves (Lukefahr et al. 1971); thus the glabrous condition has both direct effects in reducing *Heliothis* densities and indirect effects in facilitating *Heliothis* egg parasitization by *Trichogramma pretiosum* (Riley) (Hymenoptera: Trichogrammatidae) and predation by early-instar larvae of *Chrysopa rufilabris* (Burmeister) (Neuroptera: Chrysopidae) (Treacy et al. 1985). Other combinations of phytophagous and entomophagous insects associated with cotton are affected differently by the glabrous condition (see Schuster and Calderon 1986 for a recent review). This suggests that the effect of glabrous leaves on overall cotton fiber or seed yield (or fitness) may be difficult to predict from its effect on individual herbivore–natural enemy combinations.

Population Models

The previous section reviewed some of the many ways in which prey reared on different host plants have been shown to vary in their risk of being discovered and discussed variation in quality for utilization by natural enemies, although, in most cases, the consequences of such variation at the population levels were not examined explicitly. In order to better understand the consequences of host-plant variation on population dynamics of herbivore and natural enemy, it may be helpful to review some of the obvious interactions between host-plant resistance and natural enemies. The following graphical models depict the variation in equilibrium population density of herbivores as a function of host-plant suitability in both the presence and absence of natural enemies. These models may also be useful in determining when the use of most plant resistance is compatible with the use of biological control agents in applied systems (Duffey and Bloem 1986). For simplicity of presentation, the equilibrium density of herbivores in figure 12.1 is shown to decline linearly with increasing level of host-plant resistance in the absence of natural enemies in all cases.

A purely additive relationship between host-plant resistance and natural

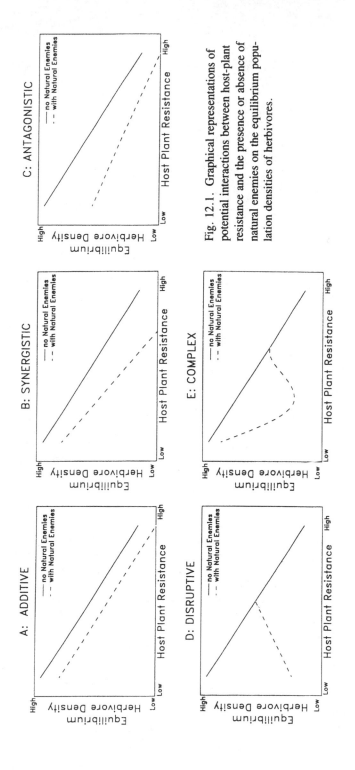

Fig. 12.1. Graphical representations of potential interactions between host-plant resistance and the presence or absence of natural enemies on the equilibrium population densities of herbivores.

enemies is shown in figure 12.1A. The incremental numerical reduction in equilibrium herbivore density caused by natural enemies is independent of that caused by host-plant resistance and uniform at all levels of host-plant resistance. The expected equilibrium pest density due to host-plant resistance and natural enemies can be predicted simply from the combined effects of both acting independently. Although this relationship may be the most restrictive in that it precludes any biological interaction, several cases will be presented in following sections demonstrating the independence of herbivore mortality from host-plant resistance and natural enemies.

A simple synergistic interaction is shown in figure 12.1B. The incremental reduction in equilibrium herbivore density caused by natural enemies is relatively greater at high host-plant resistance levels than at low. Obviously, host plant resistance and biological control would be compatible tactics in pest management if their interaction followed a synergistic model. Indeed, a synergistic interaction would be ideal. One possible mechanism for the increased susceptibility to natural enemies of herbivores on resistant plants might be that the activity of an allelochemical responsible for resistance to the herbivore imposes a metabolic stress sufficient to inhibit the herbivore's defensive responses. Another mechanism may be that the resistant plants release higher concentrations of volatile chemicals attractive to natural enemies than plants more suitable for herbivore growth and survival.

A simple, mildly antagonistic interaction is shown in figure 12.1C. Although the reduction in equilibrium herbivore density due to host-plant resistance and natural enemies is less than would be calculated if the interaction were additive, this mildly antagonistic interaction may also be a compatible interaction in pest management, because the numerical reduction in equilibrium pest density due to both factors is greater than either alone. This is probably not the case for the more severe antagonism shown in figure 12.1D.

When the antagonism is more severe (fig. 12.1D), the reduction in equilibrium herbivore density once caused by natural enemies is now replaced by that of host-plant resistance. This is probably the expectation when natural enemies are more susceptible than herbivores to plant resistance mechanisms (e.g., Campbell and Duffey 1979; Obrycki and Tauber 1984). Under this model, host plant resistance and biological control would be incompatible at low to intermediate host-plant resistance levels, and there would be essentially no biological control at high host-plant resistance levels owing to the high mortality suffered by natural enemies. The equilibrium density of herbivores on highly resistant plants in the presence of natural enemies is not expected to exceed the density when natural enemies are absent. Such a situation would occur only if natural enemies in some way facilitate herbivore survival on highly resistant plants.

More complex interdependent patterns of mortality factors probably occur. For example, host-plant resistance and natural enemies could act synergistically at low host-plant resistance levels and antagonistically at medium to high host-plant resistance levels (fig. 12.1E). Such an interaction may occur, for example, where herbivores and their natural enemies not only have different sensitivities to plant factors mediating host-plant resistance but also have different thresholds or mechanisms of exposure. In this case, the level of compatibility varies with the level of host-plant resistance. Several examples will be presented illustrating compatible interactions between host-plant resistance and biological control agents on moderately resistant cultivars but antagonistic interactions on highly resistant cultivars.

Influence of Host-Plant Resistance on Natural Enemy Populations

Applied Systems

Studies exploring the potential interaction between resistant crop varieties and natural enemies in a number of crops illustrate the graphical models described above (table 12.1). Antagonistic interactions were found in 6 of the 16 cases (37.5%) involving parasitoids, while a synergistic relationship was found in only 2 of the 16 cases (12.5%). No clear interactions (i.e., additive relationships) were found in 5 (31.2%). In the remaining 3 cases, the form of the relationship varied with the resistance level of the cultivars employed. Interestingly, host-plant resistance in soybean was responsible for 4 of the 6 clearly antagonistic interactions (see also Orr and Boethel 1986), including reductions in survival and fecundity of an egg parasitoid when host eggs were produced by adult stinkbugs feeding on resistant soybean (Orr et al. 1985).

At least in rice, host-plant resistance appears to be quite compatible with the biological control provided by generalist predators, as most studies to date indicate an additive relationship between host-plant resistance and predation (Myint et al. 1986; Salim and Heinrichs 1986). In at least two studies, however, mild antixenosis caused increased movement of prey and presumably facilitated their discovery and capture. Brown rice plant hoppers, *Nilaparvata lugens* (Stål) (Homoptera: Delphacidae), suffered heavier predation by the spider, *Lycosa pseudoannulata* (Bosenberg et Strand) Araneida: Lycosidae) on nonpreferred rice cultivars (Kartohardjono and Heinrichs 1984), and fall armyworms were more heavily attacked by *Orius insidiosus* (Hemiptera: Anthocoridae) (Say) on nonpreferred maize varieties (Isenhour et al. 1989).

In contrast, two other hemipteran predators, *Geocoris punctipes* (Say) (Hemiptera: Lygaeidae) and *Podisius maculiventris* (Say) Hemiptera: Pentatomidae) were negatively affected by resistance to caterpillars in soybean. *Geocoris*, which normally feeds on plant material to acquire water, exhibited increased development time and increased nymphal mortality when its diet included foliage from soybean cultivars resistant to the velvetbean caterpillar, *Anticarsia gemmatalis* Hübner (Lepidoptera: Noctuidae), and the soybean

TABLE 12.1.
Effect of host-plant resistance in cultivated crops on natural enemies of selected insect pests

Pest	Natural enemy	Resistant plant	Mode of resistance	Effect on natural enemy	Comments	References
Parasitoids						
Pseudoplusia includens	*Copidosoma truncatellum*	Soybean (*Glycine max* L.)	Reduced survival and growth	Antagonistic	Increased parasitoid development time	Orr and Boethel 1985
Pseudoplusia includens	*Microplitis demolitor*	Soybean	Lengthened development time, reduced weight	Antagonistic	Reduced parasitoid survival, reduced number of available prey	Yanes and Boethel 1983
Pseudoplusia includens	*Voria ruralis*	Soybean	Reduced survival	Syn./ant., varied with resistance level	Reduced parasitoid puparia per host, reduced pupal weight, increased host death prior to parasitoid emergence	Grant and Shepard 1985
Epilachna varivestis	*Pediobius foveolatus*	Soybean	Lengthened development time, reduced survival, reduced egg production	Antagonistic	Possibly a net compatibility of BC with HPR	Kauffman and Flanders 1985
Epilachna varivestis	*Pediobius foveolatus*	Soybean	Reduced survival, growth, and weight	Add./ant.	Male-biased sex ratio on one cultivar, no effects on others	Dover et al. 1987
Heliothis sp.	*Microplitis croceipes*	Soybean	Antibiosis, reduced growth	Syn./ant. varied with resistance level	Increased development time, reduced parasitoid adult weight	Powell and Lambert 1984
Nezara viridula	*Telenomus chloropus*	Soybean	Mild antibiosis and antixenosis	Antagonistic	Reduced survival, reduced total fecundity of egg parasitoid	Orr et al. 1985

Pest	Natural enemy	Crop	Plant resistance effect	Interaction	Effect on natural enemy	Reference
Spodoptera frugiperda	Campoletis sonorensis	Maize (Zea mays L.)	Reduced growth	Antagonistic	Increased parasitoid development time	Isenhour and Wiseman 1989
Spodoptera frugiperda	Campoletis sp.	Maize	Antixenosis	Synergistic	Increased parasitization rate	Pair et al. 1986
Four Lepidoptera species	Apanteles marginiventris, and others	Soybean	Reduced survival, reduced growth, lengthened development time	Additive	No differences in parasitism levels across cultivars	McCutcheon and Turnipseed 1981
Schizaphis graminum	Lysiphlebus testaceipes	Barley (Hordium vulgare L.), Sorghum (Sorghum bicolor L.)	Reduced population growth, reduced size	Antagonistic	Fewer and smaller mummies, but not disruptive	Starks et al. 1972
Schizaphis graminum	Lysiphlebus testaceipes	Oat (Avena sativa L.)	Reduced population growth	Additive	No effect on parasitoid survival, sex ratio, or development time	Salto et al. 1983
Aphis gossypii	Lysiphlebus testaceipes, and others	Cantaloupe (Cucumis melo L.)	Antibiosis and/or antixenosis	Additive	Parasitization level independent of aphid density across cultivars	Kennedy et al. 1975
Acyrthosiphon pisum	various parasitoids and predators	Alfalfa (Medicago sativa L.)	Reduced population growth	Additive	No variation in parasitism rates or predator numbers	Pimentel and Wheeler 1973
Myzus persicae	Aphidius matricariae	Chrysanthemum (Chrysanthemum sp.)	Reduced population growth	Additive	No effect of density on searching behavior or apparent effect on parasitoid population growth	Wyatt 1970

TABLE 12.1.
(*continued*)

Pest	Natural enemy	Resistant plant	Mode of resistance	Effect on natural enemy	Comments	References
Bruchus pisorum	*Eupteromalus leguminis*	Pea (*Pisum sativum* L.)	Antixenosis	Synergistic	Increased parasitization	Annis and O'Keefe 1987
Predators						
Nephotettix virescens	*Lycosa pseudoannulata*	Rice (*Oryza sativa* L.)	Reduced oviposition and survival	Additive	Predatory spider	Myint et al. 1986
Nephotettix virescens	*Cyrtorhinus lividipennis*	Rice	Reduced oviposition and survival	Additive	Predatory bug	Myint et al. 1986
Sogatella furcifera	*Lycosa pseudoannulata*	Rice	Reduced survival	Additive	All predators cause ca. 30% additional mortality on all varieties.	Salim and Heinrichs 1986
Sogatella furcifera	*Cyrtorhinus lividipennis*	Rice	Reduced survival	Additive	All predators cause ca. 30% additional mortality on all varieties.	Salim and Heinrichs 1986
Sogatella furcifera	*Harmonia octomaculata*	Rice	Reduced survival	Additive	All predators cause ca. 30% additional mortality on all varieties.	Salim and Heinrichs 1986
Sogatella furcifera	*Paederus fucipes*	Rice	Reduced survival	Additive	All predators cause ca. 30% additional mortality on all varieties.	Salim and Heinrichs 1986
Nilaparvata lugens	*Lycosa pseudoannulata* and others	Rice	Mild antixenosis, reduced survival	Synergistic	Increased predation rate due to increased prey movement	Kartohardjono and Heinrichs 1984
Anticarsia gemmatalis	*Geocoris punctipes*	Soybean	Reduced growth rate	Antagonistic	Reduced growth rate and survival	Rogers and Sullivan 1986
Pseudoplusia includens	*Geocoris punctipes*	Soybean		Antagonistic	Reduced growth rate	Rogers and Sullivan 1986

Herbivore	Natural enemy / pathogen	Plant	Effect	Interaction	Effect on natural enemy	Reference
Pseudoplusia includens	*Podisius maculiventris*	Soybean	Lengthened development time, reduced growth	Antagonistic	Similar effects on predator as on host	Orr and Boethel 1986
Spodoptera frugiperda	*Orius insidiosus*	Maize	Antibiosis, antixenosis	Synergistic	Increased predation	Isenhour et al. 1989
Heliothis zea	*Orius insidiosus*	Maize	Reduced growth	Synergistic	Increased predation	Isenhour et al. 1989
Pathogens						
Spodoptera exigua	*Bacillus thuringiensis*	Celery (*Apium graveolens* L.)	Antibiosis, reduced growth rate	Synergistic	Increased dose-specific mortality	Meade 1989
Trichoplusia ni	*Bacillus thuringiensis*	Celery	Antibiosis, reduced growth rate	Synergistic	Increased dose-specific mortality	Meade 1989
Heliothis zea	*Bacillus thuringiensis*	Soybean	Reduced growth rate	Synergistic	Increased dose-specific mortality	Kea et al. 1978
Heliothis zea	*Bacillus thuringiensis*	Soybean	Reduced survival	Synergistic	Death at earlier age	J. V. Bell 1978
Heliothis virescens	*Bacillus thuringiensis*	Cotton (*Gossypium hirsutum* L.)	Antixenosis	Synergistic	Increased host mortality	Schuster et al. 1983
Heliothis zea	*Nomuraea rileyi*	Soybean	Reduced survival	Synergistic	Death at earlier age	J. V. Bell 1978
Four Lepidoptera species	*Nomuraea rileyi*	Soybean	Antibiosis	Additive	No effect on other pathogens	Gilreath et al. 1986

looper, *Pseudoplusia includens* (Walker) (Lepidoptera: Noctuidae). Similar increases in mortality and development time occurred when predators were reared
on caterpillars that were reared on resistant soybean cultivars (Rogers and Sullivan 1986). Preimaginal development of *P. maculiventris* closely paralleled the
effect of host-plant resistance on *P. includens* reared on resistant and susceptible
soybean cultivars; development time was increased and growth rates were reduced on foliage from resistant plants compared to susceptible plants (Orr and
Boethel 1986). In the latter study, the effects of host-plant resistance extended
to the fourth trophic level in that the reproductive capacity of the egg parasitoid,
Telenomus podisi Ashmead (Hymenoptera: Sceleonidae), was reduced when it
was reared from eggs of *P. maculiventris* whose prey was *P. includens* reared on
resistant soybeans as compared to susceptible soybeans (Orr and Boethel
1986). The net outcome of this four-trophic-level interaction on the overall fitness of any of the individual participants under field conditions remains to be
ascertained. In summary, as with many of the parasitoids, and in contrast to
most other crops, antibiosis to the insect herbivores of soybean appeared to
interact negatively with the predators of those insect herbivores.

Most studies to date indicate a general compatibility between host-plant
resistance and biological control by pathogens. Host-plant resistance of maize
to *H. zea* is due to the presence of unidentified compounds in the silk of resistant varieties. In laboratory studies, larvae treated with a nuclear polyhedrosis
virus (NPV), then fed artificial diets containing silk from different cultivars,
suffered greater viral mortality when fed resistant silk than when fed susceptible
silk (Hamm and Wiseman 1986). In a related field test, viral mortality was least
on the most susceptible variety, and Hamm and Wiseman (1986) concluded that
the susceptibility of *H. zea* larvae to NPV was inversely related to the growth
and vigor of the larvae, which itself was directly related to the level of host-
plant susceptibility. Beach and Todd (1988) also found an NPV infecting *P.
includens* to be compatible with soybean host-plant resistance in that the virus
caused a nearly equivalent percentage of additional mortality on both resistant
and susceptible cultivars.

In the laboratory, the fungal pathogen *Nomuraea rileyi* (Farlow) Sampson
caused *H. zea* larvae to die at an earlier age on resistant, compared to susceptible, soybean cultivars, although the pathogen caused 100% mortality of larvae
reared on all cultivars (J. V. Bell 1978). However, Gilreath et al. (1986) found
no consistent variation in the incidence of *N. rileyi* or any other pathogen in *H.
zea*, *P. includens*, *A. gemmatalis*, or the green cloverworm, *Plathypena scabra*
(F.) (Lepidoptera: Noctuidae), collected from four soybean cultivars in the
field.

Unlike the other pathogens mentioned, vegetative growth of *Bacillus thuringiensis* (Berliner) (*Bt*) is not necessary for death of the insect. *Bt* spores and
endotoxin normally enter via feeding, and in most susceptible insects death is
caused directly by ingested endotoxin. High midgut pH inhibits spore germi-

nation. After partial cleavage by the insect's proteases, the activated toxin causes a swelling of the midgut epithelial cells, disruption of ion transport, and eventual separation and breakdown of the midgut epithelium. Such disruption leads to ion and pH imbalances in the hemolymph, total body paralysis, and eventual death. Spores then germinate within the hemolymph (Aronson et al. 1986). Thus, the effect of host-plant allelochemicals on solubilization and activation of the endotoxin are at least as important, if not more so, as the direct effects of those allelochemicals on *Bt* vegetative growth and sporulation.

In nearly all of the cases examined, *Bt* interacts synergistically with host-plant resistance, in that host insects reared on resistant plants are themselves more susceptible to *Bt* (table 12.1). For example, *H. zea* fed a resistant soybean genotype had a 40% lower LD_{50} to a commercial formulation of *Bt* than did insects reared on a susceptible genotype (Kea et al. 1978). J. V. Bell (1978) also found that *H. zea* suffered higher total mortality and succumbed at an earlier age when reared on a resistant compared to a susceptible soybean genotype. Similarly, Schuster et al. (1983) found *Bt* to be more efficacious for *H. virescens* suppression when insects were reared on a high-tannin cotton cultivar than when reared on a low-tannin cultivar. More recently, Meade (1989) compared the susceptibility of *Trichoplusia ni* (Hübner) (Lepidoptera: Noctuidae) and *Spodoptera exigua* (Hübner) (Lepidoptera: Noctuidae) to measured doses of *Bt* when reared on genotypes of celery that differed in suitability for insect survival, growth, and efficiency of conversion into insect biomass. Both insect species were affected similarly by variation among celery genotypes. Surviving insects reared on the plant genotype least suitable for insect growth and survival were themselves most susceptible to *Bt*. The mechanisms for increased susceptibility to *Bt* of insects reared on resistant plant genotypes are unknown.

It is difficult to translate results of short-term studies on the effects of host plants on particular life history parameters of natural enemies into changes in herbivore–natural enemy population dynamics. A negative impact on survivorship, growth, or fecundity of a particular natural enemy does not always lead to a disruption of biological control of the herbivore population. For example, while plant resistance led to a reduction in the size and number of parasitoids from greenbugs, overall plant damage was least and greenbug populations were smallest on resistant varieties in the presence of parasitoids (Starks et al. 1972).

Similarly, Kauffman and Flanders (1985) reported an increase in development time, reduced survival, and reduced reproduction of *Pediobius foveolatus* Crawford (Hymenoptera: Eulophidae) when reared on Mexican bean beetles, *Epilachna varivestis* Mulsant (Coleoptera: Coccinellidae), while on resistant soybeans, the population growth potential of the parasitoid was reduced less than was that of the host. Thus, despite the negative impact on individual life history parameters, the intrinsic rate of increase of the parasitoid was greatest relative to that of its host on the resistant cultivar. Even though host-plant resistance and biological control might be judged antagonistic at the individual level, they are expected to be compatible at the population level in this system. Using a modeling approach based upon energy flow, Gutierrez (1986) also concluded that, in general, the net effect of host-plant resistance would be beneficial to natural enemies if host-plant resistance suppressed herbivore populations more than natural enemy populations. Taken together, these studies illustrate that the compatibility of host-plant resistance and biological control cannot be predicted simply from measuring the effect of host-plant resistance on life history parameters or population growth of the natural enemy. Corresponding data on the herbivore are equally essential.

Natural Systems

The most relevant research on the effect of genetic variation among populations of undomesticated plants on the success of natural enemies comes from studies of plant galls. In general, gall makers inhabiting large galls are less susceptible to parasitoids owing to limitations on the length of the parasitoids' ovipositors. To the extent that gall diameter is under partial genetic control of the host plant, there exists the possibility that gall makers and their parasitoids may impose competing selection on plants for gall size. In goldenrod (*Solidago altissima* L.), the size of galls induced by *Eurosta solidaginis* (Fitch) (Diptera: Tephritidae), is a heritable character of both insect and plant genotype (Weis and Abrahamson 1986), but variation among plant genotype for gall size (and ultimate parasitoid success) would not, in general, lead to differential reproduction of plant genotypes (Weis and Abrahamson 1985). Thus, in this system, the effect of parasitoids as a selective agent favoring gall makers that induce larger galls was more significant than was the effect of parasitoids reducing herbivore damage to plants. An additional complicating factor in this system is that avian predators more readily attack large galls, and the advantage for gall makers to induce large galls appears to be partially (but not completely) offset by their increased risk of predation (Weis and Abrahamson 1986).

Price and Clancy (1986a) and Clancy and Price (1987) similarly implicated genetic variation among clones of willow, *S. lasiolepis*, in the size of galls induced by stem-galling sawflies. In the first study, mean gall diameter induced by *Euura lasiolepis* Smith (Hymenoptera: Tenthredinidae) differed consistently among willow clones over a five-year study. Sawflies in smaller galls were more susceptible to parasitization by *Pteromalus* sp. (Hymenoptera: Pteromalidae), and the percentage of parasitism by *Pteromalus* was inversely related to gall diameter. In the second study, mean gall diameter induced by another tenthredinid, *Pontania* sp., also differed significantly among willow clones. In contrast to the previous example, it was the sawflies in the more rapidly growing and larger galls that were more susceptible to parasitism. Individuals of another leaf-folding sawfly (*Phyllocolpa* sp.) attacking these willow species also consistently differed in their susceptibility to parasites and other sources of mortality, depending upon which plant clone they attacked (Fritz and Nobel 1990).

The question as to whether the variation in susceptibility of sawflies to their natural enemies induced by willow clones ultimately influences clone fitness has not yet been examined, however.

Nevertheless these studies clearly document that genetic variation among plants influences the interaction between herbivores and their natural enemies, and such variation may play an important role in determining both the species composition and densities of insects associated with particular plant genotypes (see also Fritz, this volume). Unfortunately, however, none of the "natural" tritrophic systems studied to date contribute much toward understanding the dynamic processes in ecological and evolutionary time that culminated in the present-day observable patterns.

Genetic Implications of Tritrophic Interactions

Natural Enemy Enhancement: A Mechanism of Host-Plant Resistance?

Most studies reviewed here have tended to emphasize the role of plant genetic variation on the herbivore–natural enemy interaction with little regard either for the factors initially responsible for that variation or for the consequences of subsequent genetic change within and among plant populations. Clearly, strong arguments can be made that genetic variation within plant species may substantially affect the evolution and maintenance of herbivore–natural enemy relationships. For example, differential susceptibility to natural enemies among plant genotypes may be expected to impose natural selection on herbivores for enemy-free space, that is, to identify and selectively oviposit on plant genotypes where mortality of their progeny due to natural enemies would be minimized. Simultaneously, we might also expect plant genotypes providing enemy-free space for their herbivores to be at a selective disadvantage; natural selection should act on plants to minimize the quantity of enemy-free space they provide.

Plant genetic variation may also impose substantial heterogeneity such that any attempts to model the dynamics of herbivore and natural enemy populations must be restricted to using individual plant genotypes as the largest sampling unit (Price and Clancy 1986a). On the other hand, it is not clear from the data presently available that natural enemies have been the primary agent of natural selection for any plant characteristics facilitating their own success.

The absence of any clear examples of natural enemies of herbivores imposing selection on plants is not surprising given the relatively small number of ecological studies addressing questions of this type. Applied systems, while valuable for demonstrating the potential range of plant responses, cannot, of course, address this question, since the "resistant" plant varieties examined were selected purely for the magnitude of their direct effects on herbivores; all effects on the herbivores' natural enemies are incidental. The few natural systems studied so far also have focused more on the effect of preexisting plant variation on herbivore–natural enemy interactions than on the effect of herbivores' natural enemies on plant fitness.

Such data are difficult to collect, even with the most tractable of systems. Indeed, it is difficult enough to demonstrate the importance of herbivores on plant evolution even in the absence of natural enemies (e.g., Jermy 1984). While asexually reproducing plants provide several advantages in isolating the genetic component of variation from the total phenotypic variation, asexual plants present problems in attempting to determine variation in fitness among clones, especially for long-lived species. Measurements of clonal growth or persistence over a few seasons may not be an unbiased predictor of fitness over the entire lifetime of that clone. The relatively diffuse nature of coevolution expected of most plant-herbivore interactions (i.e., the evolution of a particular trait in one or more species in response to a trait or suite of traits in several other species; see Fox 1981; Futuyma 1983a; Strong et al. 1984; Fritz, this volume) is no doubt repeated in an even greater level of complexity in most tritrophic interactions as well.

Unlike other mechanisms of host-plant resistance that act directly against herbivores, the expression of resistance via natural enemies may be variable; no selective advantage accrues to a plant producing high concentrations of potential synomones if there are no natural enemies locally available to respond to them. Thus, the intensity and fidelity of natural selection for such plant characteristics may be less, in some situations, than for plant characteristics acting directly upon the herbivores; realized levels of plant resistance based upon enhancing natural enemies may be less uniform than if based upon more direct mechanisms. While natural enemies can indeed respond to and exploit plant variation, the importance of natural enemies of herbivores as effective agents of natural selection on plant genotypes within species has not yet been ascertained for any tritrophic association.

Evolutionary Trajectories of Tritrophic Interactions

The very nature of tritrophic interactions implies the potential for genetic variation at one level to affect the ecological and evolutionary relationships between the other two. The net effect of host-plant resistance on the dynamics of herbivore populations will depend upon the shape of the herbivore–natural enemy interaction (fig. 12.1), and variation in the shape of these interactions may have diverging implications for the subsequent evolution of host selection behavior of herbivores and of natural enemies.

It would seem reasonable that there would be always a strong selective advantage for herbivores to select susceptible plants when the tritrophic interaction follows either the additive or synergistic models (figs. 12.1A and 12.1B). Presumably, herbivores that avoided antibiotic plants would benefit directly and also indirectly by reducing mortality from natural enemies.

A prediction is more difficult if the tritrophic interaction follows an antagonistic model (fig. 12.1C), however. If host-plant resistance disrupts biological control, then herbivores may be under conflicting selection pressures, depending upon the relative strength of host-plant resistance and natural enemies as sources of mortality. Under predictably high mortality from natural enemies, herbivores may realize highest fitness by seeking out and utilizing moderately resistant plants (or plant parts), provided that the direct reductions in herbivore survival and reproduction due to host-plant resistance are more than offset by the gains in avoiding mortality from natural enemies. In the absence of natural enemies, however, one would expect herbivores to discriminate against all but the most suitable host plants.

Such a mechanism could easily promote intraspecific variation among plant populations in host-plant use by herbivores as a function of the variation in occurrence and abundance of natural enemies. This scenario is, of course, a variation on the theme of enemy-free space, except that the space is provided by moderate host-plant resistance that coincidentally produces poor-quality hosts for natural enemies.

It is also reasonable that natural selection would promote positive genetic correlations in plants between independent factors that both directly and indirectly reduce their risk to herbivores. Disruptive interactions (fig. 12.1D) eventually should be selected against, especially where herbivore mortality from natural enemies is high and predictable. However, where natural enemies are a less predictable source of herbivore mortality, then natural selection may favor efficient, direct host-plant resistance mechanisms regardless of their effects on natural enemies. Therefore, the observation of negative correlations may not be surprising either in relatively new tritrophic associations where plant, herbivore, and natural enemies have had little opportunity for adaptation, or in which host-plant resistance is the more effective mechanism of plant protection. Disruptive tritrophic interactions may be more common in newly-introduced plant species or new, resistant crop varieties (e.g., soybean) than in associations with longer evolutionary histories.

Responses to selection are functions of the intensity of selection and the heritability of the traits being selected. Predictions of responses to selection require knowledge of both. The evolutionary trajectory of any tritrophic interaction depends upon the magnitude of additive genetic variation in traits mediating that interaction at all levels, and the relative intensity of selection by one level on each of the other two. Although there is a growing body of literature documenting that genetic variation in plants can influence herbivore–natural enemy interactions, the genetic mechanisms associated with the evolution, maintenance, and further development of tritrophic interactions have not been explored. Previous chapters of this book have pointed out the value of quantitative genetics in better understanding the dynamics of plant-herbivore associations (Kennedy and Barbour; Berenbaum and Zangerl; Simms and Rausher;

Karban). The application of quantitative genetics techniques also will permit us to quantify levels of genetic variation in host plants, herbivores, and natural enemies and measure selection intensities at each level. Without such information, predictions of the evolutionary trajectory of any tritrophic interaction cannot be made.

Acknowledgments

I thank R. S. Fritz, E. L. Simms, R. F. Luck, T. Meade, and an anonymous reviewer for their critical comments on previous drafts of this chapter.

PART FOUR

Evolution of Plant Resistance

13

Selective Impact of Herbivores

Robert J. Marquis

Plants are subject to predictable, systematic and directional pressures from pathogens and herbivores which participate with them in an incessant evolutionary dance.—D. Levin 1976b

The manner and degree to which individual plants interact with the cooccurring community of potential herbivores vary greatly. Damage varies among and within plant species in the same habitat, across habitats, both latitudinally and altitudinally, and between years. Some individuals and some species escape damage altogether or essentially so, while others are constantly chewed, scraped, bored, mined, sucked, galled, deflowered, uprooted, or otherwise maimed. The starting and usually unstated assumption for much of plant-herbivore theory is that these interactions are and have been detrimental to plant fitness, and have led to the evolution of traits that lower the probability of attack. Numerous plant characteristics are hypothesized to have arisen as evolutionary responses to selection by herbivores (table 13.1). Theories of host-plant specificity (e.g., Bernays and Graham 1988; Thompson 1988b), coevolution among plants and their herbivores (e.g., Ehrlich and Raven 1964; Breedlove and Ehrlich 1972; Southwood 1972; Benson et al. 1975; Berenbaum 1983; Spencer 1988b; but see Futuyma 1983a), and distribution of secondary compounds among plant parts (e.g., McKey 1979) and plant species (e.g., Feeny 1976; Rhoades and Cates 1976; Coley et al. 1985; Southwood et al. 1986) all depend on our understanding of selection by herbivores for plant resistance.

Despite this imperative, theories of plant-herbivore interactions in natural environments are based more on observations of general patterns of plant resistance than on actual analysis of the process of natural selection as affected by herbivores. Certainly the almost universal distribution of chemical and physical feeding deterrents in the plant kingdom is at first glance compelling evidence

TABLE 13.1
**Plant characters hypothesized to have evolved in response to selection pressure
by herbivores**

Plant character	References	
Constitutive production of secondary compounds	Fraenkel 1959 Ehrlich and Raven 1964 Southwood 1972 Dolinger et al. 1973 Benson et al. 1975 Janzen 1975b Feeny 1976 D. A. Levin 1976a Rhoades and Cates 1976 Edmunds and Alstad 1978 Edwards and Wratten 1980 Owen and Wiegert 1981	Langenheim et al. 1982 Herrera 1982 Bryant, Wieland et al. 1983 Gulmon and Mooney 1986 Southwood et al. 1986 Bazzaz et al. 1987 Loehle 1988 Tuomi, Niemelä, Chapin et al. 1988 Macedo and Langenheim 1989 Tallamy and Krischik 1989 Weckerly et al. 1989
Inductive production of secondary metabolites	Rhoades 1979, 1983 Bryant, Weiland et al. 1985	Tallamy and Krischik 1989
Differential distribution of secondary metabolites among tissue types and fruit ages	McKey 1974a McKey 1979	Herrera 1982
Tissue nutritive quality	Southwood 1972 Feeny 1976	Rhoades and Cates 1976 Neuvonen and Haukioja 1984
Plant trichomes	Gilbert 1971	Levin 1973
Seed and fruit morphological characters	Janzen 1969, 1971, 1975a Smith 1970	Bradford and Smith 1977
Seed size	Janzen 1969	
Seed crop size	Janzen 1969	
Selective seed abortion	Herrera 1984	Fernandes and Whitham 1989
Terrestrial plant life history	van der Meijden et al. 1988	
Algal life history and morphology	Lubchenco and Cubit 1980	
Age of first reproduction	Janzen 1975a	
Mast flowering and fruiting	Beattie et al. 1973 Janzen 1976b, 1978c	Inouye and Taylor 1979 Silvertown 1980
Intraseason flowering time	Augspurger 1981	Evans et al. 1989
Intraseason fruiting time	Thompson and Willson 1979 Johnson and Slobodchikoff 1979 Herrera 1982	
Pattern of fruit development	Herrera 1982 Janzen 1982	
Intercalary meristems, perenniality, prostrate growth form in Gramineae	Hyder 1972 Owen and Wiegert 1981	McNaughton 1984

TABLE 13.1
(continued)

Plant character	References	
Silica bodies in Gramineae	Stebbins 1981 Herrera 1985	McNaughton and Tarrants 1983 McNaughton et al. 1985
Food bodies	Janzen 1966	
Domatia	Lundstroem 1887 Janzen 1966	Benson 1985
Extrafloral or floral nectaries	Delpino 1874, in Bentley 1977 Janzen 1977	Tilman 1978 McNaughton 1983b
Butterfly egg mimics	Williams and Gilbert 1981	
Ability to compensate for tissue loss following herbivory	McNaughton 1983b McNaughton and Chapin 1985	Paige and Whitham 1987 van der Meijden et al. 1988
Leaf and plant mimicry	Gilbert 1975 Barlow and Wiens 1977	Ehleringer et al. 1986 Atkinson and Greenwood 1989
Growth pattern of clonal herbs	Edwards 1984	Southwood 1987
Seed disperal pattern	Janzen 1970 Connell 1971	Howe and Smallwood 1982
Serotiny	Zammit and Westoby 1988	
Leaf flushing times and leaf growth rates	Janzen 1975b	Aide 1988
Tissue toughness	Greenwood and Atkinson 1977	Coley 1987
Reduced visual apparency	Atkinson and Greenwood 1989	
Divaricating growth form	Greenwood and Atkinson 1977	Atkinson and Greenwood 1989
Spines and thorns	Janzen 1975b	Greenwood and Atkinson 1977
Root-shoot ratios	van der Meijden et al. 1988	
Plant growth rates	Westoby 1989	
Diel growth patterns in tropical seaweeds	Hay et al. 1988	
Germination time	Southwood 1972	
Leaf water content	Scriber 1977	
Phenology of leaf abscission	Owen 1978	
Tree habit in Galapagos cacti	Dawson 1966	

that these plant attributes have evolved in response to past selection. There is much evidence that secondary compounds (e.g., Chambliss and Jones 1966; Harley and Thorsteinson 1967; Parr and Thurston 1972; Ikeda et al. 1977; Berenbaum 1978; Blau et al. 1978; Raman et al. 1979; Bryant, Wieland et al. 1983; Miller and Feeny 1983; Martin and Martin 1984; Reichardt et al. 1984; Dreyer et al. 1985; Jogia et al. 1989; Potter and Kimmerer 1989) and plant physical characteristics (e.g., Levin 1973; Johnson 1975; Stipanovic 1983; Cooper and Owen-Smith 1986; Southwood 1986; Duffey 1986; Doss et al. 1987) do decrease herbivore feeding, growth, and survival. In examples where plants have escaped their herbivores on island habitats, absence of secondary compounds and other traits potentially related to resistance suggests that herbivores have been selective agents on mainland plant populations (Carlquist 1970; Janzen 1973c, 1975a; Rickson 1977) or populations on other islands (Schoener 1988), and that these traits are metabolically costly and are selected against in the absence of herbivores. In addition, various grass species show intraspecific differences in growth form and response to defoliation consistent with known past herbivory regimes (Kemp 1937; Bradshaw 1960; Lodge 1962; Mahmoud et al. 1975; Detling and Painter 1983; Scott and Whalley 1984; Carman 1985; Carman and Briske 1985; see also Bryant et al. 1989). Thus, there is a positive association between presence of herbivores and differences in response to damage, growth form, and the presence of resistance traits in plants.

However, problems arise in interpreting whether specific plant traits are resistance traits because of alternative selective agents (e.g., pathogens, Alexander, this volume) or alternative functions of the same trait (e.g., leaf hairs simultaneously reducing evapotranspiration and decreasing the ability of the herbivore to attack underlying leaf tissue: Reader 1979; Lieberman and Lieberman 1984; Coughenour 1985; Marquis 1987). Furthermore, verification of past selection is difficult if not impossible. This makes geographic patterns of variation in secondary compounds (e.g., Dolinger et al. 1973; Sturgeon 1979) difficult to interpret (Futuyma 1983a; but see Bryant et al. 1989). Because herbivore abundance varies greatly both temporally and spatially, we cannot measure present selection and reasonably assume that it represents past selection intensity. We can, however, measure the process of selection imposed by herbivores on plants occurring today.

Ten years ago, most evidence for the effects of herbivore feeding on plant growth and reproduction had come from the agricultural and forestry literature (Hanson et al. 1931; Chester 1950; Ellison 1960; Jameson 1963; Kulman 1971; Bardner and Fletcher 1974; Krischik and Denno 1983). Since then, the effects of herbivores on plants have been studied in a number of natural systems (see reviews by Crawley 1985; Belsky 1986; Hendrix 1988; Louda et al. 1988). The range of often diametrically opposed opinions expressed in recent literature (table 13.2) indicates that these new data have not produced a consensus about the selective impact of herbivores in natural systems. Whence comes this vari-

TABLE 13.2
Statements in recent literature concerning the role of herbivores as selective agents

Janzen 1979, 331:
"The answer to Why do 'all the good things which an animal likes have the wrong sort of swallow or too many spikes'? . . . is that the herbivores selected the plants to be that way."

Futuyma 1983a, 210:
"Although there is ample evidence that secondary compounds provide protection against herbivory, there is less evidence that they evolve rapidly in response to changes in insect fauna. This is surely not for lack of intense selection; insects clearly can have a severe impact on plant survival and reproduction, especially when, as is often the case, they attack meristems or reproductive structures."

Jermy 1984, 614:
"It is most likely, however, that insect attacks seldom if ever reduce plant fitness in space and time so extensively that they represent overall active selection factors that significantly influence plant evolution."

McNaughton 1983b, 657:
"Plants have been subjected to intense and recurrent natural selection to reduce herbivore impact upon them and to compensate for attacks when defenses have been breached."

Strong et al. 1984, 214:
"Indeed, the data suggest that many, probably the majority, of the species of insects on most host plants are too rare most of the time individually to significantly affect the fitness of their hosts."

Crawley 1985, 163:
"It does seem, however, that plant compensation is not fully effective against low-density herbivore populations and, therefore, that slight differences in the level of defoliation between plants may cause important differences in plant fitness."

Labeyrie 1987, 4:
"Maybe, then, we entomologists, shall realize that damage caused by insects is not necessarily of great consequence for all kinds of plants . . . Thus, the influence of the destruction of fruit and seeds on the evolution of the number of ovules per fruit and the number of fruits per plant, seems to me of no use for perennial plants, for which sexual reproduction only occasionally participates in the keeping up and renewal of the population. So, some speeches about coevolution often become the expression of an excessive romanticism."

Bernays and Graham 1988, 888:
"The relative importance of arthropods in plant survival is also unknown. Despite the great diversity of phytophagous insect species, density if commonly low, and most are extremely rare in relation to their host plants. They would thus exert little influence on the relative abundance of different plant genotypes. The major impact comes from relatively few species."

Fox 1988, 907:
"There is abundant evidence that [the statement that herbivores exert significant selective pressures on their food plants] is often true in spite of the seeming rarity of many herbivores or chronically low consumption rates."

Hendrix 1988, 246:
"One important selection pressure faced by all or nearly all plants results from the removal of vegetative and reproductive structures by herbivores."

Rausher 1988b, 898–899:
"The only way to support or refute the contention that herbivores commonly exert important selection pressures on resistance traits is to measure selection and determine its causes directly. Such attempts are beginning to be made and are suggesting that such selection may in fact be quite common."

Schultz 1988b, 896:
"The potential selective impact of herbivory on naturally occurring plants has been measured very few times, all within the past 5 yr. Because the quantitative genetics of the plant species must be known to infer selective impact, even these studies are preliminary; nonetheless, each suggests that even small amounts of damage can have significant fitness effects."

ation in opinion—from the differential reading, acceptance, and/or interpretation of the evidence?

Reviewer viewpoint (table 13.2) varies depending on the author's criteria for demonstration of evolutionary impact by herbivores on plants. For some, evolutionary impact is equated with the amount of tissue consumed; because herbivore population sizes are often small, individual herbivore species consume only small amounts of plant tissue, and thus are thought to have little impact (Jermy 1984; Strong et al. 1984; Bernays and Graham 1988). For others (Janzen 1979; McNaughton 1983b), distribution patterns of presumed resistance characters and growth responses to herbivory, in combination with the fossil record, provide overwhelming evidence that herbivores have affected the evolution of plants. Others see substantial evidence in the literature for phenotypic selection by herbivores (which is then apparently equated with evolutionary response) despite low damage levels and compensating mechanisms following damage (Crawley 1985; Fox 1988; Hendrix 1988). Finally, others (Futuyma 1983a; Rausher 1988b; Schultz 1988a) propose that although selection may be occurring continuously, evidence for selection alone does not reveal anything about the nature of the evolutionary response by the target plant populations.

The evolution of many plant characters (table 13.1) is difficult to explain except in the context of past selection by herbivores, yet our understanding of the population-level processes that have led to the evolution of these characters is limited. For herbivores to impose natural selection on resistance traits in host-plant populations, both phenotypic selection on variable traits and a genetic response to selection must occur (Falconer 1981; Lande and Arnold 1983; Endler 1986). The microevolutionary process leading to a change in host-plant traits, i.e., natural selection by herbivores and an observed response in the subsequent generation in one or more characters related to resistance, has not been reported for any natural system.

Criteria for Documenting Selection and Evolutionary Impact

Natural selection resulting from herbivore attack occurs when relative plant fitness as affected by damage is correlated with a particular plant trait or set of traits. Three conditions are necessary to demonstrate selection by an herbivore on plant traits: (1) intraspecific variation in damage; (2) intraspecific variation in a plant trait or set of traits that is correlated with damage; and (3) correlation between the trait (and damage as well) and fitness. In order for there to be evolutionary response to selection there must be additive genetic variance for the plant trait, and thus for damage. Many studies document a subset of the necessary conditions for selection, but few include most or all. Some studies (Simms and Rausher 1987, 1989; Marquis 1990) have considered the plant trait under selection to be overall resistance to damage, and thus synonymous with the amount of damage suffered in the presence of herbivores. Endler (1986) lists 11 plant-herbivore systems as examples of "direct demonstration of natural selection." In all cases listed by Endler (1986) except *Trifolium repens*, evidence for differential effects of tissue loss to herbivores on one or more fitness characters is missing.

Individual plant fitness is defined as total lifetime seed production, taking into account seed quality as it affects the probability of seedling establishment (e.g., Hendrix and Trapp 1989). Fitness is difficult to measure, especially for perennial plant species or if seed production as a pollen donor is to be included in total seed production. Understanding the effects of herbivores on long-lived perennials requires long-term studies; herbivore-caused effects on fitness traits over the short term cannot necessarily be extrapolated over the lifetime of a plant because damage often varies greatly from year to year (e.g., Price and Willson 1979; Barbosa and Schultz 1987).

I will first discuss the nature of evidence available for three related phenomena, intraspecific variation in damage, association of damage with plant traits, and the genotypic and environmental components underlying variability in both damage and correlated traits. I will then consider the range of effects of herbivores on plant fitness, and methodological problems associated with measuring them. Finally, I will briefly discuss results from studies in which most or all of the criteria (as outlined above) have been met in order to demonstrate an evolutionary impact by herbivores. I use the term "herbivore" in the broad sense to include both vertebrates and invertebrates that attack seeds (seed predators; Janzen 1971) and/or vegetative tissue.

Empirical Evidence

Components of Intraspecific Variation in Damage

Within a single plant population individuals often vary greatly in amount of herbivore-caused damage, both to seeds before dispersal (e.g., Newton 1967; Janzen 1975d; Mattson 1978; Moore 1978; De Steven 1981, 1983; Scurlock et al. 1982; Evans et al. 1989) and to vegetative tissue (e.g., Edmunds and Alstad 1978; Solomon 1981; McDonald 1981; Coley 1983; Langenheim and Hall 1983; Louda and Rodman 1983a; Seastedt et al. 1983; Marquis 1987; Núñez-Farfán and Dirzo 1988; Tsingalia 1989; van der Meijden et al. 1988). Damage can vary significantly between neighboring populations (e.g., Janzen 1975d; Hare 1980; Marquis 1987; Abrahamson et al. 1989; Anderson 1989) and over both short (Janzen 1978b) and extended environmental gradients (Louda 1983). Distribution patterns of damage among individuals vary greatly depending on the system: in some plant species, the distribution of damage is strongly skewed, with most individuals either suffering heavy attack (e.g., De Steven 1983) or escaping attack (e.g., Janzen 1978b; Sacchi et al. 1988), whereas in other species damage varies widely within a population (e.g., Janzen 1977,

1984; Moore 1978; Anderson 1989). Thus, in some systems, herbivores may be selecting against a very few individuals, while in others, the entire population may be under selection, but of varying intensity among different individuals.

The genetic response to selection will depend on the relative importance of plant genotype versus environment in determining level of damage. Plant genotype has been shown to affect numbers of herbivores per plant (Hare and Futuyma 1978; Dritschilo et al. 1979; Moran 1981; Service 1984b, Maddox and Cappuccino 1986; Karban 1987a; Fritz et al. 1987b; Fritz and Price 1988), herbivore community composition on a given plant (Maddox and Root 1987; Fritz and Price 1988; Fritz 1990a), pupal mortality in feeding trials (Perry and Pitman 1983), and amount of damage suffered by a plant (Debarr et al. 1972; Dimock et al. 1976; Linhart et al. 1981; Compton, Beesley et al. 1983; Åhman 1984; Marquis 1984, 1990; Berenbaum et al. 1986; Simms and Rausher 1987, 1989; McCrea and Abrahamson 1987; Sacchi et al. 1988; Silen et al. 1986; Anderson et al. 1989; Schowalter and Haverty 1989).

Several factors intrinsic to the plant, and therefore potentially under genetic control, are known to correlate with the level of attack (table 13.3). These factors include physical properties (e.g., quantity and morphology of leaf pubescence), chemical properties, and plant phenology. A few studies have demonstrated that genotypes with less damage also have higher levels of putative resistance traits (e.g., either secondary chemistry: Dirzo and Harper 1982a; Horrill and Richards 1986, Berenbaum et al. 1986; or morphological characters: Fritz and Price 1988; Sacchi et al. 1988). However, putative resistance traits do not always correlate in the expected direction with damage level (e.g., Lincoln and Langenheim 1979; Haukioja, Niemelä et al. 1985) or may show no correlation at all with damage (e.g., Tempel 1983; Greig-Smith and Wilson 1985). Consequently, caution should be exercised when attributing resistance benefits to untested plant traits (see also Dirzo 1985).

There is also a strong environmental component to the amount of damage suffered by individual plants (table 13.4). Plant traits that influence damage level, particularly secondary chemical composition, can vary with the environment in which a plant is growing (light: James 1950; Langenheim et al. 1981; Larsson, Wiren et al. 1986; Waterman et al. 1984; soil nutrients: Brown et al. 1984; Gershenzon 1984; water: Gershenzon 1984) and with the extent of previous damage (e.g., Rhoades 1979; McNaughton and Tarrants 1983; Schultz 1988b; Young 1987). Plant physical properties that change along environmental gradients can also cause variation in damage (e.g., leaf pubescence: Ezcurra et al. 1987; stem thickness: Tscharntke 1987).

The influence of the environment on level of damage is likely complex. For example, microenvironmental variation along gradients or across sharp boundaries influences damage levels by altering host-plant quality and apparency (e.g., MacGarvin et al. 1986; Harrison 1987), or by affecting the herbivore

TABLE 13.3
Plant characters shown to be correlative with or causes of within-site, intraspecific variation in herbivore damage

Plant character	Damage effect	Genetic influence
Plant flowering phenology	Solomon 1981 De Steven 1981 Berenbaum et al. 1986	Cooper 1960 Hanover 1966a Paterniani 1969 Stern and Roche 1974 Eck et al. 1975 Murfet 1977 Carey 1983 Berenbaum et al. 1986
Plant vegetative phenology	Ågren 1987 Sholes and Beatty 1987 Williams and Bowers 1987 Aide 1988 Collinge and Louda 1989	Stubblebine et al. 1978
Constitutive secondary chemistry	Jones 1962 Cooper-Driver and Swain 1976 Rice et al. 1978 Connally et al. 1980 Farentinos et al. 1981 Dirzo and Harger 1982a Compton, Beesley et al. 1983 Louda and Rodman 1983a Greig-Smith and Wilson 1985	Hanover 1966a Klun et al. 1970 Matzinger et al. 1972 Sprague and Dahms 1972 Gallun et al. 1975 Hefendehl and Murray 1976 Maxwell and Jennings 1980 Lamberti et al. 1983 Berenbaum et al. 1986
Leaf pubsecence	Doss et al. 1987 Schoener 1988	Southwood 1986
Stem thickness	Tscharntke 1987	
Plant size	Thompson 1978 Rausher et al. 1981 Williams 1983 Courtney and Manzur 1985 McCrea and Abrahamson 1985 Young 1985 Sullivan et al. 1986 Sullivan and Sullivan 1986 Forsberg 1987 Basey et al. 1988 Tiritilli and Thompson 1988	
Shoot and leaf length	Craig et al. 1986 Fritz and Price 1988 Fritz and Nobel 1989	Fritz and Price 1988 Fritz and Nobel 1989
Fruit size	Hare 1980	

TABLE 13.3
(continued)

Plant character	Damage effect	Genetic influence
Tissue nutrient concentration	Cook et al. 1978 McClure 1980b Greig-Smith and Wilson 1985	
Plant gender	Bawa and Opler 1978 Danell, Huss-Danell et al. 1985 Lovett Doust and Lovett Doust 1985 Ågren 1987 Elmqvist et al. 1988	

itself (e.g., behavior: Moore et al. 1988) or by a combination of both (Collinge and Louda 1988). Response to variation in a single factor in host-plant quality can vary among herbivore species (e.g., Hare and Futuyma 1978; Carroll and Hoffman 1980). Even the direction in which resource availability influences plant traits is not readily predictable. Microsites with low resource availability often produce stressed individuals (as measured by growth rate) which are less well defended than neighbors. In some cases, these plants suffer greater attack (Young 1985), but in other cases their quality to herbivores is reduced and they suffer less damage (Harrison 1987). High resource availability may also have multiple outcomes. For example, plants may be better defended and thus experience less damage, or they may produce more nutritious tissue or just more tissue, and thus sustain more damage. The known influence of environment on one herbivore species cannot necessarily be extrapolated to all species of an herbivore fauna; response to a single factor in host-plant quality often varies among herbivore species (Hare and Futuyma 1978; Carroll and Hoffman 1980). Furthermore, climatic changes (e.g., drought) potentially may change the preference of the herbivores themselves irrespective of changes in the host population through effects on availability of other resources (e.g., water: Young 1985) or alternative hosts.

In some cases it may be difficult to differentiate genotypic and environmental influences on resistance traits. For example, in the case of induction of secondary compounds (e.g., Rhoades 1979; Baldwin and Schultz 1983; Haukioja and Neuvonen 1985; Schultz 1988b), we might expect genetic variation (Zangerl and Berenbaum 1990) for characteristics associated with the induction process (e.g., threshold damage level, rate of induction, final level of induction, and relaxation time). But because the induction process is triggered by extrinsic factors, subsequent damage may depend on past damage that may or may not have been influenced by genotype. Genotype and environment may interact to determine level of infestation (Maddox and Cappuccino 1986; Schowalter and Haverty 1989), while host genotype may influence the level of attack by only a

TABLE 13.4
Environmental factors shown to be correlated with or causes of within-site, intraspecific variation in herbivore damage

Environmental factor	References
Light	Collinge and Louda 1988
	Harrison 1987
	Lincoln and Langeheim 1979
	Lincoln and Mooney 1984
	Moore et al. 1988
Soil moisture content	Lewis 1984
	Bernays and Lewis 1986
	Mattson and Haack 1987a
	Oluomi-Sadeghi et al. 1988
Soil acidity	Bink 1986
Associated plant species	Atsatt and O'Dowd 1976
	Thomas 1986
	Fenner 1987
	Holmes and Jepson-Innes 1989
Soil nutrient content	Stark 1965
	Onuf 1978
	Onuf et al. 1977
	McClure 1980b
	Rhoades 1983
Density of conspecifics	Stanton 1983
	Åhman 1984
	Duggan 1985
	McLain 1984
	Paulissen 1987
Tide level	Foster 1984

portion of the total herbivore fauna of a plant species (Woods and Clark 1979; Simms and Rausher 1989; Marquis 1990). Even for the same system the relative influence of genotype versus environment may change over time (Marquis 1990). Any temporal change in a number of factors (e.g., abundance of individual herbivore species, soil nutrient and water content, light level, and abundance of alternative host-plant species) could swamp out a genotype effect, or in contrast, increase the importance of genotype versus environment compared to some previous period.

The above studies demonstrate that both genotype and environment can influence plant traits that in turn affect the level of damage sustained by individual plants. Thus, there is no a priori reason to assume that individual variation in damage has a genetic basis. For plant species that host more than one type of herbivore (the majority of species), certain herbivore species may respond to environmentally based variation in their host plant while other herbivore species may respond to genotypically based resistance traits. Whether in fact genotype has a significant effect on total damage will depend on the extent of losses caused by each group. In order to understand how genotype and environment interact, field experiments are necessary in which both environmental factors and plant genotype are simultaneously manipulated, and individual herbivore species responses are monitored.

Fitness Consequences of Herbivory

Herbivores can damage plant tissue at any stage of the plant's life cycle. The consequences of such damage may vary with the plant's life stage, however. For example, just one or two bites may kill a seed or seedling, whereas loss of an equal amount of leaf tissue may have little or no measurable effect on an adult plant. Because the mean amount of leaf area lost to folivores is generally 5–15%, it has been suggested that herbivores have little selective impact. This belief ignores two important facts: first, rarely, if ever, do all plants in a population consistently experience the same level of damage. Selection can potentially occur in any population regardless of the mean damage level as long as there is variation in damage level around that mean. Second, even relatively small amounts of damage to certain tissue can cause mortality (e.g., Clark and Clark 1985; De Steven and Putz 1985) or decreases in growth (e.g., Dirzo 1984), or negatively affect seed production (e.g., Archer and Tieszen 1980). Thus, the selective impact of herbivory is not strictly determined by mean damage level suffered by a population of plants, but instead by: (1) variance in damage among individuals (Fox and Morrow 1986); (2) the relationship between fitness and damage; (3) the tissue attacked; and (4) the plant species being considered, as even closely related species of the same growth form living in the same habitat vary in their response to defoliation (Hodgkinson and Baas Becking 1977; Bentley et al. 1980; Caldwell et al. 1981; Becker 1983; Stalter and Serrao 1983; Heichel and Turner 1984; Coughenour et al. 1985a, 1985b).

Herbivores consume both vegetative and reproductive tissue, and seeds may be consumed both before and after dispersal. The effect on a parent plant's fitness of an herbivore attack depends on whether that attack is against the parent plant before seed maturation, or against the seeds themselves after they have reached maturation (Janzen 1971). Prior to some time near final seed maturation, some plants can respond to damage by reallocating resources to compensate for loss of either vegetative or reproductive tissue. Compensation may be measured as an increase in growth, flowering, and seed production by the remaining parts of damaged plants compared to undamaged control plants (Belsky 1986). Compensation following attack can occur in a number of ways (intrinsic mechanisms, sensu McNaughton 1983a): shifts in sex determination of uneaten flowers (Hendrix and Trapp 1981; Danell, Huss-Danell et al. 1985), increased photosynthesis in uneaten leaf material (Wareing et al. 1968; Hodgkinson et al. 1972; Alderfer and Eagles 1976; Detling et al. 1979; Painter and Detling 1981; Heichel and Turner 1983; von Caemmerer and Farquhar 1984), increased size of new leaves (Danell et al. 1985b), increased chlorophyll content of subsequently produced leaves (Danell, Elmqvist et al. 1985), increased nutrient uptake by new or remaining tissue (Ruess 1984; Danell, Elmqvist et al. 1985; Forno and Semple 1987), early bud break of subsequently produced leaves (Heichel and Turner 1976), changes in resource allocation to adjust shoot/root ratios (Nowack and Caldwell 1984), and delayed senescence in remaining tissue (Bardner 1968). After seed maturation, there is no possibility of compensation. Every mature seed destroyed is a seed for which there is no means of compensation.

There is often an unknown relationship between damage to vegetative tissue and seed and pollen production. Thus, the consequences of attack on vegetative tissue for fitness are less clear than those of direct damage to developing or mature seeds. For both types of damage, reduced seed production is assumed to decrease the potential number of successful offspring. Indirect effects of folivory on male fitness through decreased pollen production or pollen quality have either not been observed (Sacchi et al. 1988; Hendrix and Trapp 1989) or not been tested. When folivory alters mean seed mass (e.g., Hendrix 1979; Bentley et al. 1980), germination time (Hendrix and Trapp 1981; Hendrix 1984), viability (Marquis 1984; Ellison and Thompson 1987), and secondary chemistry of seeds (Janzen 1976a), the consequences of differential attack are far from straightforward. In the following sections, I discuss separately the known fitness impacts of damage to vegetative versus reproductive tissue.

Damage to Vegetative Tissue

Opinion varies about the degree to which plants can compensate for or tolerate loss of vegetative tissue, and therefore about the impact of nonseed predators on plant fitness (e.g., Belsky 1986; McNaughton 1986). Compensatory responses allow the plant to recoup, at least in the short term. The long-term consequences of compensatory growth in perennial plants are usually not known, however (see Belsky 1986). On a limited budget, growth and/or nutrients allocated to compensatory growth must negatively affect future growth beyond the initial compensatory response, assuming equal efficiencies of resource use before and after attack. Long-term effects may differ from short-term effects (Olson and Richards 1988), and herbivory may be beneficial to certain fitness components but detrimental to overall plant fitness (e.g., positive effect on photosynthetic rate but overall decrease in growth: Ingham and Detling 1986). For example, when moose (*Alces alces*) browse twigs of birch (*Betula* spp.), leaves of browsed shoots are larger, have more chlorophyll and remain on the plant longer than those of unbrowsed shoots, resulting in more growth in browsed than unbrowsed shoots (Danell, Huss-Danell et al. 1985). The overall effect is detrimental, however, because browsing results in reduction in the number of present and future flower buds and directs nutrients away from flowering to shoot growth (Danell, Huss-Danell et al. 1985; Bergström and Danell 1987). Most studies of compensatory responses have been with clonal grasses (e.g.,

Archer and Tieszen 1980; McNaughton et al. 1983; Ruess et al. 1983; Wallace et al. 1984, 1985; Pitelka and Ashmun 1985; Georgiadis and McNaughton 1988; Ruess 1988). Unfortunately, effects of herbivory and the compensation itself on both growth and reproduction have not been measured in these studies, making it difficult to interpret the results with respect to effects on plant fitness.

Conflicting opinions as to herbivore effects have also arisen because the nature of the evidence varies depending on whether it is from agricultural/laboratory studies or natural systems. The environmental context of a study is important because the outcome of an herbivore attack or experimental removal of leaf area is influenced by the abiotic environment in which a plant is growing. Light level (Dirzo 1984; Danell, Huss-Danell et al. 1985), nutrient concentration of the substrate (Ruess et al. 1983; Julien and Bourne 1986; Benner 1988; Maschinski and Whitham 1989; Mihaliak and Lincoln 1989), yearly variation in weather (Archer and Tieszen 1980), soil moisture level (Wallace et al. 1984), and the level of exploitative competition (Windle and Franz 1979; Bentley and Whittaker 1979; Lee and Bazzaz 1980; Archer and Detling 1984; Fowler and Rausher 1985; Parker and Salzman 1985; Cottam et al. 1986; Banyikwa 1988; Louda et al. 1988; Hartnett 1989) all influence the effect of leaf herbivory on plant fitness parameters. Overcompensation or increased growth response to tissue loss relative to growth of undamaged plants, when shown to occur, has been much more frequent for plants in cultivation and in greenhouse and laboratory studies (McNaughton 1983a, 1983b; Solomon 1983) than in natural systems. Because of the obvious environmental differences between natural habitats and artificial environments, results from one environment cannot be necessarily extrapolated to the other. Further methodological problems in studying the effects of herbivory on plant growth and reproduction are discussed by Hendrix (1988), Strauss (1988), and Brown and Allen (1989).

Most studies in natural habitats demonstrate that the long-term effects of folivores on plant growth and reproduction are negative (Crawley 1983; Belsky 1986; Verkaar 1986; Hendrix 1988). Folivory can cause greater mortality (Kulman 1971; Waloff and Richards 1977; Rausher and Feeny 1980; Parker and Root 1981; Delph 1986; Thomas 1986), decreased seed production (Waloff and Richards 1977; Kinsman and Platt 1984; Marquis 1984; Stamp 1984; Crawley 1985; Edwards 1985; Paulissen 1987), reduced leaf, stem, or aboveground biomass production (Morrow and LaMarche 1978; Archer and Detling 1984; Louda 1984; Marquis 1984; Parker and Salzman 1985; Wallace 1987), branch death in woody plants (Heichel and Turner 1984), and decreased root growth (Hodgkinson and Baas Becking 1977; Chapin and Slack 1979; Whittaker 1982; Archer and Tieszen 1983; Ganskopp 1988; Zimmerman and Pyke 1988). Less obvious effects include decreased mycorrhizal infection (Wallace 1987), decreased seed viability (Marquis 1984; Elmqvist et al. 1987) and seed size (Janzen 1976a; Bentley et al. 1980; Willson and Price 1980), delayed flowering time (Collins and Aitken 1970; Hartnett and Abrahamson 1979; Islam and

Crawley 1983; Kinsman and Platt 1984; Rai and Tripathi 1985; Marquis 1988b), decreased photosynthesis in remaining tissue (Hewett 1977; Woledge 1977; Johnson et al. 1983; Whittaker 1984), decreased size of regrowth leaves (Duncan and Hodson 1958; Heichel and Turner 1976; Bergström and Danell 1987; Tuomi et al. 1989), and change in sex expression of developing flowers (Hendrix and Trapp 1981; Spears and May 1988). In plants with strong apical dominance, destruction of apical meristems can lead to either mortality (Young 1985; De Steven and Putz 1985) or increased branching due to release of dormant axillary buds (Archer and Tieszen 1980; Whitham and Mopper 1985; Julien and Bourne 1986; Paige and Whitham 1987; Benner 1988; Room 1988). The effect on fitness of an attack on vegetative tissue depends on its timing relative to the timing of resource accumulation, allocation for future growth and reproduction, and environmental constraints on delayed flowering (McCarty and Price 1942; Miller and Donart 1981; Gregory and Wargo 1986; Islam and Crawley 1983). Because detrimental effects may not necessarily occur immediately, long-term studies are necessary. Other factors that influence the effect of leaf area loss include the absolute amount of leaf area removed (Jackston 1980; Marquis 1984), the position and/or age of the leaf removed (Stickler and Pauli 1961; Kulman 1971; Pereira 1978; Marquis 1988a), plant gender (Ågren 1987; Lovett Doust and Lovett Doust 1985; Elmqvist and Gardfjell 1988), and plant size (Marquis 1984; McCrea and Abrahamson 1985; Fenner 1987; Mendoza et al. 1987; Lubbers and Lechowicz 1989).

In contrast, examples from natural systems suggesting beneficial effects of herbivore damage on total seed production are extremely limited (Inouye 1982; Paige and Whitham 1987; Maschinski and Whitham 1989). All three investigations were of herbaceous species in which the level of sexual reproduction is constrained by plant architecture; damage leads to production of multiple flowering stalks which together produce more seeds than the single stalks of control plants. The results from these studies are intriguing, but the studies themselves are not without problems. Inouye's study of *Jurinea mollis* (1982) is not experimental, and therefore suffers from lack of randomization of experimental plants among treatments (see below). In the case of both *Ipomopsis aggregata* (Paige and Whitham 1987) and *I. arizonica* (Maschinski and Whitham 1989), treatments were not necessarily applied at random because deer and elk "chose" plants for the natural herbivory treatment. When overcompensation was observed in *I. arizonica* (Maschinski and Whitham 1989), it was only under certain limited conditions (an unspecified level of fertilization in the absence of competition). Clearly, more studies are needed of plants of this growth form. The inescapable question is why mechanisms to overrule apical dominance, resulting in multiple-stalk production, have not evolved in plant species of this growth form independently of damage by their herbivores. Shrubs and trees with many meristems may show a similar response if terminal bud removal releases lateral meristems to result in increased flowering compared to undamaged plants. The problem with woody plants, which is avoided by studying short-lived species like the monocarpic herbs mentioned above, is the difficulty of extrapolating short-term effects over the lifetime of the plant. In studies reported thus far, the effects of terminal bud grazing in shrubs have been always negative (e.g., Davis 1967; Harper 1977; Bergström and Danell 1987; Roundy and Ruyle 1989), but again, controlled experiments are limited.

Evidence for the impact of folivores on fitness in woody plants comes mainly from studies of the effects of insect outbreaks on temperate trees (e.g., Franklin 1970; Rafes 1970; Kulman 1971; Greaves 1966, 1967; Mazanec 1966, 1974; Giese and Knauer 1977; Schultz and Allen 1977; Alfaro 1982; Schowalter et al. 1986; Filip and Parker 1987; Sullivan and Wyse 1987; Benoit and Blais 1988). From one-third to complete defoliation can cause significant decreases in basal area accumulation and height growth, with mortality occurring subsequent to higher levels of foliage loss (Kulman 1971; MacLean and Ostaff 1989). Furthermore, experiments indicate that chronically high damage levels in Australian *Eucalyptus* forests decrease diameter growth (Greaves 1966, 1967; Morrow and LaMarche 1978). Attack by sucking insects can reduce shoot growth (Foster 1984; Sutton 1984) and trunk diameter growth (Dixon 1971a, 1971b).

Unfortunately, methodological problems dictate that results from most previous studies of folivory in woody plants be interpreted with caution. Usually the effects of herbivory are determined by comparing the growth rings of attacked versus unattacked trees. Because treatments (various levels of defoliation) are not applied at random, both genetic and environmental influences on growth and reproduction may be confounded with treatment effects. In addition to their methodological problems, these studies also fail to indicate the effects of chronic low levels of foliage loss (5–15%) in temperate tree species (Barbosa and Schultz 1987; Landsberg and Ohmart 1989; but see Crawley 1985). On the one hand, because of the known compensatory ability of many plant species, it seems that trees, which appear to have a very large resource base, should be able to compensate for most small losses. However, because carbohydrate movement is often restricted in plants (Watson and Casper 1984), relatively small amounts of folivory concentrated around reproductive structures could have significant detrimental effects on overall seed production (Janzen 1976a; Stephenson 1980; Marquis 1988a; Tuomi, Niemalä et al. 1988; Tuomi, Nisula et al. 1988). These results suggest that at least in some woody species, constraints on resource movement in plants limit the level of compensation following attack because the entire resource base is not available for reallocation to all parts of the plant. In contrast, there appears to be much more interspecific variability in the degree of physiological integration among herbaceous species than woody species (Watson and Casper 1984; Thomas and Watson 1988; Garrish and Lee 1989; Shea and Watson 1989).

Our limited understanding of the effects of folivory greatly surpasses our knowledge of how herbivory on roots affects plant fitness. Root herbivory and its effects have been little studied for the obvious reason of the difficulty entailed (Richards 1984; Brown and Gange 1989). Nevertheless, roots are attacked by a wide range of organisms, including fossorial mammals (D. C. Anderson 1987; Cantor and Whitham 1989), insects (e.g., Ueckert 1979; Müller et al. 1989; Sites and Phillips 1989), and nematodes (N. L. Stanton 1983; Ingham and Detling 1986). High concentrations of secondary chemicals, morphological features such as spinous coverings on corms of geophytic Iridaceae in South Africa (Lovegrove and Jarvis 1986), and impacts on plant distribution (Reichman and Smith 1985; Reichman 1988; Cantor and Whitham 1989; Reichman and Jarvis 1989) suggest that, as with aboveground parts, underground herbivores have had a selective effect on root characteristics. Karban (1980, 1982) has shown that periodic cicadas, which feed on young roots as nymphs, significantly reduce tree diameter increments. Simulated snow-goose damage on *Scirpus americana* rhizomes caused negative effects on all growth characters measured (Giroux and Bedard 1987). Because of our lack of knowledge of herbivore interactions with roots, which represent one-third or more of a plant's biomass, we are far from the goal of quantifying total herbivore impact on plants.

Damage to Reproductive Tissue

Flower parts or entire flowers are commonly eaten (Le Pelley 1932; Carlson 1964; DeBarr 1969; Dolinger et al. 1973; Bawa and Opler 1978; Boscher 1979; Inouye and Taylor 1979; Bertin 1982; Gross and Werner 1983; Edwards 1984; Crawley and Nachapong 1985; Berenbaum et al. 1986; Scharpf and Koerber 1986; Zammit and Hood 1986; Armbruster and Mziray 1987; Horvitz et al. 1987; Kirk 1987; A. N. Anderson 1987; Ågren 1988; Zwölfer 1988). Although destruction of ovaries and entire flowers is likely to have a direct negative effect on seed production (e.g., Berenbaum et al. 1986; Anderson 1989; Molau et al. 1989), the effects on pollinator visitation of damage limited to corollas, calices, and colored bracts have not been investigated. In turn, differential damage to male versus female reproductive parts has been observed (Bawa and Opler 1978), but the effects of pollen consumption (Kirk 1987) or male flower consumption on male fitness have not been considered. Compensation following attack on reproductive parts may be a frequent occurrence. An increase in the number of floral primordia which mature fruits compared to control plants has been observed following floral bud removal in agricultural species (Dale 1959; Smith and Bass 1972) and one natural system, *Pastinaca sativa* (Hendrix and Trapp 1981).

It is not known whether selection on the parent plant continues to occur once seeds have been dispersed. Seed predation varies with the distance seeds are dispersed from the parent plant (Wright 1983; Clark and Clark 1984; Howe et al. 1985), with the density of those seeds (Clark and Clark 1984), and among years at the same site (Schupp 1988). Although certain individual plants may consistently suffer more damage at the adult stage, it is not clear how often seed predation varies among progenies of different parent plants. Only one study has shown that a postdispersal seed predator differentially feeds on seeds from different parent plants, based on palatability (Thompson 1985). The potential for selection is great because postdispersal predation can be up to 100% of seeds dispersed per day, varying both among plant species and across habitats (e.g., Janzen 1971; Brown et al. 1975; Sork and Boucher 1977; Reichman 1979; Abramsky 1980; Heithaus 1981; Wright 1983; De Steven and Putz 1984; Mittelbach and Gross 1984; O'Dowd and Gill 1984; Kjellson 1985; Louda and Zedler 1985; A. N. Anderson 1987, 1989; Sork 1987; Zammit and Westoby 1988; Smith et al. 1989).

Estimation of the impact of predispersal seed predation on total seed output is direct. In contrast, the relationship between the amount of vegetative tissue removed and reduction in seed production is indirect and in most species unknown. However, we cannot necessarily measure seed predation alone and ignore other forms of herbivory for a particular system. Attacks on vegetative tissue clearly can have significant impacts on plant fitness and could confound estimates of herbivore impact based solely on measurement of attack on reproductive tissue. At this time, we do not know the relationship between level of attack on vegetative tissue and that on reproductive tissue for individual plants.

Case Studies

I review below the most complete studies available that demonstrate an evolutionary impact on plants by herbivores, i.e., studies that show that plants vary in damage level, that amount of damage is correlated with a plant trait, and that the plant trait is heritable and correlated with fitness. Only Berenbaum et al. (1986) demonstrate selection and potential for genetic response according to the above criteria. Simms and Rausher (1989) and Marquis (1990) document how variation in damage (not a trait which can be measured independent of the presence of herbivores) was related to both genotype and fitness. Three systems (*Piper arieianum, Salix lasiolepis,* and *Solidago altissima*) involved perennial plant species, for which components of fitness rather than lifetime fitness were measured. Effects of herbivores on fitness of *Trifolium repens* have been measured as survivorship of seedlings, but only in greenhouse plantings. Together, these studies provide an intriguing glimpse at the nature of the process of evolution of resistance traits in plants, and strong incentive for further work in other systems.

Ipomoea purpurea *and Its Leaf and Floral Herbivores*

The leaves of the annual morning glory *Ipomoea purpurea* (Convolvulaceae), a common weed in agricultural fields, are attacked by three chrysomelid (Coleoptera) beetles, *Chaetocnema confinis, Deloyala guttata,* and *Metriona bicolor.*

Flowers and developing fruits are eaten by the corn earworm, *Heliothus zea* (Lepidoptera: Noctuidae). Rausher and Simms (1989) found directional selection for decreased resistance (increased damage) to *C. confinis,* and increased resistance to all other species, based on large-scale plantings of half- and full-sib families in plowed fields. No evidence for stabilizing selection was found which would otherwise explain observed intermediate levels of damage. Seed production from these same plots was on average 20% greater than for plants growing in paired pesticide-sprayed plots in which additional members of the same genetic families were planted (Simms and Rausher 1989). However, genetic covariances between seed number in sprayed plots and amount of damage by any of the insect species in control plots were not significant, suggesting that there is no cost of resistance in this plant species (Simms and Rausher 1989). It would seem that "cost of resistance" refers here to the cost due to additional resistance conferred based on change in some already existing system, rather than the cost of having some resistance system versus having no resistance at all. Clarification of this point must await examination of the basis for heritable differences in damage level in *I. purpurea.*

Pastinaca sativa *and* Depressaria pastinacella

Flowers and young fruits of the biennial wild parsnip (*Pastinaca sativa*) are attacked in both North America and Europe by the parsnip webworm (Lepidoptera: Oecophoridae: *Depressaria pastinacella*). Attack is variable within populations and can result in almost total loss of lifetime fitness (Thompson and Price 1977; Berenbaum et al. 1986). Based on families that consisted of mixtures of half- and full-sibs, Berenbaum et al. (1986) have shown that the number of primary umbel seeds lost to the webworm in the field was heritable, and that this damage variation was genetically correlated with heritable variation in specific allelochemicals in the developing fruits. Because the allelochemical basis for differential damage was heritable, a genetic response in the ensuing generation would be expected, but this was not measured. Variation in various furanocoumarins accounted for the majority of differences in attack. A very important insight from this investigation is that selection for one resistance factor would simultaneously decrease another because of negative genetic correlations. Thus, evolutionary response to selection by the parsnip webworm is constrained.

Piper arieianum *and Its Folivores*

The evergreen shrub *Piper arieianum* (Piperaceae) is attacked by at least 90 species of folivores in the Atlantic lowland wet forest of Costa Rica (Marquis 1991). Damage to individual plants varies greatly within and among populations (Marquis 1984, 1987). Experimental leaf area removal has shown that 10% loss of leaf area can result in significant decreases in growth and seed production (Marquis, in press). Finally, cloning experiments suggest a genotypic basis to differences in damage, with certain genotypes consistently demonstrating lower damage than others over time. Individual herbivore species preferentially damaged certain genotypes more than others (Marquis 1990; see also Linhart 1989; Linhart et al. 1989). However, preferences for specific genotypes varied among herbivore species. For two of four time periods, there were significant negative correlations between total leaf damage per clone and growth, suggesting that selection for resistance to leaf damage had occurred during the course of the experiment (Marquis 1990). It thus appears that the intensity of selection varies over the lifetime of perennial species. The specific traits under selection have not been identified.

Salix lasiolepis *and* Euura lasiolepis

Young shoots of arroyo willow (Salicaceae: *Salix lasiolepis*) in northern Arizona are attacked by the shoot-galling sawfly, *Euura lasiolepis,* resulting in significant differential decreases in reproductive bud formation among different clones (Sacchi et al. 1988). Common garden experiments using clones demonstrated that differences in attack, and by inference damage, in unmanipulated populations have a genetic basis (Fritz and Price 1988; Sacchi et al. 1988). Eighty percent of plants surveyed in natural populations had few stem galls (<0.5 galls/shoot), resulting in loss of 5% or less of total seed production. However, 7.5% of the population experienced 20% or more loss of seed production to herbivores compared to unattacked plants. Differential attack among clones is positively correlated with mean shoot length, accounting for 39–82% variation in gall density among clones (Craig et al. 1986; Fritz and Price 1988).

Solidago altissima *and* Eurosta solidaginis

Flowering stalks of the clonal, perennial herb *Solidago altissima* (a subspecies of *S. canadensis*) are attacked by at least three different galling insects. Each of the three galling species reduces stem, rhizome, flower, and seed production of individual ramets (Hartnett and Abrahamson 1979), although effect on total genet fitness has not been documented. Clones growing naturally vary greatly in the percentage of ramets attacked by one of the galling insect species, *Eurosta solidaginis* (Tephritidae) (Weis and Abrahamson 1986; McCrea and Abrahamson 1987; Anderson et al. 1989), and cloning experiments conducted in common gardens have shown that there is genetic influence on attack by this species (McCrea and Abrahamson 1987; Abrahamson et al. 1988; Anderson et al. 1989). Ramets growing faster at the time of oviposition were more likely to be galled than slower growing ramets (Anderson et al. 1989).

Trifolium repens *and Various Leaf Herbivores*

Selection by herbivores can be difficult to demonstrate even when a particular system has undergone rather thorough investigation. Both the genetic basis and the mechanism for resistance are known in the cyanogenic stoloniferous herb, white clover (Leguminosae: *Trifolium repens*). Hydrogen cyanide is released in damaged leaves of those plants with a dominant allele at both the cyanogenic glucoside locus and the nonlinked hydrolysis enzyme locus (Corkill 1942; Atwood and Sullivan 1943). Despite the fact that hydrogen cyanide is detrimental to the growth of various species of slugs (major herbivores in the field) and affects slug distribution (Dirzo and Harper 1982a), white clover populations are mixtures of cyanogenic and acyanogenic morphs (Jones 1966; Dirzo and Harper 1982a). Some but not all herbivore species selectively graze acyanogenic morphs in the field (Dirzo and Harper 1982b). However, cyanogenesis appears to have a cost, as cyanogenic forms are more susceptible to frost damage and soil moisture stress (Foulds and Grime 1972a, 1972b; Foulds and Young 1977), flower less (Foulds and Grime 1972b; Kakes 1989), and are poorer competitors (Dirzo 1984). Spatial heterogeneity in grazing by different herbivore species and the differential responses of different genotypes to other selective factors (in this case freezing) potentially account for the maintenance of polymorphism within single populations (Dirzo and Harper 1982b). It is not clear at what stage in the life history of the plant selection may be occurring. Although seedlings may differentially survive herbivory in the greenhouse owing to their chemical phenotype (Horrill and Richards 1986; Kakes 1989; but see Ennos 1981a), adult plants are stoloniferous, potentially allowing susceptible ramets, once established, to spread vegetatively. The situation is further complicated by the fact that not all leaves of a given genet express the same phenotype (Till 1987), thus resulting in a mosaic of phenotypes within genets and perhaps further slowing the selection process.

Conclusions

Insufficient evidence is available to delineate the generality and strength of selection exerted by herbivores for resistance traits in plants. There are a limited number of natural systems that have demonstrated the prerequisites for natural selection. Variation in damage among individuals of the same species in the same location appears to be universal; loss of leaf area is detrimental to overall fitness; and in all systems investigated thus far differential damage is at least in part genetically determined. However, the amount of tissue removed and number of seeds destroyed are not high among all plant species (e.g., Coley 1983; Marquis and Braker 1992) nor do equivalent, absolute amounts of tissue loss have equivalent effects on fitness of different plant species. It would be very premature to conclude for any system that herbivores do not influence fitness; for no single system has there been investigation of the influence of herbivores at all life stages of the plant: seeds, seedlings, juveniles, and adults. Lack of impact in one stage does not necessarily rule out impact in other stages. We particularly lack knowledge of the influence of parental genotype on postdispersal seed predation and attack on young seedlings, and the effects of leaf damage on viable pollen production and root growth and functioning.

To define the selective impact of herbivores and predict the evolutionary response of plants to that selection requires that we consider the myriad of selective factors that plants face (Dirzo 1984; Fenner 1987; Sacchi et al. 1988). Because the level of both folivory and seed predation can vary dramatically from year to year at the same site (Hare and Futuyma 1978; Mattson 1978; Price and Willson 1980; Sholes 1981; De Steven 1983; Mack and Pyke 1984; Young 1985; Auld 1986; Barbosa and Schultz 1987; Elmqvist et al. 1988; Klinkhamer et al. 1988; Pasek and Dix 1988; St. Pierre 1989), among sites within the same year (e.g., Hare and Futuyma 1978; Auld 1986; Ågren 1987), and among years for the same individuals (Coley 1983; but see Alstad and Edmunds 1983; Danell, Huss-Danell et al. 1985; Duggan 1985; Crawley 1987), selection by herbivores is not likely to be "incessant" or "predictable," as proposed by Levin (1976b). Other selective pressures likely conflict with those imposed by herbivores (e.g., Dirzo and Harper 1982b; Dirzo 1984; Foster 1986), thus constraining the response of plants to herbivore selection pressure when it does occur. In addition, genetic response may be constrained by negative genetic correlations among resistance characters themselves (Berenbaum et al. 1986).

Herbivores often significantly influence plant population dynamics, limiting plant species distribution, recruitment, and community composition (Louda 1982, 1983; Watkinson 1986; Verkaar 1987; Crawley 1989). However, significant population effects do not necessarily imply significant selection pressures. Nor does lack of influence on plant abundance (e.g., Labeyrie and Hossaert 1985) rule out selection. Seed and seedling predation may have no influence on plant abundance if densities are not reduced below levels otherwise set by competition, limitation of microsites, and other sources of mortality (Duggan 1985; Watkinson 1986). However, differential pre- (e.g., Molau et al. 1989) and post-dispersal seed predation could determine which genotypes are available at the time of seedling establishment. If so, and if traits that favor escape from seed predation conflict with those that favor establishment, then herbivores would directly influence genotypes which do establish. In at least some plant species competitive ability (Windle and Franz 1979; Dirzo 1984) and growth (Coley 1986) are inversely correlated with resistance in seedlings or small plants. The relationships between resistance to attack as a seed and resistance, growth rate, and competitive ability as a seedling are unknown.

The effects of herbivory prior to seed dispersal can be ameliorated by short-term compensatory responses, but the long-term costs of compensation are rarely known. For increased compensatory ability to be an evolutionary response to herbivory, there must be additive genetic variation for the various mechanisms of compensation. Although intraspecific variation in compensatory response to herbivory is known (Giertych 1970; Mahmoud et al. 1975; Detling and Painter 1983; Carman and Briske 1985; Oesterheld and Mc-

Naughton 1988; Polley and Detling 1988; Marquis, in press), there have been no genetic analyses of variation in intraspecific response to tissue loss for wild species. These studies are needed, but will be difficult because of the necessity of controlling for the physiological state of experimental plants at the time of tissue removal. In maize, genetic variation (based on full-sib analysis) exists for both antibiosis and tolerance (ability to grow and reproduce once damage has occurred) (Ortega et al. 1980). We might expect the same to occur in natural plant populations.

Our understanding of the selection process will continue to be constrained in two ways. First, it is difficult in the field to define precisely the physiological response to tissue loss and the level at which various effects of that loss occur. Field conditions automatically introduce a level of variability in response by experimental plants that is difficult to control. For example, it may be possible to control for plant size in choosing experimental plants in order to control for resource base available to the plants. However, one must consider that previous defoliation history and local variation in resource availability may also influence the outcome of tissue removal, but that these features are not as easily controlled as plant size. Moving the experiment to a greenhouse reduces the variability introduced by the environment but changes the environment itself, thus making it difficult to extrapolate results back to the natural environment. Nonetheless, experiments under controlled environments are valuable in that they reveal the range of possible responses by plants to defoliation. Second, precise definition of the genetic versus the environmental influence on resistance requires outplanting of large families of known parentage (e.g., Simms and Rausher 1987). However, host population structure (Fenner 1987), associated species (Thomas 1986; Holmes and Jepson-Innes 1989), and population size (Stanton 1983) can affect pattern of attack. Thus, our understanding of the evolution of host resistance in those plant species that occur only in small populations may always be limited.

The above constraints do not prevent us from learning much about the nature of herbivory as a selective force. Recent studies have demonstrated that previous methods of estimating leaf damage (Lowman 1984) and predispersal seed predation (Anderson 1988) may markedly underestimate losses to herbivores. The almost universal distribution of plant characters that confer resistance against herbivores would suggest (and has led many authors to conclude) that herbivores have been a potent force in the evolutionary history of most plant species. The nature of the selection process itself remains largely a mystery.

Certain intriguing systems have received almost no attention at all. Despite the fact that herbivore impact is considered by some to be greater in tropical than in extratropical habitats (e.g., Baker 1970; Levin 1976a; Langenheim 1984), studies of herbivore effects on tropical plants in general are extremely limited, and studies of defoliation effects on tropical tree species are essentially absent (Marquis and Braker 1992). One of the most thoroughly described and perhaps simplest (fewest number of participants) tropical systems involves passion-flower vines (*Passiflora*) and their heliconiine butterfly and chrysomelid flea beetle herbivores. The herbivore species which attack a given *Passiflora* are known for many locations (Benson et al. 1975; Benson 1978) and their natural history is well understood (e.g., Benson 1978; Brown 1981; Smiley 1986). In addition, many of the secondary chemistry and physical defenses have been described (Gilbert 1971; Williams and Gilbert 1981; Spencer 1988a). Despite this wealth of background information, we know next to nothing about the impacts of the insects on *Passiflora* fitness in the wild (Williams and Gilbert 1981). A similar example involves *Eucalyptus* systems in Australia. These systems show both generally high damage and high variance in damage (Fox and Morrow 1983, 1986; Morrow and Fox 1989). Does this mean that natural selection and the associated evolutionary response is occurring continually in *Eucalyptus* systems? Something is known of the traits responsible for differential damage (Fox and Macauley 1977; Morrow and Fox 1980), and differential damage does result in differential growth (e.g., Morrow and LaMarche 1978). These systems lack only a concentrated effort, including a genetic analysis of resistance, on a few species.

We need studies that document natural variation in damage levels at all life stages and couple the fitness effects of damage to heritable plant characters. Annuals, biennials, and short-lived perennials are the obvious starting candidates for study species because of the difficulty of measuring lifetime fitness of long-lived perennials. However, perenniality is characteristic of the majority of plant species and should not be ignored. Long-term studies of variation in damage levels and consistency of damage among monitored individuals, coupled with experimental manipulation of herbivory loads, will provide much-needed understanding of the nature of selection in perennial plant species.

Summary

A large number of plant traits provide protection against herbivores. These resistance traits are ubiquitous, and may account for a significant portion of a plant's resource budget. In the case of chemical resistance, secondary compounds vary greatly in structure and concentration from plant species to plant species even among closely related taxa. Although these resistance traits provide protection against many or most of the potential herbivores in a habitat, every plant species in its native habitat suffers damage from at least one herbivore species some time in the plant's life cycle. A basic assumption found in much of the plant-herbivore literature is that herbivores, through their negative impact on plant fitness, have been the causative agents for the evolution of these plant resistance traits.

Resistance traits have been demonstrated to decrease herbivore feeding, growth, and survival. However, for any one system, we have only limited understanding of the nature of the microevolutionary process that leads to changes in plant population resistance over time. We do know that intrapopulational variation in damage is universal, that in almost every case damage has a negative impact on plant fitness components, and that a genotypic basis exists for traits that determine the amount of damage. For a few systems, each of these points has been examined in sufficient detail to demonstrate that herbivores selected for changes in resistance during the course of the study. No studies have attempted to measure an actual response to selection in nature. In only two systems (wild parsnip and the parsnip webworm, and white clover and its folivores) have actual resistance traits been identified and studied in the context of the selective impact of herbivores. Until we identify the plant traits responsible for resistance (no easy task), we will not be able to determine whether and how herbivores select for changes in those traits.

Although selection by herbivores for changes in plant resistance has been observed in natural systems, these selection pressures are neither as continuous nor as pervasive as once proposed. Absolute and relative herbivore species abundances vary greatly both spatially and temporally, as does the resultant damage. Thus, selection on resistance traits is not constant in magnitude. In turn, the abiotic environment can affect the level of resistance, diminishing the influence of the genotype. Thus, even when herbivore damage is high and highly variable among plants, differences in damage may not necessarily reflect underlying differences in plant genotypes. Finally, the nature of selection imposed by herbivores likely varies both spatially and temporally owing to changing relative abundances of different herbivore species. Different herbivore species have been shown to prefer different genotypes within the same plant population. The probable result is that herbivores promote maintenance of genetic diversity and a diversity of resistant phenotypes resulting from these genotypes.

Evolutionary changes in plant resistance traits are likely to be constrained by detrimental changes in other resistance traits, decreased growth rate and competitive ability, and decreased ability to tolerate environmental stresses other than herbivory. We have just begun to define the nature of these costs of evolutionary changes in resistance. And yet, knowledge of the constraints on these evolutionary changes is critical if we are to understand the potential of herbivores to effect change in resistance across generations.

Acknowledgments

I thank Laurel Fox, Bob Fritz, Carol Kelly, Stephen Mulkey, Ellen Simms, Victoria Sork, and one anonymous reviewer for their careful criticism and discussion of the ideas presented here. Partial support during writing of the manuscript came from National Science Foundation grant BSF-8600207.

14

Evolution of Disease Resistance in Natural Plant Populations

Helen M. Alexander

The concept that plants differ in their vulnerability to pathogens is self-evident to agricultural scientists, for whom breeding for resistance is a major component of crop disease control. Consequently, there is a large literature on the physiological and biochemical mechanisms of disease resistance (Deverall 1977; Wood 1982; Misaghi 1982), the inheritance of resistance (Nelson 1973; Day 1974; Sidhu and Webster 1977; Ellingboe 1985), and the epidemiological effects of resistance on pathogen populations (Day 1978; Wolfe 1983; Mundt and Browning 1985). The existence of disease resistance traits in natural plant populations has also not been questioned, for wild relatives of crop plants have always been a major source of resistance genes for breeding (Leppik 1970; Wahl et al. 1984; Burdon and Jarosz 1989). However, until recently, little effort had been spent on quantifying the levels of disease resistance in natural populations, much less on understanding their ecological or evolutionary significance. This situation is changing: numerous recent reviews and symposia (Burdon and Shattock 1980; Burdon 1982, 1985, 1987b, 1989; Dinoor and Eshed 1984; Alexander 1988; Augspurger 1989) have revealed that disease processes in nature are being investigated by a diverse group of scientists. These scientists include plant pathologists and foresters, whose interests range from searching for new resistance genes to understanding the effect of genetic diversity in natural populations on disease spread (Dinus 1974; Browning 1974, 1980; Dinoor and Eshed 1984). Ecologists and evolutionary biologists have also joined in this study, intrigued both by the potential importance of disease in nature and the realization that pathogens have been neglected in most studies of plant population biology (Harper 1977; Burdon 1987b). Empirical research has also been stimulated by theories suggesting that pathogens (and other "pests") may play a role in the maintenance of host genetic variation and the evolution and maintenance of sex (Haldane 1949; Clarke 1976; Jaenike 1978; Hamilton 1980; Rice 1983a; Seger and Hamilton 1988; Hamilton et al. 1990).

This chapter will critically evaluate this expanding body of literature to

326

understand the importance of pathogens in the process of evolution by natural selection in natural plant populations. I shall seek answers to three questions, which together largely determine the role of disease in plant evolution: (1) Do plants vary in resistance? (2) Is this variation heritable? (3) Is there a negative relationship between disease and plant fitness? In an agricultural context, responses to these questions are straightforward: heritable variation in disease levels is the basis for breeding resistant crop varieties, with the impetus for their development being yield losses due to disease. With natural populations, the situation is more complex. Although on a general level, one could answer "yes" to each of the questions posed above, it is surprisingly unlikely for all three topics to be addressed rigorously in any one study of a plant/pathogen interaction. This situation is partly a result of the diversity of scientists in this field, with different perspectives and objectives in their work. This breadth of focus also means that a variety of approaches have been used to study these questions, which can lead to different interpretations. For this reason, this chapter will devote considerable space to definitions and methodologies used in studying plant disease in natural populations, as well as to the results of the research. Out of necessity, most examples will concern plant/fungal interactions simply because these are much better studied than interactions with other types of pathogens.

Variation in Resistance

Definitions of Resistance

Perhaps the broadest definition of resistance is "the ability of an organism to withstand or oppose the operation of or to lessen or overcome the effects of an injurious or pathogen factor" (Committee on Technical Words 1940; Federation of British Plant Pathologists 1973); such a definition can include both the effect of the plant on retarding pathogen growth and the ability of the plant either to avoid disease or to function despite high disease levels. Burdon (1987b) has separated these two components of resistance into "active" and "passive" mechanisms. Active mechanisms involve a direct physiological or biochemical interaction between plant and pathogen. A classic example is the hypersensitive reaction, where rapid death of plant cells at the site of infection (visible as a necrotic fleck) is associated with inhibited pathogen growth (Kiraly 1980). Passive resistance results when the disease has little effect on the plant because of plant traits that are largely independent of the presence of the pathogen; these traits, however, may be active in an evolutionary sense if they resulted from past selection imposed by pathogens. In the passive mechanism of disease avoidance, genetically controlled plant traits reduce the probability of contact between plant and pathogen. An agricultural example involves ergot infection, where grain crops that are self-pollinated have low disease levels, while male-sterile varieties with flowers open for longer periods can have high disease levels (Puranik and Mathre 1971). In addition, traits such as timing of seed germination or flowering may affect whether susceptible plant tissue is exposed to a pathogen. The other passive mechanism, disease tolerance, refers to the ability of individual plants to be relatively unaffected (i.e., to maintain high plant reproduction or yield) despite heavy disease levels (Schafer 1971).

Active mechanisms of resistance are further subdivided by Burdon (1987b), depending on whether resistance is functional against certain pathogen phenotypes but ineffective against other pathogen variants (race-specific resistance) or whether resistance is effective regardless of the identity of the pathogen isolate (race-nonspecific resistance). These definitions parallel the older terms of vertical and horizontal resistance used by Van der Plank (1963), who also emphasized that these categories reflected single-gene versus polygenic control of the resistance trait. Although this generalization is often correct, inheritance mechanisms should not be assumed without appropriate crossing studies. Other authors have additionally proposed that these two modes of resistance affect different epidemiological parameters, with race-specific resistance reducing the amount of initial inoculum through discrimination against certain pathogen types while race-nonspecific resistance reduces rates of disease development for all strains (Browning et al. 1977). Although these subdivisions can be useful, resistance phenotypes cannot always easily be classified, especially with genetically variable natural plant populations.

The variety of concepts associated with the word "resistance" leads to confusion because they are measured in different ways. Active mechanisms of disease resistance are often quantified by observing the type of disease symptom on the plant: for example, does the interaction of a specific plant and pathogen lead to an actively sporulating lesion or a necrotic fleck? In contrast, the disease-avoidance mode of passive resistance may be more a measure of probability of infection: for example, do early- or late-flowering plants have a greater likelihood of infection? Lastly, disease tolerance measures a completely different attribute: do plants with similar levels of disease expression have similar probabilities of survival and reproduction?

In this chapter, variation in resistance will be considered broadly as consistent and predictable differences among plant phenotypes in their levels of disease. Active modes of resistance and the disease-avoidance mechanism of passive resistance are included, since all focus on how individual plants in the population vary in disease levels. This definition of resistance is therefore broader than is practically applied by most plant pathologists, who focus on the active resistance mechanisms. Such a strategy is understandable in crop breeding, since a cultivar that is physiologically resistant may be less risky than one that is physiologically susceptible but disease-free in the field because of disease avoidance. In studying natural populations, however, any heritable trait that leads to low disease levels may be equally effective at influencing plant fitness and should be considered, even if selection on the trait (such as flowering time) is clearly caused by many factors independent of pathogens. Disease tolerance will not be considered a mechanism of resistance in this paper, since it does not affect disease expression on the plant. This important and largely neglected concept will instead be discussed in the section on fitness effects of disease.

Methods for Measuring Variation in Resistance

Variation within natural populations in disease levels is extremely common; however, much of this variation is due simply to chance effects resulting from the limited spatial distribution of inoculum. Such variability may be important for epidemiological processes, but only consistent and predictable differences among phenotypes in disease levels or symptoms (i.e., resistance) are of interest in studies of natural selection. Thus, a manipulative approach is needed to measure resistance, where a random sample of individuals from a plant population is experimentally assayed for disease resistance. The first step of sampling plants is rarely done with a completely random sampling scheme; however, unless a clear bias is evident (that is, favoring or avoiding disease-free plants), problems are not likely. The difficulty arises in deciding how to measure the sampled plants' resistance to disease. Two general approaches have been used, either measuring disease levels after plants have been directly exposed to the pathogen by inoculation or measuring disease levels in the field on randomized experimental plants exposed to natural disease spread. The choice of methods depends on the goals of the study: (1) does the researcher want to measure only active resistance modes or are all forms of resistance, including disease avoidance, of interest? and (2) is the objective to characterize details of particular types of resistance (such as inheritance or effectiveness against different pathogen isolates) or to study the actual disease levels experienced under field conditions? Inoculation studies will be preferable if the first alternatives are the answers to the above questions, while field transplant studies have advantages for the latter choices. Clearly, use of both approaches will give the greatest understanding.

Inoculation Experiments

Inoculation involves the experimental introduction of pathogens to plants under conditions favorable for disease development. Usually performed in the greenhouse, the identity of the pathogen isolate used and the specific amount of inoculum and the method by which it is introduced to the plant can all be controlled. Since variation within treatments is reduced, only a small number of plants per family or clone may be needed to assay resistance for a particular plant/pathogen combination. Often seedlings or even excised leaves are used to increase the number of plant populations or isolates that can be tested. A common application of the inoculation approach is to inoculate a series of host lines with a series of pathogen isolates, producing a two-dimensional matrix with a unique pattern depending on the resistance and virulence of the hosts and pathogens used (Burdon 1987a). A major strength of this method is that it reveals variation in pathogen virulence and the interaction between resistance and virulence, as well as describing host resistance variation. This method is an extension of the agricultural concept of a differential series (Flor 1956; Roelfs 1985), which is used to detect pathogen variation in virulence.

The primary disadvantage of the inoculation approach is the possibility that the disease expression measured may not be representative of that experienced in the field. If susceptibility changes with plant developmental stage, resistance expressed in seedlings may not be correlated with resistance of mature plants in the field. Differences in growth conditions in the greenhouse and field may also affect susceptibility. A classic problem is the choice of pathogen isolate: use of single genotype isolates is essential to understand the genetics of plant/pathogen interactions, but studies based on only a few isolates may not reflect the inoculum diversity in the field. Isolates chosen because of unique virulence characteristics may differentiate among plants and detect many resistance factors, but may not be representative of what the natural plant population actually experiences. Therefore pathogen isolates should ideally be collected from the same site as the plant samples, with knowledge of their frequency in nature. Lastly, the disease-avoidance resistance mechanism must also be considered: if certain phenotypes are unlikely to encounter disease in the field, classifying the plant as susceptible based on forced contact with the pathogen in the laboratory may not be ecologically relevant.

Field Transplant Experiments

The alternative approach is a field transplant experiment, where plants are transplanted into a field site using a randomized design, and disease levels are measured after natural spread of inoculum. This approach follows many current studies in plant population biology, where the trait of interest (in this case, resistance) is studied with experimental plants ("phytometers," Antonovics and Primack 1982) growing under natural conditions. The ideal site of the field experiment is the population where plants were collected, so plants will be exposed to the same inoculum quantity and types they would have experienced naturally. In addition, using a combination of inoculation and natural disease spread approaches, studies can be performed in an experimental garden setting, where a subset of plants are inoculated as a source of disease spread; again choices need to be made of the number and origin of pathogen isolates. The disease nursery used in agriculture is an example of this approach, where spreader rows are inoculated.

The advantage of field transplant studies is that plants are grown under more natural conditions than in most inoculation studies, and spread of inoculum (including propagule placement and quantity) occurs similarly to that in nature and under comparable environmental conditions. Since disease spread is not manipulated, both active and passive (disease-avoidance) mechanisms can operate. This approach is particularly important for vector-borne diseases, since vector behavior may have a large effect on which plants become infected.

The disadvantages of the approach are the same as the advantages of the inoculation study. First, field studies are very time-consuming; thus the number of plants or plant/pathogen combinations that can be tested is limited. It is also unusual for studies to be repeated over many years, so results are very dependent on the environmental conditions and inoculum present in a single field season. Further, the exact identity of the pathogen isolate(s) responsible for infection, the quantity of inoculum, or details of the mode of infection are often not known. Consequently, genetic interpretation of the plant/pathogen interaction may not be feasible. Use of molecular markers may eventually solve part of this problem. If pathogen genotypes identified by markers can be assayed either in nature or in an experimental garden, patterns of infection of plant phenotypes (or genotypes) by certain pathogen genotypes could be studied in situ.

Evidence for Variation in Resistance

Although resistance variation in natural populations has only recently been studied by ecologists and evolutionary biologists, this research has a long history in plant pathology and forestry. Doolittle (1954), for instance, summarized how wild tomato species have useful resistance for diverse diseases caused by fungi, bacteria, viruses, and nematodes. Screening tests of collections suggest that resistance is widespread. For example, 32% of 446 collections of wild oats in Israel had some plants resistant to crown rust (Wahl 1970), 49% of accessions of wild emmer wheat were resistant to powdery mildew (Moseman et al. 1984), different populations and species of wild sunflowers had from 0% to 100% plants resistant to sunflower rust (Zimmer and Rehder 1976), and inoculation of loblolly pine seedlings originating from five regions across the natural range of the species had from 56% to 79% infection by fusiform rust (Powers and Matthews 1980); similar data has been summarized in Wahl et al. (1978), Segal et al. (1980), and Wahl et al. (1984). Since a major goal of these studies was to find useful disease resistance for breeding, it is reasonable to ask whether the plant sampling techniques used led to representative measurements of the resistance variation present in the natural population. Methods vary greatly, ranging from random collections to sampling where a high frequency of resistance was expected. Dinoor's (1970) analysis of sampling methods showed that collecting seed from either known resistant wild oat plants or from locations where resistance was found in the past can lead to three- to fourfold increases in the frequency of rust resistance reported, as compared to random sampling methods.

Many of the studies of wild relatives of crop plants use both inoculation and field transplant studies (specifically, disease nurseries) to assess resistance. This dual approach is common in agriculture, where inoculation studies are used to screen for resistance and in detailed genetic studies, but field trials test whether the resistance is truly effective under field conditions. Results of these two approaches are often correlated: analysis of data in Wahl (1970), for example, revealed a significant positive correlation between percentage of wild oat locations with crown rust resistance based on seedling inoculation and adult plant field-resistance trials. However, Dinoor (1970) found that relying only on seedling inoculation tests in the same system meant that resistance effective in adult plants was not uncovered in up to 40% of the cases where direct studies would have revealed its existence.

Aside from researchers who primarily screen crop progenitors for useful resistance genes for breeding, J. J. Burdon was the first to focus his research on describing the resistance structure of natural plant populations. Concentrating his efforts on the inoculation method of assessing resistance, he and others have surveyed several systems, often finding considerable resistance variation both between and within plant populations. In a broad-scale study, Burdon and Marshall (1981b) measured variation in resistance to a single rust isolate within populations, between populations, and between species of *Glycine*. This study is important since it not only revealed threefold variation among species in qualitative resistance (i.e., presence or absence of sporulating pustules), but also documented differences in more quantitative measurements of resistance (the number, developmental rate, and size of the fungal lesions). Similar variation was observed within populations. A larger number of fungal isolates were used in a study of three species of wild oats in Australia, where Burdon, Oates et al. (1983) tested plants against four races of each of two rust pathogens; these races were geographically widespread and chosen because of known differences in pathogenicity. Variation in resistance was again found within and between populations and between plant species. In addition, there were distinct geographic patterns in resistance to *Puccinia coronata*, with areas more conducive to rust development having higher resistance. Similarly, Jarosz's (1984) study of powdery mildew on *Phlox* found correlations between habitat shading and resistance levels after inoculating two mildew isolates on ten taxa collected over a broad geographic area. By using eight powdery mildew isolates chosen to differ in pathogenicity, Harry and Clarke (1986) showed that race-specific resistance was common in inbred lines of *Senecio vulgaris* derived from a range of localities. More than half of the host lines were resistant to at least one isolate.

Variation in resistance has also been found in studies that focus on the detailed resistance structure of one or a few populations. Fifty clones of *Trifolium repens* collected from a single field varied in resistance to one isolate of each of two foliar fungal pathogens; the frequency of resistance to one pathogen was normally distributed while the other was skewed in favor of resistant types

(Burdon 1980). Miles and Lenné (1984) also found continuous variation in infection of families from a population of *Stylosanthes guianensis* following inoculation of isolates of *Colletotrichum gloeosporioides*; the sampling scheme (avoiding plants with low disease levels), however, complicates interpretation. Clear race-specific differences in resistance were discovered in two populations of *Glycine canescens*, where individuals varied from complete resistance to nearly complete susceptibility in their reactions to nine pathogen races chosen to cover a range of pathogenicity; one site was more resistant than the other (Burdon 1987a). A similar inoculation approach has been used in studies of the flax rust resistance structure of wild flax populations (Burdon and Jarosz 1988; Burdon and Jarosz 1991); this work is particularly important since the frequencies of pathogen isolates used in the study are known in the natural plant populations. Knowledge of the geographic patterns of plant resistance and fungal virulence revealed that some very virulent races had limited distributions.

The inoculation studies described above are exciting because of the breadth of information they have revealed about plant resistance in natural populations. Active resistance mechanisms similar to those found in agricultural crops have been shown to be widespread both within and between natural populations. In many cases, resistance is clearly dependent on the identity of the pathogen isolate used, giving the race-specific type of resistance found in many crop systems. Authors of inoculation studies usually readily admit, however, that they are studying only a subset of the possible types of resistance (e.g., Burdon 1987a). One then must ask how well these studies describe variation in resistance important in the field (Dinoor and Eshed 1984; Alexander 1988). Field transplant studies have revealed factors (both genetic and environmental) not explored by inoculation studies. In Dinoor and Eshed (1987), randomly collected wild barley from five locations in Israel were planted in field plots, with spreader rows in each plot inoculated with a different single powdery mildew isolate. Frequency distributions of disease levels in the populations changed depending on the shading treatment (sun versus shade) and the identity of cultures. Although analyses were not presented, it appeared that the degree to which shading increased disease severity may depend on the isolate. Such a genotype × environment interaction could have a large effect on the actual selection experienced in wild barley populations.

In contrast to Dinoor and Eshed's (1987) study, most field studies have little or no data on the identity of the pathogen inoculum and thus are of limited use for phenotypic or genetic analysis of the interaction between the plant and pathogen. Their value lies in showing that plant resistance variation is effective under natural conditions. In Alexander et al. (1984) and Parker (1985), for example, maternal families (seed collected from single plants) were transplanted into randomized locations in natural populations, and variation in resistance was found with respect to the natural inoculum supply. Parker (1985) further discovered that families of *Amphicarpaea bracteata* transplanted into their native population (the "LD site") were much more prone to infection by *Synchytrium decipiens* than those transplanted into populations 1 km or more away, despite heavy infection on plants native to these distant sites. Parker's results from his field experiments were consistent with inoculation studies with a single fungal strain from the LD site. This highly inbred plant may have genetic constraints that limit evolution of disease resistance (see Parker, this volume).

Resistance is particularly difficult to study with vector-transmitted diseases. Berryman (1972) illustrates the complexity of bark beetle/fungal associations in coniferous forests and shows how disease resistance cannot be examined independently of the insect vectors. Vectors can influence the host/pathogen interaction at many levels. The differential action of ribonucleases in beetle regurgitant determined successful transmission of viruses, even when all were capable of infection with mechanical inoculation (Gergerich and Scott 1988). Work by De Nooij (1988) revealed that weevils that transmit an inflorescence disease in *Plantago lanceolata* show preference for feeding on certain host genotypes; thus natural variation in disease levels may partly reflect vector behavior. Although an inoculation study documented continuous variation in disease levels for clones from three populations, De Nooij and Van Damme (1988a) cautioned that relative susceptibility can be realistically assessed only with field trials. In a study of anther smut infection of *Silene alba*, a pollinator-transmitted disease, male genotypes with high flower production rates were more likely to become infected in a field experiment (Alexander 1989). A disease-avoidance mechanism may exist that would not be detected in an inoculation study: low disease levels of genotypes with low flower production may partly be due to a reduced chance of successful contact between plant and fungus.

Field-based quantitative genetics experiments are a strong approach for studying within-population variation in resistance; this method specifically allows an estimation of the additive genetic variation in resistance available for natural selection (see below). An example is work by Kinloch and Stonecypher (1969) which revealed large differences in fusiform rust levels among families of loblolly pine; family means ranged from 0.5 to 10.8 galls per tree. Families had been planted into sites less than 5 km from the stands that included the parent trees and had been exposed to natural inoculum or enhanced inoculum at a rust nursery. Similarly, in a field study designed to measure additive genetic variation for herbivore resistance in the annual *Ipomoea purpurea* (Rausher and Simms 1989), Simms (unpublished data) also found family variation in disease levels of both *Colletotrichum dematium* var. *ipomoeae* and *Coleosporium* (sp).

Since most field studies focus on within-population variation at only one or a few sites in only one or a few years, a relevant question is whether results can be extrapolated over time or to other populations. An example is Dirzo and Harper's (1982b) study of transplanted clover, where rust levels were highly correlated with the cyanogenesis polymorphism. However, no relationship was observed between rust levels and cyanide production in a nearby unmanipulated field, illustrating the dangers of conclusions based on a single study site. Studies of evolution of disease resistance are particularly vulnerable to problems of generalization simply because pathogen populations are biotic entities that may vary considerably from site to site.

Future Approaches to Measuring Variation in Resistance

Studies on variation in resistance in natural populations thus have focused on two extremes: either extensive data from inoculation studies of within- and between-population variation in resistance, often over a broad geographic scale, or localized field studies of variation in disease levels after natural disease spread. Both kinds of data are needed and interesting. However, the resistance that is being measured is unlikely to be the same: inoculation studies often focus only on active disease mechanisms with little knowledge of their effectiveness in nature, while field transplant studies measure resistance in a broad sense, yet often have a "black box" knowledge of the genetic structure of the pathogen population and the modes of infection. A solution is to return to the dual approach of combining greenhouse and field studies that is commonly used in agriculture: correlations between resistance patterns measured in the greenhouse and those expressed in the field must be explored and researchers using field studies should characterize the natural pathogen populations that interact with their plant populations.

Inheritance of Resistance

Methods for Determining Inheritance of Resistance

A great deal of phenotypic variation in disease resistance exists within natural populations. Because of sampling methods and experimental designs, however, relatively few studies have rigorously demonstrated *genetic* variation for resistance. For example, in comparisons of disease levels of field-collected seed families, biases in estimating genetic variation can result from dominance, maternal environmental effects, and lack of information on paternity (Mitchell-Olds and Rutledge 1986; Simms and Rausher, this volume). Comparisons of cloned genotypes have similar problems. To study the inheritance of resistance, crosses must be performed, followed by analysis of the disease resistance of progeny.

Two variations on this approach exist, differing largely in whether disease resistance is expected to be under the control of one or a few major genes or many genes of small effect. In the former case, plants that showed interesting resistance traits in previous studies are crossed or selfed, and their progeny are inoculated with appropriate pathogen isolates. The ratios of susceptible and resistant progeny are compared with the Mendelian ratios expected if resistance is controlled by a small number of major genes. The alternative approach focuses on quantitative genetic variation by crossing a large number of parents that will be representative of the population in question and scoring progeny for resistance after either inoculation or exposure to disease in a natural population. The narrow-sense heritability of disease resistance can then be determined from statistical analyses that partition variation among and within families (Falconer 1981; Simms and Rausher, this volume), indicating the proportion of the genetic variation which is additive, and therefore the potential for response to selection. Even with the same organisms, the methodology used for genetic analysis (for example, field quantitative genetics studies versus an inoculation matrix with host and pathogen lines in the greenhouse) may affect whether or not continuous or discrete disease resistance classes are apparent.

Evidence Supporting Inheritance of Resistance

Inheritance of resistance is often analyzed in plants that display race-specific resistance to particular pathogens, that is, inoculation of plants with some isolates produces resistant reactions while inoculation with others results in susceptible reactions. By crossing inbred plants of *Senecio vulgaris* with known disease reactions to five mildew isolates and inoculating the F_1 progeny with the same isolates, Harry and Clarke (1987) found evidence for single dominant resistance genes in the majority of cases, although resistance to one isolate was recessive in two crosses. Through analysis of F_2 progeny, many of the resistance genes were shown to be linked. Evidence for dominant resistance genes also appears in the study by Burdon (1987a), who crossed wild collected plants of *Glycine canescens* from two sites with a universally susceptible plant, and then tested the F_2 progeny with one to two races of the rust *Phakopsora pachyrhizi* for which the wild parent was resistant. In most cases, F_2 progeny fit ratios of 3:1, 15:1, or 63:1 resistant:susceptible plants, indicating the presence of one, two, or three dominant resistant genes in the original parent plant. In a population of a related host species, *G. argyrea*, variation in resistance to *P. pachyrhizi* was controlled by only a single locus; again the resistance allele was dominant (Jarosz and Burdon 1990). A final example of dominant resistance was found by Parker (1988c), where variation among siblings of *Amphicarpaea bracteata* in resistance to one isolate of *Synchytrium decipiens* was shown to be due to segregation of a single dominant gene.

In contrast to the above studies, the goal of quantitative genetics experiments is to document whether additive genetic variation for resistance exists in the population as a whole by studying disease levels within and among families. The best studies of this type use families created by controlled crosses. In Kinloch and Stonecypher (1969), comparisons of natural disease levels of pine seedlings resulting from either wind pollination or controlled pollination produced heritability estimates of 0.65 to 0.85. The data are consistent with a population segregating at several polymorphic loci. In the study of the annual *Ipomoea purpurea*, Simms (unpublished data) found heritable variation in disease levels for both *Colletotrichum dematium* var. *ipomoeae* and *Coleosporium* (sp). Comparison of families suggests, however, very different patterns of resistance to the two diseases. For rust infection, progeny of two families were nearly completely disease-free, suggesting the presence of a dominant resistance gene in these parents. In contrast, disease levels of *C. dematium* var. *ipomoeae* varied continuously among families, which may suggest that several genes control inheritance. Continuous variation could also result, however, when only a few loci were involved, if each locus had more than two alleles.

Future Studies of Inheritance of Resistance

The paucity of studies of inheritance of resistance in natural systems makes generalizations difficult. Much more work is needed, especially because of the controversy already existing in the literature on the origins of the gene-for-gene patterns seen in many agricultural systems. This theory, proposed by Flor (1956), states that a single gene in the host is complementary with a single gene in the pathogen, and is based on extensive inoculation trials and crosses of flax and flax rust. Resistance was usually dominant and virulence was usually recessive. Many agricultural systems have fit this model, especially those involving obligate pathogens (Barrett 1985). However, Day (1974), Barrett (1985), and others have argued that gene-for-gene interactions may be partly an agricultural artifact. They contend that dominant resistance is the easiest resistance for breeders to screen. Moreover, they warn that for resistance to be worthy of agricultural exploitation, virulence to overcome it must be rare, as is likely with a rare recessive gene. In nature, therefore, gene-for-gene interactions will be only one extreme of a continuum of types of genetic interactions between host and pathogen. Browning (1981), Burdon (1987a, 1987b), and others have instead argued that the gene-for-gene system is likely to be prevalent in nature, since, for example, resistance genes used in crop breeding are nearly always derived from wild crop progenitors. The simple, dominant resistance shown to be effective for biotrophic pathogens by Harry and Clarke (1987), Burdon (1987a), Jarosz and Burdon (1990), and Parker (1988c) supports this view, although parallel genetic studies of pathogen virulence, needed to test the gene-for-gene theory, have not been done in any natural systems. Little work has been done for other types of pathogens. Actually both sides on this argument are likely to be right. Inoculation studies may reveal specific interactions controlled by a few loci, but if field studies were done, traits affecting plant growth form and phenology (likely to be controlled by many genes) as well as quantitatively inherited resistance may also be shown to determine disease levels. In addition to the work needed on direct inheritance of resistance, particular attention must also be given to genetic correlations between resistance and other plant traits. Genetic correlations between traits can result from pleiotropy or linkage disequilibrium; Parker (1988a) discusses the latter with associations between disease resistance and allozyme loci in *Amphicarpaea bracteata*.

Effects of Disease on Plant Fitness

Given the agricultural data on the effect of disease on plant yield (James 1974; James and Teng 1979) and our general sense that pathogens must be damaging to their hosts, there is a clear expectation that disease will reduce plant fitness. Unfortunately, it appears that this belief has led to very limited measurement of the effects of disease on plant fitness in natural populations. Lack of knowledge of disease effects has been a serious impediment in interpreting the quite impressive data on variation in disease resistance in natural populations: without quantitative information on the effect of disease on fitness, how often does natural selection for disease resistance occur?

Definitions of Fitness

The concept of fitness has often been interpreted differently by population biologists and plant pathologists, leading to confusion both conceptually and in semantics (Antonovics and Alexander 1989). Fitness is defined here as the expected contribution of a phenotype to subsequent generations. In general, the focus is on the relative fitness of a phenotype compared to contributions of other phenotypes (Roughgarden 1979; Endler 1986). Operationally, plant fitness is often described in terms of its component parts (fitness components), which include such factors as the number of flowers produced per plant, number of seeds per flower, and seed size. As discussed by Primack and Antonovics (1981) and Antonovics and Alexander (1989), there are clear similarities between these fitness components and commonly used yield components in agriculture, such as heads per tiller, grains per head, and grain size. Consequently, data on crop yield losses provide valuable information on the potential effects of disease on plant fitness. It must be remembered, however, that the focus is on the individual plant, in the case of fitness, and on the crop (a population of individuals), in the case of yield (Antonovics and Alexander 1989). Crop yield can stay constant regardless of whether all plants have equal contributions of seeds per plant (equal fitness) or some plants produce many more seeds than others, leading to very unequal contributions (unequal fitness). Yield loss data are also measured in an economic context: thus a 5% yield loss due to disease may be significant in economic terms, but must be judged relative to other factors affecting fitness in a natural population.

Methods of Measuring Effects of Disease on Plant Fitness

A common way to measure fitness effects of disease in nature is a descriptive approach, comparing survival and reproduction for naturally growing plants in an undisturbed population that vary in disease levels. Such an approach has the advantage of correlating fitness components with actual disease levels in the field, with plants exposed to the natural inoculum, both in terms of spore quantity and genetic composition. Both the abiotic stresses of the natural environment and the biotic stresses of plant competition, herbivores, and other diseases will also be present. An implicit assumption of this approach, however, is that diseased and healthy plants differ in fitness only because of the presence of the pathogen: if differential infection exists owing to plant genotype or because of environmental factors affecting plant vigor, true estimates of pathogen effects will not be obtained.

In experimental studies, disease status can be controlled by either addition of inoculum (inoculation experiments) or removal or prevention of infection (chemical treatments). Experimental studies can also control plant phenotype or genotype, allowing parallel measurements of variation in disease levels and the effects of the pathogen on fitness: correlations of genetic variation in disease levels and fitness provide information on selection for resistance. Field inoculation studies are preferred over greenhouse studies for measuring fitness effects, since plant (and pathogen) growth and reproduction are so clearly promoted by the environmental conditions and lack of competition in the greenhouse. Further, the existence of resistance types that are effective in specific environments (for example, temperature-sensitive resistance; van Dijk et al. 1988) suggests the potential importance of genotype-environment interactions, with the possibility of different selection regimes in different environments. Chemical treatments to prevent or remove naturally occurring disease in the field has the advantage of allowing natural dispersal of pathogen propagules, while still manipulating the disease status of individual plants. There are potential disadvantages, however, because many pesticides affect plant growth and other characters (Paul et al. 1989).

Evidence of Effects of Disease on Plant Fitness

Burdon and Shattock (1980) outlined how pathogens may be important at different stages of the life cycle of plants, with differing effects on fitness. Unfortunately, the developmental stages where pathogens may have the greatest effect, those of seeds in the soil and germinating seedlings, have received the least study. For example, pathogens may play important roles in determining seed longevity in seed banks, but the only data available are of effects of pathogens on stored grain in agriculture, where environmental conditions are very different (Tuite and Foster 1979). Our knowledge of the effects of pathogens on seedling mortality has been greatly enhanced by the descriptive and experimental studies of Augspurger and colleagues (Augspurger 1983; Augspurger and Kelly 1984; Augspurger 1990), who have shown that damping-off diseases can lead to death in a large proportion of seedlings establishing under the canopy of tropical trees. This disease-induced mortality occurs very rapidly, and was observed only by weekly censuses of marked plants.

Most studies of disease have involved established plants that can be easily marked and followed over their lifetimes and pathogens that are easily identified and scored (e.g., fungi infecting aboveground parts). By following young plants of *Amphicarpaea bracteata*, Parker (1986) found that infection of *Synchytrium decipiens* was associated with decreased survivorship and reduced reproduction. Interestingly, at one site, fitness decreased linearly with increased infection while at another site there appeared to be a threshold effect, such that any disease above a certain level resulted in essentially zero fitness. A study by Alexander and Burdon (1984) revealed how developmental stage at time of infection can determine how disease affects fitness: downy mildew and white rust infection led to high mortality of plants of *Capsella bursa-pastoris* infected early in development, while infection of flowering plants in the summer had no effect on survivorship and a moderate effect on reproduction. Field inoculation studies by Paul and Ayres (1986a, 1986b, 1987) also showed how plants infected with rust at the same developmental stages but experiencing different environments (summer versus winter) can have different patterns of disease-induced mortality and reduced reproduction.

Fitness effects of disease may not be immediately evident. Even in annuals, where fitness is often measured by seed production in a single generation, disease effects may be underestimated if seeds from diseased plants are reduced in size, affecting success in subsequent generations (Burdon and Jarosz 1988). These year-to-year effects are obviously particularly important in perennials: for example, rust infection of *Linum marginale* has little effect on fecundity during the year of infection, but was correlated with increased overwintering mortality (Burdon and Jarosz 1988).

Pathogens that infect the reproductive parts of plants may have very interesting effects on fitness. In some cases, pathogens may be primarily drawing nutrients away from seed production or altering the inflorescence structure. Examples include two diseases of *Plantago lanceolata*, where infection of the inflorescence stalk (De Nooij and Van der Aa 1987) or the inflorescence itself (Alexander 1982) reduces seed set. In other cases, pathogens actually replace plant reproductive parts with spore-producing organs. In ergots of grasses, single ovules are replaced by the fungal sclerotia, leading to potential losses in seed number. Gray, Drury, and Raybould (1990), however, suggest that in the case of ergot infection of the clonal marsh grass *Spartina anglica*, the disease has little effect on the plant population biology. The systemic anther smut diseases of plants in the Caryophyllaceae lead to the production of spore-producing anthers in sterile flowers (Baker 1947; Alexander and Antonovics 1988; Jennersten 1988). Endophyte infections of grasses provide an interesting continuum between parasitism and mutualism, which makes fitness effects difficult to interpret. In many cases, infection causes partial or complete sterilization yet competitively vigorous plants (Bradshaw 1959; Clay 1984, 1986), suggesting poor fitness in an evolutionary sense but the potential for ecological dominance in a community. With grasses infected by the *Acremonium* endophytes, however, plants have both full seed set and enhanced vegetative growth. Although seeds are infected, the fungus is acting as a mutualist, not a pathogen (Clay 1988).

Although reasonably good estimates of fitness of annual plants can be made based on single-season measurements of seed production, data on fitness components in perennial plants are particularly difficult to interpret. Yearly differences in reproduction may exist and, depending on the nature of the disease, infection of a plant in one year may or may not be correlated with infection in a later year (Alexander 1984). Further, as noted above with *Linum marginale*, there may be a lag time between occurrence of infection and its effect on plant growth or reproduction. One useful approach is to incorporate demographic data into population projection models to predict how the numbers of healthy and diseased plants may change over time, thus giving an indication of the fitness effects of disease. Alexander (1982; summarized in Antonovics and Alexander [1989]) used such a model to analyze the effects of an inflorescence pathogen on *Plantago lanceolata,* and found that the fitnesses of diseased and healthy plants were very similar. Even though seed production was negatively correlated with disease intensity (Alexander 1982), the result of similar fitness was not unexpected, since plants with many inflorescences appeared more likely to become infected (Alexander et al. 1984).

As pointed out by Antonovics and Alexander (1989), integration of demographic data with field experiments would be more powerful, as has been done in plant/herbivore studies (Rausher and Feeny 1980). In disease studies, a useful approach would be to compare the demography of natural populations, with and without a fungal disease, with pathogen exposure controlled by fungicide application. Foster (1964; cited in Sagar 1970) found that fungicide treatments increased the chance of seedling establishment in *Bellis perennis*. Further, in an agriculture example, the relative frequency of ryegrass and clover in a pasture was a function of interspecific competition that varied depending on whether crown rust of ryegrass was controlled by a fungicide application (Latch and Lancashire 1970). Similarly, experimental introduction of pathogens in biocontrol programs provides fascinating data on the effects of disease on plant fitness. The use of the rust *Puccinia chondrillina* to control the skeleton weed (*Chondrilla juncea*) in Australia has become a classic example. Burdon and colleagues (1981) illustrated both how the frequency of a susceptible skeleton weed biotype has decreased following introduction of the rust and, in greenhouse studies, how competitive interactions between biotypes depend on the presence or absence of the disease (Burdon et al. 1984).

An effective way to understand how the effects of disease on plant fitness may lead to selection for resistance involves studies that concurrently measure

variation in disease levels and assess effects of disease on survival and reproduction of plants of known genetic relationship. This information cannot be
obtained from unmanipulated natural populations, and there are problems in
directly translating results from greenhouse inoculation studies to the field.
Field transplant studies could provide these data, but often no data have been
recorded on plant fitness. In Alexander's (1989) field experiment with 20
cloned genotypes of *Silene alba* exposed to a smut fungus that sterilizes flowers, data were collected on average flower production as well as disease levels
of the genotypes over the two-year experiment. In the case of male genotypes,
the most susceptible genotypes were actually the ones with highest healthy
flower production. As discussed earlier, this finding is probably explained by
the higher likelihood of infection of plants with many flowers, which have
many potential targets of infection for this systemic pathogen.

The approach used by Rausher and Simms (1989) to assess the potential
for natural selection for resistance and to measure costs of resistance in plant/
herbivore systems has also yielded interesting data on natural selection for disease resistance. As discussed previously, plants from a diallel cross of *Ipomoea
purpurea* planted in a field site were scored for two pathogens as well as for
several herbivore species, and fitness of the annual plant was measured by seed
production. For *Colletotrichum dematium* var. *ipomoeae,* a negative relationship was observed between seed production and leaf damage: coupled with the
genetic variation observed in disease levels, this suggests that natural selection
for resistance to this pathogen could be occurring. In the case of the rust, *Coleosporium* (sp.), genetic variation was observed in disease levels but a positive
phenotypic relationship between rust intensity and seed production was actually
observed. No explanation for this result is apparent, but it illustrates why parallel measurements of genetic variation and disease effects on fitness are needed
before postulating selection for resistance. Further, although no correlation was
found between pathogen and herbivore levels in *Ipomoea purpurea,* there is the
potential for negative or positive correlations between the incidence of different
fungal and insect species and their effect on plant fitness. Clay (1986), for example, reviews interactions between endophyte infections and herbivore damage. Although it is logistically easiest to study interactions between a single
host and pathogen, clearly a more complex set of species is interacting.

A further reason for combining studies of plant genetic variation and disease effects on fitness is to elucidate the role of disease "tolerance" in natural
populations. Although disease tolerance is often invoked as important in nature
(Browning 1974; Schmidt 1978), few studies have focused on the degree to
which plants can continue to grow and reproduce with disease. Whether or not
tolerance can evolve in response to natural selection depends on whether it is a
genetically variable trait, as is suggested by the existence of agricultural cultivars which exhibit the same disease levels but differ in yields (Schafer 1971).
An inoculation study of *S. decipiens* infection of *A. bracteata* is one of few that

provide some evidence for this phenomenon in a natural population. Parker (1986) suggests that families vary in their sensitivity to disease damage since the relationship between number of sori per plant and plant size differed among families.

Future Studies of Effects of Disease on Plant Fitness

There is a need for more quantitative studies measuring disease effects on fitness and, especially, the relationship between fitness and resistance under natural conditions. Without these data, the role of diverse types of pathogens as selective agents cannot be known. Descriptive studies of disease effects on plant fitness will be strengthened by long-term studies of plant populations, combining data on disease incidence with information on plant survival and fertility. Such studies will help determine the most likely scenario for a particular host-pathogen interaction: that is, do pathogens exert a constant selective pressure over time or do they have minimal impact on fitness most years and then occasionally devastate the plant population in the rare year when the environment is particularly conducive to infection? Factors that may alter the answer to this question include the dependence of infection on environmental conditions, the longevity of pathogen spores in the environment, vector relationships, and whether infection is systemic or nonsystemic.

Further, long-term studies should allow documentation of whether resistance traits increase, decrease, or remain at a constant frequency in populations. Currently, the best studies of this type are the observed reduction in frequency of biotype A of *Chondrilla juncea* following introduction of a rust biocontrol agent (Burdon et al. 1981) and the increased resistance observed in *Anthoxanthum odoratum* exposed to long-term treatments conducive to disease development (Snaydon and Davies 1972). However, other studies suggest more complex scenarios. For example, even for the relatively simple, two-biotype genetic structure of a population of *Amphicarpaea bracteata* (Parker 1988c), the biotype known to be resistant to at least one isolate of the fungus *Synchytrium decipiens* has been decreasing in frequency (Parker, this volume). The cause of this change is unknown: is it due to the action of other pathogens or herbivores or simply a case where traits unrelated to disease have a stronger impact on fitness? Clearly, we should not take too narrow a view in studying plant/pathogen interactions and assume that the presence of a certain disease will always have a major (or even minor) effect on fitness. Particularly useful approaches will be to consider the joint effects of several pathogens and herbivores on fitness and the interactions between disease and plant competition.

Conclusions

Review of the literature reveals evidence for all three factors necessary for the evolution of plant resistance by natural selection: variation in resistance, inheritance of resistance, and a negative effect of disease on plant fitness. Yet it is rare for all three factors to be covered in analysis of one plant/pathogen interaction. These gaps in our knowledge exist partly because the study of disease in nature is still a young, developing field, but they also reflect the research approaches being used. Scientific investigations in evolutionary biology usually involve compromises between precision, realism, and generality (Levins 1966; Harper 1982) and studies of plant/pathogen interactions are no exception. To understand the complexity of resistance variation in a population, the controls available in an inoculation study are invaluable. Particularly important is the ability to compare reaction types of different combinations of plants and pathogens and to incorporate detailed genetic analyses of the inheritance of the trait. Control, however, can lead to a reduction in realism: is the variation detected the same as that which determines disease levels in nature? Fitness measurements of greenhouse inoculated plants are also unlikely to measure realistically the effects of disease. In contrast, in a field transplant experiment, there can be simultaneous measurement of genetic variation in resistance and its relationship to plant fitness under realistic environmental conditions. Because of the lack of controls on pathogen inoculum, however, the results cannot be generalized beyond the time and place of the experiment. Clearly, since no one approach is ideal, the most information will be obtained by analysis of the same plant/pathogen interaction with many approaches.

The concept of generality also arises when one considers what organisms have been our model systems for the study of disease resistance in nature. The literature is clearly biased toward easily observed and diagnosed fungi that infect leaves and flowers, yet these are only a subset of the vast array of disease-causing organisms. Soil-borne diseases and diseases caused by bacteria and viruses will provide important comparisons. Viruses are particularly intriguing, since they appear to be commonly present in plants (MacClement and Richards 1956) despite few obvious symptoms; this raises the troubling possibility that considerable variation in fitness among individuals in nature or among cloned plants in experiments may be due to hidden viral infections (Harper 1990).

Future work on evolution of disease resistance must go in many directions. Long-term studies, both in natural and experimental populations, can show whether changes in resistance actually occur. The paradox arises that although we expect selection for resistance, plant populations are often highly variable for resistance. Is this resistance variation often neutral, and only under selection in an unusual year highly conducive to infection? Do rates of disease spread differ in uniform versus genetically variable plant populations? Are there fitness costs of resistance? How do pathogens evolve in response to changes in host resistance? As is clear in chapters by Simms and Parker in this book, these questions are ripe for investigation, yet can be approached only once there is a solid framework of knowledge on genetic variation in disease levels in nature and the effects of pathogens on plant fitness.

15

Disease and Plant Population Genetic Structure

Matthew A. Parker

In recent years, pathogens have been blamed as insidious selective agents with manifold effects on the genetic structure of plant populations. In agricultural systems, pathogens have a well-deserved reputation for evolutionary volatility. Pathogen counteradaptation to plant defenses is often so rapid that plant breeders must constantly search for new sources of resistance genes to incorporate into crops (Stakman et al. 1943; Barrett 1981). It is tempting to generalize this agricultural experience to natural communities, and to assume that rapid genetic change and coevolutionary instability may also characterize plant-pathogen interactions in nature. However, evolutionary dynamics of natural systems are only beginning to be examined.

In this chapter, I consider three issues about the impact of disease-causing microorganisms on evolution in plant populations: (1) pathogens as causes of genetic polymorphism, (2) effects of nonrandom associations between disease resistance genes and alleles affecting other plant traits, and (3) pathogens as selective agents favoring sexual reproduction. The focus throughout is on evolution at one or a small number of major loci controlling plant disease reaction. Major genes with qualitative effects on resistance are a prominent feature of plant-pathogen interactions (Day 1974; Burdon 1987a), although polygenic variation may also contribute in many cases as well (see chapters by Alexander and by Simms and Rausher, this volume). Ultimately, a comprehensive picture of plant adaptation to pathogens will require an integrated analysis of both major gene and quantitative variation. The emphasis here on evolution at a small number of major resistance loci will, I hope, contribute to this synthesis.

Disease Resistance Polymorphism in Plants

One of the most distinctive features of natural plant populations is the staggering prevalence of genetic polymorphism for disease resistance (Wahl 1970; Zimmer and Rehder 1976; Dickey and Levy 1979; Burdon 1980; Segal et al.

1980; Delwiche and Williams 1981; Burdon, Oates et al. 1983; Jarosz 1984; Nevo et al. 1984; Harry and Clarke 1986, 1987; Parker 1988a, 1988c, Alexander 1989; reviewed in Burdon 1987a). While this variation suggests a highly dynamic coevolutionary interaction between plants and their pathogens, the specific evolutionary processes responsible for these polymorphisms are poorly understood.

Observed resistance polymorphisms are often assumed to be a stable outcome of some form of balancing selection (Person 1966; Browning 1974; Harlan 1976; Burdon 1982; Dinoor and Eshed 1984). Yet alternative theories about evolutionary mechanisms are not often articulated, and hypothesis testing about causes of polymorphism has rarely been a major goal in empirical studies (Parker 1990). In principle, observed polymorphisms could represent either an equilibrium or a nonequilibrium situation (Leonard and Czochor 1980). I first review two basic equilibrium models for maintenance of resistance polymorphisms, and then consider whether nonequilibrium explanations are required to account for patterns of polymorphism in nature.

Two Models of Balanced Polymorphism

Single-locus population genetic models have shown that two basic selective processes can produce a stable polymorphism for disease resistance. First, if resistance genes are deleterious in the absence of pathogen attack, then a polymorphism can be maintained without assuming any genetic heterogeneity among pathogens (Gillespie 1975; Clarke 1976; Anderson and May 1982). Alternatively, if pathogens are genetically heterogeneous, with each specialized on a different class of host genotypes, then a resistance polymorphism can be maintained by frequency-dependent selection, regardless of any cost associated with resistance genes (Clarke 1976; Lewis 1981). These two mechanisms are thus conceptually distinct theories about causes of resistance polymorphisms. I will call these two theories the "cost of resistance" hypothesis and the "genotypic interaction" hypothesis, respectively.

Costs of Resistance

For the purposes of this chapter, the cost of disease resistance can be defined as the increment by which the fitness of a resistant host falls below that of susceptible hosts in the absence of pathogen attack (see Simms, this volume, for a complete discussion, including alternative definitions). If variation in resistance is controlled by alternative alleles at a single locus, and if alleles conferring resistance also cause inferior plant performance under disease-free conditions, then a reversal in relative fitness of different genotypes in the presence and absence of disease can generate a stable polymorphism (Gillespie 1975; Clarke 1976; Anderson and May 1982). One key condition in these models is that infection must not be universal, for if all susceptibles become infected each generation, resistant hosts will be uniformly more fit. The exact conditions necessary for polymorphism depend on the specific mechanisms assumed to generate variation in infection. In the three theoretical studies cited, disease incidence is assumed to be inversely related to the frequency of resistant hosts in the population. Even without frequency-dependent disease incidence, models of multiple-niche selection (Hedrick 1986) imply that when resistance is costly, spatial variation in environmental factors controlling disease incidence could also promote polymorphism. Variation in disease incidence is widespread in natural environments (e.g., Augspurger and Kelley 1984; Jarosz and Burdon 1988). Hence, this could be a common process promoting polymorphism, as long as the basic premise that resistance entails fitness costs is accurate.

Van der Plank (1975) and Harlan (1976) have argued that in principle, resistance costs should be common, because plant traits that contribute to optimal function under disease-free conditions are likely to differ from those necessary for optimal defense against pathogens. However, empirical evidence documenting costs associated with resistance genes is remarkably scant. In crop plants, isolated cases of a negative association between resistance and plant yield have been reported (e.g., Chaplin 1970; Simons 1979). However, some resistance alleles actually increase yield under disease-free conditions (Brinkman and Frey 1977). Also, yield effects of a particular resistance allele can vary significantly in different genetic backgrounds (Frey and Browning 1971). For natural plant populations, data on resistance costs are only beginning to emerge. Burdon and Muller (1987) found differences in germination behavior and fecundity among lines of *Avena fatua* that differed in resistance to several races of *Puccinia coronata*. However, relative plant performance was reversed in different environments: susceptible plants performed better in glasshouse experiments, and did worse in field experiments. Genotype-environment interactions may thus be important in analyzing costs associated with resistance genes.

Burdon and Muller (1987) also note that in highly inbred plants such as *A. fatua*, chance correlations may exist between alleles for resistance and many other traits (see discussion of gametic disequilibrium below). Hence, variation in fitness between resistant and susceptible plant lineages sampled from natural populations may in fact be due to fortuitous differences in genetic background. Using the annual legume *Amphicarpaea bracteata*, I performed a study of resistance costs designed to minimize bias associated with genetic background effects (Parker 1990). Selfed F1 progeny of 20 naturally outcrossed seeds were screened for disease resistance using laboratory inoculation tests, and three progeny groups (families) were found to be segregating for a major disease-resistance gene. Within each family, two lines homozygous for resistance and two lines homozygous for susceptibility were selected. Selfed progeny of each line were then grown in a disease-free common garden in order to compare plant performance. For loci unlinked to the resistance factor, this procedure randomized alleles across the resistant and susceptible lines within each family. The distribution of seed biomass per plant overlapped broadly for resistant and susceptible individuals (fig. 15.1). However, analysis of variance indicated that resistant plants had significantly higher mean seed biomass overall ($p < 0.05$). This result is the opposite of that predicted by the cost-of-resistance hypothesis. Hence, for these plants, there is no evidence that genes for disease resistance are harmful to plant fitness in a pathogen-free environment.

Overall, the limited data from natural plant populations together with the agricultural studies suggest that large fitness costs are not a general attribute of genes for disease resistance. Future studies may yet identify bona fide examples of costly resistance genes. Nevertheless, it is unlikely that the cost-of-resistance theory can provide a general explanation for resistance polymorphism, because the magnitude of costs can be altered by environmental conditions (Burdon and Muller 1987), or by changes in genetic background (Simms, this volume). Thus, it is hard to view this as a consistent factor promoting polymorphism in nature.

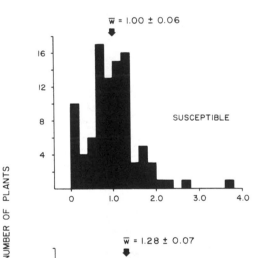

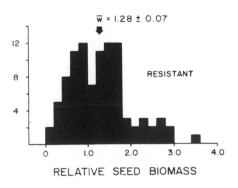

Fig. 15.1 Distribution of seed biomass per plant in *Amphicarpaea bracteata,* for six lines homozygous for disease susceptibility and six lines homozygous for resistance (n = 16 plants per line). All seed biomass values were scaled so that the mean among susceptible homozygotes was 1.00 (1 S.E. is also shown for each disease-resistance category).

Genotypic Interactions between Host and Pathogen

In genotypic interaction models, each pathogen is assumed to be specialized on a limited range of host genotypes. The relative fitness of different host genotypes will then depend on the genetic composition of the pathogen population. Haldane (1949) argued that this may lead to host polymorphism as a result of frequency-dependent disease impact. If a given type of host becomes common, this will select for a corresponding increase in the abundance of pathogen genotypes virulent on that host. Consequently, for any given host allele affecting disease reaction, there should be an inverse relationship between frequency and fitness (Haldane 1949; Clarke 1976; Lewis 1981; Levin 1983).

Several models motivated by Flor's (1956) gene-for-gene theory constitute a special class of genotypic interaction models. These assume a particular form of interaction in which hosts having a given resistance allele are vulnerable to attack by any pathogens with an appropriate matching virulence gene. Virulence genes are assumed to occur at many separate loci, making it possible to combine all of them together in a single, universally virulent pathogen genotype that can attack every host genotype. In order for polymorphism to exist in these systems, there must be some mechanism preventing fixation of this universally virulent pathogen, i.e., a fitness penalty for carrying excess virulence genes (Jayakar 1970; Leonard 1977; Levin 1983). In addition, polymorphism in the host requires overdominance at the host resistance locus (Mode 1958), or a cost associated with resistance alleles (Leonard 1977). Leonard's (1977) model thus combines both types of balancing selection distinguished above. Typically, the joint polymorphic equilibrium is locally unstable in gene-for-gene models, leading to sustained cycling of allele frequencies for many combinations of parameter values (Jayakar 1970; Leonard and Czochor 1980; Levin 1983). For finite populations in the real world, these dynamics may result in stochastic loss of polymorphism during time intervals when alleles have cycled to low frequency (Barrett 1988; Seger and Hamilton 1988).

There is strong empirical support for the basic premise of genotypic interaction models that pathogens are genetically heterogeneous in their virulence toward different host genotypes (e.g., Oates et al. 1983; Harry and Clarke 1986; Burdon 1987a; Burdon and Jarosz 1988; Parker 1988c). However, the exact structure of interactions among plant and pathogen genotypes that co-occur within local populations remains poorly studied in most cases. This is an important area for future research, because not all forms of genotypic interactions will lead to a balanced polymorphism. For example, if the pathogen population is dominated by widely virulent races that are able to attack every host genotype in a local area, then disease impact on each host type will not necessarily decrease as it becomes rare. Hence, the maintenance of host genetic polymorphism by frequency-dependent selection is less likely in cases where plants are exploited by broadly virulent pathogens rather than by narrow specialists (e.g., Burdon and Jarosz 1988; Parker 1988c, 1989). Likewise, this mechanism cannot account for polymorphism in cases where certain plant genotypes exhibit resistance to all local pathogens (e.g., Gerechter-Amitai and Stubbs 1970; Harry and Clarke 1986). If universally resistant hosts appear in a population, nothing would prevent them from replacing all hosts that are in any way susceptible, unless resistance is costly. But in that case, polymorphism would be attributable to these costs, and would have nothing to do with genotypic interactions.

To summarize, there is abundant evidence for specific interactions between plant and pathogen genotypes, but the structure of interactions is not always compatible with that necessary to generate a stable polymorphism via frequency-dependent selection. If widely virulent pathogens are prevalent, or if certain plant genotypes are found to possess resistance toward the entire pathogen population, then it is unlikely that observed polymorphisms can be explained by Haldane's (1949) mechanism or related theories. There is a great need for more detailed studies on the structure of genotypic interactions in nature to clarify how commonly this process may contribute to polymorphism.

Nonequilibrium Hypotheses for Polymorphism

To date, most empirical studies have not been sufficiently detailed to allow strong inferences about causes of resistance polymorphisms. But in certain cases, available information suggests that neither of the two simple models of balancing selection outlined above can account for observed patterns of genetic variation. While one could always devise more complicated models of balancing selection, an alternative type of explanation also deserves consideration. Observed polymorphisms may not be due to a stable balance of selective forces, but may instead represent a nonequilibrium situation.

One possibility is that observed polymorphisms are transient. Particular resistance alleles may be currently in the process of becoming fixed, but have not yet replaced their alternate alleles. Long-term studies demonstrating a consistent trend in allele frequency through time (e.g., Saghai Maroof et al. 1983) would tend to support this hypothesis. Yet even if a trend is found, interpretation is still problematical. Genotypic interaction models commonly show a tendency for sustained allele frequency oscillations (Leonard and Czochor 1980; Levin 1983; Seger 1988), so short-term trends do not necessarily imply ultimate fixation. Also, trends could be due to genetic hitchhiking caused by selection at other, unobserved loci (see below).

Another possibility is that observed polymorphisms may be a result of natural selection in the recent past, but are not currently being maintained by selection. For example, suppose that a new gene for disease resistance enters a plant population and begins to increase in frequency. This will cause selection on pathogens in favor of any variants able to attack hosts with the new resistance allele. If a new pathogen genotype appears that is virulent on plants both with and without the new resistance allele, this pathogen will rapidly become fixed, as long as it is fit in other respects (Leonard 1977). At this point, the new resistance allele becomes selectively neutral, and its subsequent evolutionary dynamics will be controlled by genetic drift. If pathogens counteradapt to new host resistance genes very rapidly, we may often expect to see such genes occurring at moderate frequencies within populations, even though they do not confer any protection against most existing pathogen genotypes. A possible example is discussed in Parker (1988c).

Testing nonequilibrium hypotheses about polymorphism is inherently more difficult than testing equilibrium theories, because they imply that the current state of a population may be determined in large part by historical factors (e.g., time since introduction of a new allele). Nonequilibrium situations can best be identified through a combination of long-term studies and perturbation experiments. For example, if a resistance allele is not being actively maintained by selection at its current frequency, then one would expect no return to current values after an experimental alteration of allele frequency. Multigeneration perturbation studies of this type would be a powerful tool in analyzing causes of resistance polymorphism, and they deserve much more attention from plant population biologists.

Disequilibrium between Alleles for Resistance and Other Traits

In analyzing the causes of any single-locus polymorphism, there is always the problem of determining whether selection is acting directly on that locus, or on associated alleles at other, unidentified loci (Ennos 1983). A case in point concerns the analysis of costs associated with resistance genes. If resistant plants are found to have inferior fitness under disease-free conditions, this could be due either to direct pleiotropic effects of resistance alleles, or to a nonrandom association between resistance alleles and deleterious alleles at other loci (Burdon and Muller 1987). Likewise, if disease resistance increases in frequency in a population through time (e.g., Saghai Maroof et al. 1983), this need not imply direct selection by pathogens. Such changes might instead be caused by hitchhiking, if resistance alleles are in disequilibrium with other selected loci. Hence, multilocus associations may be a critical factor influencing the evolutionary dynamics of disease resistance.

Gametic disequilibrium between alleles at different loci can be generated by any of the agents of evolution (mutation, migration, genetic drift, selection). Once disequilibrium is present in a population, several factors can reduce the rate of recombinational decay of disequilibrium, including linkage, asexual reproduction, and inbreeding (Hedrick et al. 1978). Whenever one or more of these three factors are present, nonrandom allelic associations may persist in a population long enough to have significant evolutionary consequences. For natural plant populations, little is known about linkage relationships of disease resistance loci (but see Harry and Clarke 1987). In domesticated plants, close linkage between different resistance loci has been found in many cases (Flor 1956; Hooker and Saxena 1971; Day et al. 1983). When such linkage exists, certain resistance alleles may be affected by genetic hitchhiking. If a particular resistance allele becomes neutral owing to pathogen counteradaptation (as discussed above), it may still hitchhike to high frequency if associated with resistance alleles at linked loci that are being favored by selection. Hence, linkage and hitchhiking may help preserve polymorphism for resistance alleles that are no longer effective toward existing pathogen genotypes in a local area. This might help explain why natural plant populations are so often polymorphic for resistance when exposed to arbitrary pathogen genotypes that may be quite different from those currently prevalent in their native habitat (Zimmer and Rehder 1976; Burdon, Oates et al. 1983; Harry and Clarke 1986; Parker 1988a).

Asexual reproduction is widespread among plants, and may thus be a major factor influencing multilocus population structure. Because recruitment of new genotypes is rare in many clonal plant populations, chance events at the time of clone establishment could result in fortuitous yet persistent associations between disease resistance and other traits. For example, Harberd (1961, 1967) and Copony and Barnes (1974) observed that clones commonly differed both in disease resistance and in several morphological and phenological characters. In some cases, abundant or widely distributed clones exhibited extreme susceptibility to certain pathogens. The causes of successful clonal proliferation are not known in these systems, and may simply be due to chance historical factors (e.g., opportunistic colonization following disturbances). Yet it appears that differential proliferation of clones, for whatever reason, may have resulted in a correlated, nonadaptive increase in the frequency of disease susceptibility within some of these populations.

Alternatively, if pathogens exert strong selection on an asexual plant population, this can cause correlated changes in the frequency of other traits associated with disease resistance. The history of biological control of *Chondrilla juncea* (Compositae) by the rust *Puccinia chondrillina* in Australia provides an excellent example. *C. juncea* is an obligate apomict, and three distinct forms occur in Australia that differ in several morphological traits and also at allozyme loci (Burdon et al. 1980). A single strain of the rust specialized on one form of *C. juncea* was released in 1971, and it soon caused extensive disease epidemics (Cullen et al. 1973). The two other forms of *C. juncea* remained uninfected and rapidly increased in abundance (Burdon et al. 1981). Thus, owing to differential impact on these asexual host lineages, pathogens greatly altered the prevalence of a whole set of morphological characters and enzyme variants.

Inbreeding is a very important factor shaping the multilocus genetic structure of many plant populations (Allard 1975). Strong correlations of allelic state across loci are common in natural populations of self-pollinated plants (Allard et al. 1972; Brown et al. 1977; Hamrick and Holden 1979; Parker

1988a). In highly self-pollinated annuals, a number of seemingly arbitrary traits have been found to be associated with disease resistance, including alleles at particular enzyme loci (Nevo et al. 1984; Nevo et al. 1986; Parker 1988a), germination behavior (Burdon and Muller 1987), and visible traits such as leaf shape, flower color, and seed size (Parker 1988a, 1991). To the extent that pathogens cause selection at resistance loci, nonadaptive evolution of these other traits may potentially occur as a correlated response.

Likewise, these interlocus associations could potentially interfere with the evolution of disease resistance in highly inbred plants. If natural selection on other traits is stronger than that exerted by pathogens at resistance loci, resistance may actually decrease in frequency, at least temporarily. A possible example of this was observed in one population of *A. bracteata*, where strong disequilibrium existed between resistance variants and many other traits (Parker 1988a). In this population, the frequency of plants with resistance toward local pathogens decreased from 41% to 16% over a two-year period (Parker 1991). The specific targets of selection are not known in this case.

To summarize, a number of interesting complexities emerge in cases where recombination is restricted by linkage, asexual reproduction, or inbreeding. Selection by pathogens at disease resistance loci may influence the evolution of functionally unrelated traits, and likewise, selection on other traits could constrain the evolution of disease resistance. However, restriction of recombination may also result in more fundamental problems for plants in their attempts to cope with pathogens. I next examine the contention that selection for increased genetic recombination in plants may be a universal outcome of pathogen attack.

Pathogens and Sex in Plants

The hypothesis that pathogens may be responsible for the evolution of sex in their hosts has recently received much theoretical attention (Levin 1975; Jaenike 1978; Bremermann 1980; Hamilton 1980; Hutson and Law 1981; Bell 1982; Tooby 1982; Rice 1983a; reviewed in Seger and Hamilton 1988). Despite a broad range of analytical approaches, certain underlying assumptions and arguments recur in most of this literature. It is assumed that pathogens have the capacity for rapid evolution (primarily because of their short generation times), and that selection on pathogens will favor those adapted to the most common phenotypes in the host population. Thus, a frequency-dependent fitness deterioration is expected for common host phenotypes as pathogens continually adapt to their defenses. If these common host phenotypes reproduce asexually, their genetically identical progeny will be born into a world full of disease-causing organisms that are preadapted for their destruction (Hutson and Law 1981). Hosts that reproduce sexually may thus have a selective advantage: by lowering the genetic similarity between parents and sexual progeny, recombination may increase the chance for escape from pathogens adapted to the parent (Tooby

1982; Rice 1983a). Although these arguments are usually developed for the case of sexual versus asexual reproduction, they extend naturally to other types of reproductive choices. For example, in plants that can reproduce by either outcrossing or self-fertilization, pathogens should select for outcrossing, all other things being equal.

I will evaluate the pathogen attack/host recombination hypothesis from two perspectives. First, how reasonable are the main biological assumptions of the theoretical models? Second, from a comparative perspective, how much diversity exists in the form of selection by pathogens on plant reproductive behavior? It is obvious that pathogens can exploit plants in a wide variety of ways, and that the resulting selective consequences for plants may be quite diverse. Therefore, I will examine several biological attributes that are likely to influence the selective impact of pathogens on plant sexuality.

Biological Assumptions

The theoretical models have identified a number of conditions that must exist in order for pathogen attack to promote host sexuality. First, to compensate for the inherent fitness advantage of asexual reproduction, selection on plants caused by pathogen attack must be intense (Hamilton 1980). Cases of severe disease epidemics dramatize the potential for strong pathogen impact on plant reproductive success (Cullen et al. 1973; Hepting 1974; Weste 1974). However, a strong demographic impact need not imply intense selection. A number of methodological problems arise in quantifying the actual intensity of selection exerted by pathogens within natural plant populations (see Alexander, this volume). As more data accumulate, it is not unlikely that we will find wide heterogeneity in pathogen impact: variation over time within a population, variation among populations within a species (Parker 1986), variation among closely related plant species exploited by a common pathogen (Jarosz 1984), or even at higher taxonomic levels (Kenneth and Palti 1984). To the extent that asexual forms have a strong inherent advantage in the absence of pathogens (Lewis 1987), this implies that pathogen attack may not always be a sufficiently potent selective agent to prevent replacement of sexuals by asexuals on an ecological time scale.

A second key assumption common to most models is that pathogens will adapt rapidly to common host phenotypes, resulting in frequency-dependent disease impact (Hamilton 1980; Hutson and Law 1981; Bell 1985). Evidence for frequency-dependent impact in natural plant populations is currently scant (Barrett 1988). For agricultural systems, there are numerous examples of rapid pathogen adaptation to widespread, abundant host phenotypes (Adams et al. 1971; Barrett 1981). Yet the vast geographic scale at which genetically uniform crop varieties may be exposed to pathogens has few parallels in nature. The likelihood of intense frequency-dependent disease impact may be much lower in natural plant communities because of their complex spatial structure, with particular host phenotypes occurring discontinuously in varying mixtures with other host phenotypes and with nonhost plants. As yet, there are no longitudinal studies in natural communities that document differential erosion of resistance for rare versus common plant phenotypes within a given population.

One indirect way to examine this problem is to perform transplant experiments. If local pathogen populations have adapted to locally common host genotypes, then novel plant genotypes introduced into a site at low frequency should escape infection. Experiments of this type have so far been performed with only two plant species. In *Amphicarpaea bracteata* (Leguminosae), transplant studies indicate that pathogens may in fact be preferentially adapted to locally common host genotypes (Parker 1985). However, differential attack on locally common host genotypes was not observed in experiments with *Podophyllum peltatum* (Parker 1989). In general, we might expect that the extent of frequency-dependent selection will depend on a number of factors (e.g., generation length, migration structure, level of host genetic diversity, pathogen mating system, see Groth and Christ, this volume) that will vary widely among different plant-pathogen interactions. Thus, for natural communities, it is not yet clear how often we can expect to see intense frequency-dependent disease impact.

A third class of assumptions concerns the nature of specificity between host and pathogen genotypes. Most models assume that interactions between host and pathogen are highly specific, so that pathogen genotypes virulent on any one type of host are not successful in attacking most other hosts (Jaenike 1978; Hamilton 1980; Tooby 1982; Rice 1983a). This assumption appears to be necessary to generate strong frequency-dependent selection on hosts. If each pathogen is virulent on a wide range of host genotypes, then disease impact on any given host genotype will not necessarily decrease as it becomes rare in the population. More specifically, consider the widely studied case where the host is a haploid with two diallelic loci controlling disease reaction (Hamilton 1980; Hutson and Law 1981; Bell and Maynard Smith 1987). There are four possible host genotypes (AB, Ab, aB, and ab). Selection for host recombination seems to work best if each pathogen is adapted to only one host genotype. For example, if pathogens adapted to host AB are equally successful on hosts Ab and aB, then host AB cannot gain any benefit from recombination. Thus, in order for host recombination to be favored, a single allele substitution at one locus must significantly alter the identity of pathogens to which a host is susceptible. The more specialized pathogens are, the more likely it is that this condition can be met.

Data from natural populations reveal that pathogens are rarely as specialized as assumed in these simple models. In most cases, each pathogen is able to attack a wide range of host genotypes, and each host is susceptible to many different pathogens (Eshed and Dinoor 1981; Harry and Clarke 1986; Burdon 1987b; Burdon and Jarosz 1988; Parker 1988c, 1989; Alexander 1989). This lack of strict specificity is perhaps not surprising: in natural communities, pathogens will generally have to contend with many different host genotypes in successive generations. As a result, pathogens with broad host ranges may often be more successful than specialists (Barrett 1983). Overall, this lack of extreme specificity reduces the likelihood that plants can lessen their progeny's disease burden by reproducing sexually.

While there is little evidence for extreme pathogen specificity of the type assumed in the simplest genetic models, selection for host recombination can still occur under certain less restrictive conditions. For any two host loci controlling disease reaction, the set of pathogens adapted to two-locus allele combinations in one phase (e.g., "coupling phase" hosts: AB, ab) must be poorly adapted to alternate host allele combinations ("repulsion phase" types: Ab, aB). If this condition is met, then host-pathogen coevolution can easily lead to sustained oscillation in the sign of the coefficient of gametic disequilibrium among host loci. Host genes promoting recombination are likely to spread when selection on the host alternately favors positive and negative associations across loci (Hamilton 1980; Hutson and Law 1981; Felsenstein 1988). This requires that coupling- and repulsion-phase host genotypes be susceptible to different groups of pathogens.

For natural plant populations, there is little genetic data available on how commonly this situation might exist. However, if the genetics of plant-pathogen interactions in nature are in any way parallel to the well-known gene-for-gene systems found in domesticated plants, then this condition is unlikely. In the classic study of flax and flax rust that formed the basis for the gene-for-gene model, Flor (1956) found that host lines carrying different multilocus allele combinations often overlapped greatly in the identity of pathogen races to which they were susceptible. To cite one specific example, pairs of flax varieties carrying two-locus allele combinations in opposite phase are not vulnerable to completely distinct sets of pathogens (table 15.1). Instead, one coupling-

TABLE 15.1
Susceptibility of four varieties of cultivated flax to different types of flax rust, *Melampsora lini*

	Flax variety	Plant genotype	Pathogen type			
			1	2	3	4
Repulsion phase	Polk	NNpp	R	S	R	S
	Koto	nnPP	S	R	R	S
Coupling phase	Redwood	NNPP	R	R	R	S
	Winona	nnpp	S	S	S	S

Sources: Flor 1955, 1956.
Note: The four host varieties are homozygous for different allele combinations at two linked loci that control disease reaction, designated "N" and "P." The four pathogen types represent the four basic disease-reaction patterns observed in tests of numerous pathogen isolates on these four varieties (R = host resistant; S = host susceptible).

phase type (Winona) is susceptible to all pathogens that are virulent on host varieties with repulsion-phase allele combinations; the other coupling-phase host line (Redwood) is generally resistant. The single type of pathogen that can attack Redwood is also virulent on hosts with all other allele combinations. If all of these genotypes were allowed to coevolve together, one might expect these interactions to result in directional selection on plants for the particular allele combination that confers resistance to the greatest number of pathogen types (e.g., Redwood). Selection on pathogens should then result in fixation of genotypes most successful on the prevailing host variety (e.g., pathogen type 4). Hence, this pattern of interaction will not select for host recombination, because pathogen attack will not cause a sustained oscillation in the coefficient of disequilibrium among host loci (i.e., NNpp and nnPP favored in some generations, and nnpp and NNPP favored in other generations). Thus, the structure of genotypic specificity found in agricultural gene-for-gene systems deviates considerably from that assumed in models for the evolution of recombination.

It has been suggested that gene-for-gene systems may be an artifact of crop plant breeding procedures (Day et al. 1983; Barrett 1985), and may thus not be typical of plant-pathogen interactions in natural communities. Compared to domesticated plants, natural plant populations probably have a more complex array of disease resistance mechanisms (Segal et al. 1980; Burdon 1982; Heath 1987; Alexander, this volume). Nevertheless, recent studies of natural plant populations have identified patterns of resistance phenotype structure and inheritance that are not inconsistent with the gene-for-gene interactions seen among crop plants (e.g., Harry and Clarke 1986, 1987; Burdon 1987b; Parker 1988c). If gene-for-gene interactions of the type illustrated in table 15.1 prove to be common in nature, this will be a serious difficulty for the theory that pathogens are responsible for sex in plants.

One final issue in evaluating the theoretical models is the assumption that mating occurs at random among all sexual members of the host population (Hamilton 1980; Hutson and Law 1981). In natural populations, the capacity for sex to produce genetically different offspring will depend on the spatial structure and kinship of individuals who actually exchange genes. In most plant species, the spatial scale of pollen movement tends to be highly restricted (Levin and Kerster 1974). Thus, plant populations may often be subdivided into small, genetically homogeneous neighborhoods (Wright 1946; Turner et al. 1982). If neighboring plants tend to be genetically similar owing to genetic drift, and if most mating occurs among neighbors, then outcrossing may not reduce the genetic similarity of parents and offspring enough to alter disease impact substantially. Thus, in natural plant populations, ecological constraints on pollen dispersal may limit the potential benefit of sexual reproduction in reducing pathogen impact.

To summarize, models have helped to clarify the conditions necessary for pathogens to act as selective agents favoring sex in plants. The models generally require that pathogens exert strong and consistent selection on plants, that pathogens adapt rapidly to common host phenotypes, that interactions between host genotypes and pathogen genotypes be highly specific, and that sexual forms mate randomly. The biology of plant-pathogen interactions in natural communities quite often appears to deviate from one or more of these assumptions. It is thus unlikely that this theory, in its present form, can be a robust explanation for the widespread existence of sexual reproduction among plants.

Pathogen Impact on Plant Reproduction: A Comparative View

Pathogens are quite diverse in their manner of exploiting plants, particularly with respect to their mode of transmission and the stage of the host's life cycle attacked. Consequently, the specific form of selection by pathogens on plant reproductive behavior is likely to be rather variable. I will consider four basic issues in this survey: (1) flowers as infection sites, (2) vertical disease transmission to clonal progeny, (3) developmental variation in disease resistance, and (4) nonadaptive modification of host reproductive phenotype by systemic pathogens.

Flowers as Infection Sites

In many cases, flowers serve as specific sites for invasion by pathogens (Batts 1955; Campbell 1958; Buddenhagen and Elsasser 1962; Sands and McIntyre 1977). In effect, these pathogens act like venereal diseases, parasitizing the mating process in plants. Pathogens may sometimes be transmitted by the same animal vectors that provide pollen transport services to plants (e.g., Schroth et al. 1974; Jennersten 1983). Floral infection has two potential selective consequences for plants. First, plants may experience selection to minimize exposure to these pathogens by shifting their reproductive effort to vegetative propagules (corms, bulbs, stolons, rhizomes, tillers, etc.). If individual plants within a population vary in their relative allocation to sexual versus vegetative reproduction, and if flowering increases the chance of exposure to damaging pathogens, then plants with increased allocation to vegetative reproduction may contribute more propagules to future generations.

A second possible consequence of floral infection would be selection for changes in floral morphology or phenology. Rapid postfertilization changes in floral traits to reduce pollinator attraction (e.g., Gori 1983) or, in dioecious taxa, rapid dehiscence of male flowers following pollen donation (e.g., Alexander 1989) are traits that might help avoid exposure to florally transmitted pathogens. In cultivated barley, the trait of cleistogamy provides effective resistance to infection by the loose smut pathogen (*Ustilago nuda*). This pathogen invades plants through open flowers, so barley varieties that produce only closed, self-fertilized cleistogamous flowers escape infection. In barley breeding programs, there has been extensive artificial selection for cleistogamy to control this disease (Pedersen 1960). Natural selection for cleistogamy might presumably occur for similar reasons. Genetic variation in relative allocation to cleistogamous versus open (chasmogamous) flowers has been documented for several plants (Clay 1982; Schemske 1984). If floral infection is a serious problem, selection may favor individual plants that produce a relatively higher proportion of cleistogamous flowers. This example is important because it illustrates how pathogens may indirectly alter plant mating systems, in this case, potentially causing a shift toward self-fertilization.

Vertical Disease Transmission to Clonal Progeny

Direct transmission of pathogens from parents to offspring is extremely common in vegetative propagules (Agrios 1978). This is one basic disadvantage of most forms of asexual reproduction in plants. Disease transmission in seeds or pollen has been documented (Bennett 1969; Mathre 1978), but appears to be fairly infrequent for most types of plant pathogens. Embryos lack direct vascular connections with their maternal parent, and may thus be physically isolated from pathogens that have colonized maternal tissue in many cases.

If there is a higher probability of disease transmission to asexual progeny, this might favor plants with an increased allocation to sexual reproduction. For example, plants of the herbaceous perennial *Arisaema triphyllum* that are infected with the systemic rust *Uromyces ari-triphylli* invariably transmit this disease to their vegetative propagules (corm buds). However, pollen and seed from infected plants are viable and free of infection (Parker 1987). Hence, the optimal allocation to vegetative versus sexual reproduction will be quite different in environments where this disease is common, compared to areas lacking this pathogen.

Differences in the probability of vertical diseases transmission to sexual and vegetative propagules may often have more to do with developmental patterns than with relative genetic similarity between progeny and parent. For example, in citrus, seeds may contain nucellar embryos that are genetically identical to the parent tree. Yet these seeds are typically free of viruses that have infected the parent, unlike vegetatively derived propagules (Frost 1948). Thus, seeds might in some cases be viewed simply as a morphological adaptation for escape from parental pathogens. The usual link between seed formation and genetic recombination may be incidental to this effect.

Developmental Variation in Disease Resistance

Sexual and vegetative propagules often differ greatly in size, morphology, and developmental patterns. This can result in unequal disease resistance toward parental pathogens, even when there is no direct vertical disease transmission. For example, in the perennial herb *Podophyllum peltatum*, ramets produced asexually via rhizome branching are resistant to a specialist rust, *Puccinia po-*

dophylli, while sexual progeny derived from seedling recruitment are highly susceptible (Parker 1988b). The difference in disease resistance appears to be due to size-related variation in a morphological structure that protects emerging shoots from contact with pathogen spores in the soil. New ramets generated clonally by rhizome branching are much larger when first formed, and are thus more effectively protected from exposure to pathogens. The net effect of differential disease impact on sexual progeny may be to select for an increased allocation to vegetative reproduction in this plant (Parker 1988b). This type of developmental variation in resistance could be fairly common. Plant vulnerability to damping-off fungi is strongly affected by developmental stage, with newly germinated seedlings being most susceptible in many cases (Neher et al. 1987). Vegetative propagules may often be larger or more physiologically mature (e.g., more lignified) when they first encounter such pathogens (Populer 1978). Thus, vegetative reproduction may sometimes enable plants to bypass a stage in the life cycle (young seedlings) that is inherently susceptible to certain types of disease.

Modification of Host Phenotype by Systemic Pathogens

Certain ascomycete endophytes of grasses can have striking effects on the reproductive behavior of their hosts. For example, infection of *Agrostis tenuis* by *Epichloe typhina* causes complete suppression of flowering and a corresponding increase in the rate of vegetative spread (Bradshaw 1959). Since infected plants invariably transmit pathogens to vegetative propagules (tillers), this alteration of host reproduction could be beneficial to the pathogen, although it is probably maladaptive for the host. In *Danthonia spicata,* uninfected plants have a mixed mating system, with individuals producing both chasmogamous and cleistogamous flowers. *D. spicata* plants infected with *Atkinsonella hypoxylon* produce cleistogamous flowers only (Clay 1984). Seeds from these cleistogamous flowers are viable, and are also infected by *A. hypoxylon* at high frequency (Clay and Jones 1984). Thus, infected plants can still reproduce sexually, but are reproductively isolated from the rest of the population as a result of enforced inbreeding.

Both of these cases represent extreme manipulation of host reproductive behavior by the pathogen. The overall consequences for the evolution of the plant's mating system are likely to be complex, because of the high frequency of vertical disease transmission, the reproductive isolation between infected and healthy plants, and the fact that endophytes may protect their hosts from herbivore consumption by producing poisonous alkaloids (Clay 1986). To the extent that endophytes increase the ecological success of their hosts in environments where herbivores are abundant (Clay 1988), the loss of opportunity for sex or outcrossing can be viewed as a cost payed by plants in exchange for the benefit of protection from herbivory.

Summary: Pathogens and Plant Reproduction

Together, these examples highlight the diversity of pathogen impact on plant reproductive behavior. Flowering can make a plant vulnerable to infection with certain groups of pathogens, which might be expected to favor vegetative reproduction. But once a plant is infected, it can be disadvantageous to reproduce vegetatively, because pathogens are often transmitted to clonal progeny. Leaving aside disease transmission, there can be basic developmental differences between sexual and vegetative propagules that strongly affect their relative disease resistance. Finally, some pathogens may directly manipulate plant reproduction in a way that may be maladaptive for the host.

This apparent diversity of selective outcomes contrasts with the dominant theme of the theoretical literature, where there is a near consensus that pathogen attack will promote outcrossed sexuality among hosts. The comparative studies suggest that it may be difficult to achieve a decisive experimental test of the proposition that pathogens are the main agents responsible for sex in plants. As pointed out by Bierzychudek (1987), evaluating a particular theory about the benefits of sex requires more than experiments on a model system chosen to closely match the theory's assumptions. Regardless of what happens in any ideal model system, the core problem is to judge the accuracy and generality of the assumptions themselves in mirroring the diverse natural world.

Progress in evaluating the pathogen attack/host recombination hypothesis will require further work in several specific areas. Theoretically, we need to know the probable effects of increased complexity. For example, if a large number of loci influence disease resistance, can host recombination be favored under a broader range of conditions than is observed with two-locus models (Seger 1988; Seger and Hamilton 1988)? When one pathogen fails to promote host recombination because of an "incorrect" form of genotypic specificity (e.g., table 15.1), will the introduction of other pathogen taxa to the system generally have the effect of favoring recombination? Empirically, we need comparative studies that more accurately address theoretical concerns: what forms of plant-pathogen genotypic specificity are typical in nature? Do natural plant populations commonly exhibit a rapid oscillation in the sign of disequilibrium among host loci owing to selection by pathogens? Is frequency-dependent disease impact as intense or consistent as imagined in most models? In summary, I have argued that in its present form, the pathogen attack/host recombination hypothesis may not be particularly robust when applied to natural plant populations. However, theoretical exploration of this idea is still incomplete, and it remains a hypothesis that deserves serious attention from plant population biologists.

Conclusions

In organizing this review, I have considered pathogen impact on polymorphism and on sex in plants as separate issues. However, there is a close relationship between these two topics (Tooby 1982; Bell 1985). Sex can only modify progeny genotype arrays, and will thus only be visible to natural selection, when genetic diversity is present within the pool of mating individuals. Hence, the magnitude of genetic polymorphism generated by host-pathogen coevolution has been a key concern in evaluating theories of pathogen-mediated selection for recombination (Seger and Hamilton 1988).

Nevertheless, it is important to distinguish between polymorphism arising from different selective mechanisms. With respect to the dichotomy adopted here, of cost-of-resistance versus genotypic-interaction models, only polymorphisms generated by the latter mechanism are significant for the evolution of sex. The processes invoked in genotypic interaction models for polymorphism are identical to those in theories of pathogen-mediated evolution of sex (frequency-dependent selection on hosts due to a heterogeneous pathogen population, with each pathogen specialized on a particular class of host genotypes). In contrast, the cost-of-resistance theory makes no assumptions about specific genotypic interactions between host and pathogen, and can operate even if the pathogen population is genetically uniform. Thus, polymorphisms maintained by a cost-of-resistance mechanism are not likely to furnish the appropriate genetic raw material to allow selection for host recombination. In conclusion, discriminating between the two main theories for maintenance of disease-resistance polymorphism is not only an important problem in its own right, but is also critical for analyzing the role of pathogens in the evolution of plant mating systems.

The widespread genetic variation for both disease resistance and pathogen virulence in nature, together with the potential for intense selection, suggests that the evolutionary interaction between plants and pathogens may often be highly dynamic. These attributes make plant-pathogen interactions promising systems for examining many fundamental problems in coevolutionary biology. Despite the numerous gaps in our current knowledge, the work reviewed here makes it clear that pathogens can no longer be ignored as key agents shaping the genetic structure of natural plant populations.

Acknowledgments

I am grateful to Ellen Simms and Helen Alexander for comments and suggestions during the writing of this chapter. Financial support for research described here was provided by National Science Foundation grant BSR-8717222.

16

Theory and Pattern in Plant Defense Allocation

Arthur R. Zangerl and Fahkri A. Bazzaz

Optimal-Defense Theory

Plants are subject to attack by a wide diversity of organisms ranging from bacteria and viruses to large mammals. Plants have, however, acquired a similarly wide array of defenses against attack, including physical defenses such as thorns and trichomes and numerous chemical defenses. In this chapter, we will examine factors that govern within-plant defense allocation. Particular emphasis will be placed on chemical defenses that reduce or prevent damage by herbivores.

Optimal-defense theory as originally outlined by McKey (1974a) and later elaborated by Rhoades (1979) has three main tenets. First, defense is assumed to have a cost; resources allocated to defense cannot simultaneously be allocated to other functions. Second, the fitness impact associated with loss of a given quantity of tissue depends on the characteristics of the tissue; not all plant parts are equally valuable to the plant. Third, some plant parts or tissues are more likely to be attacked than other plant parts or tissues. Given these three tenets, defenses should be preferentially allocated to more valuable parts with high probability of attack. In practice, empirical tests of this theory are few, however, owing to the difficulty of measuring those very tenets that have such great logical appeal in theory.

The Cost of Defense

At the heart of the assumption that chemical defense has a cost is the axiom that materials and energy allocated to a particular function cannot be allocated simultaneously to another function (but see later discussion of multiple function compounds). The direct cost of defense is equivalent to the amount of resources devoted to production and maintenance of the defense. The ultimate and more relevant cost in fitness terms, however, is the indirect cost of diverting resources

363

from other important functions. The difficulty of measuring these costs is formidable. The first, and perhaps most difficult, step is to identify the function or
functions of a particular plant constituent. For example, Seigler and Price
(1976) argued that secondary chemical pools may be little more than temporary
repositories for carbon skeletons and that any defensive value of these compounds is incidental. Costs of chemicals that have more than one function must
be spread proportionately over the benefits derived from each function. If all of
the functions of a putative defense can be successfully identified and quantified,
cost can be estimated in a variety of ways (also see Simms, this volume).

Trade-offs between defense and productivity are generally assumed to exist
in crop systems. Bottrell and Adkisson (1977) suggested that the emergence of
synthetic organic insecticides in the 1940s and 1950s prompted a shift in cotton
breeding programs from yield maximization in environments with pests to yield
maximization in pest-free environments. The resultant varieties were far more
insecticide-dependent than ancestral stocks. The same trade-off between yield
maximization and resistance is indicated for agronomic systems generally (Pimentel 1976). Although the crop-breeding literature provides much information on consequences of selection for desired traits on other desirable traits
(e.g., the trade-off between yield and nutritional value; see Axtel 1981), there
are few detailed studies of the effects of breeding for high yield on pest resistance (see Krischik and Denno 1983). As the environmental and other problems
associated with pesticide use increase, enhanced efforts to breed for increased
resistance should provide a wealth of information on the relationship between
defense and yield.

Direct Costs

One way to quantify the direct cost of compounds or structures associated with
resistance is to estimate their biosynthetic cost (e.g., Penning de Vries 1975a;
Chew and Rodman 1979). For example, Gulmon and Mooney (1986) estimated
that the cost of several potentially defensive compounds ranged from 2.5 to 5.0
gCO_2 g^{-1} product. However, measures of this kind can underestimate overall
cost, because they may not account for costs associated with specialized containment structures, autotoxicity-related repair, or transport of precursors and
products within the plant. Moreover, these estimates ignore the indirect cost of
diversion of resources to defense from resource acquisition itself; diversion of
nitrogen, for example, to alkaloid synthesis from nitrogen-dependent photosynthesis will enhance defense but reduce growth.

Indirect Costs

Correlative data relating defense investment to plant fitness or growth without
herbivory are easily obtained and integrate all of the indirect costs associated
with defense. Their obvious drawback is that they reflect correlation and not
causation. Surprisingly, however, few studies provide evidence of any correla-

tion. In table 16.1 we list significant correlations ranging from -0.382 to -0.848 between measures of defense and plant performance. An interesting feature of these data is that all but two (Coley 1986; Zangerl, unpublished data) implicate a genetically controlled cost of resistance or defense. The negative correlation between alkaloid content and yield in tobacco is one that has been repeatedly observed (Matzinger et al. 1989). By selecting simultaneously for both high tobacco yield and high alkaloid content, Matzinger et al. (1989) were able to reduce a genetic correlation of -0.67 between yield and alkaloid content to -0.26 after only two cycles of selection. However, it is questionable whether natural selection is consistent enough to significantly reduce genetic correlations in wild populations.

The relationship between allelochemical content and growth and fitness in wild parsnip is all the more interesting in that the cost of defensive compounds present at concentrations of less than 1% should not be measurable (Gulmon and Mooney 1986), yet in wild parsnip there are negative correlations (table 16.1). Because furanocoumarins are localized in specialized oil tubes (Ladygina et al. 1970), the cost of containment together with the cost of production is probably much higher than would be expected based upon the concentration of the chemicals.

Negative correlations between defense and fitness or fitness components are not always found. Simms and Rausher (1987) found that variation in fitness costs was not associated with genetically variable resistance to insects in *Ipomoea purpurea*. In later work with the same plant, Simms and Rausher (1989) found genetic variation for resistance to four types of insect herbivores. However, genetic covariances between resistance measured in control plots and seed production measured in insecticide-treated plots were not different from zero, indicating no cost of resistance.

An alternative to examining continuous variation in defense investment and plant performance is to compare two genotypes or phenotypes that are differentially resistant. Foulds and Grime (1972b) found that acyanogenic genotypes of *Trifolium repens* produced nearly twice as much dry matter as cyanogenic genotypes and were far more likely to flower by the end of the season. Cyanogenesis in this species is attributed to genes at two loci. The Ac gene is dominant at one locus and is responsible for the production of the cyanogenic glucosides linamarin and lotaustralin. The Li gene is dominant at the second locus and is responsible for the presence of linamarase, a β-glucosidase. Kakes (1989) recently reexamined the role of the Ac and Li genes in *Trifolium repens* as they relate to flower production. The two loci are not linked, but disequilibrium has been observed. Kakes minimized nonlinkage disequilibrium by controlled crosses and measured flower production in offspring, segregating for both loci. Offspring containing the Ac gene produced only a half to a third as much floral biomass as plants without cyanogenic glucosides. The Li gene had no significant effect on flower production. Kakes estimated that the biosynthetic

TABLE 16.1
Costs of defense as estimated by correlations between defense level and fitness components

Species	Defense	Concentration (%)	Fitness component	Correlation	References
Nicotiana tabacum	alkaloid	(2.36 – 5.46)	yield	– .648	Nielsen et al. 1985
Nicotiana tabacum	alkaloid	(2.11 – 8.14)	yield	– .848	Vandenberg and Matzinger 1970
Cecropia peltata	tannins	(1.00 – 6.00)	leaf production	– .520	Coley 1986
Gossypium hirsutum	resistance	(na)	lint yield	– .727	Wilson 1987
Pinus monticola	terpenes	(na)	growth	– .383	Hanover 1966a
Pastinaca sativa	furanocoumarin	(0.022 – 0.054)	growth	– .382	Zangerl, unpublished data
Pastinaca sativa	bergapten	(0.15)[a]	flowers	– .632	Berenbaum et al. 1986
Prunus hybrids	9 phenolic compounds	(na)	growth	– .23 to – .92	Nachit and Feucht 1983

Note: na = not available.
[a]Mean concentration of seed bergapten (from Nitao and Zangerl 1986)

cost of glucoside production in Ac plants was approximately 5 kJ, far less than the estimated 130 kJ equivalent loss in flower production, but cautioned that not all costs of glucoside production could be determined. Cates (1975) showed that forms of *Asarum caudatum* resistant to slugs produced fewer seeds than palatable forms in the absence of slugs, suggesting a trade-off between resistance and seed production. However, this cost was more than compensated by the benefit of resistance when slugs were present. Windle and Franz (1979) grew two cultivars of barley (*Hordeum vulgare*) differentially resistant to greenbugs (*Schizaphis graminum*), in pure and mixed stands with and without aphids. The susceptible cultivar was a superior competitor in the absence of aphids but inferior when aphids were present. Castro et al. (1988) also found that in the absence of herbivores, the greenbug-susceptible genotype of barley produced almost twice as much aboveground biomass as the resistant genotype.

Costs that are merely associated with defense may actually reflect costs of a character that is correlated with, but not functionally related to, defense. Measurement of cost in inducible defense systems can provide more definitive results if induction is restricted to a single defense chemical. The difficulty with this approach is that it is never certain that an inducing agent increases only the level of defense. Attempts to measure costs in inducible systems have yielded mixed results. Smedegaard-Petersen and Stolen (1981) examined induction of barley resistance by compatible (nonvirulent) powdery mildew fungus and found that the fungus, while not producing any disease symptoms, significantly reduced grain yield. But Brown (1988), attempting a similar experiment in which tomatoes were injected with chitin (an inducer of proteinase inhibitors), did not find differences in seed production between treated and control plants.

Another potential approach to estimating defense cost is to employ specific chemical inhibitors or inducers of defense compounds. The advantage of this approach is that damage to the plant, if any, would be minimal. Growth or seed production of plants treated with specific inhibitors or inducers of compounds could be compared to controls—differences would ostensibly reflect the cost of defense production and deployment. Inhibitors and inducers have already been used to define the roles of certain compounds with putative defense roles. Aminooxyacetic acid was used to block synthesis of phenolics in an effort to assess the importance of phenolics in resistance of birch, *Betula pendula*, to insect feeding (Hartley 1988); phenolic production was lowered but feeding preference was unaffected. Similarly, Moesta and Grisebach (1982), using L-2-Aminooxy-3-phenylpropionic acid to inhibit glyceollin accumulation in soybean, *Gylcine max*, found that resistance to the fungal pathogen *Phytopthora megasperma* f. sp. *glycinea* was greatly reduced. Gossypol, an allelochemical of cotton, *Gossypium hirsutum*, is induced by volatile constituents from leaves infected with *Aspergillus flavus* (Zeringue 1987), and aspirin and related hydroxy-benzoic acids inhibit wound responses of tomato, *Lycopersicon esculentum* (Doherty et al. 1988). Most recently, Baldwin et al. (1990) employed auxin as an inhibitor of alkaloid induction in tobacco to estimate the cost of alkaloid defense. Plants were damaged with and without application of auxin to the damaged tissues and were compared to appropriate controls. Damaged, uninhibited plants produced 2.5 times more alkaloid and significantly less seed mass than inhibited or undamaged controls. The estimated cost of a 1% dw increase in alkaloid content was equivalent to a 5.4 g decline in seed production. Baldwin cautions that the induction of other factors in addition to alkaloids may account for some of this cost. If the targets of these inhibitors and inducers are restricted to defense (an important, if not readily satisfied condition), convenient, noninvasive techniques may be available to provide wound-free estimates of defense cost.

Perhaps the ultimate approach to measuring costs may soon be available. Genetic engineering techniques might be used to introduce gene modifiers whose only function would be to increase synthesis of a particular defense. The resulting costs could then be measured in a uniform genetic background.

While the notion that there must be a cost, direct or indirect, to any defense is attractive, studies of the magnitude of these costs are few. The simplest explanation for the paucity of data indicating a cost of defense is that the cost, while real, is slight and therefore undetectable or biologically irrelevant. Alternatively, the costs are large but detection is hampered by complex interactions between defense traits that are phenotypically or genetically correlated. While some biosynthetically related defense compounds are positively correlated with one another in populations (Berenbaum et al. 1986), others are negatively correlated. For example, tannin production is negatively correlated with cyanide production in *Lotus corniculatus* (Haskins and Gorz 1986) and leucocyanidin is negatively correlated with biosynthetically related dhurrin in *Sorghum bicolor* (Ross and Jones 1983). An increased cost associated with an increase in one allelochemical might be offset by a decreased cost associated with a decline in the other allelochemical. In these systems, then, an attempt to establish a cost associated with any one defense compound is likely to fail even though one may exist.

The environmental context in which cost is measured can also have an important and confounding influence. If the resources available for defense are abundant and those for other functions are limiting, defenses may accumulate as overflow metabolism with no real cost (e.g., Mihaliak and Lincoln 1985; Mole et al. 1988). If, however, resources for defense as well as other functions are limiting, defense could exact a large cost. Although cost of a defense is generally measured in the absence of herbivores (there is a presumable net benefit in the presence of herbivores), this cost may be grossly underestimated in correlated multiple defense systems when herbivores are present. For example, if two defense chemicals are negatively correlated with one another and each is effective against a different suite of herbivores, the cost of increasing chemical defense may include an additional indirect cost of reduced protection afforded by the other chemical.

While there do appear to be significant costs associated with at least some chemical defenses, much more information is required before the magnitude of costs in simple systems and in complex systems with multiple defenses can be adequately evaluated. To the extent that cost is a fundamental component of differential defense allocation, limited progress can be made in evaluating defense theory until costs are demonstrated and compared to costs of other plant activities.

Determinants of Within-Plant Defense Allocation

The optimum allocation of defense to parts within plants should reflect overlying patterns of attack probability and value to the plant. Attack probability is a function of the likelihood of discovery and feeding after the plant part is located. The value of the part to the plant is usually a function of the direct and indirect fitness costs associated with the loss of the part. Although there is great intuitive appeal to the notion that some plant parts are more likely to be attacked than others, the difficulties of defining and measuring probability of attack are sufficiently great (e.g., apparency) that this aspect is frequently omitted in interpretations of defense patterns. Even if probability of attack can be measured, only those herbivores currently able to utilize the plant will be measured and not the wide variety of herbivores that are excluded by existing defense allocations. Estimates of the fitness value of plant parts are easier to obtain but are commonly available only for leaves.

Probability of Attack

Ideally, to understand the existing pattern of defense allocation in a plant, probabilities of attack need to be determined in the absence of any defenses. Because the ideal is unattainable for most plants, an alternative approach can be used to infer probabilities of attack. Herbivores require fairly large quantities of water, nitrogen, and carbohydrates for growth and reproduction. Insofar as herbivore growth is limited by supplies of these materials, selection should favor consumption of those plant parts that provide ample supplies of these materials. A crude prediction of attack probability can then be obtained by estimating the nutritional value of plant parts, ignoring defenses.

Probabilities of attack inferred from nutritional profiles of plants alone are likely to be inaccurate and must be tempered by consideration of other factors. Tissue toughness, for example, reduces net nutritional value of plant parts but may be dictated more by requirements of structural strength than by defense. Competition among herbivores may lead to resource partitioning, in which some species are forced to specialize on less nutritionally rewarding plant parts.

Another important but poorly studied factor influencing probability of attack is apparency (Feeny 1976). Here, "apparency" is used in a temporal sense; the longer a plant part is available, the more likely an herbivore will find it. Thus, more apparent tissues will have higher probabilities of being attacked. In essence, apparency becomes an issue of optimal foraging by herbivores; when does the cost of searching exceed the benefit of finding and utilizing ephemeral tissues? If the cost/benefit is unfavorable, herbivores should not attempt to exploit ephemeral tissues.

Value of Plant Parts

Clearly all plant parts contribute to fitness in a plant, yet it can be argued that, gram for gram, the loss of a particular part will have a more negative effect on plant fitness than the loss of another part. Similarly, the loss of a particular organ can have a more negative effect on fitness if removed at one stage of development than at another (Mooney and Gulman 1982). The most often used method of determining the fitness value of a plant part is to remove the part and observe the effect of its absence on fitness (McKey 1979; Krishik and Denno 1983). While this approach has merit, no investigators have systematically removed equal amounts of each plant part in any species to assess the relative importance gram for gram of leaves, stems, roots, flowers, and fruits. In the absence of such information, generalizations might be based upon the ease of replacement of plant parts. In the simplest example, vegetative parts are more easily replaced than reproductive parts. Because plants are plastic and have indeterminate growth forms with many dormant meristems, lost vegetative parts generally can be replaced. However, for monocarpic species especially, reproduction is a one-time opportunity. A loss of any fruits in these species could represent an irreversible reduction in fitness. Thus, as plants switch from resource acquisition, an activity that plants spend most of their life cycle engaged in, to a short period of reproduction, the stakes and therefore the impacts of losses to herbivores increase.

Leaves

Raupp and Denno (1983) reviewed the literature on leaf age and suitability for herbivores and concluded that young leaves were nutritionally more suitable for herbivores but that the abundance of herbivores did not always parallel leaf age. Other factors, including chemical defense and predators, modify the actual distributions of herbivores (Strong et al. 1984). Hartnett and Bazzaz (1984) examined the distribution of aphids in relation to leaf age and leaf contribution to new leaf production. They observed that aphids concentrated on stems near young leaves whose potential contribution to new leaf production is greatest.

Fully expanded leaves operating at maximum photosynthetic rate are probably the most nutritionally rewarding for herbivores. Generally, maximum photosynthetic capacity is positively correlated with nitrogen content (Field and

Mooney 1986) and negatively correlated with low water potentials (Boyer 1982). Thus, leaves operating at maximum photosynthetic rate will be rich in nitrogen and water, resources that are important for herbivore growth. In a review of leaf value and defense relative to other leaves within a plant, Krischik and Denno (1983) concluded that not all leaves are equally valuable to a plant; terminal (younger) leaves are more valuable than basal (older) leaves and also tend to be better defended than basal leaves, although there were a number of exceptions.

While young leaves appear to be more valuable to herbivores and may be better defended, the value of leaf tissue generally relative to other tissues in the plant is difficult to ascertain. The primary function of leaves is carbon fixation, yet removal of leaf tissue constitutes a loss of nutrients as well as photosynthetic capacity. A sizable literature examines the effects of removal of leaves of differing age on fitness (see Krischik and Denno 1983) and negative effects have been documented even under natural field conditions (Marquis 1984). An alternative method of assessing leaf value that does not involve damage is leaf demography. By following leaf births and deaths, and knowing the age-specific photosynthetic rate, life tables can be constructed and used to estimate the contribution of each leaf to new leaf production. Studies employing this technique also indicate that young, fully expanded leaves are the most valuable in terms of productivity (Bazzaz 1984; Hartnett and Bazzaz 1984). Leaf demographic techniques, however, may overestimate leaf value if remaining leaves can compensate for lost tissue by enhanced photosynthesis. Numerous examples of compensatory growth involving foliage replacement can be cited (see McNaughton 1983a; Maschinski and Whitham 1989; but see also Belsky 1986).

In addition to changes in attack probability and value of leaves over their lifetimes, positional effects of leaves may also be important in determining value. Marquis (1988a) demonstrated in *Acer pennsylvanicum* that removal of 25% of the area from leaves subtending infructescences significantly reduced seed number in those infructescences, while removal of area from leaves near, but not subtending, the infructescence had no effect. The density of plants can also influence leaf value. At low density, individuals of *Abutilon theophrasti* with 75% of their leaf area removed and in the presence of undefoliated neighbors are not adversely affected. However, at high density a 75% reduction in leaf area cuts biomass and seed production by half (Lee and Bazzaz 1980).

Seeds and Fruits

For a different reason, seeds and fruits should also have high probabilities of attack. While these structures are not primarily active in resource acquisition, they are repositories for a rich supply of energy and nutrients intended for use by developing embryos. Moreover, the nutrients are in concentrated form, presumably to minimize seed size and weight. As such, these concentrated packets of complex carbohydrates and proteins should be nutritionally rewarding for herbivores that consume them. These structures are also extremely valuable to the plant, representing the closest approximation to fitness within a generation. A considerable effort in many cases is required to achieve pollination (i.e., nectar and pollen production, large showy petals), and for monocarpic species the consequences of fruit and seed losses are especially severe, because there is no future opportunity to compensate for such losses. One factor that lessens attack probability is the decline in water content as the fruit or seed ripens. Whether this drying process is attributable to dormancy or defense, the low water content of seeds is a formidable obstacle even to specialized frugivores (Baker and Loschiavo 1987). In general, however, fruits and seeds should have both high probabilities of attack and high fitness value.

Roots

The nutrient acquisition and storage functions of roots vary widely among species. In those species with tap roots, storage is clearly an important function and, in those species, roots should be good sources of nutrition for herbivores. There is some question, however, as to the availability of these resources owing to their location in a solid substrate. Clearly, foraging efficiency by many herbivores is greatly reduced by the energy-intensive effort associated with digging around in the soil. However, there remains an abundant supply of well-adapted nematodes, soil pathogens, and arthropods that flourish in this medium and pose a significant threat to plants. Roots, like leaves, can be readily replaced. Therefore, their defense needs based on value and attack probabilities are perhaps comparable or slightly less than those of leaves. Unfortunately, root chemistry is often neglected in the study of plant/animal interactions.

Stems

The primary function of stems is to support leaves and reproductive structures. With few exceptions (notably species in the Cactaceae), stems do not contribute greatly to gross photosynthesis in comparison to leaves. They are, however, a vital connection between the root and the leaves and without them, neither the shoot nor the root could survive. Moreover, a rather small loss of stem tissue to herbivores could cause the plant to topple, with a loss of all the leaves and reproductive structures above the breakpoint. A section of stem is then considerably more valuable than a comparable amount of leaves, roots, or even seeds and fruits. Consequently, the integrity of every centimeter of stem is critical, although breaks higher up the stem will have progressively less impact. While the value of stems is high, their attack probabilities may be low. The premium placed on high structural strength leads to tissue toughness that reduces the nutritional quality of stem material for chewing herbivores. In cases where stem material persists for very long periods (e.g., in trees) however, the attack probability may be high.

Predictions of Within-Plant Defense Allocation

We suggest that fruits, seeds, and to a lesser extent, flowers, are more heavily defended than vegetative plant parts. The basis for this prediction is that plant function consists primarily of two activities: resource acquisition and reproduction. Because resource acquisition is a means to effect reproduction and because reproductive organs generally constitute a small proportion of plant biomass and an attractive reward to herbivores, seeds and fruits should be defended most heavily. Although much attention has been paid to the importance of allelochemicals in seeds (E. A. Bell 1978; Janzen 1978a; Janzen et al. 1986), little is known about the defense of seeds relative to the other organs.

Predictions of defense allocation among vegetative parts are less easily made. Plant growth is most efficient when a balanced supply of resources is available (Bloom et al. 1985). To that extent, mineral nutrition (primarily a function of the roots) and carbon acquisition (a function of the shoots) are equally important functions carried out by equally valuable parts. The amount of resources allocated to mineral nutrition and carbon acquisition, however, varies widely among species and habitats and can lead to differences in the value of a unit of tissue (see later discussion on root/shoot ratios). Thus, consistent predictions of defense allocation between root and shoot cannot be made on the basis of value alone; generalities, if any exist, are more likely related to differences in probability of attack. While arguments can be made that roots are protected because they are surrounded by soil, we would reply that many soil organisms are well adapted to utilize roots, and that therefore no consistent predictions can be made for differential defense of roots and shoots. Stems, despite their high value to the plant, are probably poor fare for most herbivores, especially compared to nutrient-rich leaves and roots, because they consist predominantly of nutritionally deficient material (e.g., lignins and cellulose). Therefore, we would expect that stems generally require less chemical defense relative to leaves, roots, and reproductive structures.

A major impediment to testing such a general pattern is that the chemistry of individual plant parts is rarely examined. Janzen noted in 1979 that "herbivores do not eat Latin binomials"; they eat plant parts. Ten years hence, the amount of information on within-plant defense distribution is still minuscule within the vast literature on natural products. Although we do not profess to have exhaustively searched the literature, we have assembled a collection of studies in which at least two plant parts were analyzed (table 16.2). Chemical concentrations of fruits and seeds are in every case nearly the same or greater than those of leaves. In some cases the levels are severalfold higher (e.g., in *Lupinus* and *Pastinaca*). In cases where data are available for both leaves and stems, leaf concentrations are generally similar to stem concentrations with some higher and some lower. Similarly, the patterns for stems versus roots and leaves versus roots are ambiguous. Fruits and seeds, however, are almost al-

TABLE 16.2
Distribution of chemical constituents with potential or demonstrated allelochemical activity among plant parts

Species	Constituent	Units	Root	Stem	Leaves	Flowers	Seeds/fruits	Reference
Asclepias eriocarpa	cardenolides	mg/g	nd	nd	3.7	5.5	3.4	Isman et al. 1977
Asclepias eriosa	cardenolides	mg/g	nd	nd	3.0	0.9	2.4	Isman et al. 1977
Asclepias fascicularis	cardenolides	mg/g	nd	nd	.2	0.7	0.3	Isman et al. 1977
Asclepias vestita	cardenolides	mg/g	nd	nd	6.8	2.6	7.1	Isman et al. 1977
Erica australis	phenols	mg/g	1.7	0.8	1.3	2.5		Carballeira 1980
Asclepias eriocarpa	cardenolide	mg/g	2.5	4.5	3.7–4.5	2.6	5.0	Nelson et al. 1981
Pinus ponderosa	-pinene + subinene	a	24.8	21.5	34.0	nd	nd	Radwan et al. 1982
Cardamine cordifolia	glucosinolates	mg/g	1.11	.07	.32	nd	nd	Louda and Rodman 1983b
Petteria ramentacea	alkaloids	mg/g	nd	.16–1.2	.39	nd	1.8	Wink and Witte 1985b
Chichorium intybus	chicorin	% dw	0	nd	.02–.08	1.75[b]	nd	Rees and Harborne 1985
Chichorium intybus	8-deoxylactucin	% dw	.04–.29	nd	.04–.15	nd	nd	Rees and Harborne 1985
Lupinus polyphyllus	alkaloids	nmoles/kg	1	4–15	4–20	2–8[c]	100–150	Wink 1985
Cytisus scoparius	alkaloids	nmoles/kg	.1	8–20	1–5	nd	8	Wink 1985
Pastinaca sativa	furanocoumarins	mg/g	.01–.14	.35–1.43	.21–2.16	2.26–3.11	8.68	Berenbaum 1981b; Zangerl 1986; Nitao and Zangerl 1986
Calotropis procera	alkaloids	mg/g	nd	3.93	1.65	nd	3.26	Erdman 1983
Citrus sinensis	caffeine	μg/g	nd	nd	6.0	62.0	nd	Stewart 1985
Crotalaria spectabilis	alkaloid	% dw	nd	nd	0–1.52	nd	2.32–5.35	Johnson et al. 1985
Gossypium hirsutum	gossypol	%	nd	nd	nd	.44–1.28	1.2–3.6	Lukefahr and Houghtaling 1969
Solanum tuberosum	solanine	mg/g	.05–.38	.03–.06	.51–.62	1.58–3.54	nd	Lampitt et al. 1943
Brassica compestris	progoitrin	μg/g	.13	nd	nd	nd	2–20	Tookey et al. 1980
Heteromeles arbutifolia	phenolics	%[d]	.3	.3–1.3	2.8–5.4	nd	nd	Mooney and Chu 1974
Heteromeles arbutifolia	cyanogens	%[c]	nd	nd	7	nd	2	Dement and Mooney 1974
Heteromeles arbutifolia	tannins	%[c]	nd	nd	12	nd	24	Dement and Mooney 1974
Brassica napus var. napobrassica York cultivar	glucosinolate	μmoles/g	10.3	nd	7.8	nd	136.6	Jürges and Röbbelen 1990
Seestem cultivar	glucosinolate	μmoles/g	8.1	nd	3.5	nd	99.7	Jürges and Röbbelen 1980
Hyoscyamus niger	alkaloid	%	nd	.01–.025	.04–.08	.07–.10	.07–.10	Singh and Sharma 1977
Hyoscyamus muticus	alkaloid	%	nd	.4–.60	.5–1.4	.6–1.15	.9–1.3	Singh and Sharma 1977
Phalaris arundinacea	gramine	μg/g	nd	10	270–300	nd	nd	Coulman et al. 1977
Phalaris arundinacea	hordenine	μg/g	nd	380	830–870	nd	nd	Coulman et al. 1977

Note: nd = not determined.

[a]Electronic units/kg. [b]February sampling period. [c]Maximum values estimated from graphical data. [d]Petals. [e]Stamens, ovaries, and petals from graphical data.

ways greater or equal to stem or root concentrations, although few such comparisons are available.

Fine-Scale Patterns of Defense Allocation

We have thus far concentrated on defense allocation patterns among major plant parts. An infinite variety of optimal defense patterns might exist within any one of these structures. Thanks primarily to food chemists whose interest is in safeguarding food quality, copious information is available on the distribution of potential allelochemicals within plant parts. Rather than attempt to survey this extensive literature, we will use a few examples to illustrate a general pattern. Frequently, distributions of potentially defensive compounds within plant parts are highly nonrandom. The outer peel of an onion (Carmen Hybrid), for example, contains 500 times more quercetin than the inner rings, and contains all of the kaempferol (Bilyk et al. 1984). The alkaloid content of potato skins is anywhere from 5 to 50 times higher than that of the flesh (Sizer et al. 1980; Maga 1980) and usually higher than 20mg/100g, an amount considered safe for human consumption (Sizer et al. 1980). Similarly, purine alkaloid concentrations in tea seed coats, *Camellia sinensis,* are 20 times higher than in the cotyledons, and coffee seed coats (*Coffea arabica*) contain 70 times more purine alkaloid than the cotyledons (Suzuki and Waller 1987).

Similar examples of fine-scale defense allocation can also be found in wild species. Langenheim et al. (1978) investigated resin distribution within the tropical legume *Hymenaea*. They found that resin-bearing rings, pockets, or cavities are located at leaf margins, near the epidermis in young stems, near the cambium in older stems, and toward the outside of the fruit. Furanocoumarins are located only in the seed coat, not inside the seeds of wild parsnip (Berenbaum and Zangerl 1986).

All of these patterns appear to emphasize a line of first defense near the outside of the plant part. If one assumes that the internal tissues are the ones that are most valuable to the plant, a defense allocation that minimizes the chances of an herbivore even getting close to those tissues is advantageous. Thus, at this localized level, defense and value are not always spatially congruent, but defense may nonetheless be "optimized."

The Defense of Sex

Recent interest in patterns of male and female reproductive effort have led to studies of differential susceptibility and defense of male and female functions. In studies that have examined dioecious plants, males appear to be more heavily damaged than females. For example, Danell, Elmqvist, et al. (1985) found that bark-eating voles consistently remove a higher percentage of the bark of males than of females in *Salix myrsinifolia-phylicifolia,* and herbivores consistently removed a higher proportion of leaf area in male *Rubus chamaemorus* (Ågren 1987). However, no measures of chemical defenses were made in either of these studies. Elmqvist and Gardfjell (1988) showed that male and female plants of *Silene dioica* responded differently to manual defoliation; females experienced less mortality following defoliation and, being larger plants, they were better able to compensate for losses. In a preference test with the same species, a generalist snail, *Arianta arbustorum*, showed a clear preference for male foliage; but again, no measurements of defense investment were made. Whether or not this trend of higher male susceptibility reflects a general pattern of reduced defense allocation in males is yet to be determined.

Within monoecious plants, data on male and female susceptibility and defense are even more scarce. Wink (1985) measured alkaloid content of *Lupinus polyphyllus* pollen and carpels and found slightly higher concentrations in pollen (8 mmoles/kg for pollen versus 5 mmoles/kg for carpels), but seeds contained far more alkaloid (100–150 mmoles/kg). *Heliothis virescens*, a major insect pest on cotton (*Gossypium hirsutum*), utilizes cotton anthers as a major food source. Anther color variants, cream and yellow, differentially affect insect growth—yellow anthers suppress insect growth by 15% (Hanney et al. 1979). Hanney (1980) later found that yellow anthers contained more gossypol (1.09% versus 0.87% dw), an allelochemical known to interfere with growth of *H. virescens*.

The study of optimal sex defense should prove to be a very interesting and challenging endeavor. Approaches to studying defense of physiologically autonomous sexes in dioecious plants will likely differ both conceptually and mechanistically from approaches applied to sex defense in physiologically integrated monoecious plants. Moreover, the value of each sex at any particular stage will be more difficult to measure, at least in obligately outcrossing species. And of course there will be important variations in sex defense among species; for example, species that utilize pollen consumers for pollen dispersal may not defend their pollen at all.

The Role of Inducible Defenses in Optimal Defense Theory

Although the primary focus of this chapter has been spatial allocation of defenses, temporal allocation patterns can be equally important. The widespread distribution of inducible secondary compounds (Bailey and Mansfield 1982) indicates that facultative defense may be an integral part of most species' defense repertoire. Evidence is accumulating that induced defenses affect herbivores (e.g., Haukioja and Niemelä 1977; Tallamy 1985; Edwards et al. 1986; Baldwin 1988; Gibberd et al. 1988; Zangerl 1990), however, there is valid scepticism with regard to methodology and interpretation of many studies (Fowler and Lawton 1985; Haukioja and Neuvonen 1985). Induced defenses may either augment a minimum level of defense (the minimum can be called the "constitutive" level), or they may put into place a defense that was not previously present. Ostensibly, the primary benefit of inducible defenses is that they incur cost as a function of actual attack, not probability of attack. Thus, defenses and defense-associated costs are increased only when the benefits of defense can be realized. The drawback of an induced defense system is that there is a lag time between attack and deployment of the defense. In some chemical defense the lag is as little as four hours (Dixon et al. 1986) but in others, such as mechanical defenses (thorns), the lag can be several weeks (Young 1987). During this lag period, the plant remains at risk. Thus, while induced defenses vary as a function of actual attack, reliance upon this type of defense should vary depending upon the length of the lag period and probability of attack. Populations that have a high probability of attack and a long lag period for defense induction should not rely on induced defense, but rather on a more potent constitutive defense. Selection in populations with low probabilities of attack, however, should favor reliance on inducible defense, thereby avoiding costs of a defense that, for most of the time, would provide no benefit (fig. 16.1).

A Case Study in Optimal Defense Patterns: Furanocoumarins in Wild Parsnip

Wild parsnip (*Pastinaca sativa*) is well suited for study of optimal-defense patterns. It is a facultative biennial species that produces a group of biosynthetically related compounds known as furanocoumarins (fig. 16.2). These compounds are toxic or deterrent to a variety of organisms (Murray et al. 1982) and are distributed throughout the plant (Berenbaum 1981b; and see table 16.2). Significant amounts of genetic variation have been found in natural populations for all five of these compounds (Berenbaum et al. 1986; Zangerl et al. 1989) and negative genetic correlations between some of the compounds and flower

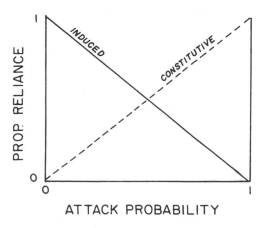

Fig. 16.1 Hypothetical relationship between type of defense and probability of attack.

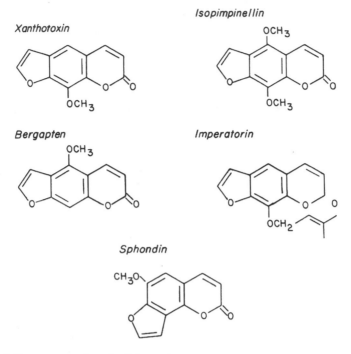

Fig. 16.2 Furanocoumarins of wild parsnip.

number have been documented (Berenbaum et al. 1986). Negative phenotypic correlations between furanocoumarin concentration and seedling growth rate have also been demonstrated (table 16.1). Although there appear to be costs to furanocoumarin defense, there is also evidence of benefits in terms of resistance to both specialists and generalists. Angelicin, an angular furanocoumarin of wild parsnip, reduces fecundity of *Papilio polyxenes,* black swallowtail, a specialist on parsnip leaves and flowering parts (Berenbaum 1981b). Resistance of parsnip to another specialist, *Depressaria pastinacella,* parsnip webworm, is associated with bergapten (Berenbaum et al. 1986), a furanocoumarin that interferes with food utilization (Berenbaum et al. 1989). *Trichoplusia ni,* the cabbage looper, is a generalist species that occasionally feeds on wild parsnip and is negatively affected by furanocoumarins both in the plant and in artificial diets (Zangerl 1990). A summary of parsnip defense attributes is presented in table 16.3.

By closely examining several aspects of furanocoumarin allocation in wild parsnip, we can provide a number of useful insights into defense optimization strategies. Some of the insights will have general applicability and others will reflect peculiarities of wild parsnip.

TABLE 16.3
Defense characteristics of wild parsnip

Attribute	Reference
Costs of furanocoumarin defense	
Negative genetic correlations between bergapten, xanthotoxin, isopimpinellin, sphondin, and flower number	Berenbaum et al. 1986
Negative phenotypic correlation between total furanocoumarin content and growth	see table 16.1
Genetic variation in furanocoumarin chemistry	
All furanocoumarins have significant heritabilities	Berenbaum et al. 1986
Genetic variation for inducibility of certain furanocoumarins	Zangerl et al. 1989 Zangerl and Berenbaum 1990
Furanocoumarin induction	
Flowers/fruits not induced by damage	Nitao 1988
Leaf furanocoumarins induced by damage	Zangerl 1990
Furanocoumarin localization	
In oil tubes	Ladygina et al. 1970
Modulators of furanocoumarin efficacy	
Myristicin, and cooccurring synergist	Berenbaum and Neal 1987
Solar UV: enhances toxicity to insects	Berenbaum 1978
Within-plant distribution of furanocoumarins	
Fruits > flowers > leaves, stems > roots	see table 16.2
Resource limitations on furanocoumarin production	
Both light and nutrients limit production	Zangerl and Berenbaum 1986
Compensatory reproduction	
Able to redirect resources from damaged inflorescences to new inflorescences	Hendrix 1979
Effects of furanocoumarins on insects that utilize wild parsnip	
Depressaria pastinacella: reduced growth and consumption associated with bergapten in artificial diet	Berenbaum et al. 1989
Papilio polyxenes: reduced growth and fecundity on leaves augmented with angelicin	Berenbaum and Feeny 1981
Trichoplusia ni: reduced growth, and consumption, and feeding efficiency on artificial diet containing xanthotoxin	Zangerl 1990

Parsnip Flower and Fruit Defense

With furanocoumarin concentrations 800 times higher than in roots and four times higher than in leaves (table 16.2), fruits are by far the most heavily defended parts of the wild parsnip with respect to this class of chemicals. This pattern is consistent with earlier predictions; however this overall pattern masks a subtle interplay between defense and compensatory reproduction. Parsnip flowers and fruits are frequently attacked by parsnip webworm. In some cases, the primary umbel, which is the first to flower, is completely destroyed by this insect (Berenbaum et al. 1986). Hendrix (1979) discovered that plants could compensate for losses in the primary umbel by producing more inflorescences that flower later (primarily tertiary and higher-order umbels). In the end, seed number is conserved but seed size is slightly reduced.

The ability to compensate for reproductive losses seemed, at first, at odds with the observation that furanocoumarin concentrations steadily increase as an umbel progresses from buds to ripe fruit (fig. 16.3). Nitao and Zangerl (1986) hypothesized that the ability to compensate for lost reproductive parts depends upon the timing of defloration, and therefore defense should vary with development stage. Initially, investment in an umbel is small; buds weigh only 0.57 mg, but the ripe fruit weighs 4.45 mg (fig. 16.3). The loss of an umbel in bud thus represents a minor loss of resources compared to the loss of the same umbel with well-developed fruits. If compensatory reproduction involves reallocation of unused resources from damaged umbels to production of more umbels flowering later, the ability to compensate should decline as development advances.

This hypothesis was tested by manually removing all the floral units in the primary umbel at different stages of development for both large and small plants. The ability to compensate for lost reproduction in the primary umbel was slightly stage-dependent in large plants and wholly stage-dependent in small plants (fig. 16.3). Thus, compensatory reproduction is not sufficient to recoup losses under all circumstances, and a stronger defense is required for more developed parts (e.g., fruits) whose value to the plant is greater than that of less developed parts (e.g., buds).

Parsnip Leaf Defense

Wild parsnip produces compound leaves that for most of the life cycle are arranged in a rosette. The rosette arrangement allows an investigator to determine

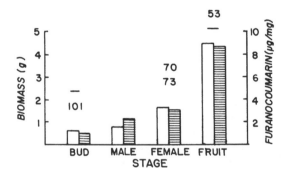

Fig. 16.3 Increases in biomass (open bars) and furanocoumarin concentration (hatched bars) with floral and fruit development in wild parsnip. The top number above each pair of bars represents the percentage of control seed production by small plants whose primary umbel was deflorated at that particular stage of development. The bottom number represents the percentage of control seed production by large plants whose primary umbel was deflorated at the stage of development. Dashes indicate absence of data. Significant treatment effects were found for stage-dependent defloration in small plants but not in large plants (From Nitao and Zangerl 1986).

precisely the relative age of leaves and effect of age on photosynthetic rate and defense. Young leaves, not fully expanded, are frequently net carbon sinks (Zangerl 1986). Shortly after full expansion, leaves operate at maximum photosynthetic rate but leaves age, and carbon gain declines and eventually ceases (Zangerl 1986). Clearly, the value of a leaf is proportional to its net carbon gain. Yet, despite changes in net photosynthetic rate with age, furanocoumarin defense is inconsistently correlated with photosynthetic rate (table 16.4). The only generality applicable to all plants is that dead leaves are not defended.

This seemingly suboptimal defense allocation among leaves does not extend to allocations within leaflets. Oil tubes, which appear to be the primary repository of furanocoumarins throughout the plant, are exclusively associated with veins in leaves (Warning 1934). Furanocoumarin allocation to veins within leaflets follows a distinct pattern common to all individuals. Roughly half of all the furanocoumarin in a leaflet is located in the primary vein and one third is in the secondary veins, despite the fact that the primary and secondary veins make up only 8% and 11% respectively of total leaflet biomass (table 16.5). Breaks in veins caused by an herbivore will affect both leaflet support and transport of materials; however it is the location of the break that will largely determine its impact. Likely impacts from variation in break locations can be predicted. The portion of a parsnip leaflet isolated by specific breakpoints is presented in figures 16.4A and B. While there is some secondary movement of sap around breaks via the network of fine veins, the efficiency of these secondary routes is presumably lower than in the intact system. The furanocoumarin content of veins at the same breakpoints is presented in figure 16.4C. If defense allocation

TABLE 16.4
Correlations between photosynthetic rate and furanocoumarin concentration of leaves within five wild parsnip plants

	Plant				
	1	2	3	4	5
Correlation	−.028	.846*	.523	.385	−.642*

Source: Zangerl 1986.
*$p < .05$.

TABLE 16.5
Distribution of furanocoumarin and biomass within a wild parsnip leaflet

Leaflet part	Furanocoumarin (ug/mg)	Leaflet furanocoumarin (%)	Biomass (mg)	Leaflet biomass (%)
Primary vein	8.1	43	2.7	8
Secondary veins	5.6	37	3.4	11
Remainder	0.4	20	26.2	81

382 Arthur R. Zangerl and Fahkri A. Bazzaz

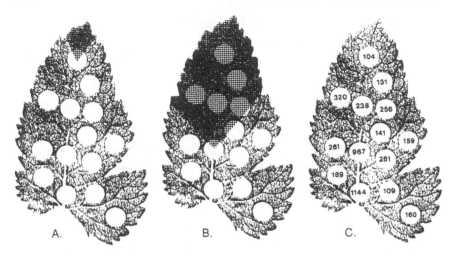

Fig. 16.4 (A) Impact area estimation for a midvein break in terminal section of midvein and (B) for first third of midvein. Impact areas were estimated in this way for all the locations where leaf disks were removed. (C) Furanocoumarin content (ng) of leaf disks at four locations along the midvein, at secondary veins, and at higher-order veins.

within the leaflet is optimal, defense investments should be correlated with potential impact and is in fact highly correlated in this case (fig. 16.5). Thus, defense within leaflets is closely related to the value of tissues.

Trichoplusia ni, by virtue of its behavioral flexibility, is able to circumvent this leaflet defense for part of its larval growth cycle. First-instar larvae avoid even the smallest veins of wild parsnip. The benefit of the defense to the plant is that first-instar damage is limited to the area of tissue consumed. Later instars, however, are unable to avoid the smallest veins but continue to avoid large veins. Thus, the effect of herbivory is continually minimized by confining damage to those parts of the leaf which have the least impact on leaf function. Although the defense strategy does not prevent damage by *T. ni,* it does minimize the impact of damage. An alternate hypothesis is that veins are avoided because they are tougher than the rest of the leaf tissues. *Trichoplusia ni* is capable of penetrating vein tissue and facultatively trenches across leaves (including veins) of other species as a means of disabling pressurized defense systems (Doussard, 1990, personal communication). *Trichoplusia ni* occasionally trenches across parsnip leaflets, but on two occasions when two penultimate-instar larvae caged on leaflets did cut across parsnip midveins, both larvae died within 24 hours (Zangerl, 1989, personal observation).

Within-leaflet and between-leaf patterns of defense allocation in parsnip illustrate the importance of considering scale in interpretation of defense allocations. Gross-scale differences in the distribution of defense compounds almost certainly affect large herbivores, simply because large herbivores sample

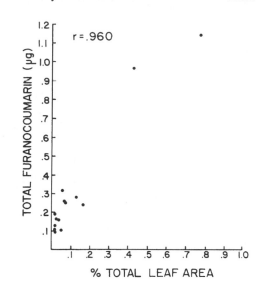

Fig. 16.5 Relationship between impact area associated with break positions in leaflet vascular distribution system and defense allocated to those positions.

larger amounts of tissue. However, the fine-scale distribution of furanocoumarins within leaflets is irrelevant to a large grazer because it does not sample at a fine scale. Conversely, large-scale distributions of defense (e.g., amounts per leaf) are irrelevant to small herbivores if there is defense localization. Such within-leaflet patterns of defense have relevance to insect feeding patterns but no relevance to feeding by an ungulate, for example. To our knowledge, no large animals feed on parsnip leaves, but there are several foliage-feeding insects on wild parsnip. In a very real sense, optimal defense patterns must take into account herbivore bite size. The impact of a single insect bite of a parsnip leaflet can be as small as the amount of photosynthetic tissue removed by the bite itself or as large as the total photosynthetic value of the leaflet if the midvein is severed at the base.

Another tool in the parsnip leaf defense arsenal is the ability to increase production of furanocoumarins in response to damage. Both manual and insect damage can induce furanocoumarin production, and for the first 48 hours following damage, induction appears to be restricted to the damaged leaflet (table 16.6). Growth of *T. ni* larvae on damaged leaflets with induced furanocoumarins was 70% less than growth on adjacent intact leaflets (comparable induced and uninduced levels of xanthotoxin in artificial diet have similar impacts on larval growth) (table 16.7). If defense induction is primarily a cost-saving strategy, its occurrence should vary with overall probability of attack as predicted earlier (fig. 16.1). Populations that routinely experience low probabilities of attack should rely more on inducible defense, while populations with higher probability of attack should rely more upon constitutive defense. Two central Illinois populations differ significantly in their frequencies of attack by leaf- and

TABLE 16.6
Increase in furanocoumarin concentration 48 hours after manual damage and feeding by
Trichoplusia ni

	Factor increase	
Furanocoumarin	Insect-damaged[a]	Manually damaged[b]
Imperatorin	2.38	2.58
Bergapten	3.20	2.76
Isopimpinellin	3.67	2.10
Xanthotoxin	2.81	2.50
Sphondin	3.68	3.18

Source: Zangerl 1990.

Note: All increases were significant, $p < .05$.

[a]10 First-instar larvae were permitted to feed for 48 hours on one leaflet. At the end of the feeding period the damaged leaflet and opposing intact leaflet were compared for furanocoumarin content. The number of leaflet pairs compared was 10.

[b]Fifty pin pricks were made in each of nine leaflets. After 48 hours the damaged leaflets were compared to intact opposite leaflets.

TABLE 16.7
**Growth of *Trichoplusia ni* on damaged and intact leaflets and on artificial diets containing
constitutive or induced levels of xanthotoxin**

	n	Growth (RGR mg/mg/d)
Intact leaflets	14	0.40
Manually damaged leaflets[a]	14	0.12
0.06% fw xanthotoxin diet	10	0.54
0.24% fw xanthotoxin diet	10	0[b]

Source: Zangerl 1990.

[a]Furanocoumarin content of damaged leaflets was 1.74 times higher than that of intact leaflets.

[b]Larval consumption rates were lower but not significantly different from the 0.06% xanthotoxin treatment.

fruit-feeding herbivores (table 16.8). The habitat associated with high levels of attack is heavily shaded and is therefore marginally suitable for the shade-intolerant parsnip. The other habitat is virtually open with no overstory. Individuals from both populations were grown from seed in the greenhouse for constitutive and damage-induced levels of furanocoumarin. Although tested in a common environment, individuals from the population with high probability of attack had significantly greater constitutive amounts of two furanocoumarins while individuals from the population with low probability of attack tended to have greater inducibility (table 16.8). Both constitutive and induced levels of furanocoumarin exhibit significant levels of genetic variation in both of these populations (Zangerl and Berenbaum 1990) and therefore differences between the populations in inducibility likely reflect evolved responses to differences in probability of attack.

TABLE 16.8

Constitutive and inducible furanocoumarin defense in two populations of wild parsnip with different probabilities of attack

	n	Race Street	Phillips Tract	P
Frequency of damaged plants in the field				
Plants with parsnip webworms[a]	50	100%	52%	<.001
Plants with leaf damage[b]	440	41%	5%	<.001
Furanocoumarin allocation in greenhouse plants				
Constitutive levels				
Xanthotoxin	287	1.95 mg/g	1.67 mg/g	<.007
Bergapten	287	0.25 mg/g	0.18 mg/g	<.001
Damage-induced increase				
Xanthotoxin	287	32.6%	50.0%	.079
Bergapten	287	58.9%	81.9%	.083

Source: Zangerl and Berenbaum 1990.

[a]First 50 plants in a two-meter wide transect were collected and examined for parsnip webworm damage in summer 1985. Significance level from g-test of independence.

[b]Presence or absence of leaf damage scored for 440 plants in plots at each locality in spring 1988. Significance level from g-test.

The Resource Availability/Defense Hypothesis

In recent years, a body of theory has developed that relates defense allocation to resource availability and the indirect cost of defense (Bryant, Chapin et al. 1983a; Coley 1983, 1988; Coley at al. 1985; Chapin et al. 1986; Gulmon and Mooney 1986; Bazzaz et al. 1987). We are calling these ideas collectively the "resource availability/defense hypothesis." The hypothesis reflects a different approach to interpretation of defense patterns and bears discussion. Certain refinements to the hypothesis involving effects of variation in the relative availabilities of carbon and nutrients (Bryant, Chapin et al. 1983a) and nutritional suitability for herbivores (Bryant et al. 1987) do not change the central premise and will not be discussed. The essence of the hypothesis is that selection in resource-rich habitats favors plants with high growth rates. High growth rates are achieved by producing inexpensive leaves that can be quickly and economically replaced as the canopy moves higher. In contrast, plants in resource-poor habitats are characterized by slow growth and long-lived leaves. Leaf replacement is much more costly in these habitats and therefore defense investments must be higher to avoid leaf losses. Moreover, because selection in resource-rich habitats favors plants with high growth rates, the indirect cost of defense would place defended plants at a competitive disadvantage compared to undefended plants. A summary of these ideas is presented in table 16.9. While this hypothesis is based on comparisons between species, some insights can be gained by examining these ideas in a within-plant optimal-defense context.

Although not explicitly stated, most of these associations can be interpreted in traditional optimal-defense terms. For example, plants with expensive

TABLE 16.9
Correlates of defense investment

Low defense investment	High defense investment
Fast growth	Slow growth
High regrowth potential	Low regrowth potential
Cheap, short-lived leaves	Expensive, long-lived leaves
Nutrient-rich habitats	Nutrient-poor habitats
Selection for competition	Selection for stress tolerance
High indirect cost of defense	Low indirect cost of defense

leaves will suffer more from a loss of leaf area than will plants losing an equivalent area of easily replaced leaves; expensive leaves should then be better defended. With three exceptions, all of the correlates of defense in table 16.9 can be recast in terms of value and probability of attack (table 16.10).

The remaining correlates of defense are not well understood in optimal-defense terms. These are the indirect costs of defense, nutrient availability in the habitat, and selection for competition versus stress tolerance. Proponents of these relationships argue that in nutrient-rich habitats, plants that invest heavily in mechanisms that maximize growth will outcompete plants that divert resources to defense (Bryant, Chapin et al. 1983a; Gulmon and Mooney 1986). Thus, the benefit of defense is more than offset by the indirect cost of reduced growth and competition. Gulmon and Mooney (1986) modeled the indirect cost of defense as a function of potential growth rate and concluded that the indirect cost of defense is greater for fast-growing plants. They did not however, examine the cost of defense in the context of its benefit. Coley et al. (1985) constructed a similar model including effects of herbivory but assumed a specific nonlinear relationship between defense investment and the cost and benefit of defense, which if assumed otherwise could easily change the outcome of the model.

We propose an alternative model based upon leaf demographic techniques (Bazzaz and Harper 1977) that incorporates an indirect cost for both defense and herbivory. Leaf production is calculated as follows:

TABLE 16.10
Correlates of defense explained in opitmal-defense terms

Correlate	Optimal-defense terms	Defense investment
Expensive leaves	Highly valuable	High
Cheap leaves	Less valuable	Low
Long-lived leaves	Higher probability of attack	High
Short-lived leaves	Lower probability of attack	Low
High regrowth potential	Less valuable	Low
Low regrowth potential	More valuable	High

(1) $$dL/dt = L \times H \times (B - C) + L \times H,$$

where L is the number of leaves on the plant, B is the leaf-specific birth rate (leaves/leaf/week), C is the number of leaf equivalents allocated to defense (leaves/leaf/week) and H is the proportion of leaf equivalents surviving herbivory owing to defense. The first term in the equation estimates the indirect costs of herbivory and defense in terms of reduced production of new leaves. The second term estimates an additional direct cost of herbivory in terms of leaf area immediately lost. An assumption in this model is that there is no leaf turnover, a valid assumption for short time periods or for species with long-lived leaves.

This model was used to examine the relative costs and benefits of defense for undefended and defended plants using data from Coley (1986) for *Cecropia peltata*, a neotropical tree (table 16.11). Figure 16.6 shows the cost of defense, the costs of herbivory, and the net cost or benefit of defense in the defended and undefended plants. The cost of allocating 6% of leaf biomass to defense is a 33% reduction in growth after 18 months (fig. 16.6A). The cost associated with a modest herbivore damage rate (0.49% of leaf area each week), assuming no benefit from defense investment, is a 31% reduction in both defended and un-

TABLE 16.11
Estimates of leaf birth rate, defense cost, and benefit for *Cecropia peltata*

Cost of tannin defense

Defense investment (mg/g tannin)	Leaf production in 18 months	Birth rate (B) (leaves/leaf/week)[a]
0	55[b]	0.051
60	35	0.057

Proportion of leaf equivalents attributable to cost of defense, $C = 0.057 - 0.051 = 0.006$

Benefit of tannin defense

Defense investment (mg/g tannin)	Leaf area removed in 10 days[c] (%)	Leaf area surviving (H) per week[d]
0	0.70	0.9951
60	0.40	0.9972

Source: Coley 1986

Note: Coley measured tannin content and leaf production after an eighteen-month period of growth. Tannin content ranged from approximately 10 mg/g to 60 mg/g and leaf production ranged from 30 to 58 leaves. These data were used to estimate the cost of defense in a plant with no tannin and in one with the maximum observed tannin concentration of 60 mg/g. Because growth is an iterative process, maximum birth rate (B, leaves/leaf week) was calculated by setting H to unity in equation (1) and varying B until the final observed leaf production after 72 weeks for an undefended plant was obtained. The cost of defense (C in leaves/leaf/week) was determined by subtracting B for a defended plant from B for the undefended plant. Both measures of B were calculated for a plant beginning with one leaf.

[a]Estimated by setting H to unity substituting B values until final leaf production rate was attained after 18 months.

[b]Extrapolated from regression in Coley.

[c]Obtained from linear extrapolation of high- and low-tannin groups, mean 44 and 21 mg/g, respectively. Plants were placed in field to expose them to natural infestation.

[d]Leaf loss rates were calculated on a per week basis and subtracted from unity.

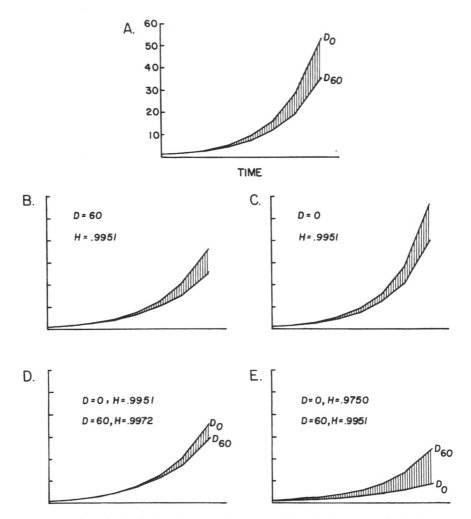

Fig. 16.6 Growth simulations incorporating cost and benefit of defense. (A) Direct and indirect cost of defense is indicated by shaded difference in growth rate between defended plant (D_{60}) and undefended plant (D_0). (B) Direct and indirect cost of herbivory for a defended (slower-growing plant) assuming no benefits of defense versus (C) the same costs of herbivory in an undefended plant. (D) The difference in growth between a defended plant and an undefended plant with both cost of herbivory and benefit of defense included; in this example a moderate differential in defense benefit is assumed based upon field observations by Coley (1986). (E) Difference in growth between defended and undefended plants as in D but assuming a greater differential in defense benefit consistent with feeding damage observed by Coley in experiments with a generalist feeder.

defended plants but a greater absolute reduction of 16 versus 11 leaf equivalents in the undefended plant (figs. 16.6B and C). Thus, the indirect cost of herbivory in absolute terms is greater for the undefended, faster-growing plant. In figure 16.6D, the natural field herbivory rates observed by Coley (1986) for high- and low-tannin groups were used to estimate the benefit of defense. At this rate of herbivory the cost of defense allocation is greater than the benefit of defense and the undefended plant's net growth is higher than that of the defended plant. These rates of herbivory are probably low, however, because Coley measured natural herbivory by moving potted plants to the field for a 10-day observation period. No doubt, herbivores required some time to colonize the plants. In a separate laboratory experiment *Spodoptera latifascia* were placed on both high-tannin (mean 44 mg/g) and low-tannin (mean 21 mg/g) plants (Coley 1986). The rate of leaf loss to herbivory in the low-tannin group was four times higher than the loss in the high-tannin group. Incorporating this fourfold difference in benefit into the model (even though the difference in benefit should be greater because we are comparing undefended with heavily defended plants), the benefit of defense far outweighs its costs (fig. 16.6E).

These simulations demonstrate that the net cost or benefit of a defense is dependent upon the degree of herbivore damage. They also demonstrate that the indirect cost of herbivory must be considered as well as the indirect cost of defense. Within a habitat the same factors that influence relative success of individuals within a population should also influence relative species success. Thus, the argument that selection for growth in resource-rich habitats tends to limit defense investment needs further analysis. At issue is the implicit assumption in resource-rich habitats that a plant's competitors are sustantially free from herbivory despite their having little or no defense and that the plant cannot afford to invest in defense lest it be outcompeted. If carbon acquisition is the limiting factor in resource-rich habitats, then losses to herbivores of the carbon acquisition tissues are bound to have profound effects upon growth and competition. If, indeed, there are differences in defense investment associated with different habitats, then more likely there are differences in probability of attack or in unit value of tissues. The apparency hypothesis (Feeny 1976) is one explanation for higher probability of attack in long-lived plants or in plants with long-lived parts. If plants or parts of plants are more likely to be found by herbivores in low-nutrient habitats, plants will have to maintain high constitutive defense levels. If in high-nutrient habitats, plants and plant parts are short-lived, then plants may avoid significant damage with inducible defenses.

An argument can also be made that plant parts vary in their value between habitats. Bloom et al. (1985) argue that plants equalize the marginal product/unit cost of operations; that is to say, plants will allocate more resources to leaf production if light is limiting or to roots if nutrients are limiting so that an optimal balance of resources is available for maximum growth. If light or carbon is the limiting factor, plants should increase the amount of aboveground

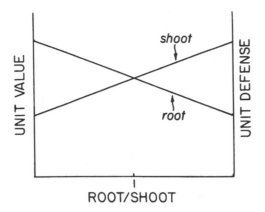

Fig. 16.7 Hypothetical relationship between root and shoot allocation and unit defense and value of roots and shoots.

tissue. Thus while roots in toto are equally valuable to shoots in toto, the total quantity of shoot material is greater and the unit value is less. This reasoning can be applied to explain apparently lower defense investment in high-nutrient environments, but only for shoots; the unit value and therefore defense of roots in these environments should be greater. At the level of whole plants, there is no basis for arguing that some plants are more valuable than others. Because plants alter root/shoot ratios to balance resource acquisition, the unit values of roots and shoots vary accordingly. Thus, we propose an extension of optimal-defense theory to explain defense allocation in terms of unit value of resource acquisition investment (fig. 16.7). Shoots in plants with high root/shoot values have a higher unit value and should have more defense per unit weight if all else is equal; unit shoot value and defense will decline however as root/shoot ratio decreases. The empirical basis for the resource availability/defense hypothesis seems to rest on patterns of leaf defense rather than whole-plant defense allocation patterns. To evaluate the hypothesis properly, studies are needed that measure defense in both shoots and roots as a function of resource availability in contrasting habitats.

Epilogue

While some progress has been made toward evaluation of defense theory, the field is clearly very much in its infancy, as evidenced by the rather small number of hypotheses generated. To facilitate the resolution of some apparent exceptions to the optimal-defense theory we propose the following hypotheses:

1. Spatial congruity between tissue value and defense investment need not occur to achieve optimal defense allocation (e.g., first lines of defense).

2. Inducible defenses are a cost-saving strategy that may be employed by organisms in all habitats; however, populations with high probabilities of attack should rely less upon induction for protection.
3. The value of roots and shoots to a plant is the same; however the level of defense allocated to a unit quantity of root is inversely proportional to the biomass ratio of roots to shoots, assuming equal probabilities of attack.
4. The scale of defense distribution must be interpreted in the context of the herbivore. For example, fine-scale distribution of defenses is probably relevant only for small herbivores.

At the same time that these and other hypotheses are pursued, continued studies of the cost of defense are very much needed. To the extent that any cost is associated with defense, plants should optimize defense allocation among plant parts. However, the relative magnitudes of cost and benefit of defense become far more important in evaluating the validity of the habitat quality/ defense investment hypothesis. More information is also needed to evaluate cost/benefit dynamics in correlated multiple defense systems. Needless to say, there is a great deal to learn about defense optimization, but in optimistic terms, there is a great deal of interesting research awaiting those willing to undertake it.

Acknowledgments

For comments on the manuscript, we thank M. Berenbaum, E. Fajer, K. Hoy-Burgess, M. Cohen, L. Wolfe, R. Maurasio, and an anonymous reviewer. The research on wild parsnip was supported by National Science Foundation grants BSR 86-06015 and BSR 88-18205 to M. Berenbaum and A. Zangerl. Thanks also to Robert Fritz and Ellen Simms for their patience and encouragement.

17

Costs of Plant Resistance to Herbivory

Ellen L. Simms

Because herbivory can greatly reduce plant growth and reproduction (see Marquis, this volume), it may significantly influence the distribution and abundance of plants in communities (Harper 1977; Hay 1981, 1984; Gaines and Lubchenco 1982; Crawley 1983, 1989; Power and Matthews 1983; Hockey and Branch 1984; Hay and Fenical 1988). Moreover, variation among plants in susceptibility to herbivory may influence the composition and organization of consumer communities as well (Krischik and Denno 1983; Price 1983a; Strong et al. 1984; Fritz, this volume; Karban, this volume). Understanding how plants evolve resistance to herbivory is therefore a requisite for understanding many aspects of both plant and animal ecology.

Both optimality and coevolutionary theories have been developed to explain the evolution of plant resistance. In models based on each type of theory, costs of resistance are important determinants of equilibrium levels of resistance to herbivory. Therefore, my objectives in this chapter are threefold. First, I will examine how costs of resistance influence theoretical predictions of the evolution of plant resistance to herbivores. Because mathematical treatment of the coevolution of plants and herbivores is scanty, this section will also include a brief review of models of plant-pathogen and host-parasite coevolution and an assessment of their applicability to plant-herbivore coevolution. I will then summarize the biological mechanisms by which costs of resistance to herbivores might be incurred and review the empirical evidence available regarding the frequency and magnitude of costs of resistance. Finally, I will examine some hypotheses that suggest why resistance might not be costly in some natural populations.

Theoretical Foundations

Various authors have argued that the degree of resistance to herbivory characteristic of a particular plant population reflects a compromise between the benefits of reduced herbivory and the costs of resistance (Janzen 1973a, 1973b; Feeny 1970, 1975, 1976; Whittaker and Feeny 1971; Jones 1972; Rehr, Bell et al. 1973; Rehr, Janzen et al. 1973; Rehr, Feeny et al. 1973; McKey 1974a; Cates 1975; Rhoades 1979, 1983, 1985; Lubchenco and Gaines 1981; Mooney and Gulmon 1982; Futuyma 1983a; Krischik and Denno 1983; Coley et al. 1985; Gulmon and Mooney 1986; Rosenthal 1986; Kakes 1987; Simms and Rausher 1987; Loehle 1988; Pimentel 1988; Fagerström 1989; Zangerl and Bazzaz, this volume). Because most optimality arguments are rooted in plant physiological ecology, costs and benefits are often envisioned in terms of resources diverted from other fitness-enhancing functions within a plant, such as photosynthesis, growth, and reproduction.

Most authors postulate that variation among species and populations in degree of resistance reflects underlying variation in the costs and benefits of resistance. Such variation may be produced by any of several mechanisms. For example, the severity of herbivore pressure in an environment will determine to a large extent the benefits associated with resistance. Increased herbivore pressure presumably increases the benefit of defenses and thus favors increased allocation of resources to resistance (Rhoades 1983). In contrast, the cost of a particular defense may be determined in part by the kinds and amounts of resources that must be diverted to it (Janzen 1974; McKey 1979; Dirzo 1984). Consequently, defenses that require limiting resources should be more costly than those that utilize common resources. For example, it has been argued that carbon-based chemical defenses are less costly under nitrogen-poor conditions but more costly when light availability limits carbon fixation (McKey 1979; Bryant, Chapin et al. 1983a; Bazzaz et al. 1987).

Developing in parallel with the optimality theory described above is a related body of theory devoted to understanding the coevolution of plants and their herbivores (Ehrlich and Raven 1964). Whereas optimality theory treats herbivores as qualitatively equivalent to abiotic environmental factors, coevolutionary theory considers the evolution of plant defenses in the context of coevolving herbivore populations.

Mathematical models can be used to examine the importance of costs of resistance in some types of coevolutionary interactions. Although few mathematical models describe plant-herbivore coevolution, several models have been developed to examine particular components of plant-herbivore coevolution. For example, Łomnicki (1974) modeled the interaction between insect population dynamics and the evolution of host-plant resistance to determine whether plant resistance alone could regulate insect populations. He found that plant resistance would evolve to an intermediate equilibrium and could regulate herbivorous insect populations, but only if resistance was costly in terms of plant fitness.

Others have used graphical or numerical cost-benefit models to predict the evolution of continuously varying polygenic resistance traits (Fagerström et al.

1987; Simms and Rausher 1987; Fagerström 1989). These models show that under certain conditions (when cost is a linear or concave-upward function of allocation to resistance and benefit approaches an asymptote), stabilizing selection can maintain the population at an intermediate optimal level of resistance.

Except for the models described above, which do not treat coevolution in the strict sense (that is, reciprocal evolutionary responses of plants and herbivores), the coevolution of plants and herbivores has not been modeled mathematically. In contrast, there is a rich literature of mathematical models of two related types of antagonistic interactions, those between plants and pathogens and between hosts and parasites. These models may provide insight into the role of costs in the process of coevolution between plants and herbivores.

Flor's (1956, 1971) elegant explanation of the relatively simple complementary gene-for-gene mode of genetic control of plant resistance and pathogen virulence in the flax-rust interaction inspired early mathematical modeling of plant-pathogen coevolution. Mode (1958) used a theoretical framework established by Wright (1955) to model coevolution between diploid, randomly mating plants and pathogens. By assuming that host fitness varies inversely with pathogen fitness (making the average fitness of each host genotype dependent upon the frequency of pathogen genotypes and the average fitness of each pathogen genotype dependent upon the frequency of host genotypes), Mode demonstrated that frequency-dependent selection produces a stable equilibrium in which virulence and resistance each exhibit a balanced polymorphism, but only if the pathogen experiences a cost of virulence. The model predicts that at equilibrium most plants in a population are defended against at least some pathogens but some plants are susceptible to all pathogens. Hence, the population appears phenotypically variable and the average level of resistance in the population is lower than it would be were the resistance allele fixed. Although Mode's (1958) model does not incorporate cost of resistance, subsequent models that do (Leonard 1977; Leonard and Czocher 1978, 1980) produce similar predictions. However, the stabilities of the polymorphic equilibria in these models are sensitive to time lags (Fleming 1980; Levin 1983; Gould 1988).

Finally, when neither trait is costly, simple genetic models predict that a different mechanism will produce polymorphism for resistance with different phenotypic results. Jayakar (1970) demonstrated that for diploid hosts and parasites with recessive resistance and virulence traits that have no cost, coevolution causes fixation of the virulence allele in the parasite population, after which the now ineffective host resistance allele persists in neutral polymorphism. Leonard (1977) reached a similar conclusion for a diploid host and haploid pathogen. Thus, at equilibrium the average level of resistance in the population is zero, a result differing from that predicted by optimality theory, which presumes that without costs resistance will go to fixation, resulting in a maximally resistant population.

Clearly, the equilibrium level of resistance resulting from coevolution between plants and pathogens is substantially influenced by the presence or absence of resistance and virulence costs. How well do these models and their results apply to coevolution between plants and herbivores? Only if the genetic and ecological features of herbivore biology are similar to those of pathogens and parasites, and if their interactions with hosts are equivalent, will host-parasite and plant-pathogen models be applicable to plant-herbivore coevolution.

Comparing Herbivores with Pathogens and Parasites

One important assumption of most simple genetic models of plant-pathogen interaction is that encounters between plants and pathogens are random. That is, the probability with which each plant encounters a particular pathogen genotype is a function of the frequency of that genotype. While this assumption may be valid for certain highly mobile herbivores, such as grazers (see Pollard, this volume), and for the initial phase of colonization by some herbivores that are assumed to rain from above onto plants (e.g., aphids), it seems unlikely that it will apply to less mobile herbivores or those that hatch, mature, mate, and oviposit on the same plant. Incorporating density-dependent effects into models allows the relaxation of the random encounter assumption, and host-parasite models that include density-dependent interactions suggest that mechanisms other than cost can also maintain intermediate levels of resistance (Levin and Udovic 1977).

Another concern in applying plant-pathogen models to plant-herbivore coevolution is the mode of genetic control of virulence and resistance traits. Most plant-pathogen models assume dominant single-gene diallelic resistance and recessive multilocus virulence (based on Flor's [1956, 1971] description of the flax-rust interaction). Although gene-for-gene interactions have been demonstrated in Hessian fly–wheat interactions (Caldwell et al. 1966; Gallun and Hatchett 1968; Gallun 1972), neither the genetic mechanisms controlling host-plant resistance to herbivores nor those involved in herbivore virulence are understood in most systems. Moreover, many authors argue that gene-for-gene interactions are the exception rather than the rule (Robinson 1976; Gould 1983; Claridge and Den Hollander 1983; Wilhoit, this volume). Consequently, for many plant-herbivore interactions, coevolution might best be modeled using a quantitative genetic approach (Simms and Rausher, this volume). Quantitative models would also be more appropriate for modeling many plant-pathogen interactions which may in fact be characterized by quantitative resistance and virulence (Burdon 1987a; Alexander, this volume).

There are available no quantitative models of pairwise antagonistic coevolution. However, quantitative optimality models suggest that while costs of resistance might maintain an intermediate level of resistance, they will not by themselves maintain large amounts of genetic variation. Stabilizing selection on polygenic resistance, due to the benefit of deterring herbivores and the fitness cost of defense, will lead to an intermediate equilibrium level of resistance, but at evolutionary equilibrium, genetic variance will drop to zero (Robertson 1956) or a low value maintained by a balance among mutation, selection, and genetic drift (Lande 1976).

Intraspecific variability in virulence and among herbivores may alter this prediction, however. The expression of a trait in two different environments can be treated as two separate traits (Falconer 1981). Hence, evolution of resistance to an herbivore population with two virulence types may be treated as the evolution of phenotypic plasticity in response to an environmental variable whose value is determined by the proportion of each type of herbivore found on a plant. A model by Hastings and Hom (1989) suggests that the number of polymorphic loci maintained by stabilizing selection on a polygenic trait may be as large as the number of available environments. In other words, the number of different combinations of virulence genotypes to which plants are exposed may determine the number of polymorphic resistance loci and hence the amount of genetic variation in resistance present in a plant population. The model suggests that stabilizing selection (due to changes in the balance between costs and benefits of resistance caused by interactions with numerous herbivore genotypes) could maintain an intermediate level of resistance and a high level of genetic variation even when resistance is polygenically controlled.

One problem in applying models of the evolution of phenotypic plasticity to the evolution of plant resistance lies in their failure to incorporate frequency-dependent selection. If coevolution occurs between plants and herbivores, the frequency with which plants encounter different states of the herbivore environment will change as the herbivores evolve in response to evolutionary change in plant resistance. Until frequency dependence is incorporated into quantitative models, few conclusions can be drawn about the importance of plant-herbivore interactions in maintaining genetic variation in natural plant populations. Of course, the effect of costs of quantitative resistance traits on evolutionary equilibria resulting from such interactions will also remain unknown.

In summary, although models of coevolution between plants and pathogens may not be used directly to describe the interactions between plants and herbivores, they provide a framework for understanding how costs of resistance influence both the evolution of resistance to herbivores and the importance of the process in maintaining genetic variation in natural plant populations.

Mechanisms by Which Costs May Be Incurred

Until recently, there has been little empirical evidence available with which to evaluate the assumption that resistance is costly to plants (Janzen 1979; Fox 1981; Gould 1983, 1988). This deficiency may be attributed in part to confusion about how costs can be measured. To develop methods for empirically detecting and measuring costs of resistance, one must first understand the various mechanisms by which costs may arise.

Allocation Costs

Costs were first envisioned as internal trade-offs in the allocation of resources (Chew and Rodman 1979). Rhoades (1979) and others argued that within the plant, a limited quantity of resources (usually energy or nutrients) are optimally allocated to maximize plant fitness. Allocation to defenses diverts resources from other functions, which presumably disrupts otherwise optimal patterns of allocation and thereby reduces fitness. Although some authors have termed costs due to this mechanism "physiological costs" (Estes and Steinberg 1988; Gould 1988) or "physiological constraints" (Bryant et al. 1988), I prefer the term "allocation cost," which distinguishes this cost from other types of physiological cost, which will be described shortly. This concept of cost has been historically associated with chemical defenses and, in this context, includes costs of production (Orians and Janzen 1974; Chew and Rodman 1979; Coley et al. 1985; Gulmon and Mooney 1986), sequestration (Orians and Janzen 1974; McKey 1979), maintenance (Burbott and Loomis 1969; Waller and Nowacki 1978), and marginal costs (Bloom et al. 1985; Zangerl and Bazzaz, this volume).

Critics of this concept note that many chemicals involved in plant resistance appear to have other functions within the plant (Muller 1969; Robinson 1974; Mothes 1976; Seigler and Price 1976; Seigler 1977; Jones 1979) and may simply be pleiotropic effects of genes controlling traits that evolved in response to selection by other environmental stresses such as competing plants (Olney 1968; Whittaker and Feeny 1971), ultraviolet radiation (Lee and Lowry 1980), nutrient deficiency (Lieberman and Lieberman 1984), or drought (Rhoades 1977). For example, although alkaloid-containing lupins experience less herbivory from insects, molluscs, and vertebrates than do non–alkaloid-containing lupins (Wink 1984b), alkaloids may also function in nitrogen storage (Wink and Witte 1985a), making their allocation costs negligible. Furthermore, because alkaloids also inhibit pathogen infection (Wink 1984a; Wippich and Wink 1985) and growth of neighboring plants (Wink 1983), any costs of alkaloid production should be charged in part to these functions.

Self-Toxicity

Many chemicals that protect plants from herbivores by the mechanism of antibiosis (Painter 1951) are also toxic to plants (McKey 1974a, 1979; Orians and Janzen 1974; Rhoades and Cates 1976; Gulmon and Chu 1981, Zucker 1983), creating the potential for costs due to self-toxicity. For example, cyanogenic morphs of *Trifolium* sp. and *Lotus* sp. are rare in colder climates, and self-toxicity caused by freezing damage is thought to have constrained their distribution (Daday 1958, 1965; Foulds and Young 1977; but see Compton, New-some et al. 1983, and Ellis et al. 1977b for contrasting evidence). Moreover, Lieberei et al. (1989) have shown that hydrogen cyanide production by some clones of the rubber tree (*Hevea brasilensis*) inhibits induced defense against a fungal pathogen. A similar mechanism may cause susceptibility of highly cyanogenic flax varieties to *Colletotrichum lini* (Lüdtke and Hahn 1953).

Biochemical and Developmental Mechanisms

When a new resistance trait arises by mutation, disruption of biochemical pathways may also produce costs. For example, the ability to catalyze a new substrate may be achieved at the expense of the ability to metabolize the normal substrate. Catabolic imbalance and catabolic repression may also disrupt normal metabolism (Mortlock 1982). Although such biochemical costs have received little consideration in the plant-herbivore literature (but see Chew and Rodman 1979; Berenbaum and Zangerl 1988), they are prominent in discussions of the evolution of herbicide-resistance (Holt et al. 1983; Hirschberg et al. 1987; Gressel 1987) and insecticide-resistance (Plapp 1976; 1986; Tsukamoto 1983; Devonshire 1975).

Ecological Mechanisms

Finally, some defenses may involve ecological trade-offs (Futuyma 1983; Gould 1983a; Simms and Rauscher 1987, 1989). Most plant species host many herbivorous species (e.g., Strong et al. 1984; Maddox and Root 1990; Fritz, this volume), and in some plant species genotypes resistant to one type of herbivore are susceptible to other types (Kinsman 1982; Da Costa and Jones 1971; but see Fritz and Price 1988; Maddox and Root 1990; Zangerl and Bazzaz, this volume). In such cases, allocating resources to defense against the first type of herbivore would enhance fitness by reducing its herbivory, but would reduce fitness by the extent to which damage caused by other herbivores was increased.

These trade-offs may be directly mediated by the plant, such as when traits that confer resistance to one herbivore are constituents of feeding or oviposition attractants used by other herbivores ((Da Costa and Jones 1971; Beck and Schoonhoven 1980; Blau et al. 1978). On the other hand, trade-offs driven by genetic variation in resistance among plants may be mediated by interactions among other species. For example, traits that confer resistance to one herbivore may increase susceptibility to a second herbivore that would otherwise be competitively excluded by the first herbivore (Fritz et al. 1986; Faeth 1988; Fritz, this volume). Such interactions may span more than one trophic level. Herbivores feeding on susceptible genotypes may be subject to higher levels of predation than those on resistant genotypes (Ferguson and Metcalf 1985; Campbell and Duffey 1979; Duffey 1980, Hare, this volume; Fritz, this volume).

Ecological trade-offs may also occur between phenotypes within an herbivore species. These cause the type of costs incorporated by Clarke (1976) into his first coevolutionary model. Ecological costs arising from specific resistance to particular herbivore phenotypes are equivalent to those described by Parker (this volume) as genetic specificity.

Empirical Detection of Costs

Definition of Resistance

The diversity of mechanisms by which resistance costs may be incurred dictates that costs must be detected by more than one method. Before considering methods for measuring costs of resistance one must first define resistance and operationalize that definition. Two methods have been employed to determine level of resistance.

Specific-Trait Approach

If the mechanism of defense is well understood and due to a specific trait, then the level of plant resistance may be defined by the status of that trait. Such a trait can be measured by any of a number of qualitative or quantitative assays. For example, if glandular trichomes provide the major defense, then resistance may be measured by determining trichome density and the volume of the glandular exudate. This method of defining resistance, termed the specific-trait approach, is powerful because it provides a definitive link between the genotype and the resistance phenotype. However, because defenses against some herbivores may act as feeding or oviposition stimulants for other herbivores, plants possessing a "resistance" trait may experience more damage than those without it. Moreover, resistance against a single species of herbivore may involve several interacting mechanisms, and focusing on one may preclude understanding the entire picture. If the investigator is interested in the overall costs of reducing vulnerability to herbivory, or if the mechanisms of resistance are unknown, the bioassay approach may also be used to define resistance.

Bioassay Approach

The bioassay approach (Tingey 1986; Lewis and van Emden 1986; Waller and Jones 1989) involves defining the level of resistance as (1) the inverse of the number of herbivores per plant (or some plant part), (2) the inverse of the feeding damage per herbivore, or (3) the complement of the proportion of plant tissue lost to herbivory. Resistance defined by bioassay must indicate the reduction in herbivory experienced by a plant as a result of any (unknown) resistance traits that it possesses. It is generally not appropriate to measure herbivore growth rate, pupal mass, or other measures of herbivore fitness, because these observations do not indicate the direct effect of resistance on individual plants. An exception to this rule arises when the entire population of herbivores occupies a single plant. In this case, reducing herbivore fitness reduces herbivore population growth rate and benefits the individual plant on which the population resides.

Because resistance must be distinguished from components of plant fitness, measures of tissue lost to herbivory are usually standardized and expressed as a proportion of the total plant tissue (Painter 1958; Beck 1965; Carter 1973; Horber 1980; Simms and Rausher 1987, 1989). Proportionalizing the measure of damage is particularly important when the tissue being used by the herbivore is a major component of fitness. For example, highly productive plants may appear less resistant because they lose more seeds to predators than plants that produce few seeds. In this case, resistance can be uncoupled from fitness by expressing seed destruction as a proportion of the total number of seeds produced by the plant. Proportionalizing the measure of damage is not appropriate, however, when the investigator controls the number of herbivores on each plant, as is the case in many laboratory experiments. Finally, because the ultimate goal is to understand how plant resistance traits evolve in nature, bioassays should be performed with herbivores that might actually impose selection on the plant population being studied.

The advantage of the bioassay approach to defining resistance is that it can be used even when the mechanisms of resistance are unknown or poorly understood. Its major shortcoming is that the resistance mechanism remains unknown.

Methods of Detecting Costs

Obviously, the methods by which costs of resistance may be detected will be dictated to some extent by the way in which resistance is defined. The units in which costs are measured will also constrain the methodology.

Tables 17.1 and 17.3 through 17.7 summarize a number of studies that provide empirical evidence regarding costs of resistance to herbivory. An effort was made to be comprehensive, but it is likely that many more examples exist in the agricultural literature. The studies in these tables illustrate the diversity of methods for detecting costs of resistance, but can be broadly subdivided into two major classes on the basis of the units in which costs are measured.

Measurement in Resource Units

One class of methods involves attempts to calculate allocation costs in terms of some limiting resource (e.g., carbon, energy, nitrogen) (table 17.1). Chew and Rodman (1979) and Mooney and Gulmon (1982) used the ideas of Penning de Vries et al. (1974) to calculate costs in units of energy and material resources. After McDermitt and Loomis (1981) streamlined the calculations suggested by Penning de Vries (1974, 1975b) and Penning de Vries et al. (1974), Gulmon and Mooney (1986) used them to estimate the direct costs in terms of g CO_2/g

TABLE 17.1
Studies providing evidence regarding costs of resistance to herbivory in terms of resources

Species	Units of measure	Conclusions	Citations
Theoretical calculations in terms of units of resources allocated to defenses			
Various	Chemical energy	Costly[a]	Chew and Rodman 1979
Diplacus aurantiacus	% of lifetime growth	Costly[a]	Mooney and Gulmon 1982
Theoretical	% of leaf growth	Costly[a]	Gulmon and Mooney 1986
Empirical measures in terms of resources			
Dioclea megacarpa	Total seed nitrogen	At least 30% involved in resistance	Rosenthal 1977
Cicer arietinum	Water	Significant water loss through glandular trichomes	Lauter and Munns 1986

[a]Because all organic chemicals contain carbon, by this definition they are necessarily costly.

of several defensive compounds. The results and limitations of these studies are discussed by Zangerl and Bazzaz (this volume).

Another method for detecting costs is to observe differential allocation of resources to the defense of different plant parts that presumably differ in their contribution to plant fitness (Mooney and Gulmon 1982). In a review of several studies, Krischik and Denno (1983) observed that young leaves, which are generally believed to be more valuable, are better defended than older leaves. Hay et al. (1988) reached similar conclusions but Zangerl (1986) found no evidence for correlations between photosynthetic rates and furanocoumarin concentrations in leaves of *Pastinaca sativa* that differed in age.

Finally, Pimentel (1988) cites a study of water loss from chickpea trichomes (Lauter and Munns 1986) as evidence that resistance is costly. While Lauter and Munns demonstrated substantial water loss via glandular exudates, which is likely to be detrimental to chickpeas growing in dry habitats, they emphasized that it is not clear what function the glandular trichomes and their exudates perform for the plant. Consequently, this study does not provide clear evidence that resistance to herbivory is costly.

While measuring costs in units of resources is interesting and useful for understanding physiological trade-offs involved in plant defense, it cannot be used to understand constraints on the evolution of resistance (Simms and Rausher 1987). Moreover, this method can be used only to detect costs of resistance traits whose mechanisms are known and well characterized. Another major drawback is that detecting costs by these methods ignores other causes of costs, such as ecological trade-offs. Finally, pleiotropic benefits of defense traits can reduce or eliminate the fitness costs of the resistance functions they perform. It is difficult to imagine how this complication could be incorporated into calculations of costs in terms of resource units.

Measurement in Units of Fitness Components

In recognition of the evolutionary nature of optimality arguments and the consequent need to measure costs and benefits in terms of fitness, optimality theory has recently been recast in terms of plant growth (Coley et al. 1985; Fagerström et al. 1987; Fagerström 1989), with the explicit assumption that plant growth is a major component of fitness in long-lived iteroparous plants. In the case of semelparous plants, it is feasible to estimate fitness by measuring other components as well, such as survival and seed production. In either case, determining the cost of possessing a resistance trait is equivalent to measuring selection acting on that trait.

Consider a simple numerical model of the evolution of a resistance trait controlled by a single diallelic locus with a dominant allele, A, coding for resistance. For simplicity, assume that environment has no effect on the resistance phenotype (e.g., resistance is not inducible). This locus produces two phenotypes: phenotype 1 (aa), which is susceptible, and phenotype 2 (AA and Aa), which is resistant. Each phenotype can occur in two different environments: one in which herbivores are absent (environment I) and one in which herbivores are present at some natural density (environment II). The fitness values associated with the four combinations of phenotype and environment are expressed in table 17.2 relative to the fitness of the susceptible phenotype in the herbivore-free environment ($W_{1,1}$). In the herbivore-free environment the resistant phenotype will not generally benefit from its defenses; but the cost of resistance remains and its fitness will be less than that of the susceptible phenotype.

The difference

$$(1) \qquad\qquad W_{1,1} - W_{2,1} = C$$

represents the cost associated with resistance (C).

In general, the fitness of the susceptible phenotype in the herbivore-free environment will be greater than its fitness in the environment with herbivores because of the detrimental effect of herbivory. The difference between these two fitnesses is defined as the cost of herbivory, H, i.e.,

TABLE 17.2
Calculations of costs and benefits of resistance that exhibits single-gene inheritance

Environment	Phenotype	
	Nonresistant	Resistant
Herbivores absent	$W_{1,1}$	$W_{1,1} - C$
Herbivores present	$W_{1,1} - H$	$W_{1,1} - (H - B) - C$

Note: $W_{1,1}$ = fitness of the most susceptible phenotype in the environment without herbivores. H = cost of herbivory (reduction of fitness due to herbivory). C = cost of resistance (reduction of fitness due to defenses). B = benefit of resistance (amount by which defenses reduce loss of fitness due to herbivory).

(2) $$H = W_{1,I} - W_{1,II}.$$

If herbivory were beneficial to plant fitness, then H would have a negative value and the additive inverse of H would represent the fitness benefit of herbivory.

The fitness of the resistant phenotype in the environment with herbivores will be affected to some extent by herbivory (assuming that the defense is not perfect), which is called K. Its fitness will also be reduced by C, the cost associated with resistance. Thus,

(3) $$W_{2,II} = W_{1,I} - C - K.$$

In most cases, the value of K will range from 0 (perfect resistance) to H (ineffective resistance). If herbivory were beneficial, K would have a negative value. Finally, the benefit associated with resistance may be defined as

(4) $$B = H - K,$$

where $0 \leq B \leq H$. Then

(5) $$W_{2,II} = W_{1,I} - C - (H - B).$$

In this form the benefit appears as the amount by which the detrimental effects of herbivory on fitness, H, are reduced by resistance.

Comparisons among Populations

These definitions illustrate how costs of resistance may be detected by comparing levels of defense in different populations of species polymorphic for resistance (table 17.3). These geographical comparisons correspond to Endler's (1986) "Method One" for detecting natural selection. In such studies, it is assumed that if natural selection occurs, then geographic variation in herbivore density (the selective factor) will produce parallel geographic variation in resistance traits. Specifically, in an environment with herbivores, the selection pressure by herbivores (reflected in H, the cost of herbivory) will outweigh selection against resistance due to its costs (C), resulting in the spread of the resistant phenotype. In contrast, in the absence of herbivores, selection against resistance due to its costs will reduce the frequency of the resistant phenotype.

In both geographic comparison studies cited in table 17.3, resistance is reduced or absent in the absence of herbivores, suggesting that resistance might be costly. Several qualifications must accompany this conclusion, however. One drawback of geographical comparisons is that without knowing the ancestral state of the resistance trait, it is impossible to determine what process has caused an observed gradient in resistance. If susceptibility is the ancestral state, then the gradient may be due to selection by herbivores for resistance (cost of

TABLE 17.3
Studies providing evidence regarding costs of resistance to herbivory in natural plant populations

Geographic comparisons within a single species between populations in sympatry and allopatry with herbivores

Species	No. of populations	Pattern	Conclusions	Citations
Hymenaea courbariil	2	Resistance neg. corr. with herbivore presence	Resistance costly	Janzen 1975a
Asarum caudatum	14	Freq. (unpalat. plants) neg. corr. with slug pop. density	Resistance costly	Cates 1975

Comparisons of chemical defenses between species and without ant defenses.

Taxon	No. of species	Pattern	Conclusions	Citations
Acacia	9	Non-ant acacias contain HCN, but ant acacias do not.	HCN is costly	Rehr, Feeny et al. 1973
Ipomoea	19	Indole alkaloids are no less numerous in spp. with defense nectaries than in spp. without such nectaries.	Indole alkaloids not costly	Steward and Keeler 1988
Cecropia	5	Mullerian bodies are not produced in absence of ants.	Mullerian bodies costly	Janzen 1973c

herbivory and concomitant benefit of resistance). By contrast, if resistance is the ancestral state, then the gradient may be due to selection against resistance (cost of resistance). Geographic comparisons are also subject to the limitations described by Endler for detecting selection by Method One. In particular, if only a few localities have been sampled (as is true with Janzen 1975a), then it is likely that some significant differences in frequencies of resistant phenotypes between environments will appear by chance alone. Furthermore, differences among populations in unrelated environmental factors may produce a negative correlation between herbivore pressure and defenses. Carefully designed geographical comparisons can provide important circumstantial evidence for costs of resistance. They can also be corroborated by experimental manipulations (e.g., Cates 1975). However, because biases against negative results may have prevented the publication of studies that failed to show a correlation between resistance and herbivore pressure, we cannot conclude that resistance will always be costly.

Comparisons among Species

Comparisons among species have also been used to test for costs of resistance (table 17.3). For example, that certain resistance traits found in most Central

American acacias are absent in those species in which ants protect the trees from herbivory (Rehr, Feeny et al. 1973; Janzen 1973c) suggests that these traits are costly. In contrast, Steward and Keeler (1988) found that among members of the genus *Ipomoea,* the presence of extrafloral nectaries was not correlated with the number of alkaloid or mechanical defenses a species produced. Because extrafloral nectaries have been shown to function in ant-mediated antiherbivore defense of leaves (Keeler 1977) and flowers and seeds (Keeler 1980; Beckmann and Stucky 1981) in this genus, Keeler's results suggest three alternative hypotheses: (1) neither alkaloids nor mechanical defenses are costly, (2) ants provide inadequate defenses to *Ipomoea,* or (3) both ants and alkaloids provide positive benefits to *Ipomoea* that outweigh their costs.

These studies depend upon the same theoretical framework as comparisons among populations within a species. However, conclusions drawn from simple comparisons among species are even more tenuous because the histories of selection forces responsible for variation in resistance levels among species and the genetic mechanisms underlying resistance are likely to vary more than those of different populations within a species.

A more productive approach is to incorporate the phylogenetic history of the species being compared, to provide an estimate of the number of times each trait has evolved and the order in which the traits evolved (Donoghue 1989). For example, consider a hypothetical lineage in which species possess either ant or chemical defense, but never both. Furthermore, suppose that the species all share a common ancestor which possessed chemical defense and lacked ant defense (chemical defense is ancestral). Although the hypothetical lineage could have arisen via several phylogenetic pathways, to illustrate my point I will describe two possible paths. One lineage might have experienced several phylogenetic events in which acquisition of ant defense was followed by loss of chemical defense. This pattern would suggest that chemical defense is costly. By contrast, an alternative lineage might have experienced some phylogenetic events like those just described, but it also experienced an equal number of events in which loss of chemical defense was followed by later acquisition of ant defense. The latter pattern would reveal no information about the costs of chemical defenses. No phylogenetic studies have been performed to detect costs of resistance to herbivory.

Comparisons within Populations

Discrete Polymorphisms

Studies that compare the fitnesses of different resistance phenotypes from a single polymorphic population in a common herbivore-free environment provide the best evidence we have on the frequency and magnitude of costs of resistance. Many of these experiments have been done on crop species (table 17.4). For example, Windle and Franz (1979) compared growth in monoculture

TABLE 17.4
Studies providing evidence regarding costs of resistance to herbivory in fitness units

Species	Component of fitness	Conclusions	Citations
Nicotiana tabacum	Leaf yield	Nicotine costly	Matzinger et al. 1960
Nicotiana tabacum	Leaf yield	Nicotine costly	Matzinger and Mann 1964
Nicotiana tabacum	Leaf yield	Nicotine costly	Legg et al. 1965
	Plant height	Nicotine not costly	
	Days to flower	Nicotine not costly	
Nicotiana tabacum	Leaf yield	Nicotine costly	Vandenberg and Matzinger 1970
Nicotiana tabacum	Leaf yield	Nicotine costly	Matzinger and Mann 1971
Nicotiana tabacum	Leaf yield	Nicotine costly	Matzinger et al. 1972
Nicotiana tabacum	Leaf yield	Nicotine not costly	Matzinger et al. 1989
Hordeum vulgare	Aboveground biomass	Greenbug resist. not costly	Windle and Franz 1979
	Competitive ability	Greenbug resist. not costly	
Hordeum vulgare	Biomass at 25 days	Greenbug tolerance costly	Castro et al. 1988
	No. of leaf primordia	Greenbug tolerance not costly	
Glycine max	Seed yield	Mexican beetle resistance costly	Tester 1977
Glycine max	Seed yield	Resistance to herbivores costly	Mebrahtu 1985
Zea mays	Ear yield	Resist. to Eur. corn borer costly	Klenke 1985
Vigna unguiculata	Seed weight	Resist. to herbivores not costly	Bosque-Perez et al. 1987

Note: Comparisons are among genotypes within crop species.

and the competition abilities of two barley cultivars that were genetically identical (isogenic) except for a locus controlling both tolerance and antibiosis to the greenbug *Schizaphis graminum*. Although the results of this paper have been cited as evidence for costs of resistance (e.g., Berenbaum et al. 1986; Simms and Rausher 1987; Brown 1988), when examined in detail the picture is not so clear. Greenbugs reduced plant growth in the susceptible cultivar, but infestation actually increased aboveground biomass in the resistant cultivar. Thus, the resistance allele not only benefits the plant by reducing herbivory but may in fact increase plant growth in response to herbivory (overcompensation). Furthermore, at seven weeks (when the experiment was terminated), growth of resistant plants in the absence of greenbugs was equal to that of susceptible plants in similar conditions, which suggests that resistance is not costly. Finally, although in the absence of greenbugs resistant plants were initially poorer competitors than susceptible plants, this competitive differential disappeared by seven weeks of age. Consequently, if anything, this study suggests that resistance at this locus may not be costly.

One particularly interesting group of studies involves cyanogenesis and includes both geographic surveys and examinations of polymorphism within single populations (table 17.5). Although cyanogenesis occurs in over 800 plant species spanning 100 plant families (Eyjolfsson 1970; Gibbs 1974; Conn 1979), the vast majority of studies on this trait have focused on two herbaceous perennial legumes: *Trifolium repens* and *Lotus corniculatus*.

Cyanogenesis in these two legumes is largely controlled by two unlinked loci (Corkill 1942; Atwood and Sullivan 1943; Nass 1972). Cyanogenic glucoside synthesis is determined by the incompletely dominant allele at the *AC/ac* locus, and production of the hydrolytic enzyme b-glucoside linamarase is determined by the incompletely dominant allele at the *Li/li* locus (Hughes and Stirling 1982; Hughes et al. 1984; Maher and Hughes 1973). Although inheritance may be more complex, in general only those plants possessing both the cyanogenic glucoside and the hydrolytic enzyme (*Ac-Li-*) release hydrogen cyanide when damaged. However, in some herbivores hydrogen cyanide is released from cyanogenic glucosides by gut enzymes, which make the *Ac-lili* genotypes

TABLE 17.5
Studies regarding costs of cyanogenesis in *Trifolium repens* and *Lotus corniculata*

Geographic comparisons

Species	No.	Gradient	Pattern	Citations
Trifolium repens	6	Elevation	Freq. (HCN) negatively correlated	Daday 1954b
T. repens		Latitude	Freq. (HCN) positively correlated	Daday 1954a
T. repens	51	Jan. mean temp.	Freq. (HCN) positively correlated	Daday 1958
T. repens	4	Moisture	Freq. (*Ac*) positively correlated	Foulds and Grime 1972a
T. repens	17	Elevation	Freq. (HCN) negatively correlated	de Araújo 1976
T. repens	22	Elevation	Freq. (HCN) positively correlated	Till-Bottraud et al. 1988
T. repens	19	Latitude	No specific trend*	Kakes 1987
Lotus corniculatus	3	Coast-inland (200 m)	Freq. (HCN) lower near sea	Jones 1962 cited in Ellis et al. 1977c
L. corniculatus	3	Coast-inland (200 m)	Freq. (*Li*) pos. corr. with moisture	Ellis et al. 1977c
L. corniculatus	13	Latitude (50 m)	Freq. (HCN) negatively correlated	Compton, Newsome et al. 1983
		Elevation (555 m)	Freq. (HCN) negatively correlated	
L. corniculatus	4	Moisture	Freq. (*Ac.*) positively correlated	Foulds and Grime 1972a

Table17.5
(**continued**)

Within-population comparisons

Species	Component of fitness	Pattern	Citation
T. repens	Mortality Dry mass/plant	In drought: $AC > ac$ With plenty of water: $AC < ac$ (not signif.)	Foulds and Grime 1972b
T. repens	% plants flowering No. flowers/plant Dry mass/plant Root length	HCN = non-HCN HCN = non-HCN HCN = non-HCN HCN = non-HCN	Foulds 1977
T. repens	Photosynthetic rate	Frost reduced PS in non-HCN only	Foulds and Young 1977
T. repens	Yield/pot	At high density: Li/li mixture > monoculture; in monoculture: $Li =$ li	Ennos 1981b
T. repens	Survival	HCN < non-HCN	Dirzo and Harber 1982b†
	Stolons/plant Stolon decay rate	HCN < non-HNC HCN > non-HCN	
T. repens	Vegetative dry mass‡ Flower head dry mass‡ Seedling survival	No difference among HCN genotypes $Ac < ac$ No difference among HCN genotypes	Kakes 1989
L. corniculatus	Survival under cold stress	HCN = non-HCN	Ellis et al. 1977b
L. corniculatus	No. seeds/plant Survival Plant dry mass Root length	HCN = non-HCN HCN = non-HCN HCN = non-HCN HCN = non-HCN	Foulds 1977
L. corniculatus	Photosynthetic rate	Frosting reduced PS in HCN only	Foulds and Young 1977
L. corniculatus	No. pods/plant No. pods/plant‡	$Ac > ac$ $Ac > ac$	Keymer and Ellis 1978

*Found linkage disequilibrium between Ac and Li locus, suggesting epistatic selection, perhaps favoring $AcLi$
phenotypes.
†Grazing by molluscs, sheep, and weevils throughout study.
‡With herbivores present.

effectively cyanogenic for these herbivores. It is generally believed that cyano-genesis provides resistance to herbivores, but conflicting evidence of the effects of cyanogenic glycosides on herbivores (reviewed by Hruska 1988) suggests that generalizations about their effectiveness for deterring various types of her-bivores should be drawn cautiously.

Daday (1954a, 1954b, 1958) observed that *T. repens* populations are poly-morphic at both the AC/ac and Li/li loci over large geographic areas and that cyanogenic forms are less frequent in environments with low temperatures. He proposed that stabilizing selection, due to herbivory (favoring cyanogenesis) and low temperature (favoring acyanogenic forms), actively maintains the polymorphism in these populations. These observations, and those later reported by de Araújo (1976) and Till-Bottraud et al. (1988), inspired numerous attempts to determine experimentally which environmental factors might be responsible for maintaining apparently stable cyanogenesis polymorphisms.

In a field transplant experiment, Daday (1965) observed that at a coastal site (high temperature) cyanogenic plants grew larger than acyanogenic forms and that *Ac-* phenotypes produced more flowers than did *acac* genotypes, suggesting a benefit of cyanogenic glucosides in high-temperature environments. In contrast, at a montane site (low temperature), the acyanogenic morph *acaclili* produced more flowers than did *Ac-lili* and sustained less frost damage than the *Ac-Li-* genotype, suggesting a cost of cyanogenesis in low-temperature environments. However, the genotypes did not differ in vegetative growth.

Foulds and Grime (1972a) observed a cline in cyanogenic frequencies in *T. repens* along a soil moisture gradient. They found that glucoside forms (*Ac-*) were significantly less frequent in dry soils and reasoned that if frost damage could disrupt vacuolar membranes and allow hydrolysis of cyanogenic glucosides, then drought stress might produce a similar effect (Foulds and Grime 1972b). They performed two experiments to detect selection against cyanogenic forms under soil moisture stress. First, they manipulated soil moisture availability in the field by varying soil depth. In all four soil depths used, the cyanogenic phenotype (*Ac-Li-*) produced significantly fewer flowers than any of the acyanogenic phenotypes. However, the final dry weight did not differ among phenotypes in any treatment. After five severe drought cycles in a greenhouse, *Ac* phenotypes experienced significantly more mortality than did *ac* phenotypes. These results suggest that in both field and greenhouse conditions, and in all water regimes including extreme drought stress, acyanogenic morphs are more fit than cyanogenic morphs. In contrast, Foulds (1977) found no difference in vegetative growth, germination percentage, seed production, or mortality between cyanogenic and acyanogenic phenotypes when grown in either moist or drought-stressed conditions.

Ennos (1981b) observed the growth of different phenotypes in various competitive regimes established in outdoor pots. In competition with the *Li* phenotype, *li* plants were more fit than *Li* plants, whereas there was no fitness differential when plants competed with the *li* phenotype. These results suggest that there is some cost associated with the gene producing the hydrolyzing enzyme. In contrast, in a field experiment involving cuttings of four genotypes (*AcacLili, Acaclili, acacLili,* and *acaclili*), Kakes (1989) found evidence for a cost of cyanogenic glucosides but not of the hydrolyzing enzyme.

Finally, Dirzo and Harper (1982b) transplanted replicate cuttings from five cyanogenic (*Ac-Li-*) and five acyanogenic (*acaclili*) plants into a field where they were subjected to normal levels of herbivory and pathogen attack. They found that acyanogenic plants produced leaves at a faster rate and produced more stolons than cyanogenic plants. Furthermore, surveys on a natural population indicated that cyanogenic plants produce fewer flowers than the acyanogenic morphs. Dirzo and Harper also observed that cyanogenic plants lost more stolons to decay and were more likely to be infected by a rust, *Uromyces trifolii*. These results suggested that cyanogenic morphs are less fit than acyanogenic morphs.

In summary, most studies (except Daday 1965 [in low temperature sites] and Foulds 1977) found that cyanogenic plants are less fit than acyanogenic ones. Most of these studies were performed in the field and, although Dirzo and Harper (1982b) were the only researchers who reported the activities of herbivores on their experimental plants, presumably all field-grown plants experienced at least some herbivory. Consequently, these studies suggest that even in the presence of herbivores the cyanogenic morph has reduced fitness. Not only is the trait costly, but the costs do not appear to be offset by the benefit of reduced herbivory. It is difficult to understand what selective benefit of cyanogenesis might maintain the polymorphism.

In contrast to *T. repens,* in which selection appears to act against cyanogenesis, this trait does not appear to be costly in *L. corniculatus.* Although Foulds and Grime (1972b) found a positive relationship between the frequency of glucosidase production and soil moisture, and Jones (1973) found that populations of *L. corniculatus* had a lower frequency of the *Li* allele in droughted areas than in moist places, Foulds (1977) detected no difference in germination, growth, mortality, or seed production between cyanogenic and acyanogenic phenotypes in either water-stressed or unstressed conditions. On the basis of an extensive survey of populations throughout Europe, in which few general trends were apparent, Jones (1977) concluded that there is unlikely to be a general explanation for the distribution of cyanogenesis in *L. corniculatus.* Compton, Newsome et al. (1983) found that plants with cyanogenesis were rare in the south but predominated in harsh conditions further north.

To further complicate the situation in *L. corniculatus,* Ellis et al. (1977b) found that cyanogenesis varies within plants over time. They found some "stable" plants, which did not vary, and other "unstable" plants, in which cyanogenesis varied as much over time as among plants in the population. Within the variable plants, there was no correlation between expression of cyanogenesis and the mean monthly maximum and minimum temperatures in either the same or the previous month.

In a controlled environment chamber without herbivores, Keymer and Ellis (1978) found that *Ac-lili* plants (which produce cyanogenic glycosides, but not linamarase) produced the most seed pods, followed by the cyanogenic (*Ac-Li-*) and acyanogenic (*acac Li-*) forms. The *acac lili* (acyanogenic) forms produced the fewest seed pods. These results suggest that rather than being costly in the absence of herbivory, cyanogenesis is beneficial. The ranking of phenotypes in the presence of molluscan grazers (also in growth chambers), changed in only one respect: pod production in the acyanogenic (*acaclili*) morph equaled or exceeded that of the *acacLi-* morph. In outdoor cold-frame experiments, from which grazers were excluded, *Ac-lili* plants produced the most pods, followed by *Ac-Li-*, with *acaclili* plants producing fewer than 20% as many pods as *Ac-lili* plants. These results suggest an absence of any apparent cost of cyanogenesis.

Cyanogenesis seems to have different implications for the two legumes. In general, the cyanogenic morph of *T. repens* is less fit than acyanogenic morphs. This relationship persists even in the field, where herbivores are presumably present. In contrast, the cyanogenic morph of *L. corniculatus* appears to equal or exceed the fitness of most acyanogenic forms regardless of herbivore pressure. Thus, in neither species is there clear evidence that a balance between costs of cyanogenesis in the absence of herbivores and the benefit of deterring herbivory is responsible for maintaining a polymorphism.

Quantitative Resistance

Theory. Unlike cyanogenesis, many resistance traits exhibit continuous phenotypic variation and are likely controlled by many loci, each with small effect (Berenbaum et al. 1986; Simms and Rausher, this volume). To measure costs on such quantitative traits, equation (5) must be modified to a continuous form. When resistance varies continuously, its cost is usually assumed to be a monotonically increasing function of the amount of resources allocated to defense, because as more resources are allocated to defense, fewer are available for nondefensive structures and processes that enhance survival and seed production in an herbivore-free environment (Fagerström et al. 1987; Fagerström 1989). For simplicity, this relationship is depicted as linear in figure 17.1. The benefit function, by contrast, is bounded above by H because benefits cannot exceed the detrimental effects of herbivory (perfect resistance) (Fagerström et al. 1987; Simms and Rausher 1987; Fagerström 1989). Hence, a plot of benefit as a function of allocation to resistance will plateau at some value less than or equal to H (leveling-off may occur at values less than H if no allocation can completely eliminate the reduction of fitness caused by herbivory).

The fitness of a phenotype with resource allocation R is thus

$$(6) \quad \begin{aligned} W(R) &= W(0) + B(R) - H - C(R) \\ &= [W(0) - H] + [B(R) - C(R)], \end{aligned}$$

where $W(0)$ is the fitness of the phenotype that allocates no resources to resistance (equivalent to $W_{1,1}$ in equation [5]). Consequently, the phenotype with maximum fitness corresponds to the value of R that maximizes the difference $B(R) - C(R)$. This value is depicted as $\hat{R}$ in figure 17.1. If resistance is a quan-

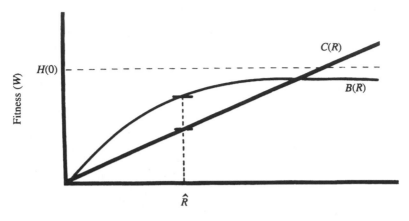

Resources Allocated to Resistance (R)

Fig. 17.1 Optimality model for the calculation of cost and benefit of resistance, when treated as a quantitative trait. The amount of fitness lost to herbivory by the most susceptible genotype is designated $H(0)$. The fitness cost of resistance, as a function of the amount of resources allocated to resistance, is $C(R)$. $B(R)$ is the fitness benefit of resistance, as a function of allocation. The point $\hat{R}$ corresponds to the genotype with maximum fitness. (Reprinted from Simms and Rausher 1987.)

titative trait controlled by many loci with small additive effects, then natural selection will move the average level of resistance in the population to $\hat{R}$ (Slatkin 1978; Lande 1976).

A corollary of this model is that at evolutionary equilibrium the slope of the cost function (marginal cost) is equal to the slope of the benefit function (marginal benefit), i.e.,

(7)   marginal benefit = dB/dR = dC/dR = marginal cost.

This result follows from the fact that $dW/dR = dB/dR - dC/dR = 0$ at the equilibrium because the equilibrium corresponds to the phenotype for which W is greatest. A second corollary of the model is that at evolutionary equilibrium the slope of fitness as a function of resistance is 0, i.e.,

(8) $dW/dR = 0$.

This model can be combined with the standard methodology of quantitative genetics (e.g., Falconer 1981; Mather and Jinks 1982) to estimate in fitness units the cost of resistance (Simms and Rausher 1987). If resistance is a quantitative character associated with a given allocation of resources, then equation (6) describes the fitness of an individual whose genotypic value is R. In an herbivore-free environment both H, the cost of herbivory, and $B(R)$, the benefit of resistance, will be zero. Consequently, in an herbivore-free environment

(9) $$W(R) = W(0) - C(R).$$

Differentiating with respect to R yields

(10) $$dW/dR = -dC/dR \quad \text{or} \quad dC/dR = -dW/dR,$$

which says that in an herbivore-free environment the marginal cost of resistance is equal to, and hence can be estimated by, the genetic regression of fitness on resistance (see Simms and Rausher, this volume). Although this method can measure absolute costs only if there are phenotypes available that exhibit no resistance, in any case, it indicates whether such costs are zero or not (assuming that the marginal costs do not plateau).

Empirical evidence in crop plants. Numerous studies provide empirical evidence regarding the costs of continuously distributed resistance traits, and many of these include crop species (table 17.4). Klenke (1985) found that selection for increased resistance to the European corn borer reduced losses in corn yield to artificial infestations. However, this reduction did not compensate for the loss in yield that accompanied selection for resistance. Klenke estimated that an 8.4% reduction in grain yield in the absence of herbivores was due to changes in gene frequency in response to selection for resistance. Interestingly, he estimated that the majority of yield loss (18.8%) was caused by inbreeding depression due to genetic drift. In other words, although there was a strong cost of resistance, only about one third of the cost appeared due to antagonistic pleiotropy. The remainder apparently resulted from linkage to other traits that reduced fitness. In natural populations, such linkage could be broken up and lost through random mating. This result suggests that estimates of costs of resistance reported from recently selected resistant cultivars might be inflated by linkage disequilibrium and emphasizes the importance of studying populations near evolutionary equilibrium.

In a study cited by Coley (1986) and Kakes (1989) as evidence for costs of resistance, Vandenberg and Matzinger (1970) compared the leaf yields of two cultivars and eight native, or primitive, strains of *Nicotiana tabacum* collected from Central and South America that varied in their nicotine and nornicotine content. Except for one primitive strain which converted nicotine to nornicotine and consequently exhibited an order of magnitude higher nornicotine content than the other lines, the strains exhibited a significant negative correlation between leaf yield and percentage nornicotine content (arcsine square-root transformed) ($r = -2289.9$, $P < 0.006$). There was also a negative correlation of leaf yield with percentage nicotine content (arcsine square-root transformed)

($r = -729.1$, $P < 0.002$). Although these results suggest that nicotine production is costly in terms of plant growth, several caveats must be attached to this conclusion. For example, some lines produced many basal shoots, but in conformity with standard practices of tobacco production, all plants were pruned to a single stalk. Consequently, the yield data probably do not accurately reflect total aboveground plant growth. Another problem is that because tobacco is grown to be smoked by nicotine addicts, it is likely that selection (conscious or otherwise) by farmers has artificially increased the nicotine content of these lines well beyond levels which natural selection by herbivores would favor. Thus, even if these compounds were costly to produce at these levels, it does not necessarily follow that resistance to herbivory is costly. Finally, Matzinger and coworkers (1989) demonstrated that simultaneous selection for increased concentrations of total alkaloid and constant yield (by means of index selection [Falconer 1981]) produced an increase in total alkaloid content without decreasing yield. Consequently, as Klenke found for corn, previously reported negative genetic correlations between leaf yield and total alkaloid content in tobacco (Matzinger et al. 1960; Matzinger and Mann 1964, 1971; Legg et al. 1965; Matzinger et al. 1972) may have been predominantly due to linkage disequilibrium, and thus do not indicate a pleiotropic cost of alkaloid production.

Tester (1977) demonstrated that two soybean cultivars resistant to Mexican bean beetle accumulated nitrogen more slowly than did two susceptible cultivars. He suggested that the higher nitrogen content of susceptible cultivars resulted in higher yields of pods. Unfortunately, because the yield data were not published, this field study does not provide definitive evidence for costs of resistance.

In another study of soybean, Mebrahtu (1985) raised the parents and progeny of a cross between a moderately resistant line and a susceptible cultivar in infested and uninfested field cages. Resistance was additively controlled with a broad-sense heritability of 43%. Under field conditions with herbivores present, the resistant progenies produced significantly higher seed yields, indicating a fitness cost of herbivory and a concomitant benefit of resistance. Significant negative genotypic correlations of leaf damage (the measure of resistance) with time to maturity, lodging, and plant height suggest that resistance exacts a fitness cost. However, resistance was also positively correlated with larger seed size.

Castro et al. (1988) provide evidence for costs of herbivore tolerance in a crop plant. Greenbug densities were the same on both susceptible and tolerant barley cultivars, indicating that resistance was not acting in the tolerant cultivar. The tolerant cultivar did not exhibit reduced biomass when infested, but infestation by greenbugs reduced the biomass of intolerant plants, indicating that herbivory is costly and that tolerance alleviated that cost. In the absence of greenbugs, the susceptible cultivar had higher biomass and greater leaf area at

25 days of age than did the tolerant cultivar, indicating a cost of tolerance. However, for one indicator of future growth, the number of leaf primordia differentiated on the apex, the relationship was reversed.

Empirical evidence from natural plant populations. Table 17.6 lists studies that provide information regarding costs of continuously distributed resistance traits from within-population comparisons in natural populations. Many of these studies report negative phenotypic correlations between herbivore resistance and components of fitness as evidence for costs of resistance. Because environmental covariance may cause the phenotypic covariance between two traits to differ from the genetic covariance in both sign and magnitude (Rausher and Simms 1989), a negative phenotypic correlation may not be indicative of a true pleiotropic cost of resistance. However, in an extensive literature survey Cheverud (1988) demonstrated that when correlations are drawn from a sample size of 40 or more, phenotypic correlation matrices are of similar sign and magnitude to their component genetic correlation matrices. Thus, phenotypic evidence for costs of resistance cannot be ruled out unconditionally.

The earliest evidence of costs from comparisons within a natural plant population comes from Hanover (1966a) who showed that monoterpene production in *Pinus monticola* is genetically controlled and that growth is negatively cor-

TABLE 17.6
Studies providing evidence regarding costs of resistance to herbivory in fitness units

Species	Component of fitness	Pattern	Citations
Pinus monticola	Growth	Negatively correlated with α-pinene	Hanover 1966a
		Neg. corr. with total terpene	
Asarum caudatum	No. seeds/fruit	Palatable plants > unpalatable plants	Cates 1975
	Plant dry weight	Palatable plants > unpalatable plants	
Pastinaca sativa	Umbel production	Neg. corr. with furanocoumarins	Berenbaum et al. 1986
Ipomoea purpurea	No. seeds/plant	No corr. with flea beetle damage	Simms and Rausher 1987
I. purpurea	No. seeds/plant	No corr. with damage by any of three types of herbivores	Simms and Rausher 1989
Betula sp.	Seedling height	Pos. corr. with vole damage	Rousi 1988
Pinus sylvestris	Growth rate	Pos. corr. with vole damage	Rousi 1989
Nicotiana sylvestris	No. seeds/plant	Neg. corr. with alkaloid conc.	Baldwin et al. 1990

Note: Comparisons are within population, in natural populations.

related with the production of some monoterpenes. Unfortunately, it is difficult to determine from the text of the paper whether these were genetic correlations.

Berenbaum et al. (1986) reported large negative genetic correlations of umbel production (a component of fitness) with the production of several types of furanocoumarins known to confer resistance to the parsnip webworm. Although their results suggest significant and large costs of resistance, umbel production was measured in the greenhouse. Genotype-by-environment interactions can cause patterns of genetic correlations measured in artificial environments to differ from those present in an organism's natural environment (Service and Rose 1985; Maddox and Cappuccino 1986). Consequently, it is not clear whether costs measured in a greenhouse will also occur in nature.

Coley (1986) measured a cost of tannin defenses in *Cecropia* seedlings as a negative correlation of leaf production during 18 months with tannin concentration of six-week-old leaves at the end of the 18-month period. Although the seedlings were the offspring of many parents, their actual parentage was unknown, making this a phenotypic regression. Coley, however, argued that the uniformity of the transplant environment left only genetic differences in growth and tannin concentration.

Rousi found that height at the close of the growing season was a good predictor of resistance to vole attack in birch seedlings; the tallest seedlings were least likely to be destroyed. If early growth rate is an important fitness component in long-lived trees, then Rousi's results suggest that resistance to vole damage is not costly to birch seedlings. Furthermore, in a study on *Pinus sylvestris*, Rousi (1989) discovered a possible positive genetic correlation between growth rate in the absence of herbivores (two years in a greenhouse) and resistance to vole damage (outdoors during the winter). Unfortunately, some families experienced greater than 25% mortality, and the relationship between survival and resistance was not reported. Nevertheless, among those families with low mortality, a positive genetic correlation between growth rate and resistance was recorded in two of four experiments. Rousi did not report the outcomes of the other two experiments; they presumably revealed no relationship. Both of these experiments suggest that costs of resistance are not universal.

In two field experiments, Simms and Rausher (1987, 1989) assessed costs of resistance to herbivory in the common morning glory (*Ipomoea purpurea*). In both studies, resistance was defined as the reciprocal of the proportion of plant tissue damaged by natural insect infestations in the field. Resistance was measured in plants from a set of half-sib families and correlated with fitness in the absence of herbivores (as estimated by measuring the seed production of plants from the same set of half-sib families but protected from herbivory by insecticides). The first experiment considered only resistance to the folivorous sweet potato flea beetle (*Chaetocnema confinis*). The experimental population in this experiment exhibited significant additive genetic variation for resistance to the flea beetle but there was no evidence that this resistance was costly (Simms and Rausher 1987). In a second experiment, conducted on an experimental population derived from the same base population as the first experiment, Simms and Rausher (1989) found significant additive genetic variation for resistance to corn earworm (*Heliothis zea*) feeding on pods, tortoise beetles (*Deloyala guttata* and *Metriona bicolor*) feeding on leaves, and generalist folivores (mainly orthopterans), but only nonadditive genetic variation for resistance to the sweet potato flea beetle. The experiment produced no evidence indicating that any of the three heritable resistances were costly.

Although in these experiments fitness was assessed in the field, the pesticide may have distorted the results by producing a genotype-by-environment interaction. To test for that possibility, E. L. Simms and M. A. Bucher raised seeds from ten fourth-generation inbred lines from the same population of morning glories in the greenhouse under two treatments. In the control treatment, plants were grown one per pot, trained up bamboo stakes, and watered daily. The second treatment consisted of the same conditions except that plants were sprayed with carbaryl (4 ml/l) every two weeks, starting three weeks after the seeds were planted. *Bacillus thuringiensis* (4 ml/l) was added to the spray mixture on alternate applications. This schedule followed that used by Simms and Rausher (1987, 1989). From each plant, all seed pods were collected after 95 days and counted. Although lines differed significantly in total seed mass ($F_{9,97} = 7.09$, $P < 0.0026$) and average seed size ($F_{9,97} = 5.65$, $P < 0.0007$), there was neither a significant treatment effect nor an interaction between line and treatment for either parameter. Consequently, it is unlikely that spraying with *B. thuringiensis* or carbaryl had any direct effect on seed production by morning glories in the field. Therefore, the absence of apparent costs of resistance in this population cannot be attributed to pesticide effects.

Another drawback to the conclusions drawn by Simms and Rausher (1987, 1989) is that they are based on the absence of negative genetic correlations between fitness in the absence of herbivores and resistance traits. Unfortunately, as pointed out by Pease and Bull (1988) and later elaborated upon by Charlesworth (1990), if trade-offs responsible for costs of traits such as resistance to herbivores involve more than two traits, then particular pairs of traits may be positively correlated. Therefore, the absence of a negative genetic correlation between seed production in the absence of herbivores and level of resistance does not rule out the possibility that these two traits may be part of a suite of traits involved in a complex functional constraint that ultimately imposes a cost of resistance that cannot be measured by this method.

More recently, Baldwin and colleagues (1990) found a significant negative phenotypic correlation between leaf alkaloid content and seed number in field-grown, undamaged wild tobacco (*Nicotiana sylvestris*). The phenotypic correlation of alkaloid content with total plant biomass was also negative but not significant. Alkaloids significantly reduce herbivory in wild tobacco (Baldwin 1988); therefore, these results suggest that resistance to herbivory in wild tobacco may be costly.

Empirical Evidence Regarding Costs of Inducible Resistance

There are at least three studies that report on the cost of resistance traits that were induced by environmental treatments (table 17.7). In one study, Brown (1988) artificially induced the production of proteinase inhibitors in tomato by injecting them with chitin and found no difference in several fitness components (including total biomass, shoot length, post-treatment growth, number of reproductive structures, number of mature fruits, number of seeds per plant, and mean seed mass) between induced and uninduced plants. Although there was a significant increase in the concentration of proteinase inhibitor in the chitin-injected plants over water-injected and uninjected plants, only one fitness component differed among treatments; germination in the water-injected plants was 95.97% whereas chitin-injected plants had 93.1% germination. However, only 91.2% of seeds from uninjected plants germinated. Consequently, Brown failed to detect a cost of resistance.

For two grass species, Cheplick and coworkers (1989) compared the growth and performance of plants infected by fungal endophytes with uninfected individuals. The relationship between clavicipitaceous endophytic fungi and grasses (and sedges) is generally considered to be mutualistic because in-

TABLE 17.7
Studies providing evidence regarding costs of resistance to herbivory in fitness units (costs of inducible resistance)

Species	Component of fitness	Pattern	Citations
Lycopersicon esculentum	Total biomass Shoot length Post-trt. growth No. mature fruits No. seeds/plant Mean seed mass	Induced plants = uninduced plants	Brown 1988
Lolium perenne (seedlings) (adults)	Biomass (in absence of herbivores) Relative growth rates (in absence of herbiv.)	Endophyte-infected plants = uninfected plants Infected plants > uninfected	Cheplick et al. 1989
Festuca arundinacea	Biomass (in absence of herbivores)	Infected plants > uninfected	
Nicotiana sylvestris	No. seeds/plant	Artificial damage that induced alkaloids < same amount of damage that did not induce alkaloids	Baldwin et al. 1990

fected host plants supply the fungus with photosynthate while ergotamine alkaloids produced by the fungus augment plant resistance to insect and mammalian herbivory (Clay 1987a; Bacon and Siegel 1988; Lewis 1988). Because endophyte-infected grasses and sedges generally outperform uninfected conspecifics under experimental conditions (Latch et al. 1985; Clay 1987b; Stovall and Clay 1988), this mechanism for resistance to herbivory may not involve costs. Several studies have shown, however, that environmentally stressful conditions (such as low light) decrease mycorrhizal infection and presumably decrease resistance (Harley 1969; Abbott and Robson 1984; Son and Smith 1988). This pattern may indicate that under stressful conditions plants starve their endosymbionts, thereby sacrificing herbivore resistance and reducing its cost. Consequently, Cheplick and coworkers wished to determine whether costs of herbivore resistance produced by endophytic infection were dependent on environmental stress. All three of their experiments were conducted in the absence of herbivores.

In their first experiment, the relative growth rates of adult *Lolium perenne* plants infected by *Acremonium lolii* (a fungal endophyte) were significantly greater than uninfected plants under nine treatments composed of three soil moistures and three nutrient concentrations. In another experiment designed to compare infected and uninfected *L. perenne* seedlings under three nutrient conditions, all seedlings were stunted at the lowest nutrient level, but the biomass of infected seedlings did not differ from uninfected ones. At intermediate and high nutrient levels, infected seedlings had significantly greater biomass. These results indicate that any costs to perennial rye of herbivore-resistance produced by endophyte infection are either undetectably small or are offset by unrelated benefits. In contrast, a similar experiment on *Festuca arundinacea* revealed that compared to uninfected seedlings, seedlings infected with *A. coenophialum* had greater biomass when nutrient availability was high but a significantly lower biomass under reduced nutrient availability. Moreover, the benefits of endophyte infection in fescue became increasingly greater with increasing nutrient availability. These results suggest that herbivore resistance produced by endophyte infection in fescue may involve an allocation cost. However, because infected and uninfected plants were obtained from different seed lots, they were not genetically equivalent, and differences in biomass may not have been due to endophyte infection.

Finally, wild tobacco plants damaged artificially or by caterpillars exhibit a significant increase in alkaloid content, but treatment with auxin at the site of damage inhibits the alkaloid response (Baldwin et al. 1990). Lifetime seed mass in plants in which alkaloid response was induced by artificial damage was significantly lower than in equally damaged plants in which the alkaloid response was inhibited by the auxin treatment. This result suggests that induction of alkaloid-based resistance is costly to plant fitness in wild tobacco.

Summary of Empirical Evidence

The bulk of the evidence reviewed here suggests that in many cases resistance to herbivory does involve fitness costs. However, results from Windle and Franz (1979), Simms and Rausher (1987, 1989), Brown (1988), Cheplick et al. (1989), and Rousi (1988, 1989) suggest that costs of resistance may not be universal. We do not have sufficient evidence to know how commonly costs constrain the evolution of plant resistance to herbivory. Clearly, further efforts must be made to answer this question more definitively. Equally clearly, the finding that costs may not be ubiquitous leads to the question of why costs might be small or absent. The remainder of this chapter will be devoted to considering this issue.

How Costs Might Evolve (A Possible Explanation for Low Costs)

Although the cost of a resistance trait is critical to theoretical predictions of its evolutionary trajectory and final equilibrium level, there has been little or no examination of the evolution of cost itself. To see how costs might evolve, consider an example from McKenzie et al. (1982), who reported evolutionary change in the cost of resistance to the pesticide diazinon in the Australian sheep blowfly, *Lucilia cuprina*. Diazinon is widely used in Australia to control sheep blowfly, whose larvae parasitize sheep. Resistance to diazinon was first reported in *L. cuprina* in 1967, about 10 years after introduction of the insecticide (Shanahan 1967), and is largely caused by an allelic substitution at a single locus (Arnold and Whitten 1976; McKenzie et al. 1980).

In 1969–70 Arnold and Whitten showed that, in the absence of insecticide, the R (resistant) allele reduced blowfly fitness (unpublished results reported in McKenzie et al. 1982). Simple single-gene models of the evolution of pesticide resistance (Comins 1977; Taylor and Georghiou 1979; Curtis et al. 1978; Tabashnik and Croft 1982; Gould 1984; May and Dobson 1986) predict that when selection for resistance is relaxed by abandoning use of the pesticide, the cost of resistance will cause the frequency of resistant genotypes to decline over time. However, by 1977–78, field surveys found *L. cuprina* populations to be close to fixation for the R allele, regardless of the recent history of pesticide usage (McKenzie et al. 1980).

McKenzie et al. (1982) provided an explanation for the conflict between theoretical expectations and empirical observations by demonstrating that the relationship between resistance and fitness in the absence of diazinon depended upon the genetic background in which resistance was expressed. McKenzie et al. (1982) inserted the susceptible allele (r) from a susceptible laboratory population into the background genome of a pure-breeding resistant population collected from the field in 1979 by repeatedly back-crossing susceptible individuals into the resistant population. They established population cages with initial frequencies of 25% resistant alleles and observed them over 10 generations to determine whether selection against the resistant allele in the absence of diazinon would cause evolution to a higher frequency of susceptible alleles. Subsequent to reaching Hardy-Weinberg equilibrium, allele frequencies did not change, indicating that the population did not evolve. Because the population was initially genetically variable for resistance, evolutionary stasis indicates that there was no directional selection against resistant genotypes and hence no cost of resistance. This conclusion was supported by a field experiment in which resistant and susceptible individuals did not differ in egg-to-adult viability when raised on live sheep.

In contrast, when the resistant allele was inserted into the susceptible background, the outcome was dramatically different. From an initial allele frequency of 0.5, these populations evolved toward higher frequencies of the susceptible allele. This result indicates that in the absence of diazinon, resistance in these populations was accompanied by fitness costs. Because the experimental populations differed only in the genetic background against which the resistance alleles were tested, the results of these two experiments indicate that the cost of resistance to diazinon depends upon interactions of alleles at the resistance locus with other portions of the genome. McKenzie et al. (1982) concluded from their results that resistance to diazinon in *L. cuprina* was costly when it first arose, but that continued use of the pesticide created the opportunity for mutation and recombination to produce occasional resistant individuals that experienced a lower cost of resistance; i.e., cost of resistance became a genetically variable trait. Natural selection on the cost of resistance then eliminated resistant individuals with high fitness costs, producing a population of resistant flies that experienced no costs of resistance.

Clearly, a similar process could operate on plants with costly resistance to herbivory. For example, suppose that when it first arises, a resistance trait is costly. If the cost does not exceed the fitness benefit conferred by reducing herbivory, the trait will spread in the plant population and the population will equilibrate at an allocation to resistance labelled $\hat{R}$ in figure 17.2A. This point represents the maximum level of fitness along this phenotypic axis, and the peak in the fitness function is maintained by balancing selection due to the costs and benefits of resistance. So long as herbivores occur in the environment, the resistance trait will presumably remain beneficial and the population will occupy the peak. However, the peak may not remain stationary. Mutation, or migration of new alleles into the population, can move the peak. In particular, any new alleles that reduce the cost of resistance without reducing its benefit will move the peak toward increased resistance, and directional selection will cause the population to progress to the new equilibrium (figure 17.2B). Given enough time, this gradual adaptive process might greatly reduce and possibly eliminate costs of resistance.

Gradual adaptation can occur as a result of four distinct mechanisms which

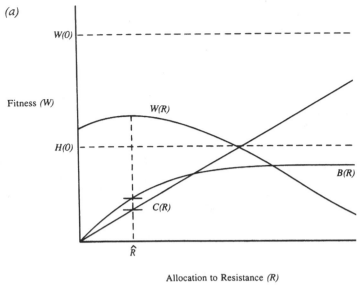

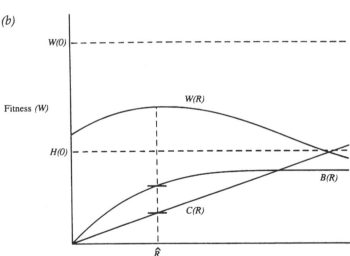

Fig. 17.2 Optimality model demonstrating the consequences for equilibrium levels of resistance of evolutionary reduction in costs of resistance. The fitness of the most susceptible genotype in the absence of herbivory is designated $W(0)$. The amount of fitness lost to herbivory by the most susceptible genotype is designated $H(0)$. The fitness cost of resistance, as a function of the amount of resources allocated to resistance is $C(R)$. $B(R)$ is the fitness benefit of resistance, as a function of allocation. The point $\hat{R}$ corresponds to the genotype with maximum fitness. (a) When resistance initially arises, cost function is steep and the equilibrium allocation to resistance is low. (b) As the cost of resistance evolves to a lower level, the model predicts that maximal fitness will be achieved by allocating more resources to resistance, and $\hat{R}$ will move to the right.

are illustrated in the following hypothetical example. Suppose that all members of a plant population possess a metabolic pathway in which the penultimate product is a chemical that deters herbivory. Further suppose that the final product of this pathway contributes in some other way to plant fitness. Noticeable resistance arises in the population when a mutation at a structural locus disrupts the function of the enzyme that converts the herbivore deterrent into the normal end product. This mutation allows a buildup of the herbivore deterrent, but a cost is incurred because the normal end product is no longer available to perform its function. Four mechanisms may change this cost of resistance. First, at this same structural locus, a new allele may arise which produces an enzyme with reduced activity. A plant with this allele could be resistant at reduced cost because the end product is still produced. Mutations at other loci could also modify the cost of resistance. For example, a new allele may arise at the locus that regulates the enzyme-producing structural gene. If this new allele reduces enzyme production only after signaled by herbivore damage, then resistance will be inducible rather than constitutive. In a plant with a functional structural gene, this mutation reduces the overall cost of resistance because the plant experiences the cost only when it is induced. Furthermore, a mutation at another structural locus might allow the plant to use the final metabolic product of the pathway more efficiently and thereby reduce the fitness cost of the original resistance trait. Obviously, this type of mutation would benefit plants possessing inducible as well as constitutive resistance. Because this mutation and the previously described regulatory mutation modify the negative fitness consequences of the original resistance allele, these mutations are called fitness modifiers (Karlin and McGregor 1974). Finally, a less costly resistance mutation may arise at an unrelated locus. For example, modification of a less critical metabolic pathway may produce an unrelated chemical compound that provides equivalent resistance but with less detriment to fitness.

Lenski (1988a, 1988b) showed that the cost to *Escherichia coli* of resistance to virus T4 can evolve by three mechanisms. In one case (Lenski 1988a), twelve resistance mutations mapped to a single locus in the *E. coli* genome. Although for three of these twelve resistance mutations fitness in the absence of the virus was only 50% that of the T4-susceptible parent strain, other mutations were less costly. For example, one mutation carried only a 35% reduction in fitness.

In the same experiment, three resistance mutations mapped to another locus. These mutant strains experienced only a 15% reduction in fitness (Lenski 1988a). Presumably, in environments with T4 virus, strains resistant at this locus will replace strains with resistance at the locus described above, thereby causing evolution to a lower cost of resistance.

In another experiment, Lenski (1988b) found that fitness modifiers could also reduce the costs of resistance to T4 in *E. coli*. Specifically, a resistant strain that originally experienced a 43% reduction in fitness evolved after 400 gener-

ations to a relative fitness of 0.86. Subsequent examination revealed that mutation at a modifier gene reduced the cost of resistance by 49%.

Of course, the preceding review of evolution of costs of resistance does not
include any examples from the literature on plant resistance to herbivory. There
is nonetheless no obvious reason to believe that the mechanisms described
above should fail to operate for these traits.

Conclusions

The debate over costs of resistance to herbivory should move beyond the basic
question of whether they exist. Clearly, costs do exist in some circumstances.
Equally clearly, they are nonexistent or at least unmeasurable in other circumstances. We should instead refine the question to ask, "Under what conditions
is herbivore-resistance costly?" The theoretical framework described above
suggests some hypotheses that need testing. Firstly, we can predict that resistance is most costly when the relationship between herbivore and plant is new
(assuming that the resistance is specific to this herbivore and had not evolved
earlier in response to previous herbivores) and when the fitness costs of herbivory are severe. Thus, inducible defenses and other modifications that may reduce resistance costs should be associated with older resistance traits. Testing
these hypotheses will require studies of plant-herbivore associations for which
we have extensive historical information. Such information is most likely to be
available for plants or herbivores recently introduced into new habitats, and
research subjects may be found in the context of biocontrol efforts, accidental
escapes, and releases of genetically engineered organisms.

The genetic, biochemical, and physiological mechanisms of resistance are
also likely to determine the initial cost of resistance as well as the extent to
which it can be modified by selection. We may find that some resistance mechanisms have very labile costs whereas others exhibit high costs that cannot be
modified. More information is needed regarding the genetic regulation of resistance traits, their biochemical pathways, and physiological effects.

Costs of resistance traits may also be experienced at different developmental stages than those during which resistance is expressed. For example, most
studies ignore the period during seed maturation and seed germination, yet herbivore resistance may have important pleiotropic effects (either beneficial or
detrimental) in overwintering seeds. We know virtually nothing about the ontogeny of resistance expression.

In addition to our need for empirical work, coevolutionary theory must be
modified to accommodate evolving costs of resistance. How might the eventual
amelioration of resistance costs influence evolutionary equilibrium levels of resistance? What effects will this phenomenon have on predictions of genetic variability? Are there circumstances in which costs might not be expected to
change? Incorporating these issues into more complex models of plant-

herbivore coevolution will provide a better understanding of how plant resistance to herbivores evolves.

Acknowledgments

I am grateful to J. Antonovics, D. Eitzman-Hougen, F. Gould, D. Pilsen, M. D. Rausher, and M. Uyenoyama for conversations that were instrumental in the development of my ideas about costs of resistance. Many of these conversations occurred while I was supported by the NSF (BSR-8507359 to M. D. Rausher and myself). Preparation of this chapter was also supported by the NSF (BSR-9196188) while I was assistant professor in the Department of Biology at Wake Forest University.

18

Toward Community Genetics

Janis Antonovics

To anyone interested in the richness and diversity of the biological world, the concept of coevolution is as seductive as it is tantalizing. If interactions among organisms are indeed a driving force for reciprocal evolution, then given the obvious fact that such interactions are often complex, it is easy and tempting to cast coevolution as a major player in producing mutual adaptations that preserve and generate species diversity.

Such thoughts perforce flutter across the mind of any sensate natural historian, whether in a Costa Rican rain forest or in an English hedgerow, where like the dappled shade in early spring they cannot but stimulate, energize, and increase the sense of wonder. But to a professional biologist, often that selfsame natural historian back from a field trip, the study of coevolution is fraught with conceptual pitfalls and with practical difficulties that are often as prosaic as they are real.

The term "coevolution" has itself been applied to rather diverse phenomena, sometimes referring to processes that are macroevolutionary and recognizable through correlated phylogenies, or sometimes referring to processes that are defined as such only if there is a reciprocal interplay of selection pressures and gene frequencies among the ecological interactants (for discussion see Thompson 1989). For the ecologist interested in species interactions, it is essential to realize that a "species" is composed of a highly heterogeneous class of individuals differing phenotypically because of the varied influences of not only environment, age, or phenology, but also genotype. And genetic variation affecting the processes involved in species interactions will change those processes both in the immediate sense and in terms of future evolutionary change. Over the past 20 years the species as a unit recognized by taxonomists has come increasingly under attack as a valid unit of ecological analysis (Birch 1960; Antonovics 1976a; Harper 1982). At best the Latin binomial and what it represents is seen as an approximation, while at worst it is seen as dangerous and completely misleading. The major impasse, and one which to some extent leads

426

us to persist in a typological characterization of the species, is that at present we have no convenient (one is tempted to say simple-minded) substitute for the species concept. The papers in this volume cite numerous instances where genotypes within a species may behave in a quite contrasting manner in response to pathogens or herbivores. Conversely, different species may be treated as homogeneous by other pathogens or herbivores and should therefore perhaps be treated as "equivalent" genotypes (Janzen 1979). However, the technical difficulties in operationally defining genotypic or guild categories serve to make the prospect of abandoning the species as an ecological unit seem like a dangerous flirtation with prospective anarchy.

Probably the simplest (minimal) framework that encapsulates the essence of these complexities is shown in figure 18.1a. This diagram, originally developed by Levin and Udovic (1977) in the context of a mathematical model of coevolving populations, illustrates the different ways in which two noninterbreeding populations may interact with regard to their gene frequencies and their numerical abundances. For the theoretician, even this simple two-species model requires one to keep track of population size and gene frequency, and if one adds in the component ecological parameters (e.g., birth rates, death rates, disease transmission coefficients), the model can rapidly become overparameterized. Yet such an analysis becomes necessary if one wants to understand the dynamics and possible equilibria of relative species abundances, population sizes, and gene frequencies.

However, even a cursory flight into biological reality tells us that this coevolutionary diagram is greatly oversimplified. For example:

1. Most systems do not involve pairwise interactions but involve multiple interactions, either at one or several trophic levels (see Fritz, this volume).
2. The various interactions presented by the arrows are likely to be highly nonlinear.
3. These interactions are likely to vary with time, and to have implicit in them delays, different generation times, or different response times.
4. The spatial scales of the interactions are also likely to be noncongruent, with hosts and parasites having different-sized genetic and ecological neighborhoods (sensu Antonovics and Levin 1980) and/or quite different metapopulation or interdemic structures.
5. There is likely to be a lot of chance or environmental variation impinging on the processes; not only will this make the results "fuzzy," but the variance itself will be an important parameter in determining differential fitness (Real 1980).

From the viewpoint of the empiricist, all these parameters and complexities could be important, yet minimally must be measured in at least two organisms, often with different life histories, different dispersal ranges, and requiring dif-

ferent rearing techniques. Failure to coordinate just the latter can mean an ig-
nominious end to an otherwise well-conceived project. Given this scenario, it
may seem that any sensible biologist would abandon the study of species interactions, brush coevolution under the rug, and resort to simpler systems, to
model evolutionary paradigms where, for example, toxic mine spoils or pol-
luted cities provide an unresponding backdrop for the evolutionary theater.
However, we often become professional biologists precisely because of our fas-
cination with the intricacies of the biological world, and we would not be scientists if the study of complexity did not have its rewards, its modest inroads, or
even its dreams of major insights.

The biological richness at the heart of species interactions, combined with
the difficulties in analyzing them explicitly at either the empirical or the theoretical level, leads to the question of whether there is indeed not a much larger
discipline which encompasses the discussions regarding the role of genetic variation in influencing species interactions and determining community structure.
During a discussion of this issue a number of years ago with Dr. Jim Collins, of

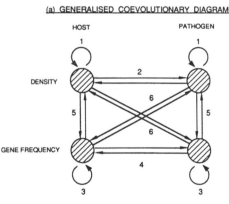

(a) GENERALISED COEVOLUTIONARY DIAGRAM

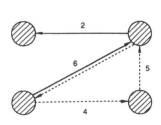

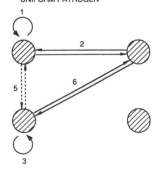

(b) INTERACTIONS OF PARTICULAR INTEREST
TO THE AGRICULTURALIST

(c) INTERACTIONS OF GENETICALLY
VARIABLE HOST WITH GENETICALLY
UNIFORM PATHOGEN

Arizona State University, he suggested that we need a new level of analysis, a new area of biology called "community genetics." This discipline would emphasize the analysis of evolutionary genetic processes that occur among interacting populations in communities. There was also an earlier excursion into this idea by Wilson (1976), who discussed ways in which diffuse coevolution could occur at the level of communities. If molecular, cellular, developmental, and population genetics, why not community genetics? Conceptually, it may not be farfetched to partition the subject of ecological genetics into subdisciplines at the population, community, and perhaps even ecosystem level (Loehle and Pechmann 1988). Above all, the concept of community genetics frees us from the overly restrictive frame of reference, the reciprocality, that coevolutionists would choose for their own discipline. If our research is no longer dependent on returning with a genuine scalp of reciprocal coevolution, we can generalize community processes in terms of interactions that occur among genotypes as individuals, and by extension use our efforts as a vehicle for further understanding when and how taxonomic characterizations should be incorporated into ecological thinking, or when they are misleading.

Within the subject of community genetics, there are two major approaches, one reductionist and the other holistic, to understanding the underlying complexity of species interactions. The approach that links population biology to community interactions is clearly reductionist, and it is this approach that is emphasized in the present volume. It is also this approach that I wish to discuss

Opposite: Fig. 18.1 (*a*) Diagram of the possible types of interactions between densities and gene frequencies for two interacting species (after Levin and Udovic 1977). Arrows: (1) Intraspecific density dependence; (2) effect of density of one population on growth rate of the other; (3) differential fitness among genotypes unrelated to conspecific density, or composition of the other species; (4) density-independent differential fitness caused by the other species; (5) intraspecific density-dependent selection; (6) density of one species affecting relative fitnesses of genotypes in the other species, or gene frequency of one species affecting the density of the other.

(*b*) The subset of interactions that are of primary interest to the agriculturalist. Crop genotype changes pathogen abundance (solid arrow 6), which affects crop yield (arrow 2). Pathogen evolves virulent genotypes (arrow 4), which changes pathogen abundance (arrow 5), which in turn affects crop yield (arrow 2) and leads to further breeding for resistant varieties (dotted arrow 6).

(*c*) The interactions between densities and gene frequencies for a genetically variable host population and a genetically uniform pathogen population. Pathogen population size is influenced by and influences host numerical (arrows 2) and gene frequency dynamics (arrows 6). Density dependence occurs in the host (arrow 1) and there is genetic variation in costs of resistance that would result in genetic change in the absence of the pathogen (arrow 3). However, there may (if costs are in female fecundity) or may not (if costs are in terms of male fecundity) be direct effects of resistance on density regulation (arrows 5).

first, and to elaborate and extend upon using results from our own studies of a natural plant-pathogen system. Later, I will discuss more holistic approaches and suggest that they may be equally important for achieving a full understanding of community genetics.

The Reductionist Approach

By focusing on the plant component in plant-herbivore, plant-pathogen interactions, this book takes a reductionist approach to the study of community genetics. In doing so, it also immediately exposes the enormous commonality between the interests of the agricultural scientist and the population biologist (Harper 1967). The focus of both disciplines has been to document and understand the nature of variation in plant resistance, and to assess the degree to which such variation affects herbivores and pathogens and thus how resistance affects the fitness or yield of the plants themselves. This focus represents a small subset of the coevolutionary diagram (fig. 18.1b). It is immediately obvious from the papers in this volume that plant populations contain an abundance of genetic variation in resistance and susceptibility: direct studies of crop plants, of their wild relatives, and of other natural populations strongly support this contention. It cannot be overemphasized how dramatic these differences can be: within one population, individuals can be totally resistant or seemingly totally susceptible. What is equally clear, however, is that beyond the immediate documentation of such variation in resistance relatively little is known about its origin, stability, and distribution. It is tempting to assume that its origin is mutational or recombinational, that its greatest expression should be in areas of greatest herbivore/pathogen abundance and that it has been amplified and is being maintained by coevolutionary interactions. However, the chapters in this book seriously question this scenario. Agriculturalists often find substantial variation in resistance to a pathogen even in populations that seemingly have never been exposed to the pathogen, at least in the recent past. The lack of a cost to resistance, or even a positive correlation with fitness, if taken at face value, implies a nonequilibrium world with perhaps resistances spreading toward possible fixation. Absence of the occurrence of "losers" in the coevolutionary process is evident from the well-known cases of large-scale effects of introduced alien pests on the abundance of natural or seminatural populations (Gibbs 1978; Cullen et al. 1973), from the success of biological control programs, and from the use of pathogens directly as weed control agents (Wilson 1969; Templeton et al. 1979; Zettler and Freeman 1972). The studies described in the present volume make clear that assumptions about the coevolutionary process are at best hypotheses still in need of critical examination for the many different types of plant herbivore/pathogen interactions that occur in nature.

In spite of the common interests of the agriculturalist and the population biologist in the phenomenon of plant resistance, there are important differences that need to be clarified so the two can continue to interact creatively. Earlier chapters (e.g., Alexander) have mentioned the conflict in any empirical science between achieving generality and achieving precision and realism. In any science, the pure and applied fields have quite different orders of priority in this regard. The pure scientist is primarily interested in generality, in theories that cast a broad sweep of understanding and unification over seemingly diverse aspects of a particular discipline. Excursions into particular systems are necessary to check the general theories and to develop new biological insights, but are to some extent secondary. Applied scientists, on the other hand, are interested in precision and realism. They have to operate within a specific real-world framework to achieve some direct predictability. Excursions into generality are secondary, necessary only to see the interconnectedness of ideas and approaches, so that these can be a stimulus to new answers for solving the particular problem at hand.

Pure and applied sciences also work on quite different time scales. The pure scientist (daily pressures of the job apart) would ideally like to be timeless, to generate truths that are eternal and irrefutable: underlying the scientific enterprise there is a belief that an objective reality exists that can eventually be understood. On the other hand, the applied scientist's goals are driven by urgencies and immediacies: achievement is more often seen and rewarded in terms of a rapid, transient success than by eternal solutions. For example, agricultural scientists want to predict immediately whether the herbivore/pathogen will reduce yield in that selfsame year so management practices can be put into force that minimize yield reduction. This is the reason for the focus on herbivore and pathogen effects expressed largely in terms of disease symptoms. Unfortunately, the concepts of resistance-susceptibility or virulence-avirulence are only tangentially related to demographic parameters of relevance in fitness estimation. Even in the agricultural literature, there needs to be frequent reminder of the fact that disease occurrence may relate only partially to yield decrement; overuse of pesticides and fungicides, in particular, may at times be motivated more by an aesthetic desire for clean crops than by economic considerations. This focus on symptoms rather than on fitness effects has led to discordant methodologies, to a noncorrespondence of concepts, and to differences in assumptions about the evolutionary process in the approaches of population biologists and applied plant pathologists (Antonovics and Alexander 1989).

Perhaps the greatest disjunction that has resulted from the focus on resistance per se has been a failure to appreciate the potential dynamical behaviors of plant-herbivore or plant-pathogen systems. Indeed, there has been a serious lack of explicit studies on how pathogens and herbivores influence the dynamics of plant populations, largely because of the focus on crop populations where numbers are preset at planting. For this reason the impact of plant pathogens and herbivores on plant population dynamics (defined as changes in numerical size of populations over successive generations) has not been a major concern of the agriculturalist, except in cases of large-scale pandemics (Klinkowski 1970). Instead the concern has been with direct yield reduction, and with management of the less controlable component of the system, the pathogen or herbivore. As a result, there has been extensive study of pathogen spread in crop populations and this has provided information on spore dispersal, predictions of crop damage, and models for disease management (Leonard and Fry 1986, vol. 1). The emphasis however has been on epidemiology, on the short-term, within-season dynamics, rather than on long-term processes of the kind that characterize pathogen behavior in natural ecological systems. In order to understand and appreciate this dynamic behavior, it is important to express disease effects in terms of fitness impacts. Indeed, except for the purposes of recognizing pathogen presence or measuring transmission rates and modes, the nature of the disease symptoms is essentially irrelevant to the ecological and evolutionary dynamics.

Of particular concern to plant pathologists has been the genetic basis of disease resistance and susceptibility, or pathogen virulence and avirulence. But again, because these measures of resistance and virulence are difficult to translate into precise fitness effects (Nelson 1979; Antonovics and Alexander 1989), the agricultural literature can be used only tangentially for generalizations regarding natural populations. Moreover, it has been suggested that particular genetic interactions of plants and their pathogens (e.g., the gene-for-gene hypothesis) may be a product of selection in agricultural circumstances, rather than a reflection of patterns that actually exist in nature (Day 1974; Barrett 1985; but see Parleviet and Zadocks 1977). Indeed, many authors have appealed to natural systems as justification for a pathogen control strategy based on genetic diversity (Browning 1974; Harlan 1976), but others have cautioned that factors other than genetic diversity may also minimize disease incidence in nature (Schmidt 1978; Alexander 1988), and that more evidence based on ecological genetic studies is needed (Dinoor and Eshed 1984).

The above reasons make it perhaps understandable, but not less remarkable, that there have been, as far as I know, no explicit models of *numerical* dynamics of plant and pathogen populations, or of the interaction of such dynamics with gene frequency change. Models of gene frequency change in plant-pathogen systems (for review, see Leonard and Czochor 1980) have demonstrated that reasonable, empirically based assumptions can lead to the maintenance of genetic polymorphism in both the pathogen and the host, and have resulted in predictions for use of multiline mixtures for disease control (Leonard 1977; Barrett 1978; Marshall and Weir 1987). Generalized models of parasite-host systems developed by May, Anderson, and coworkers (Anderson and May 1979; May and Anderson 1979) provide the major theoretical paradigms that can be applied to natural plant-pathogen systems. They have been extended only in a limited way to analysis of genetic effects and evolution of sexual systems (May and Anderson 1983). While such models capture the essential dynamical properties of the systems they describe, their application to particular classes of disease requires empirical evaluation and extension to incorporate specific biological processes.

There obviously is also a dynamic in agricultural systems; the evolution of virulent pathogens may require the development of new varieties. But the dynamics will be slower and much less self-generating, being determined more by the pace of plant breeding programs and the rate of deployment of pest management strategies than by the biology itself. Rational deployment of multilines, where the opportunity exists for resetting plant genotype frequencies in each planting season, is probably the only situation in which joint dynamics of host and pathogen become critical in an agricultural context.

Finally it may be worth remembering that for the agriculturalist, the pursuit of intraspecific resistances may soon become a secondary process: given the possibility of interspecific gene transfer, resistances can be moved in from alien sources or engineered from the genes of the pathogen itself. Understanding the mechanics of interspecific variation in resistance (why do mosses not have fungal infections, even though they have no cuticle and live in cool moist habitats?) may become more crucial to the applied enterprise than understanding intraspecific variation.

In summary, the major goal of the agriculturalist is to understand and explore that small subset of the coevolutionary diagram that relates crop genotype, pathogen pressure, and yield. This is because crop genotypes and densities are normally reset at specific values each season and because it is essential to have immediate assessment of disease symptoms so as to invoke appropriate control measures. Longer-term dynamics are more a product of agricultural practices and possibilities than of the intrinsic biology of the host and pathogen. The population biologist, however, has a greater interest in a broader exploration of the coevolutionary diagram so as to include more of the direct feedback loops that are likely to be critical to the long-term numerical and gene frequency dynamics of natural populations. The generalities of the population biologist are in turn of interest to the agriculturalist at two levels: as a guide to strategies of crop deployment, particularly in low-input agricultural systems, and as a guide to gene conservation efforts. It is remarkable that a number of crop deployment and breeding strategies (e.g., multilines, hybrid varieties) are often justified beyond their intrinsic merits on the basis of a mythology of how the natural world should operate; empirical data on how it actually does operate (e.g., the role of genetic variation in ecological success, the extent of inbreeding depression) are often lacking or fragmentary at best.

The natural extension of many of the chapters in this book is therefore to ask how all the individual components fit together to produce an ecological and evolutionary dynamic that can be used to predict distributions, abundances, adaptations, and other evolutionary trajectories. In their discussion of the expected dynamics of the coevolutionary diagram, Levin and Udovic (1977) point out that unless one makes unrealistic simplifying assumptions, the only generality to emerge is that it represents an "anything goes" situation. Populations that can coexist may not do so if there is genetic variation, and vice versa. Approach to gene frequency equilibrium may be accompanied by reduced population size. Or heterozygote deficiency may be associated with gene frequency equilibrium.

Fortunately, however, species interactions fall into various classes, each of which has distinct properties. The dynamical properties of plant-herbivore and plant-pathogen systems have not been explored sufficiently to say which types of interactions will have similar dynamics, but this book does suggest lines across which generalizations could be made. There is obviously the distinction between parasites, parasitoids, grazers, and predators: the ecological dynamics of these contrasting systems are well explored. Another distinction could be between situations where the plant is long-lived relative to the herbivore or pathogen and the reverse situation where the herbivore, for example, a mammalian grazer, is long-lived relative to the plants being grazed. In the former situation the host plants (whether at a phenotypic, genetic, or interspecific level) may be simply acting as a heterogeneous environment, and preexisting concepts and models of selection, habitat choice, and speciation in heterogeneous environments may be directly applicable to these cases (e.g., Maynard Smith 1966; Rausher 1984a). In the case of large grazers, the evolutionary forces acting on the plant community, as pointed out by Pollard (this volume), may have more direct parallels and be better described by models that are based on interspecific frequency–dependent behaviors characteristic of Mullerian (positive frequency–dependent) and Batesian (negative frequency–dependent) mimicry. Other possible distinctions might be systems with and without alternative hosts, whether pathogens are seed transmitted or not, and whether transmission is vector based or the result of wind dispersal.

To illustrate the nature of the gene frequency and numerical dynamics that can result from host-pathogen systems, I will digress into a description of some of the work we have been doing with the anther smut disease commonly found on members of the Caryophyllaceae. We in particular have been studying the anther smut (*Ustilago violacea*) of white campion (*Silene alba*).

Resistance and Susceptibility and the Dynamics of the *Ustilago-Silene* System.

This system was chosen for study because of the clear-cut effects of the pathogen on the host, its technical convenience, and its intriguing biology (Alexander and Antonovics 1988).

Silene alba or white campion (= *Melandrium album* or *Lychnis alba*, Caryophyllaceae) is a common weed of roadsides, old fields, and crops such as peas and alfalfa (McNeill 1977). It is a short-lived perennial that germinates in late summer and early spring. In growth chambers seed to flowering occurs in about six weeks; plants are dioecious and easily crossed. Under short days plants remain vegetative and can be cloned by cuttings.

Both male and female individuals of *Silene* infected by the anther smut fungus *Ustilago violacea* (Basidiomycetes, Ustilaginales) produce stamens with anthers that carry purple fungal spores. Infected females retain a rudimentary ovary but produce no seed. Except for initial stages of the infection process, usually all flowers on a plant are diseased and the individual is completely sterile (Alexander and Antonovics 1988). Otherwise infected plants appear normal, although they usually show increased flower production and a longer flowering period (Alexander 1987).

The spores (teliospores) produced in the anthers are diploid and transmitted by pollinators. Following transfer to another flower, they germinate and undergo meiosis to produce a short basidium of four haploid cells which in turn bud to produce yeastlike cells (sporidia). Fusion of sporidia of opposite mating types produces a binucleate infection hypha that penetrates the host tissue. Completely diseased plants have never been observed to recover and produce healthy flowers (Alexander and Antonovics 1988). There is no seed transmission, so newly established plants are disease-free.

The sporidial haploid stage can be maintained in liquid or solid agar culture using techniques similar to those for yeast (Cummins and Day 1977). Plants can be artificially infected at high frequencies by wounding rosettes with sporidial suspensions of mixed mating type or by soaking seedlings in such suspensions.

Our study populations are in the vicinity of Mountain Lake Biological Station, Pembroke, in southwestern Virginia. In this region, the pathogen is restricted to one host species, even though in Europe it is found on a wide array of hosts. In Virginia and other parts of the Southeast, we have found a similar anther smut infection in the fire pink, *Silene virginica*, but electrophoretic studies show it to be distinct from the anther smut on *S. alba* (Stratton 1990, personal communication).

In classical host-parasite models, the probability of an individual becoming diseased increases with the density of diseased individuals in the population. This assumes that spores are dispersed freely into the air, or that there is random encounter among mobile infected and uninfected individuals (as in animals). With pollinator-transmitted diseases, a different transmission process may be at play. It is known that pollinators increase flight distances when plants are more widely spaced (Levin 1972; Levin and Kerster 1974). As a result, given perfect "adjustment" of pollinator flight distances, disease transmission will be independent of density and dependent only on frequency of diseased individuals. A plant will receive spores if a pollinator visited a diseased plant on a prior visit, regardless of the absolute density of individuals within the population. Empirical evidence for this assumption comes from field studies (Alexander 1990b) and experiments where we independently varied the density and frequency of diseased and healthy individuals in experimental arrays (Antonovics, unpublished data). Clearly, one would not expect this assumption to hold precisely, given spore carry-over or variations in pollinator behavior. However, the frequency-dependent nature of the disease transmission process is expected in any venereal disease where mate encounter rates are relatively independent of density, or in any vector-transmitted disease where vectors actively search out hosts and have a motility that exceeds that of the host. This anther smut disease is therefore a venereal disease not only by virtue of its being transmitted by pollinators, but also because of its transmission dynamics. Models of this disease therefore have broader relevance to venereal diseases in general, particularly those like gonorrhea, syphilis, and chlamydia that greatly reduce the fertility of their human hosts. In plants, anther smut diseases infect a large range of hosts in the Caryophyllaceae (Goldschmidt 1928) and related families such as the Portulacaceae (D. Ford, 1989, personal communication). Other pollinator-transmitted diseases are caused by fungi (Leach 1940; Broadbent 1960), by bacteria (Schroth et al. 1974), and by viruses (George and Davidson 1963; Cooper, Kelley et al. 1988).

Models of frequency-dependent transmission show that the conditions for host-pathogen coexistence are far more restrictive than in the case of "normal" diseases where transmission rates depend on density (Getz and Pickering 1983). Analysis of the *Silene-Ustilago* system is simplified because the possible state transitions are far fewer than in other host-parasite systems (fig. 18.2). We assume that individuals are either diseased or disease-free, and that diseased plants are completely sterile and never recover.

If we examine the purely numerical dynamics of this system (i.e., ignore

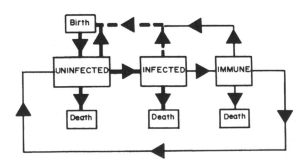

Fig. 18.2 Diagrammatic flow chart for state transitions in a directly transmitted disease (after Anderson and May 1979). The solid lines indicate state transitions considered in our theoretical models. Plants when first diseased may produce some healthy flowers, but this condition is transitory and rare: therefore this transition is indicated as a dotted line.

any genetic variation), then frequency-dependent transmission leads either to the population purging itself of the disease, or extinction of both disease and host (fig. 18.3). Moreover, unlike in the classical density-dependent transmission disease spread models (Kermack and McKendrik 1927), there is no threshold density for initial increase of the disease and therefore no critical density before epidemics occur. Initial disease spread will occur whenever the disease transmission coefficient, β, is greater than the death rate, d, of diseased individuals, regardless of density. However, if healthy individuals establish even more rapidly than the rate of disease spread, the disease becomes proportionately rarer in the population.

This simple model therefore predicts that the host and the pathogen should not stably coexist, and that the disease cannot regulate the plant population. One has either to invoke a nonequilibrium scenario, or to consider biologically realistic extensions of the model that may change these initial conclusions. The most obvious extension is to assume that the plant population is regulated by density-dependent factors that act independently of the disease itself (e.g., resource availability, safe sites for germination). Under these conditions (figs. 18.4 and 18.5), depending on the parameter values, equilibrium coexistence becomes possible (see also Alexander and Antonovics 1988 for an application of a similar model to a real population). In general, coexistence will occur if the density-dependent forces regulating the population act differentially on healthy and diseased individuals. This is very likely because density effects in plants, especially perennials, usually act most strongly at the seedling stage (Harper 1977; Shaw 1986). Yet the disease is transmitted largely at the adult flowering phase (but some seedling infection is known to occur if spores fall on young plants: Baker 1947; Antonovics and Alexander 1989; Alexander 1990b). Therefore the establishment of the healthy class is more likely to be influenced by density; diseased individuals can die only as adults and may be relatively insensitive to density. Again, this discordance in density-dependent effects on diseased and healthy plants may be generalizable to most venereal diseases, where almost by definition transmission is at an adult phase. We can therefore encapsulate the dynamics of such a system by assuming that density dependence acts solely on the birth rate of the healthy individuals, and not on the death rates of either the healthy or diseased. The result (table 18.1) is that as long as β is less than the birth rate at low density, b_0, and is greater than d, then coexistence of host and pathogen are possible.

While this may seem satisfying in that it provides an explanation of host-pathogen coexistence, it is important to calibrate this interpretation against reality. We currently have insufficient demographic data or data on transmission rates to assess whether real world populations fall in the "coexistence" region of parameter space or not. In one population that was studied intensively, coexistence was in fact not predicted (Alexander and Antonovics 1988). Moreover, the genetics of this host-pathogen system suggest that the model may be

DENSITY DEPENDENT
DISEASE TRANSMISSION

$$\frac{dX}{dt} = rX - \beta XY$$

$$\frac{dY}{dt} = \beta XY - dY$$

FREQUENCY DEPENDENT
DISEASE TRANSMISSION
(POLLINATOR)

$$\frac{dX}{dt} = rX - \beta \frac{XY}{N}$$

$$\frac{dY}{dt} = \beta \frac{XY}{N} - dY$$

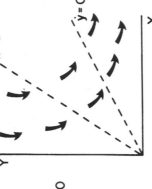

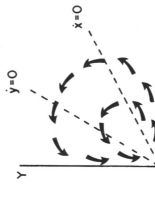

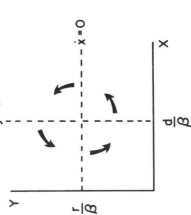

Fig. 18.3 Phase diagrams showing changes in numerical abundance of a host (X) and pathogen (Y) given density- or frequency-dependent disease transmission. Zero growth isoclines of host and pathogen are shown as dotted lines. $r =$ birth-death rate of host, $\beta =$ disease transmission rate, $d =$ death rate of host, $N = X + Y$. Model is of *Silene-Ustilago* system illustrated in fig. 18.2.

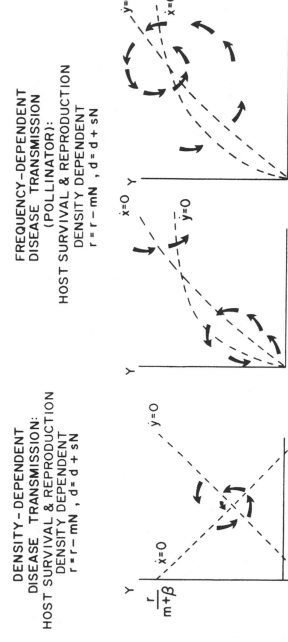

Fig. 18.4 Phase diagrams showing changes in numerical abundance of a host (*X*) and pathogen (*Y*) given density- or frequency-dependent disease transmission, but with linear density dependence of *r* and *d* on total plant density, *N*. Symbols as in fig. 18.2, with *s* and *m* = constants. Model is of *Silene-Ustilago* system illustrated in Fig. 18.3. Zero-growth isoclines for the frequency-dependent case are implicit functions in *X* and *Y*, and have been drawn only approximately.

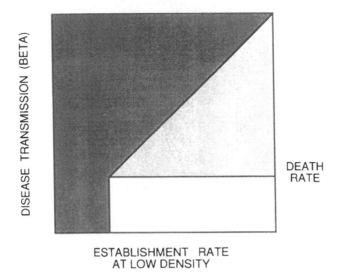

Fig. 18.5 Conditions for host-pathogen coexistence assuming density dependence of establishment rate of host. Dark shading = region of host and pathogen extinction. Light shading = region of coexistence. Unshaded = region of pathogen elimination. The model is as in the frequency-dependent case of fig. 18.4 except death rate d = constant, and $r = b/(kN + 1)$ where b = establishment rate at low (zero) density, and k = constant determining intensity of density-dependent recruitment into the healthy class.

oversimplified. Using an experimental population in which genotypes were replicated by cloning, Alexander (1989) showed that genotypes sampled as seeds from one diseased population have large extremes of resistance. The level of resistance is highly heritable, although the precise genetic basis is still under investigation (Alexander 1990a). Her studies showed that resistance in males, but curiously not in females, is genetically correlated with lower flower production and with later flowering. This suggests that there are costs to resistance if we assume that fewer flowers and later flowering result in lower reproductive success. This result is a very plausible scenario for any venereal disease: reduced mating will lead to a decreased chance of acquiring the disease but correspondingly will reduce reproductive success of those individuals if they do remain healthy.

This experimental population also contained diseased individuals produced by artificial inoculation with six fungal isolates from the same population. Initial attempts to follow the fate of these isolates and to examine whether there is

TABLE 18.1
**Components of a coevolutionary model of the *Silene-Ustilago* system that incorporates both
plant and fungal genetic variation**

Genotypes

X_1 Host genotype 1
X_2 Host genotype 2
Y_{11} Host genotype 1, fungal genotype 1
Y_{12} Host genotype 1, fungal genotype 2
Y_{21} Host genotype 2, fungal genotype 1
Y_{22} Host genotype 2, fungal genotype 2

Transmission coefficients

	Host 1		Host 2	
	Fungus 1	Fungus 2	Fungus 1	Fungus 2
Host 1	$\beta 1.11$	$\beta 1.12$	$\beta 1.21$	$\beta 1.22$
Host 2	$\beta 2.11$	$\beta 2.12$	$\beta 2.21$	$\beta 2.22$

Recursion Equations

$$X_1 = X_1 [1 + r_1 - \frac{1}{N}(\beta 1.11 Y_{11} + \beta 1.12 Y_{12} + \beta 1.21 Y_{21} + \beta 1.22 Y_{22})]$$

$$Y_{11} = \frac{X_1}{N}(\beta 1.11\ Y_{11} + \beta 1.21\ Y_{21}) - d_{11}\ Y_{11}$$

Note: X_i represents numbers of healthy individuals of genotype i, Y_{ij} represents numbers of ith genotype of host diseased with jth fungal genotype. *Beta h.ij* represents the transmission rate of the fungus to the hth healthy genotype from the ith host genotype diseased by the jth pathogen. Recursions are shown only for X_1 and Y_{11}, but are analogous for other healthy and diseased genotypes.

genetic variation in fungal virulence have not been successful because of the low level of electrophoretic variation in the population. Essentially all the populations in this region of Virginia show very little among- or within-population variation for a large number of electrophoretic markers (Stratton 1990, personal communication). We therefore at present do not know the degree of genetic variation in virulence of the pathogen; it is plausible that during the introduction of the pathogen into the United States the populations went through a severe bottleneck effect and now show relatively little variation not only electrophoretically but also with regard to genes for virulence.

The models developed for numerical dynamics could be extended to incorporate genetic variation in resistance and susceptibility and therefore to reflect the entirety of the coevolutionary diagram presented at the outset of this paper. The problem is that such a formulation is heavily overparameterized: the minimal assumption of only two host and two pathogen genotypes results in six types of individuals and eight different transmission coefficients describing disease spread from diseased to healthy individuals (table 18.1).

In our analyses to date we have made the simplifying assumption that there is no genetic variation in the fungus and that diseased plant genotypes do not show differential transmission (e.g., as a result of different pollinator visitation rates). In this way we are essentially examining the dynamics of the system assuming that the only genetic variation is in the resistance/susceptibility of the host. The model therefore still represents only half of the coevolutionary diagram (fig. 18.1C). We also assume that there are only two genotypes of the host (i.e., that the individuals are haploid); this is justified by the fact that the results of single-locus haploid models approximate those of diploid models if heterozygotes are intermediate.

To include some of the essential biological features, we have included plant genotypes with extremes of resistance and susceptibility, yet with costs to these resistances that reflect the results of the experimental studies described earlier (Alexander 1989). Thus in the simulations presented here, we have assumed widely divergent resistances, expressed by transmission coefficients of 1.0 (low resistance) and 0.1 (high resistance) for the two host genotypes, and costs that reflect the approximately twofold difference in flower production between the most resistant and the most sensitive genotypes in the experimental studies. In combination with the assumed birth and death rates, populations monomorphic for each of these genotypes would result in either population extinction for the low-resistance genotype, or failure of the disease to establish or persist for the high-resistance genotype. The results (fig. 18.6) show however that in polymorphic populations both genotypes will coexist, with an oscillatory approach to equilibrium. This shows that extremes of resistance and susceptibility can be maintained within one population and that given such variation the regions for host-pathogen coexistence are increased substantially. All our models so far have been deterministic: given the large fluctuations in population size, pathogen extinction in nature may occur owing to effects in small populations.

The costs of resistance however occur only in males, and we can modify the mating scheme in the model to reflect this. We assume that female success is unaltered (i.e., seed production of resistant and susceptible plants does not differ), but that susceptible genotypes are much more successful as males. This introduces frequency-dependent host "fitnesses" into the model: high-fecundity males are at a disproportionate advantage when they are rare, but at a disadvantage when they are common. But when they are common, disease spread will be more rapid. The resulting population dynamics show even more severe oscillations, an apparent stable limit cycle, and maintenance of polymorphism in resistance (fig. 18.7). It is important to note that if we were measuring costs solely in terms of seed production of females, then the costs would go unnoticed and we would be at a loss to explain the dynamics.

These results therefore provide insight into the ecological genetics, not only of this particular disease, but into venereal diseases in general. We can draw several important conclusions.

CHANGE IN NUMBERS OVER TIME

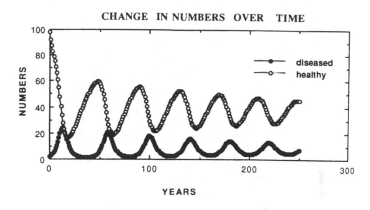

FREQUENCY OF SUSCEPTIBLE GENOTYPES

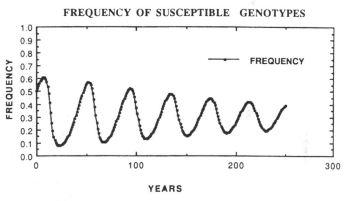

Fig. 18.6 Numerical and gene frequency dynamics of the *Silene-Ustilago* system assuming resistance costs are in terms of female fecundity. Model is as described in figure 18.5 and table 18.1, assuming density-dependent establishment and no genetic variation in the pathogen. Susceptible genotype: *beta* = 1.0, *b* = 0.9. Resistant genotype: *beta* = 0.1, *b* = 0.6; *k* = 0.02, *d* = 0.3.

1. Venereal disease dynamics can be stabilized by differential density effects on diseased and healthy plants.
2. Such effects are likely in the case of venereal diseases because their transmission is at the adult stage, so that the diseased class is likely to experience less density-dependent mortality than the healthy class which passes through a juvenile stage.
3. Resistance to venereal diseases is likely to have a direct cost whenever such greater resistance is achieved through reduced mating.
4. Variation in resistance is likely to be greater in the sex with the greatest variation in mating success; this will usually be males.

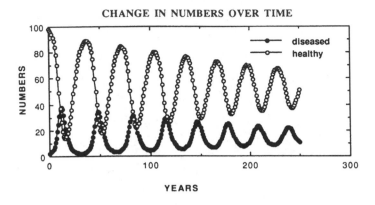

CHANGE IN NUMBERS OVER TIME

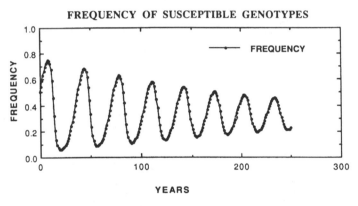

FREQUENCY OF SUSCEPTIBLE GENOTYPES

Fig. 18.7 Numerical and gene frequency dynamics of the *Silene-Ustilago* system assuming resistance costs are in terms of male fecundity. Model is as in fig. 18.6, except that relative male fitnesses are 1 and 0.667 for the susceptible and resistant genotypes. Recursions are modified to include differential male mating success.

5. These costs can permit the maintenance of large extremes of resistance and susceptibility in one population, and extend the regions for coexistence of host and pathogen.

6. Genetically variable hosts can result from the impact of genetically uniform pathogens. One does not necessarily need to invoke a coevolutionary "arms-race" to explain host-genetic variation (see also Parker, this volume).

I have focused on our study to illustrate how knowledge of plant resistance can be used to develop dynamic models that increase our understanding of the nuts and bolts of species interactions. Things that were extremely puzzling, such as the presence of highly susceptible and highly resistant individuals in the same population, no longer seem so. Conversely, such interactions immediately can be seen to have important repercussions for more general biological phenomena. For example, venereal diseases could be an important force in maintaining genetic variance in male mating success, and therefore in promoting continuing sexual selection.

The analysis of species interactions in terms of the population ecology and population genetics of the interacting components will be possible only in a limited number of cases. Even in the *Silene-Ustilago* system, it is difficult to envisage the behavior of the system with the added complexity of fungal variation in virulence. It is clear that while such reductionist approaches provide an important methodological bridge from population biology to community genetics, there may be other approaches, involving quite different types of theory and empirical goals, that also contribute to our general understanding of how genetics influences community structure.

The Holistic Approach

The reductionist is always the snob in biology (and probably in other sciences), putting his or her holistic counterpart on the defensive by proclaiming to have a more precise analysis, greater technical expertise, and therefore a surer handle on causality. Much of this snobbery is justified whenever the holist takes refuge in unsupported generalities. Harper (1982) has criticized the holistic approach to ecology because it is didactically unsound, methodologically weak (often being based solely on description), and motivated more often by emotional and intuitive feelings about nature than by evidence. For example, he describes the teaching of introductory ecology, where the first lab is a field trip to a woodland, as being equivalent to starting a chemistry class by showing them the structure of DNA. And he assails the blind belief that communities are "integrated and harmonious" because it leads to an answer that is always "safe and ignorant." However were it not for some virtues in holistic approaches, we would still all be paralyzed by physics envy, searching for individual quarks and electrons in a futile quest for biological understanding rather than discovering biologically based laws and generalities. What is critical is that higher levels of analysis have rigorously formulated questions, and that these be empirically tractable so that holistic approaches do not become a refuge for crypticism, mysticism, and evasiveness. It is also essential that holistic approaches do not violate ideas and generalizations attained by more reductionist levels. In technical terms (Rosenberg 1978) laws at higher levels should "supervene on" those at lower levels, but otherwise they can have a valid autonomy of their own.

Within the area of community genetics, there are numerous questions that do not require a case-by-case, species-by-species, year-by-year approach. Perhaps the most important of these is, What is the relationship between genetic diversity and species diversity? It is a question that can be asked within trophic levels, among trophic levels, or within particular guilds or subsets of communities. For example, we simply do not know whether the abundant demonstration of ecotypic differentiation in temperate regions is a function of the greater abundance of scientists there (the Swedish tradition!) or whether intraspecific genetic variation in some way compensates for the low level of species diversity within these regions. The corresponding prediction might be a lower level of genetic diversity within species from the tropics. Alternatively, given that genetic and ecological diversity can be maintained by similar forces (Antonovics 1976a), one might argue that there should be a positive correlation between genetic diversity and species diversity of a community. I know of very few attempts to examine this question even descriptively (an exception is presented in figure 18.8).

Obviously, this larger question of the relationship between species diversity and genetic diversity is overgeneralized and should be broken down by considering subsets of genetic diversity and subsets of community components. For example, we can ask how the diversity of resistance genes relates to the presence or absence of particular pathogens. While numerous studies bearing on this issue have been done, the answers remain equivocal largely because information addressing his issue has often been obtained indirectly as part of studies with goals other than direct assessment of community genetic structure.

Perhaps it is feasible to address other, less grandiose questions. For example, there is currently an intense debate raging about the possible role of pests and pathogens in the evolutionary maintenance and (perhaps) origin of outbreeding and sexual reproduction (Hamilton 1980). And greater outbreeding and less parthenogenesis (hence a greater potential for generating genetic variation) have been correlated with greater pathogen and pest pressure (Levin 1975; Glesener and Tilman 1978). Such observations are certainly consistent with many micromodels of biotic interactions generating and preserving variation by negative-frequency dependent selection or genetic feedback.

It is important to emphasize that holistic approaches do not preclude exper-

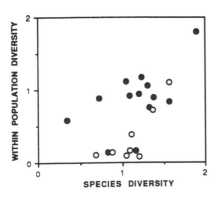

Fig. 18.8 Relationship between seed type diversity and weed species diversity among different populations of *Oryza sativa* (open circles) and *O. glaberrima* (closed circles) from rice fields in Africa. Seed types were identified on the basis of five polymorphic traits, and weed diversity was measured on the estimated biomass of 26 species. Diversity, *H*, is calculated using the Shannon-Weaver diversity index. From Morishima and Oka 1979.

imental analysis, nor is it intrinsically easier or harder to carry out experiments at different levels of analysis. Certainly there has been the lack of an experimental tradition in much of plant community ecology, but this is rapidly changing. A vast array of experiments are possible. The role of genetic variance in species abundance can be addressed by manipulating genetic variance of species in experimental communities. For example, biological control measures have generated many natural experiments, and these have shown that more outcrossing, presumably more genetically variable, species are more recalcitrant to biological control measures (Burdon and Marshall 1981a). And we have artificially manipulated the genetic variance among arrays of individuals of *Anthoxanthum odoratum* in a seminatural grassland and shown that genetic variation is a critical component of demographic success (Antonovics and Ellstrand 1984; Ellstrand and Antonovics 1985; Antonovics and Ellstrand 1985; Schmitt and Antonovics 1986; Kelley et al. 1988). In these studies, circumstantial evidence strongly implicated the role of pests or pathogens.

A major impediment to the empirical understanding of coevolutionary systems is the recognition of genetic variants. This problem occurs at two levels. The first, and easiest to resolve in the future, is the ability to follow the fate of particular individuals and their progeny in natural and experimental populations. This requires the development of easy and reliable techniques for assessing genetic identity and statistical methods for inferring parentage (Meagher 1986). The need is to develop better and better DNAometers, whether through the use of electrophoretic variants, restriction-fragment-length polymorphisms, or eventually more direct analysis of sequence information. The lack of well-developed procedures in this regard is one of the major hurdles in the development of community genetics. The second, more difficult, problem is identifying the genetic basis of particular phenotypic variants. If the genotypes underlying these variants can be identified (e.g., identification of resistance genes by RFLP markers), then direct assessment of genotypic change is far easier. At present this hardly seems feasible for most systems, but may be an important source of interaction between the agricultural scientist and the field biologist: time and effort devoted to identifying specific loci may be worth the effort from an applied viewpoint and, given the availability of appropriate probes, these same loci could be studied in related wild populations. As it is, resolving the genetic versus environmental basis for intraspecific variation will continue to require classical experimental methods involving transplants and crosses, which are time-consuming and often remove populations from their natural context.

The other major problem confronting the discipline, and one that I hope became apparent in the previous section, is that given a set of interacting factors and processes, it is almost impossible to understand their joint dynamics by superficial intuition or by casual observation. It is necessary to use theoretical (computer or analytical) models to predict the consequences of multiple nonlinear interactions. It is necessary to do experiments to isolate the nature of the particular interactions that might be involved. And finally it is necessary to calibrate the predicted outcomes against long-term studies of relative abundances of the interacting components. For no plant-based system (except perhaps in a few cases of biological control and major species introductions) do we have even the crudest long-term data on disease and host abundance.

All these problems are confounded by a too ready acceptance of ecology and evolution as sciences that can be done on the cheap. A thorough analysis of ecological and evolutionary dynamics is an extremely expensive enterprise, requiring laboratories to identify genetic markers and to rear organisms; requiring garden and growth chamber facilities for crosses and experiments; requiring computer facilities for modeling; and requiring time and personnel to handle the extensive, thorough analysis of field processes. Just as a molecular biology lab should perhaps not be without its high-speed centrifuge, its DNA synthesizer, and its attendant culture collections, so a community genetics lab should not be without is gene-detection equipment, its growth chambers, its computers, and its accessible field sites. Perhaps practicing ecologists realize these issues, but they need to be far more impatient and insistent in imparting the lesson to administrators and politicians who hold the purse strings, and who are only too willing to imagine that all an ecologist needs is a shovel, boots, and binoculars as if, by analogy, a modern molecular biologist could solve major issues in that science using a pestle and mortar, a bit of toluene, and some chromatography paper.

Conclusion

There is enormous richness and opportunity in the discipline of community genetics, the study of the genetics of species interactions and their ecological and evolutionary consequences. This opportunity presents itself at levels ranging from issues that form the focus of the present volume, namely understanding intraspecific variation in plant resistance, to well beyond. It extends to a detailed analysis of the numerical and genetic dynamics of particular systems, as outlined here for the *Silene-Ustilago* systems, and to answering quite general questions about the relationship between trophic structure and genetic structure. The reductionist aspects of the discipline have goals and methodologies almost identical to those of the agriculturalist concerned with pest-induced yield reduction. The holistic aspects also have applied repercussions: knowledge of the relationship between species diversity, abundance, and genetic diversity will provide a backdrop against which crop deployment, gene conservation, and species conservation strategies can be rationally developed. Both reductionist and holistic components will demand a combination of theoretical, experimental, and observational analysis to resolve the processes in such richly interacting systems, and to replace what is often now only a mythology by real, tempered understanding. Much of the development of the field will depend on advances in genetic analysis and monitoring, on improved DNAometers. In general, it will be a far more expensive enterprise than either pure ecology or pure genetics. The obvious but impressive message of this volume is that genetic variation at the intraspecific level can be critically important to ecological processes. But equally this volume impresses upon us the fact that our knowledge is at best fragmentary. We need to know far more about the properties of intraspecific variation in resistance, the degree to which such resistance variation is related to the presence of herbivores and pathogens, the kinds of ecological and evolutionary dynamics that this generates (or is a product of), the degree to which this affects species distributions and abundances, and the degree to which this has consequences for major evolutionary processes such as the evolution of breeding systems and speciation. These issues will form the agenda for community genetics well into the next century.

Acknowledgments

The studies described here were carried out in collaboration with Helen Alexander, Don Stratton, David Hall, and Peter Thrall. I wish to thank them for their insight and input.

References

Abbott, L. K., and A. D. Robson. 1984. The effect of VA mycorrhizae on plant growth. In C. L. Powell and D. J. Bagyaraj, eds., VA mycorrhiza, 113–130. CRC Press, Boca Raton, FL. [17]

Abrahamson, W. G. 1975. Reproductive strategies in dewberries. Ecology 56:721–726. [10]

Abrahamson, W. G., S. S. Anderson, and K. D. McCrea. 1988. Effects of manipulation of plant carbon nutrient balance on tall goldenrod resistance to a gallmaking herbivore. Oecologia 77:302–306. [13]

Abrahamson, W. G., J. F. Sattler, K. D. McCrea, and A. E. Weis. 1989. Variation in selection pressures on the goldenrod gall fly and the competitive interactions of its natural enemies. Oecologia 79:15–22. [7, 13]

Abramsky, Z. 1980. Experiments on seed predation by rodents and ants in the Israeli desert. Oecologia 57:328–332. [13]

Adams, M. W., A. H. Ellingboe, and E. C. Rossman. 1971. Biological uniformity and disease epidemics. BioScience 21:1067–1070. [15]

Adkisson, P. A., and V. A. Dyck. 1980. Resistant varieties in pest management systems. In F. G. Maxwell and P. R. Jennings, eds., Breeding plants resistant to insects, 233–251. John Wiley and Sons, New York. [12]

Ågren, J. 1987. Intersexual difference in phenology and damage by herbivores and pathogens in dioecious *Rubus chamaemorus* L. Oecologia 72:161–169. [13, 16]

———. 1988. Between-year variation in flowering and fruit set in frost-prone and frost-sheltered populations of dioecious *Rubus chamaemorus*. Oecologia 76:175–183. [13]

Agrios, G. N. 1978. Plant pathology. Academic Press, Orlando, FL. [15]

Ahmad, S., S. Govindarajan, J. M. Johnson-Cicalese, and C. R. Funk. 1987. Association of a fungal endophyte in perennial ryegrass with antibiosis to larvae of the southern armyworm, *Spodoptera erdania*. Entomologia Experimentalis et Applicata 45:287–294. [2]

Ahmad, S., J. M. Johnson-Cicalese, W. K. Dickson, and C. R. Funk. 1986. Endophyte-enhanced resistance in perennial ryegrass to the bluegrass billbug, *Spenophorus parvulus*. Entomologia Experimentalis et Applicata 41:3–10. [2]

Åhman, I. 1984. Oviposition and larval performance of *Rabdophaga terminalis* on *Salix* spp. with special consideration to bud size of host plants. Entomologia Experimentalis et Applicata 35:129–136. [13]

Aide, T. M. 1988. Herbivory as a selective agent on the timing of leaf production in a tropical understory community. Nature 336:574–575. [13]

Ainsworth, G. C. 1981. Introduction to the history of plant pathology. Cambridge University Press, Cambridge. [8]

Alderfer, R. G., and C. F. Eagles. 1976. The effect of partial defoliation on the growth and photosynthetic efficiency of bean leaves. Botanical Gazette 137:351–355. [13]

Alexander, H. M. 1982. Demography of and intraspecific variation in *Plantago lanceolata* in relation to infection by the fungus *Fusarium moniliforme* var. *subglutinans*. Ph.D. Diss., Duke University. [14]

––––––. 1984. Spatial patterns of disease induced by *Fusarium moniliforme* var. *subglutinans* in a population of *Plantago lanceolata*. Oecologia 62:141–143. [14]

––––––. 1987. Pollinator limitation in a population of *Silene alba* infected by the anthersmut fungus *Ustilago violacea*. Journal of Ecology 75:771–780. [18]

––––––. 1988. Spatial heterogeneity and disease in natural populations. In J. Jaeger, ed., Spatial components of epidemics, 144–164. Prentice-Hall, Englewood Cliffs, N.J. [14, 15, 18]

––––––. 1989. An experimental field study of anther-smut disease of *Silene alba* caused by *Ustilago violacea:* Genotypic variation and disease incidence. Evolution 43:835–847. [14, 17]

––––––. 1990a. Dynamics of plant/pathogen interactions in natural plant communities. In J. J. Burdon and S. R. Leather, eds., Pests, pathogens and plant communities, 31–45. Blackwell Scientific Publications, Oxford. [18]

––––––. 1990b. Epidemiology of anther-smut infection of *Silene alba* caused by *Ustilago violacea:* Patterns of spore deposition and disease incidence. Journal of Ecology 78:166–179. [18]

Alexander, H. M., and J. Antonovics. 1988. Disease spread and population dynamics of anther-smut infection of *Silene alba* caused by the fungus *Ustilago violacea*. Journal of Ecology 76:91–104. [14, 18]

Alexander, H. M., J. Antonovics, and M. D. Rausher. 1984. Relationship of phenotypic and genetic variation in *Plantago lanceolata* to disease caused by *Fusarium moniliforme* var. *subglutinans*. Oecologia 65:89–93. [14]

Alexander, H. M., and J. J. Burdon. 1984. The effect of disease induced by *Albugo candida* (white rust) and *Peronospora parasitica* (downy mildew) on the survival and reproduction of *Capsella bursa-pastoris* (shepherd's purse). Oecologia 64:314–318. [14]

Alexopolous, C. J., and C. W. Mims. 1979. Introductory mycology. 3d ed. John Wiley and Sons, New York. [8]

Alfaro, R. I. 1982. Tree mortality and radial growth losses by the western spruce budworm in a Douglas fir stand in British Columbia. Canadian Journal of Forest Research 12:780–787. [13]

Al-kherb, S. M., A. P. Roelfs, and J. V. Groth. 1987. Diversity for virulence in asexually reproducing population of *Puccinia coronata*. Canadian Journal of Botany 65:994–998. [8]

Allard, R. W. 1975. The mating system and microevolution. Genetics 79:115–126. [15]

Allard, R. W., G. R. Babbel, M. T. Clegg, and A. L. Kahler. 1972. Evidence for coadaptation in *Avena barbata*. Proceedings of the National Academy of Sciences USA 69:3043–3048. [15]

Alstad, D. N., and G. F. Edmunds, Jr. 1983. Adaptation, host specificity, and gene flow in the Black Pineleaf Scale. In R. F. Denno and M. S. McClure, eds., Variable plants and herbivores in natural and managed systems, 413–426. Academic Press, New York. [9, 13]

Altieri, M. A., and M. Z. Liebman. 1986. Insect, weed, and plant disease management in multiple cropping systems. In C. A. Francis, ed., Multiple cropping systems, 183–218. Macmillan, New York. [5]

Altieri, M. A., and L. L. Schmidt. 1987. Mixing broccoli cultivars reduces cabbage aphid numbers. California Agriculture 41:24–26. [5, 9]

Anderson, A. N. 1987. Effects of seed predation by ants on seedling densities at a woodland site in SE Australia. Oikos 48:171–174. [13]

———. 1988. Insect seed predators may cause far greater losses than they appear to. Oikos 52:337–340. [13]

———. 1989. Pre-dispersal seed losses to insects in species of *Leptospermum* (Myrtaceae). Australian Journal of Ecology 14:13–18. [13]

Anderson, D. C. 1987. Belowground herbivory in natural communities: A review emphasizing fossorial animals. Quarterly Review of Biology 62:261–286. [13]

Anderson, R. M., and R. M. May. 1979. Population biology of infectious diseases. I. Nature 280:361–367. [18]

———. 1982. Coevolution of hosts and parasites. Parasitology 85:411–426. [15]

Anderson, S. S., K. D. McCrea, W. G. Abrahamson, and L. M. Hartzel. 1989. Host genotype choice by the ball gallmaker *Eurosta solidaginis* (Diptera: Tephritidae). Ecology 70:1048–1054. [11, 13]

Andeweg, J. M., and J. W. DeBruyn. 1959. Breeding of non-bitter cucumbers. Euphytica 8:13–20. [2]

Angseeing, J. P. A. 1974. Selective eating of the acyanogenic form of *Trifolium repens*. Heredity 32:73–83. [10]

Angseeing, J. P. A., and W. J. Angseeing. 1973. Field observations on the cyanogenesis polymorphism in *Trifolium repens*. Heredity 31:276–282. [10]

Annis, B., and L. E. O'Keefe. 1987. Influence of pea genotype on parasitization of the pea weevil, *Bruchus pisorum* (Coleoptera: Bruchidae) by *Eupteromalus leguminis* (Hymenoptera: Pteromalidae). Environmental Entomology 16:653–655. [12]

Antonovics, J. 1976a. The input from population genetics: "The new ecological genetics." Systematic Botany 1:233–245. [18]

———. 1976b. The nature of limits to natural selection. Annals to the Missouri Botanical Garden 63:224–247. [11]

Antonovics, J., and H. M. Alexander. 1989. The concept of fitness in plant fungal pathogen systems. In K. J. Leonard and W. E. Fry, eds., Plant disease epidemiology. Vol. 2, Genetics, Resistance, and Management, 185–214. McGraw-Hill, New York. [5, 14, 18]

Antonovics, J., and N. C. Ellstrand. 1984. Experimental studies on the evolutionary significance of sexual reproduction. I. A test of the frequency dependent selection hypothesis. Evolution 38:103–115. [18]

———. 1985. The fitness of dispersed progeny: Experimental studies with *Anthoxan-*

thum odoratum. In P. Jacquard, G. Heim, and J. Antonovics, eds., Genetic differentiation and dispersal in plants, 369–381. Springer-Verlag, Berlin. [18]

Antonovics, J., N. C. Ellstrand, and R. N. Brandon. 1988. Genetic variation and environmental variation: Expectations and experiments. In L. D. Gottlieb and S. K. Jain, eds., Plant evolutionary biology, 275–303. Chapman and Hall, London. [2, 7]

Antonovics, J. and D. A. Levin. 1980. The ecological and genetic consequences of density-dependent regulation in plants. Annual Review of Ecology and Systematics 11:411–452.

Antonovics, J., and R. B. Primack. 1982. Experimental ecological genetics in *Plantago.* VI. The demography of seedling transplants of *P. lanceolata.* Journal of Ecology 70:55–75. [14]

Archer, M. 1973. The species preferences of grazing horses. Journal of the British Grassland Society 28:123–128. [10]

Archer, S., and J. K. Detling. 1984. The effects of defoliation and competition on regrowth of tillers of two North American mixed grass prairie graminoids. Oikos 43:351–357. [13]

Archer, S., and L. L. Tieszen. 1980. Growth and physiological responses of tundra plants to defoliation. Arctic and Alpine Research 12:531–552. [13]

————. 1983. Effects of simulated grazing on foliage and root production and biomass allocation in an arctic tundra sedge (*Eriophorum vaginatum*). Oecologia 58:92–102. [13]

Armbruster, W. S., and W. R. Mziray. 1987. Pollination and herbivore ecology of an African *Dalechampia* (Euphorbiaceae): Comparisons with new world species. Biotropica 19:64–73. [13]

Arnold, G. W. 1964. Factors within plant associations affecting the behaviour and performance of grazing animals. In D. J. Crisp, ed., Grazing in terrestrial and marine environments, 133–154. Blackwell, Oxford. [10]

Arnold, G. W., E. S. deBoer, and C. A. P. Boundy. 1980. The influence of odor and taste on the food preferences and food intake of sheep. Australian Journal of Agricultural Research 31:657–666. [10]

Arnold, G. W., and J. L. Hill. 1972. Chemical factors affecting selection of food plants by ruminants. In J. B. Harborne, ed., Phytochemical ecology, 71–101. Annual Proceedings of the Phytochemical Society. Academic Press, London. [10]

Arnold, J. T. A., and M. J. Whitten. 1976. The genetic basis for organophosphorus resistance in the Australian sheep blowfly, *Lucilia cuprina* (Wiedemann) (Diptera: Calliphoridae). Bulletin of Entomological Research 66:561–568. [17]

Arnold, S. J. 1981. Behavioral variation in natural populations. I. Phenotypic, genetic, and environmental correlations between chemoreceptive responses to prey in the garter snake, *Thamnophis elegans.* Evolution 35:489–509. [3, 7]

Aronson, A. I., W. Beckman, and P. Dunn. 1986. *Bacillus thuringiensis* and related insect pathogens. Microbiological Reviews 50:1–24. [12]

Arveson, J. N. 1969. Jackknifing U-statistics. Annals of Mathematical Statistics 40:2076–2100. [3]

Asay, K. H., I. T. Carlson, and C. P. Wilsie. 1968. Genetic variability in forage yield, crude protein percentage, and palatability in the reed canarygrass *Phalaris arundinacea* L. Crop Science 8:568–571. [10]

Atkins, I. M., and R. G. Dahms. 1945. Reaction of small-grain varieties to green bug attack. United States Department of Agriculture Technical Bulletin 901. [5]

Atkinson, I. A. E., and R. M. Greenwood. 1989. Relationships between moss and plants. New Zealand Journal of Ecology 12:67–96. [13]

Atkinson, W. D., and B. Shorrocks. 1981. Competition on a divided and ephemeral resource: A simulation model. Journal of Animal Ecology 50:461–471. [5, 11]

Atsatt, P. R. 1981a. Ant-dependent food plant selection by the mistletoe butterfly *Ogyris amaryllis* (Lycaenidae). Oecologia 48:60–63. [6]

———. 1981b. Lycaenid butterflies and ants: Selection for enemy-free space. American Naturalist 188:638–654. [6]

Atsatt, P. R., and D. J. O'Dowd. 1976. Plant defense guilds. Science 193:24–29. [2, 13]

Atwood, S. S., and J. T. Sullivan. 1943. Inheritance of a cyanogenic glucoside and its hydrolysing enzyme in *Trifolium repens* L. Journal of Heredity 34:311–320. [13, 17]

Augspurger, C. K. 1981. Reproductive synchrony of a tropical shrub: Experimental studies of effects of pollinators and seed predators on *Hybanthus prunifolius* (*Violaceae*). Ecology 62:775–788. [13]

———. 1983. Seed dispersal of the tropical tree, *Platypodium elegans,* and the escape of its seedlings from fungal pathogens. Journal of Ecology 71:759–771. [3, 14]

———. 1984. Seedling survival of tropical tree species: Interactions of dispersal distance, light gaps, and pathogens. Ecology 65:1705–1712. [3]

———. 1989. Impact of pathogens on natural plant populations. In A. J. Dary, M. J. Hutchings, and A. R. Watkinson, eds., Plant population ecology, 413–433. Blackwell Scientific Publications, Oxford. [14]

———. 1990. Spatial patterns of damping-off disease during seedling recruitment in tropical forests. In J. J. Burdon and S. R. Leather, eds., Pests, pathogens, and plant communities, 131–144. Blackwell Scientific, Oxford. [14]

Augspurger, C. K., and C. K. Kelly. 1984. Pathogen mortality of tropical tree seedlings: Experimental studies of the effects of dispersal distance, seedling density, and light conditions. Oecologia 61:211–217. [14, 15]

Auld, T. D. 1986. Variation in predispersal seed predation in several Australian *Acacia* spp. Oikos 47:319–326. [13]

Axtel, J. D. 1981. Breeding for improved nutritional quality. In K. J. Frey, ed., Plant breeding 2:365–432. Iowa State University, Ames. [16]

Ayers, M. P., and S. F. MacLean. 1987. Development of birch leaves and the growth-energetics of *Epirrita autumnata* (Geometridae). Ecology 68:558–568. [9]

Ayers, M. P., J. Suomela, and S. F. MacLean. 1987. Growth performance of *Epirrita autumnata* (Lepidoptera: Geometridae) in mountain birch: Tree, broods, and tree x brood interactions. Oecologia 74:450–457. [9]

Ayers, M. P., and D. L. Thomas. 1990. Alternative formulations of the mixed-model ANOVA applied to quantitative genetics. Evolution 44:221–226. [3]

Ayyangar, G. N. R., and B. W. X. Ponnaiya. 1941. The occurrence and inheritance of a bloomless sorghum. Current Science 10:408–409. [2]

Bach, C. E. 1980. Effects of plant diversity and time of colonization on an herbivore-plant interaction. Oecologia 44:319–326. [5]

———. 1984. Plant spatial pattern and herbivore population dynamics: Plant factors affecting the movement patterns of a tropical cucurbit specialist (*Acalymma innubum*). Ecology 65:175–190. [6, 11]

Bacon, C. W., and M. R. Siegel. 1988. Endophyte parasitism of tall fescue. Journal of Production Agriculture 1:45–55. [17]

Bailey, J. A., and J. W. Mansfield. 1982. Phytoalexins. John Wiley and Sons, New York. [4, 16]

Bailey, V. 1903. Sleepy grass and its effect on horses. Science 17:392–393. [10]

Baker, H. G. 1947. Infection of species of *Melandrium* by *Ustilago violacea* (Pers.) Fuckel and the transmission of the resultant disease. Annals of Botany 11:333–348. [14, 18]

———. 1970. Evolution in the tropics. Biotropica 2:101–111. [13]

Baker, J. E., and S. R. Loschiavo. 1987. Nutritional ecology of stored-product insects. In F. Slansky and J. G. Rodriquez, eds., Nutritional ecology of insects, mites, spiders, and related invertebrates, 321–344. John Wiley, New York. [16]

Baker, J. P. 1979. Electrophoretic studies on populations of *Myzus persicae* in Scotland from October to December, 1976. Annals of Applied Biology 91:159–164. [5]

Baker, P. F. 1960. Aphid behavior on healthy and yellow-virus infected sugar beet. Annals of Applied Biology 48:384–391. [2]

Baldwin, I. T. 1988. Short-term damage-induced increases in tobacco alkaloids protect plants. Oecologia 75:367–370. [11, 16, 17]

Baldwin, I. T., and J. C. Schultz. 1983. Rapid changes in tree leaf chemistry induced by damage: Evidence for communication between plants. Science 221:277–279. [13]

Baldwin, I. T., C. L. Sims, and S. E. Kean. 1990. The reproductive consequences associated with inducible alkaloidal responses in wild tobacco. Ecology 71:252–262. [16, 17]

Banks, C. J. 1957. The behaviour of individual coccinellid larvae on plants. British Journal of Animal Behaviour 5:12–24. [12]

Banyikwa, F. F. 1988. The growth response of two East African grasses to defoliation, nitrogen fertilization, and competition. Oikos 51:25–30. [13]

Barbosa, P. 1988a. Natural enemies and herbivore-plant interactions: Influence of plant allelochemicals and host specificity. In P. Barbosa and D. K. Letourneau, eds., Novel aspects of insect-plant interactions, 201–210. John Wiley and Sons, New York. [12]

———. 1988b. Some thoughts on "the evolution of host range." Ecology 69:912–915. [6]

Barbosa, P., W. Cranshaw, and J. Greenblatt. 1981. Influence of food quantity and quality on polymorphic dispersal in the gypsy moth *Lymantria dispar*. Canadian Journal of Zoology 59:293–296. [4]

Barbosa, P., and J. A. Saunders. 1985. Plant allelochemicals: Linkages between herbivores and their natural enemies. Recent Advances in Phytochemistry 19:107–137. [12]

Barbosa, P., J. A. Saunders, J. Kemper, R. Trumbule, J. Olechno, and P. Martinat. 1986. Plant allelochemicals and insect parasitoids: Effects of nicotine on *Cotesia congregata* (Say) (Hymenoptera: Braconidae) and *Hyposoter annulipes* (Cresson) (Hymenoptera: Ichneumonidae). Journal of Chemical Ecology 12:1319–1328. [12]

Barbosa, P., and J. C. Schultz. 1987. Insect outbreaks. Academic Press, New York. [1, 13]

Barbour, J. D., R. R. Farrar, Jr., and G. G. Kennedy. 1991. Interaction of fertilizer regime with host-plant resistance in tomato. Entomologia experimentalis et applicata. In press. [2]

Barbour, J. D., and G. G. Kennedy. 1991. The role of the steroidal glycoalkaloid α-tomatine in the host-plant resistance of tomato to the Colorado potato beetle. Journal of Chemical Ecology. In press. [2]

Bardner, R. 1968. Wheat bulb fly, *Leptohylemyia coarctata* Fall., and its effect on the growth and yield of wheat. Annals of Applied Biology 61:1–11. [13]

Bardner, R., and K. E. Fletcher. 1974. Insect infestations and their effects on the growth and yield of field crops: A review. Bulletin of Entomological Research 64:141–160. [13]

Barker, G. M., R. P. Pottinger, P. J. Addison, and R. A. Prestidge. 1984. Effect of *Lolium* endophyte fungus infections on behavior of adult Argentine stem weevil. New Zealand Journal of Agricultural Research 27:271–277. [10]

Barker, J. S. F., and R. H. Thomas. 1987. A quantitative genetic perspective on adaptive evolution. In V. Leoschcke, ed., Genetic constraints on adaptive evolution, 3–23. Springer-Verlag, Berlin. [3]

Barlow, B. A., and D. Wiens. 1977. Host-parasite resemblance in Australian mistletoes: The case for cryptic mimicry. Evolution 31:69–84. [13]

Barlow, R. B., and R. O. D. Dixon. 1973. Choline acetyltransferase in the nettle *Urtica dioica* L. Biochemical Journal 132:15–18. [10]

Barrett, J. A. 1978. A model of epidemic development in variety mixtures. In P. R. Scott and A. Bainbridge, eds., Plant disease epidemiology, 129–137. Blackwell, Oxford. [5, 18]

———. 1980. Pathogen evolution in multilines and variety mixtures. Zeitschrift für Pflanzenkrankheiten und Pflanzenschutz 87:383–396. [5, 8]

———. 1981. The evolutionary consequences of monoculture. In J. A. Bishop and L. M. Cook, eds., Genetic consequences of man-made change, 209–248. Academic Press, London. [8, 15]

———. 1983. Plant-fungus symbioses. In D. J. Futuyma and M. Slatkin, eds., Coevolution, 137–160. Sinauer, Sunderland, MA. [5, 15]

———. 1985. The gene-for-gene hypothesis: Parable or paradigm? In D. Rollinson and R. M. Anderson, eds., Ecology and genetics of host-parasite interactions, 215–225. Academic Press, Orlando, FL. [8, 14, 15, 18]

———. 1987. The dynamics of genes in populations. In M. S. Wolfe and C. E. Caten, eds., Populations of plant pathogens: Their dynamics and genetics, 39–54. Blackwell Scientific Publications, Oxford. [5]

———. 1988. Frequency-dependent selection in plant-fungal interactions. Philosophical Transactions of the Royal Society of London B319:473–483. [15]

Barrett, J. A., and M. S. Wolfe. 1978. Multilines and super-races: A reply. Phytopathology 68:1535–1537. [5]

Barrus, M. F. 1911. Variation of varieties of beans in their susceptibility to anthracnose. Phytopathology 1:190–195. [8]

Barton, N. H., J. S. Jones, and J. Mallet. 1988. No barriers to speciation. Nature 336:13–14. [6]

Barton, N. H., and M. Turelli. 1989. Evolutionary genetics: How little do we know? Annual Review of Ecology and Systematics 23:337–370. [3, 7]

Basey, J. M., S. H. Jenkins, and P. E. Busher. 1988. Optimal central-place foraging by beavers: Tree-size selection in relation to defensive chemicals of quaking aspen. Oecologia 76:278–282. [13]

Bate-Smith, E. C. 1972. Attractants and repellents in higher animals. In J. B. Harborne, ed., Phytochemical ecology, 45–46. Annual Proceedings of the Phytochemical Society 8:45–46. Academic Press, London. [10]

Batts, C. C. V. 1955. Infection of wheat by loose smut, *Ustilago tritici*. Nature 175:467–468. [15]

Batzli, G. O. 1983. Responses of arctic rodent populations to nutritional factors. Oikos 40:396–406. [10]

Bawa, K. S., and P. A. Opler. 1978. Why are pistillate inflorescences of *Simarouba glauca* eaten less than staminate inflorescences? Evolution 32:673–676. [13]

Bazzaz, F. A. 1984. Demographic consequences of plant physiological traits. In R. Dirzo and J. Sarukhan, eds., Perspectives on plant population ecology, 324–346. Sinauer Associates, Sunderland, MA. [16]

Bazzaz, F. A., N. R. Chiariello, P. D. Coley, and L. F. Pitelka. 1987. Allocating resources to reproduction and defense. Bioscience 37:58–67. [13, 16, 17]

Bazzaz, F. A., and J. L. Harper. 1977. Demographic analysis of the growth of *Linum usitatissimum*. New Phytologist 78:193–208. [16]

Beach, R. M., and J. W. Todd. 1988. Discrete and interactive effects of plant resistance and nuclear polyhedrosis viruses for suppression of soybean looper and velvetbean caterpillar (Lepidoptera: Noctuidae) on soybean. Journal of Economic Entomology 81:684–691. [12]

Beattie, A. J., D. E. Breedlove, and P. R. Ehrlich. 1973. The ecology of the pollinators and predators of *Frasera speciosa*. Ecology 54:81–91. [13]

Beck, S. D. 1965. Resistance of plants to insects. Annual Review of Entomology 10:207–232. [10, 17]

Beck, S. D., and L. M. Schoonhoven. 1980. Insect behavior and plant resistance. In F. G. Maxwell and P. R. Jennings, eds., Breeding plants resistant to insects, 115–135. John Wiley and Sons, New York. [11, 17]

Becker, P. 1983. Effects of insect herbivory and artificial defoliation on survival of *Shorea* seedlings. In S. L. Sutton, T. C. Whitmore, and A. C. Chadwick, eds., Tropical rain forest: Ecology and management, 241–252. Blackwell Scientific, Oxford. [13]

Becker, W. A. 1984. Manual of quantitative genetics. 4th ed. Academic Enterprises, Pullman, WA. [3]

Beckmann, R. L., Jr., and J. M. Stucky. 1981. Extrafloral nectaries and plant guarding in *Ipomoea pandurata* (L.) G. F. W. Mey. (Convolvulaceae). American Journal of Botany. 68:72–79. [17]

Beddington, J. R., C. A. Free, and J. H. Lawton. 1978. Characteristics of successful natural enemies in models of biological control of insect pests. Nature (London) 273:513–519. [12]

Beed, G. L., T. A. Brindley, and W. B. Showers. 1972. Influence of resistant corn leaf tissue on the biology of the European corn borer. Annals of the Entomological Society of America. 65:658–662. [2]

Bell, A. A., R. D. Stipanovic, G. W. Elzen, and H. J. Williams. 1987. Structural and genetic variation of natural pesticides in pigment glands of cotton (*Gossypium*). In G. R. Waller, ed., Allelochemicals: Role in agriculture and forestry, 477–490. ACS Symposium Series 330. American Chemical Society, Washington, DC. [12]

Bell, E. A. 1978. Toxins in seeds. In J. B. Harborne, ed., Biochemical aspects of plant and animal coevolution, 143–161. Academic Press, New York. [16]

Bell, G. 1982. The masterpiece of nature. University of California Press, Berkeley. [15]

———. 1985. Two theories of sex and variation. Experientia 41:1235–1245. [15]

Bell, G., and V. Koufopanou. 1986. The cost of reproduction. In Oxford surveys of evolutionary biology, vol. 3, ed. R. Dawkins and M. Ridley, 83–131. Oxford University Press, Oxford. [3, 6]

Bell, G., and J. Maynard Smith. 1987. Short-term selection for recombination among mutually antagonistic species. Nature 328:66–68. [15]

Bell, J. V. 1978. Development and mortality in bollworms fed resistant and susceptible soybean cultivars treated with *Nomuraea rileyi* or *Bacillus thuringiensis*. Journal of the Georgia Entomological Society 13:43–50. [12]

Belovsky, G. E. 1981. Food plant selection by a generalist herbivore: The moose. Ecology 62:1010–1030. [10]

Belsky, A. J. 1986. Does herbivory benefit plants? A review of the evidence. American Naturalist 127:870–892. [10, 13, 16]

———. 1987. The effects of grazing: Confounding of ecosystem, community, and organism scales. American Naturalist 129:777–783. [10]

Benner, B. L. 1988. Effects of apex removal and nutrient supplementation on branching and seed production in *Thlaspi arvense* (Brassicaceae). American Journal of Botany 75:645–651. [13]

Bennett, C. W. 1969. Seed transmission of plant viruses. Advances in Virus Research 14:221–261. [15]

Benoit, P., and R. Blais. 1988. The effects of defoliation by the larch casebearer on the radial growth of tamarack. Forestry Chronicle 142:190–192. [13]

Benson, W. W. 1985. Amazon ant-plants. In G. H. Prance and T. E. Lovejoy, eds., Key environments: Amazonia, 239–266. Pergamon Press, New York. [13]

Benson, W. W. 1978. Resource partitioning in passion vine butterflies. Evolution 32:493–518. [13]

Benson, W. W., K. S. Brown, Jr., and L. E. Gilbert. 1975. Coevolution of plants and herbivores: Passion flower butterflies. Evolution 29:659–680. [13]

Bentley, B. L. 1976. Plants bearing extrafloral nectaries and the associated ant community: Interhabitat differences in the reduction of herbivore damage. Ecology 57:815–820. [10]

———. 1977. Extrafloral nectaries and protection by pugnacious bodyguards. Annual Review of Ecology and Systematics 8:407–427. [10, 13]

Bentley, S., and J. B. Whittaker. 1979. Effects of grazing by a chrysomelid beetle, *Gastrophysa viridula*, on competition between *Rumex obtusifolius* and *Rumex crispus*. Journal of Ecology 67:79–90. [13]

Bentley, S., J. B. Whittaker, and A. J. C. Malloch. 1980. Field experiments on the effects of grazing by a chrysomelid beetle (*Gastrophysa viridula*) on seed production and quality in *Rumex obtusifolius* and *Rumex crispus*. Journal of Ecology 68:671–674. [13]

Beregovoy, V. H., K. J. Starks, and K. G. Janardan. 1988. Fecundity characteristics of the greenbug biotypes C and E cultured on different host plants. Environmental Entomology 17:59–62. [5]

Berenbaum, M. R. 1978. Toxicity of a furanocoumarin to armyworms: A case of biosynthetic escape from insect herbivores. Science 201:532–534. [13, 16]

——. 1981a. Patterns of furanocoumarin distribution and insect herbivory in the Umbelliferae: Plant chemistry and community structure. Ecology 62:1254–1266. [11]

——. 1981b. Patterns of furanocoumarin production and insect herbivory in a population of wild parsnip (*Pastinaca sativa*). Oecologia 49:236–244. [4, 16]

——. 1983. Coumarins and caterpillars: A case for coevolution. Evolution 37:163–179. [13]

——. 1984. Effects of tannin ingestion on two species of papilionid caterpillars. Entomologia Experimentalis et Applicata 34:245–250. [4]

——. 1985. Brementown revisited: Interactions among allelochemicals in plants. In G. A. Cooper-Driver, T. Swain, and E. E. Conn, eds., Chemically mediated interactions between plants and other organisms. Recent Advances in Phytochemistry 19:139–169. Plenum Press, New York. [4, 10]

——. 1986. Post-ingestive effects of phytochemicals on insects: On paracelsus and plant products. In T. A. Miller and J. Miller, eds., Insect-plant interactions, 121–153. Springer-Verlag, New York. [4]

——. 1990. Evolution of specialization in insect-umbellifer associations. Annual Review of Entomology 35:319–343. [4]

Berenbaum, M. R., and P. Feeny. 1981. Toxicity of angular furanocoumarins to swallowtail butterflies: Escalation in a coevolutionary arms race? Science 212:927–929. [16]

Berenbaum, M. R., and J. Neal. 1985. Synergism between myrisicin and xanthotoxin, a naturally co-occurring plant toxicant. Journal of Chemical Ecology 11:1349–1358. [4]

Berenbaum, M. R., and J. J. Neal. 1987. Interactions among allelochemicals and insect resistance in crop plants. In G. R. Waller, ed., Allelochemicals: Role in agriculture and forestry, 416–430. ACS Symposium Series 330. American Chemical Society, Washington, DC. [16]

Berenbaum, M. R., and A. R. Zangerl. 1986. Variation in seed furanocoumarin content within the wild parsnip (*Pastinaca sativa*). Phytochemistry 25:659–661. [16]

——. 1988. Stalemates in the coevolutionary arms race: Syntheses, synergisms, and sundry other sins. In K. C. Spencer, ed., Chemical mediation of coevolution, 113–132. American Institute of Biological Sciences, Washington, DC. [11, 17]

Berenbaum, M. R., A. R. Zangerl, and K. Lee. 1989. Chemical barriers to adaptation by a specialist herbivore. Oecologia 80:501–506. [4, 16]

Berenbaum, M. R., A. R. Zangerl, and J. K. Nitao. 1984. Furanocoumarins in seeds of wild and cultivated parsnip (*Pastinaca sativa*). Phytochemistry 23:1809–1810. [4]

——. 1986. Constraints on chemical coevolution: Wild parsnips and the parsnip webworm. Evolution 40:1215–1228. [2, 3, 4, 6, 11, 13, 16, 17]

Bergelson, J. M., and J. H. Lawton. 1988. Does foliage damage influence predation on the insect herbivores of birch? Ecology 69:434–445. [11]

Bergman, J. M., and W. M. Tingey. 1979. Aspects of interaction between plant genotypes and biological control. Bulletin of the Entomological Society of America
25:275–279. [11, 12]

Bergström, R., and K. Danell. 1987. Effects of simulated winter browsing by moose on
morphology and biomass of two birch species. Journal of Ecology 75:533–
544. [13]

Bernays, E. A. 1981. Plant tannins and insect herbivores: An appraisal. Ecological Entomology 6:353–360. [10]

————. 1986. Diet-induced head allometry among foliage-chewing insects and its importance to gramnivores. Science 231:495–497. [7]

Bernays, E. A., R. F. Chapman, J. MacDonald, and J. E. R. Salter. 1976. The degree
of oligophagy in *Locusta migratoria* L. Ecological Entomology 1:223–230. [10]

Bernays, E. A., and M. Graham. 1988. On the evolution of host specificity in phytophagous arthropods. Ecology 69:886–892. [4, 5, 6, 13]

Bernays, E. A., and A. C. Lewis. 1986. The effect of wilting on palatability of plants
to *Schistocerca gregaria*, the desert locust. Oecologia 70:132–135. [13]

Bernays, E. A., and P. Wege. 1987. Significance levels of inferential statistics and their
interpretation: A lesson from feeding deterrent experiments. Annals of the Entomological Society of America 80:9–11. [2, 4]

Berryman, A. A. 1972. Resistance of conifers to invasion by bark beetle–fungus association. BioScience 22:598–602. [14]

Bertin, R. I. 1982. The ecology of sex expression in red buckeye. Ecology 63:445–
456. [13]

Bicchi, C., A. D'Amato, C. Frattini, G. M. Nano, E. Cappelletti, R. Caniato, and R.
Filippini. 1990. Chemical diversity of the contents from the secretory structures of
Heracleum sphondylium. Phytochemistry 29:1883–1887. [4]

Bieniek, M. E., and W. F. Millington. 1986. Thorn formation in *Ulex europaeus*
in relation to environmental and endogenous factors. Botanical Gazette 129:145–
150. [10]

Bierzychudek, P. 1987. Resolving the paradox of sexual reproduction: A review of experimental tests. In S. C. Stearns, ed., The evolution of sex and its consequences,
163–174. Birkhauser Verlag, Basel. [15]

Bilyk, A., P. L. Cooper, and G. J. Sapers. 1984. Varietal differences in distribution of
quercetin and kaempferol in onion (*Allium cepa* L.) tissue. Journal of Agricultural
and Food Chemistry. 32:274–276. [16]

Bink, F. A. 1986. Acid stress in *Rumex hydrolapathum* (Polygonaceae) and its influence
on the phytophage *Lycaena dispar* (Lepidoptera: Lycaenidae). Oecologia 70:447–
451. [13]

Birch, J. C. 1960. The genetic factor in population ecology. American Naturalist 94:5–
24. [18]

Bjarko, M. E., and R. F. Line. 1988a. Heritability and number of genes controlling leaf
rust resistance in four cultivars of wheat. Phytopathology 78:457–461. [3]

————. 1988b. Quantitative determination of the gene action of leaf rust resistance in
four cultivars of wheat *Triticum aestivum*. Phytopathology 78:451–456. [3]

Blackman, R. L. 1971. Variation in the photoperiodic response within natural populations of *Myzus persicae* (Sulz.). Bulletin of Entomological Research 60:533–
546. [6]

————. 1972. The inheritance of life-cycle differences in *Myzus persicae* (Sulz.) (Hem., Aphididae). Bulletin of Entomological Research 62:281–294. [6]

Blackman, R. L., V. F. Eastop, and M. Hills. 1977. Morphological and cytological separation of *Amphorophora* Buckton (Homoptera: Aphididae) feeding on European raspberry and blackberry (*Rubus* spp.). Bulletin of Entomological Research 67:285–296. [6]

Blakley, N. 1982. Biotic unpredictability and sexual reproduction: Do aphid genotype–host genotype interactions favor aphid sexuality? Oecologia 523:396–399. [5, 6]

Blanch, P. A., J. C. Asher, and J. H. Barnett. 1981. Inheritance of pathogenicity and cultural characters of *Gaeumannomyces graminis* var. *tritici*. Transactions of the British Mycological Society 77:391–399. [8]

Blau, P. A., P. Feeny, L. Contardo, and D. S. Robson. 1978. Allylglucosinolate and herbivorous caterpillars: A contrast in toxicity and tolerance. Science 200:1296–1298. [13, 17]

Blaxter, K. L., F. W. Wainman, and R. S. Wilson. 1961. The regulation of food intake by sheep. Animal Production 3:51–61. [10]

Bloom, A. J., F. S. Chapin, III, and H. A. Mooney. 1985. Resource limitation in plants: An economic analogy. Annual Review of Ecology and Systematics 16:363–392. [16, 17]

Boethel, D. J., and R. D. Eikenbary. 1986. Interactions of plant resistance and parasitoids and predators of insects. Ellis Horwood, Chicester. [1, 11]

Bohn, G. W., A. N. Kishaba, J. A. Principe, and H. H. Toba. 1973. Tolerance to melon aphid in *Cucumis melo* L. Journal of the American Society for Horticultural Science 98:37–40. [2]

Boidin, J. 1986. Intercompatibility and the species concept in the saprobic basidiomycotina. Mycotaxon 26:319–336. [8]

Boozaya-Angoon, D., K. J. Starks, D. C. Weibel, and G. L. Teetes. 1984. Inheritance of resistance in sorghum, *Sorghum bicolor*, to the sorghum midge, *Contarinia sorghicola* (Diptera: Cecidomyiidae). Environmental Entomology 13:1531–1534. [2]

Borlaug, N. E. 1953. New approach to the breeding of wheat varieties resistant to *Puccinia graminis tritici*. Phytopathology 43:467. [5]

Boscher, J. 1979. Modified reproduction strategy of leek *Allium porrum* in response to a phytophagous insect, *Acrolepiopsis assectella*. Oikos 33:451–456. [13]

Bosque-Perez, N. A., K. W. Foster, and T. F. Leigh. 1987. Heritability of resistance in cowpea to the western plantbug. Crop Science 27:1133–1136. [4, 17]

Bottrell, D. G., and P. L. Adkisson. 1977. Cotton insect pest management. Annual Review of Entomology 22:451–481. [16]

Bowers, M. D. 1988a. Chemistry and coevolution: Iridoid glycosides, plants, and herbivorous insects. In K. Spencer, ed., Chemical mediation of coevolution, 133–165. Academic Press, New York. [6]

————. 1988b. Plant allelochemistry and mimicry. In P. Barbosa and D. K. Letourneau, eds., Novel aspects of insect-plant interactions, 273–311. John Wiley and Sons, New York. [12]

Bowers, M. D., and G. W. Puttick. 1986. Fate of ingested iridoid glycosides in lepidopteran herbivores. Journal of Chemical Ecology 12:169–178. [12]

Box, G. E. P. 1953. Non-normality tests on variances. Biometrika 40:318–335. [3]

Boyer, J. S. 1982. Plant productivity and environment. Science 218:443–448. [16]

Boygo, T., and W. A. Becker. 1963. Exact confidence intervals for genetic heritability estimated from paternal half-sib correlations. Biometrics 19:494–496. [3]

Bradford, D. F., and C. C. Smith. 1977. Seed predation and seed number in *Scheelea* palm fruits. Ecology 58:667–673. [13]

Bradshaw, A. D. 1959. Population differentiation in *Agrostis tenuis*, Sibth. II. The incidence and significance of infection by *Epichloe typhina*. New Phytologist 58:310–315. [14, 15]

———. 1960. Population differentiation in *Agrostis tenuis* Sibth. III. Populations in varied environments. New Phytologist 59:92–103. [13, 15]

———. 1965. Evolutionary significance of phenotypic plasticity in plants. Advances in Genetics 13:115–155. [7]

Brasier, C. M. 1987a. Recent genetic changes in the *Ophiostoma ulmi* population: The threat to the future of the elm. In M. S. Wolfe and C. E. Caten, eds., Populations of plant pathogens: Their dynamics and genetics, 213–226. Blackwell Scientific Publications, Oxford. [8]

———. 1987b. Some genetical aspects of necrotrophy with special reference to *Ophiostoma ulmi*. In P. R. Day and G. J. Jellis, eds., Genetics and plant pathogenesis, 297–310. Blackwell Scientific Publications, Oxford. [8]

———. 1988. *Ophiostoma ulmi*, cause of Dutch elm disease. In D. S. Ingram and P. H. Williams, eds., Genetics of plant pathogenic fungi, vol. 6, Advances in plant pathology, ed. G. S. Sidhu, 207–223. Academic Press, London. [8]

Brattsten, L., and S. Ahmad. 1986. Molecular aspects of insect-plant associations. Plenum Press, New York. [4]

Brattsten, L. B., C. F. Wilkinson, and T. Eisner. 1977. Herbivore-plant interactions: Mixed-function oxidases and secondary plant substances. Science 196:1349–1352. [4]

Breedlove, D., and P. Ehrlich. 1972. Coevolution: Patterns of legume predation by a lycaenid butterfly. Oecologia 10:99–104. [13]

Bremermann, H. J. 1980. Sex and polymorphism as strategies in host-pathogen interactions. Journal of Theoretical Biology 87:671–702. [15]

Briggs, D., and S. M. Walters. 1984. Plant variation and evolution. Cambridge University Press, Cambridge. [2]

Briggs, J. B. 1965. The distribution, abundance, and genetic relationships of four strains of the rubus aphid (*Amphorophora rubi* [Kalt.]) in relation to raspberry breeding. Journal of Horticultural Science 40:109–177. [6]

Brinkman, M. A., and K. J. Frey. 1977. Growth analysis of isoline-recurrent parent grain yield differences in oats. Crop Science 17:426–430. [15]

Broadbent, L. 1960. Dispersal of inoculum by insects and other animals, including man. In J. Y. Horsfall and A. E. Dimond, eds., Plant pathology: An advanced treatise, 3:97–135. Academic Press, New York. [18]

Bronson, C. R., and A. H. Ellingboe. 1986. The influence of four unnecessary genes for virulence on the fitness of *Erysiphe graminis* f. sp. *tritici*. Phytopathology 76:154–158. [5]

Brower, L. P., J. N. Seiber, C. J. Nelson, S. P. Lynch, and M. M. Holland. 1984. Plant-determined variation in the cardenolide content, thin-layer chromatography profiles, and emetic potency of monarch butterflies, *Danaus plexippus*, reared on the milkweed plants in California: 2. *Asclepias speciosa.* Journal of Chemical Ecology 10:601–639. [12]

Brower, L. P., J. N. Seiber, C. J. Nelson, S. P. Lynch, M. P. Hoggard, and J. P. Cohen. 1984. Plant-determined variation in the cardenolide content and thin-layer chromatography profiles, of monarch butterflies, *Danaus plexippus,* reared on milkweed plants in California. 3: *Asclepias californica.* Journal of Chemical Ecology 10:1823–1857. [12]

Brower, L. P., J. N. Seiber, C. J. Nelson, S. P. Lynch, and P. M. Tuskes. 1982. Plant-determined variation in the cardenolide content, thin-layer chromatography profiles, and emetic potency of monarch butterflies, *Danaus plexippus,* reared on the milkweed, *Asclepias eriocarpa* in California. Journal of Chemical Ecology 8:579–633. [12]

Brown, A., E. Nevo, and D. Zohary. 1977. Association of alleles at esterase loci in wild barley, *Hordeum spontaneum.* Nature 268:430–431. [15]

Brown, A. H. D., O. H. Frankel, D. R. Marshall, and J. T. Williams, eds. 1988. The use of plant genetic resources. Cambridge University Press, Cambridge. [2]

Brown, B. J., and T. F. H. Allen. 1989. The importance of scale in evaluating herbivory impacts. Oikos 54:189–194. [13]

Brown, D. G. 1988. The cost of plant defense: An experimental analysis with inducible proteinase inhibitors in tomato. Oecologia 76:467–470. [17]

Brown, J. H., J. J. Grover, D. W. Davidson, and G. A. Lieberman. 1975. A preliminary study of seed predation in desert and montane habitats. Ecology 56:987–992. [13]

Brown, K. S., Jr. 1981. The biology of *Heliconius* and related genera. Annual Review of Entomology 26:427–456. [13]

Brown, P. H., R. D. Graham, and D. J. D. Nicholas. 1984. The effects of manganese and nitrate supply on the levels of phenolics and lignin in young wheat plants. Plant and Soil 81:437–440. [13]

Brown, V. K. 1985. Insect herbivores and plant succession. Oikos 45:17–25. [11]

Brown, V. K. and A. C. Gange. 1989. Differential effects of above- and below-ground insect herbivory during early plant succession. Oikos 54:67–76. [13]

Brown, V. K., A. C. Gange, I. M. Evans, and A. L. Storr. 1987. The effect of insect herbivory on the growth and reproduction of two annual *Vicia* species at different stages in plant succession. Journal of Ecology 75:1173–1178. [13]

Brown, V. K., and P. Hyman. 1986. Successional communities of plants and phytophagous Coleoptera. Journal of Ecology 74:963–975. [11]

Brown, W. L. 1960. Ants, acacias, and browsing mammals. Ecology 41:587–592. [10]

Browning, J. A. 1974. Relevance of knowledge about natural ecosystems to development of pest management programs for agro-ecosystems. Proceedings of the American Phytopathological Society 1:191–199. [14, 15, 18]

———. 1980. Genetic protective mechanisms of plant-pathogen populations: Their coevolution and use in breeding for resistance. In M. K. Harris, ed., Biology and breeding for resistance to arthropods and pathogens in agricultural plants, 52–75. Texas Agricultural Experimental Station Miscellaneous Publication 1451. [14]

———. 1981. The agro-ecosystem–natural ecosystem dichotomy and its impact on phytopathological concepts. In J. M. Thresh, ed., Pest, pathogens and vegetation, 159–172. Pitman Books, London. [14]

Browning, J. A., and K. J. Frey. 1969. Multiline cultivars as a means of disease control. Annual Review of Phytopathology 7:355–382. [5, 8]

Browning, J. A., M. D. Simons, and E. Torres. 1977. Managing host genes: Epidemiologic and genetic concepts. In J. G. Horsfall and E. B. Cowling, eds., Plant disease, 1:191–221. Academic Press, New York. [14]

Brues, C. T. 1924. The specificity of food plants in the evolution of phytophagous insects. American Naturalist 58:127–144. [7]

Bryant, E. H., L. M. Combs, and S. A. McCommas. 1986. The effect of an experimental bottleneck upon quantitative genetic variation in the housefly. Genetics 114:1191–1211. [3]

Bryant, J. P., F. S. Chapin III, and D. R. Klein. 1983a. Carbon/nutrient balance of boreal plants in relation to vertebrate herbivory. Oikos 40:357–368. [10, 16, 17]

———. 1983b. Carbon/nutrient balance of boreal plants in the role of plant chemistry. Annual Review of Ecology and Systematics 11:261–285. [13]

Bryant, J. P., F. S. Chapin, P. Reichardt, and T. Clausen. 1985. Adaptation to resource availability as a determinant of chemical defense strategies in woody plants. In: T. Swain, G. A. Cooper-Driver, and E. E. Conn, eds., Chemically mediated interactions between plants and other organisms. Recent Advances in Phytochemistry 19:219–237. Plenum Press, New York. [10]

Bryant, J. P., T. P. Clausen, P. B. Reichardt, M. C. McCarthy, and R. A. Werner. 1987. Effect of nitrogen fertilization upon the secondary chemistry and nutritional value of quaking aspen (Populus tremuloides Michx.) leaves for the large aspen tortrix (Choristoneura conflictana [Walker]). Oecologia 73:513–517. [16]

Bryant, J. P., and P. J. Kuropat. 1980. Selection of winter forage by subarctic browsing vertebrates: The role of plant chemistry. Annual Review of Ecology and Systematics 11:261–285. [10]

Bryant, J. P., J. Tahvanainen, M. Sulkinoja, R. Julkunen-Titto, P. Reichardt, and T. Green. 1989. Biogeographic evidence for the evolution of chemical defense by boreal birch and willow against mammalian browsing. American Naturalist 134:20–34. [13]

Bryant, J. P., J. Tuomi, and P. Niemalä. 1988. Environmental constraint of constitutive and long-term inducible defenses in woody plants. In K. C. Spencer, ed., Chemical mediation of coevolution, 367–389. Academic Press, New York. [3, 17]

Bryant, J. P., G. D. Wieland, T. Clausen, and P. Kuropat. 1985. Interactions of snowshoe hare and feltleaf willow in Alaska. Ecology 66:1564–1573. [13]

Bryant, J. P., G. D. Wieland, P. B. Reichardt, V. E. Lewis, and M. C. McCarthy. 1983. Pinosylvin methyl ether deters showshoe hare feeding on green alder. Science 222:1023–1025. [13]

Buddenhagen, I. W., and T. A. Elsasser. 1962. An insect-spread bacterial wilt epiphytotic of bluggoe banana. Nature 194:164–165. [15]

Bull, J. J. 1987. Evolution of phenotypic variance. Evolution 41:303–315. [7]

Bulmer, M. G. 1980. The mathematical theory of quantitative genetics. Oxford University Press, Oxford. [3]

Bultman, T. L., and S. E. Faeth. 1988. Abundance and mortality of leaf miners on artificially shaded emory oak. Ecological Entomology 13:131–142. [6, 11]

Burbott, A. J., and W. D. Loomis. 1969. Evidence for metabolic turnover of monoterpenes in peppermint. Plant Physiology 44:173–179. [17]

Burdon, J. J. 1980. Variation in disease resistance within a population of *Trifolium re-pens*. Journal of Ecology 68:737–744. [14, 15]

———. 1982. The effect of fungal pathogens on plant communities. In E. I. Newman, ed., The plant community as a working mechanism, 99–112. Blackwell, Oxford. [14, 15]

———. 1985. Pathogens and the genetic structure of plant populations. In J. White, ed., Studies on plant demography: A festschrift for John L. Harper, 313–325. Academic Press, London. [14]

———. 1987a. Diseases and plant population biology. Cambridge University Press, Cambridge. [3, 5, 8, 14, 15, 17]

———. 1987b. Phenotypic and genetic patterns of resistance to the pathogen *Phakopsora pachyrhizi* in populations of *Glycine canescens*. Oecologia 73:257–267. [14, 15]

———. 1989. Directory of the ecological and evolutionary plant-pathogen species cooperative. Commonwealth Scientific and Industrial Research Organization, Canberra. [14]

Burdon, J. J., and G. A. Chilvers. 1977. Controlled environmental experiments on epidermic rates of barley mildew in different mixtures of barley and wheat. Oecologia 28:141–146. [5]

Burdon, J. J., R. H. Groves, and J. M. Cullen. 1981. The impact of biological control on the distribution and abundance of *Chondrilla juncea* in south-eastern Australia. Journal of Applied Ecology 18:957–966. [8, 14, 15]

Burdon, J. J., R. H. Groves, P. E. Kaye, and S. S. Speer. 1984. Competition in mixtures of susceptible and resistant genotypes of *Chondrilla juncea* differentially infected with rust. Oecologia 64:199–203. [14]

Burdon, J. J., and A. M. Jarosz. 1988. The ecological genetics of plant-pathogen interactions in natural communities. Philosophical Transactions of the Royal Society of London B321:349–363. [14, 15]

———. 1989. Wild relatives as sources of disease resistance. In A. H. D. Brown, O. H. Frankel, D. R. Marshall, and J. T. Williams, eds., The use of plant genetic resources, 280–296. Cambridge University Press, Cambridge. [14]

———. 1991. Host-pathogen interactions in natural populations of *Linum maringale* and *Melampsora lini:* I. Patterns of resistance and racial variation in a large host population. Evolution 45:205–217. [14]

Burdon, J. J., N. H. Luig, and D. R. Marshall. 1983. Isozyme uniformity and virulence variation in *Puccinia graminis* f. sp. *tritici* and *P. recondita* f. sp. *tritici* in Australia. Australian Journal of Biological Sciences 36:403–410. [8]

Burdon, J. J., and D. R. Marshall. 1981a. Biological control and the reproductive mode of weeds. Journal of Applied Ecology 18:649–658. [18]

———. 1981b. Inter- and intra-specific diversity in the disease-response of *Glycine* species to the leaf-rust fungus *Phakopsora pachyrhizi*. Journal of Ecology 69:381–390. [14]

Burdon, J. J., D. R. Marshall, and R. H. Groves. 1980. Isozyme variation in *Chondrilla juncea* in Australia. Australian Journal of Botany 28:193–198. [15]

Burdon, J. J., and W. J. Muller. 1987. Measuring the cost of resistance to *Puccinia coronata* in *Avena fatua*. Journal of Applied Ecology 24:191–200. [15]

Burdon, J. J., J. D. Oates, and D. R. Marshall. 1983. Interactions between *Avena* and

Puccinia species. I. The wild hosts: *Avena barbata* Pott ex Link, *A. fatua* L., *A. ludoviciana* Durieu. Journal of Applied Ecology 20:571–584. [14, 15]

Burdon, J. J., and R. C. Shattock. 1980. Disease in plant communities. Applied Biology 5:145–219. [8, 14]

Burdon, R. D., and R. D. Shelbourne. 1974. The use of vegetative propagules for obtaining genetic information. New Zealand Journal of Forestry Science 4:418–425. [3]

Burgess, R. S. L., and R. A. Ennos. 1987. Selective grazing of acyanogenic white clover: Variation in behaviour among populations of the slug *Deroceras reticulatum*. Oecologia 73:432–435. [10]

Burke, L. G., and J. F. Jackson. 1982. Inheritance of the glandless leaf trichome trait in *Nicotiana tabacum* L. Tobacco Science 26:51–52. [2]

Burnes, T. A., R. A. Blanchette, G. C. Wang, and D. W. French. 1988. Screening jack pine seedlings for resistance to *Endocronartium harkressii*. Plant Disease 72:614–616. [8]

Burnett, J. H. 1975. Mycogenetics. John Wiley and Sons, London. [8]

———. 1983. Speciation in fungi. Transactions of the British Mycological Society 81:1–14. [8]

Burrows, G. E., W. C. Edwards, A. J. Pollard, and R. J. Tyrl. 1985. Toxic plants of Oklahoma: Species causing phytodermatoses. Oklahoma Veterinarian 37:23–26. [10]

Busbice, D. H., R. R. Hill, Jr., and H. L. Carmahan. 1972. Genetics and breeding procedures. In C. H. Hanson, ed., Alfalfa science and technology, 283–318. American Society of Agronomy, Madison, WI. [2]

Bush, G. L. 1975a. Modes of animal speciation. Annual Review of Ecology and Systematics 6:339–364. [6]

———. 1975b. Sympatric speciation in phytophagous parasitic insects. In P. W. Price, ed., Evolutionary strategies of parasitic insects and mites, 187–206. Plenum Press, London. [6]

Bush, L., J. E. Slosser, W. D. Worrall, and G. J. Puterka. 1987. Status of greenbug biotypes in Texas. Southwestern Entomologist. 12:229–236. [5]

Caldwell, M. M., J. H. Richards, D. A. Johnson, R. S. Nowack, and R. S. Dzurec. 1981. Coping with herbivory: Photosynthetic capacity and resource allocation in two semiarid *Agropyron* bunchgrasses. Oecologia 50:14–24. [13]

Caldwell, R. M., R. Gallun, and L. Compton. 1966. Genetics and expression of resistance to Hessian fly, *Phytophaga destructor* (Say). In J. Mackey, ed., Proceedings of the Second International Wheat Genetics Symposium, 462–463. Hereditas (Suppl.) 2. [17]

Caligari, P. D. S., and K. Mather. 1975. Genotype-environment interactions. III. Interactions in *Drosophila melanogaster*. Proceedings of the Royal Society of London B191:387–411. [7]

Camazine, S. 1985. Olfactory aposematicism. Association of food toxicity with naturally occurring odor. Journal of Chemical Ecology 11:1289–1295. [10]

Camm, E. L., C. K. Wat, and G. H. N. Towers. 1976. An assessment of the roles of furanocoumarins in *Heracleum lanatum*. Canadian Journal of Botany 54:2562–2566. [4]

Campbell, A., B. D. Frazer, N. Gilbert, A. P. Gutierrez, and M. Mackauer. 1974.

Temperature requirements of some aphids and their parasites. Journal of Applied Ecology 11:419–423. [5]

Campbell, B. C., and S. S. Duffey. 1979. Tomatine and parasitic wasps: Potential incompatibility of plant antibiosis with biological control. Science 205:700–702. [11, 12, 17]

Campbell, D. R. 1989. Measurements of selection in a hermaphroditic plant: Variation in male and female pollination success. Evolution 43:318–334. [3]

Campbell, W. P. 1958. Infection of barley by *Claviceps purpurea*. Canadian Journal of Botany 36:615–619. [15]

Cantor, L. F., and T. G. Whitham. 1989. Importance of belowground herbivory: Pocket gophers may limit aspen to rock outcrop refugia. Ecology 70:962–970. [13]

Carballeira, A. 1980. Phenolic inhibitors in *Erica australis* L. and in associated soil. Journal of Chemical Ecology 6:593–596. [16]

Carey, K. 1893. Breeding system, genetic variability, and response to selection in *Plestrites* (Valerianaceae). Evolution 37:947–956. [13]

Carlquist, S. 1970. Hawaii: A natural history. Natural History Press, Garden City, NY. [13]

Carlson, E. C. 1964. Effect of flower thrips on onion seed plants and a study of their control. Journal of Economic Entomology 57:735–741. [13]

Carman, J. G. 1985. Morphological characterization and defoliation responses of selected *Schizachyrium scoparium* genotypes. American Midland Naturalist 114:37–43. [13]

Carman, J. G., and D. D. Briske. 1985. Morphologic and allozymic variation between long-term grazed and non-grazed populations of the bunch-grass *Schizachyrium scoparium* var. *frequens*. Oecologia 66:332–337. [13]

Carroll, C. R., and C. A. Hoffman. 1980. Chemical feeding deterrent mobilized in response to insect herbivory and counteradaptation by *Epilachna tredecimnotata*. Science 209:414–416. [11, 13]

Carroll, G. 1988. Fungal endophytes in stems and leaves: From latent pathogen to mutualistic symbiont. Ecology 69:2–9. [10]

Carter, W. 1973. Insects in relation to plant disease. Wiley, New York. [17]

Cartier, J. J. 1963. Varietal resistance of peas to pea aphid biotypes under field and greenhouse conditions. Journal of Economic Entomology 56:205–213. [6]

Cartier, J. J., and R. H Painter. 1956. Differential reactions of two biotypes of the corn-leaf aphid to resistant and susceptible varieties, hybrids, and selections of sorghums. Journal of Economic Entomology 49:498–508. [6]

Casagrande, R. A., and D. L. Haynes. 1976. The impact of pubescent wheat on the population dynamics of the cereal leaf beetle. Environmental Entomology 5:153–159. [12]

Castillo-Chavez, C., S. A. Levin, and F. Gould. 1988. An analytical model of physiological and behavioral adaptation to varying environments. Evolution 42:986–994. [6]

Castle, W. E. 1921. An improved method of estimating the number of genetic factors concerned in cases of blending inheritance. Science 54:223. [3]

Castro, A. M., C. P. Rumi, and H. O. Arriaga. 1988. Influence of greenbug on root growth of resistant and susceptible barley genotypes. Environmental and Experimental Botany 28:61–72. [16, 17]

Caswell, H. 1978. Predator-mediated coexistence: A nonequilibrium model. American Naturalist 112:127–154. [5]

Caten, C. E. 1987. The concept of race in plant pathology. In M. S. Wolfe and C. E. Caten, eds., Populations of plant pathogens: Their dynamics and genetics, 21–37. Blackwell Scientific Publications, Oxford [8]

Cates, R. B. 1975. The interface between slugs and wild ginger: Some evolutionary aspects. Ecology 56:391–400. [16, 17]

Cates, R. G., and G. H. Orians. 1975. Successional status and palatability of plants to generalized herbivores. Ecology 56:410–418. [10]

Cates, R. G., and R. A. Redak. 1988. Variation in terpene chemistry of Douglas-fir and its relationship to western spruce budworm success. In K. Spenser, ed., Chemical mediation of coevolution, 317–344. Academic Press, New York. [4]

Chalfant, R. B. 1965. Resistance of bunch bean varieties to the potato leafhopper and the relationship between resistance and chemical control. Journal of Chemical Ecology 58:681–682. [2]

Chambliss, O. L., and C. M. Jones. 1966. Chemical and genetic basis for insect resistance in cucurbits. Journal of Horticultural Science 89:394–405. [2, 13]

Chapin, E. C., and G. W. Schneider. 1975. Resistance to the common peach tree borer (*Samminoidea exitosa* Say) in seedlings of "Rutgers red leaf" peach. HortScience 10:400. [2]

Chapin, F. S., III, J. D. McKendrick, and D. A. Johnson. 1986. Seasonal changes in carbon fractions in Alaskan tundra plants of differing growth form: Implications for herbivory. Journal of Ecology 74:707–731. [16]

Chapin, F. S., III, and M. Slack. 1979. Effect of defoliation upon root growth, phosphate absorption, and respiration in nutrient-limited tundra graminoids. Oecologia 42:67–79. [13]

Chaplin, J. F. 1970. Associations among disease resistance, agronomic characteristics, and chemical constituents in flue-cured tobacco. Agronomy Journal 62:87–91. [15]

Chapman, I. 1826. Some observations on the Hessian fly: Written in the year 1797. Memoir of the Philadelphia Society for the Promotion of Agriculture 5:143–153. [4]

Chapman, R. F. 1977. The role of the leaf surface in food selection by acridids and other insects. Colloques Internationaux du Centre National de la Recherche Scientifique 265:133–149. [10]

———. 1982. Chemoreception: The significance of receptor numbers. Advances in Insect Physiology 16:247–356. [6]

Charlesworth, B. 1990. Optimization models, quantitative genetics, and mutation. Evolution 44:520–538. [3, 17]

Cheplick, G. P., K. Clay, and S. Marks. 1989. Interactions between infection by endophytic fungi and nutrient limitation in the grasses *Lolium perenne* and *Festuca arundinaceae*. New Phytologist 111:89–97. [17]

Chesson, P. L. 1986. Environmental variation and the coexistence of species. In J. Diamond and T. J. Case, eds., Community ecology, 240–256. Harper and Row, New York. [11]

Chester, K. S. 1950. Plant disease losses: Their appraisal and interpretation. Plant Disease Report 193 (Suppl.): 190–362. [13]

Cheverud, J. M. 1988. A comparison of genetic and phenotypic correlations. Evolution 42:958–968. [3, 17]

Chew, F. S., and J. E. Rodman. 1979. Plant resources for chemical defense. In G. A. Rosenthal and D. H. Janzen, eds., Herbivores: Their interaction with secondary plant metabolites, 271–307. Academic Press, New York. [16, 17]

Chew, R. M. 1974. Consumers as regulators of ecosystems: An alternative to energetics. Ohio Journal of Science 74:357–370. [10]

Chin, K. M., and M. S. Wolfe. 1984. The spread of *Erysiphe graminis* f. sp. *hordei* in mixtures of barley varieties. Plant Pathology 33:89–100. [5]

Christ, B. J., C. O. Person, and D. D. Pope. 1987. The genetic determination of variation in pathogenicity. In M. S. Wolfe and C. E. Caten, eds., Populations of plant pathogens: Their dynamics and genetics, 7–19. Blackwell Scientific Publications, Oxford. [8]

Clancy, K. M., P. W. Price and T. P. Craig. 1986. Life history and natural enemies of an undescribed sawfly near *Pontania pacifica* (Hymenoptera: Tenthredinidae) that forms leaf galls on arroyo willow, *Salix lasiolepis*. Annals of the Entomological Society of America 79:884–892. [11]

Clancy, K. M., and P. W. Price. 1987. Rapid herbivore growth enhances enemy attack: Sublethal plant defenses remain a paradox. Ecology 68:733–737. [12]

Claridge, M. F., and J. Den Hollander. 1983. The biotype concept and its application to insect pests of agriculture. Crop Protection 2:85–95. [5, 17]

Clark, A. G. 1987. Senescence and the genetic correlation hang-up. American Naturalist 129:932–940. [3]

Clark, D. A., and D. B. Clark. 1984. Spacing dynamics of a tropical rain forest tree: Evaluation of the Janzen-Connell model. American Naturalist 124:769–788. [13].

Clark, D. B., and D. A. Clark. 1985. Seedling dynamics of a tropical tree: Impacts of herbivory and meristem damage. Ecology 66:1884–1892. [13]

Clark, L. R., and M. J. Dallwitz. 1974. On the relative abundance of some Australia Psyllidae that coexist on *Eucalyptus blakelyi*. Australian Journal of Zoology 22:387–415. [11]

Clarke, B. 1976. The ecological genetics of host-parasite relationships. In E. R. Taylor and R. Muller, eds., Genetic aspects of host-parasite relationships, 87–103. Blackwell Scientific, Oxford. [3, 14, 15, 17]

Clarke, D. D., J. R. Bevan, and I. Crute. 1987. Genetic interactions between wild plants and their parasites. In P. R. Day and G. J. Jellis, eds., Genetics and plant pathogenesis. 195–206. Blackwell Scientific Publications, Oxford. [8]

Clay, K. 1982. Environmental and genetic determinants of cleistogamy in a natural population of the grass *Danthonia spicata*. Evolution 36:734–741. [15]

———. 1984. The effect of the fungus *Atkinsonella hypoxylon* (Clavicipitaceae) on the reproductive system and demography of the grass *Danthonia spicata*. New Phytologist 98:165–175. [14, 15]

———. 1986. Grass endophytes. In N. Fokkema and J. van den Heuvel, eds., Microbiology of the phyllosphere, 188–204. Cambridge University Press, Cambridge. [14, 15]

———. 1987a. The effect of fungi on the interaction between host plants and their herbivores. Canadian Journal of Plant Pathology 9:380–388. [1, 17]

————. 1987b. Effects of fungal endophytes on the seed and seedling biology of *Lolium perenne* and *Festuca arundinacea*. Oecologia 73:358–362. [10, 17]

————. 1988. Fungal endophytes of grasses: A defensive mutualism between plants and fungi. Ecology 69:10–16. [10, 14, 17]

————. 1990. Comparative demography of three graminoids infected by systemic clavicipitaceous fungi. Ecology 71:558–570. [10]

Clay, K., and J. P. Jones. 1984. Transmission of *Atkinsonella hypoxylon* (Clavicipitaceae) by cleistogamous seed of *Danthonia spicata* (Gramineae). Canadian Journal of Botany 62:2893–2895. [15]

Clayton, G. A., J. A. Morris, and A. Robertson. 1957. An experimental check on quantitative genetical theory. I. Short-term responses to selection. Journal of Genetics 55:131–151. [3]

Clements, F. E. 1920. Plant Indicators. Carnegie Institution of Washington, publication 290. [10]

Cockerham, C. C. 1963. Estimation of genetic variances. Statistical genetics and plant breeding. National Academy of Sciences NRC 982:53–94. [3]

Cockerham, C. C., and B. S. Weir. 1977. Quadratic analyses of reciprocal crosses. Biometrics 33:187–203. [3]

Cohen, D. 1967. Optimizing reproduction in a randomly varying environment when a correlation may exist between the condition at the time a choice has to be made and the subsequent outcome. Journal of Theoretical Biology 16:1–14. [7]

Cohen, M. G., M. R. Berenbaum, and M. A. Schuler. 1989. Induction of cytochrome P450 mediated detoxification in the black swallowtail. Journal of Chemical Ecology 15:2347–2355. [4]

Coleman, J. S. 1986. Leaf development and leaf stress: Increased susceptibility associated with sink-source transition. Tree Physiology 2:289–299. [11]

Coleman, J. S., C. G. Jones, and W. H. Smith. 1987. The effect of ozone on cottonwood-leaf rust interactions: Independence of abiotic stress, genotype, and leaf ontogeny. Canadian Journal of Botany 65:949–953. [11]

Coley, P. D. 1980. Effects of leaf age and plant life history patterns on herbivory. Nature 284:545–546. [11]

————. 1983. Intraspecific variation in herbivory in two tropical tree species. Ecology 64:426–433. [6, 11, 13, 16]

————. 1986. Costs and benefits of defense by tannins in a neotropical tree. Oecologia 70:238–241. [2, 13, 16, 17]

————. 1987. Interspecific variation in plant anti-herbivore properties: The role of habitat quality and rate of disturbance. New Phytologist 106:251–263 [13]

————. 1988. Effects of plant growth rate and leaf lifetime on the amount and type of antiherbivore defense. Oecologia 74:531–536. [16]

Coley, P. D., J. P. Bryant, and F. S. Chapin III. 1985. Resource availability and plant antiherbivore defense. Science 230:895–899. [1, 10, 11, 13, 16, 17]

Colhoun, J. 1973. Effects of environmental factors on plant disease. Annual Review of Phytopathology 11:343–364. [3]

Collier, H. O. J., and G. B. Chesher. 1956. Identification of 5-hydroxytryptamine in the sting of the nettle (*Urtica dioica*). British Journal of Pharmacology 11:186–189. [10]

Collinge, S. K., and S. M. Louda. 1988. Herbivory by leaf miners in response to experimental shading of a native crucifer. Oecologia 75:559–566. [13]

————. 1989. Influence of plant phenology on the insect herbivore/bittercress interaction. Oecologia 79:111–116. [13]

Collins, G. N., and J. H. Kempthorne. 1917. Breeding sweet corn resistant to the corn earworm. Journal of Agricultural Research 11:549–572. [4]

Collins, W. J., and Y. Aitken. 1970. The effect of leaf removal on flowering time in subterranean clover. Australian Journal of Agricultural Research 21:893–903. [13]

Comins, H. N. 1977. The development of insecticide resistance in the presence of migration. Journal of Theoretical Biology 64:177–197. [17]

Comins, H. N., and M. P. Hassell. 1987. The dynamics of predation and competition in patchy environments. Theoretical Population Biology 31:393–421. [5]

Committee on Technical Words. 1940. Report. Phytopathology 30:361–368. [14]

Compton, S. G., S. G. Beesley, and D. A. Jones. 1983. On the polymorphism of cyanogenesis in *Lotus corniculatus* L. IX. Selective herbivory in natural populations at Porthdafarch, Anglesey. Heredity 51:537–547. [10, 13]

————. 1986. On the polymorphism of cyanogenesis in *Lotus corniculatus* L. 10. Successional differences in the distribution of cyanogenic and acyanogenic plants. Journal of Natural History 20:1443–1460. [10]

Compton, S. G., and D. A. Jones. 1985. An investigation of the responses of herbivores to cyanogenesis in *Lotus corniculatus* L. Biological Journal of the Linnean Society 26:21–38. [10]

Compton, S. G., D. Newsome, and D. A. Jones. 1983. Selection for cyanogenesis in the leaves and petals of *Lotus corniculatus* L. at high latitudes. Oecologia 60:353–358. [10, 17]

Conn, E. E. 1979. Cyanide and cyanogenic glycosides. In G. A. Rosenthal and D. H. Janzen, eds., Herbivores: Their interaction with secondary plant metabolites, 387–412. Academic Press, New York. [4, 10, 17]

Connally, G. E., B. O. Ellison, J. W. Fleming, S. Geng, R. E. Kepner, W. M. Longhurst, J. H. Oh, and G. F. Russell. 1980. Deer browsing of douglas fir trees in relation to volatile terpene composition and in vitro fermentability. Forest Science 26:179–193. [13]

Connell, J. 1971. On the role of natural enemies in preventing competitive exclusion in some marine animals and in rain forest trees. In P. J. van der Boer and G. R. Gradwell, eds., Dynamics of populations, 298–310. Centre for Agricultural Publishing and Documentation, Waageningen. [13]

Conner, J. 1988. Field measurements of natural and sexual selection in the fungus beetle, *Bolitotherus cornutus*. Evolution 42:736–749. [3]

Cook, G. C., and K. M. El-Zik. 1991. Evaluation of twelve upland cotton genotypes for resistance to *Phymatotrichum omnivorum*. Plant Disease 75:56–58. [8]

Cook, J. L., M. W. Houseweart, H. M. Kulman, and L. C. Thompson. 1978. Foliar mineral differences related to sawfly defoliation of white spruce. Environmental Entomology 7:780–781. [13]

Cooper, J. L., S. E. Kelley, and P. R. Massalski. 1988. Virus-pollen interactions. Advance in Vector Research 5:221–249. [18]

Cooper, J. P. 1960. Selection for production characters in ryegrass. In R. Creed, ed., Ecological genetics and evolution, 41–44. Blackwell Scientific, Oxford. [13]

Cooper, S. M., and N. Owen-Smith. 1985. Condensed tannins deter feeding by browsing ruminants in a South African savanna. Oecologia 67:142–146. [10]

———. 1986. Effects of plant spinescence on large mammalian herbivores. Oecologia 68:446–455. [10, 13]

Cooper, S. M., N. Owen-Smith, and J. P. Bryant. 1988. Foliage acceptability to browsing ruminants in relation to seasonal changes in the leaf chemistry of woody plants in a South African savanna. Oecologia 75:336–342. [10]

Cooper-Driver, G. A., and T. Swain. 1976. Cyanogenic polymorphism in bracken in relation to herbivore predation. Nature 260:604. [10, 13]

Copony, J. A., and B. V. Barnes. 1974. Clonal variation in the incidence of Hypoxylon canker on trembling aspen. Canadian Journal of Botany 52:1475–1481. [15]

Corkill, L. 1942. Cyanogenesis in white clover (*Trifolium repens* L.). V. The inheritance of cyanogenesis. New Zealand Journal of Science and Technology B23:178–193. [13, 17]

———. 1952. Cyanogenesis in white clover (*Trifolium repens* L.). VI. Experiments with high-glucoside and glucoside-free strains. New Zealand Journal of Science and Technology A34:1–16. [10]

Cottam, D. A., J. B. Whittaker, and A. J. C. Malloch. 1986. The effects of chrysomelid beetle grazing and plant competition on the growth of *Rumex obtusifolius*. Oecologia 70:452–456. [13]

Coughenour, M. B. 1985. Graminoid responses to grazing by large herbivores: Adaptations, exaptations, and interacting processes. Annals of the Missouri Botanical Garden 72:852–863. [10, 13]

Coughenour, M. B., S. J. McNaughton, and L. L. Wallace. 1985a. Responses of an African graminoid tall-grass (*Hyparrhenia filipendula* Stupf.) to defoliation and limitations of water and nitrogen. Oecologia 68:80–86. [13]

———. 1985b. Responses of an African graminoid (*Themeda triandra* Forsk.) to frequent defoliation, nitrogen, and water: A limit to adaptations to herbivory. Oecologia 68:105–110. [13]

Coulman, B. E., D. L. Woods, and K. W. Clark. 1977. Distribution within the plant, variation with maturity, and heritability of gramine and hordenine in reed canary grass. Canadian Journal of Plant Science 57:771–777. [16]

Courtney, S. P. 1982. Coevolution of pierid butterflies and their cruciferous foodplants. IV. Host plant apparency and *Anthocharis cardamines* oviposition. Oecologia 52:258–265. [6]

———. 1984. The evolution of egg clustering by butterflies and other insects. American Naturalist 123:267–281. [6]

———. 1986. The ecology of pierid butterflies: Dynamics and interactions. Advances in Ecological Research 15:15–131. [6]

———. 1988. If it's not coevolution it must be predation? Ecology 64:910–911. [6]

Courtney, S. P., and M. I. Manzur. 1985. Fruiting and fitness in *Crataegus marogyna*: The effects of frugivores and seed predators. Oikos 44:398–406. [13]

Cox, T. S., and J. H. Hatchett. 1986. Genetic model for wheat/Hessian fly (Diptera: Cecidomyiidae) interaction: Strategies for deployment of resistance genes in wheat cultivars. Environmental Entomology 15:24–31. [5]

Craig, T. P., J. K. Itami, and P. W. Price. 1990. The window of vulnerability of a shoot-galling sawfly to attack by a parasitoid. Ecology 171:1471–1482. [6, 11]

Craig, T. P., P. W. Price, K. M. Clancy, G. L. Waring, and C. F. Sacchi. 1988. Forces preventing coevolution in the three-trophic-level system: Willow, a gall-forming herbivore, and parasitoid. In K. C. Spencer, ed., Chemical mediation of coevolution, 57–80. Academic Press, New York. [3]

Craig, T. P., P. W. Price, and J. K. Itami. 1986. Resource regulation by a stem-galling sawfly on the arroyo willow. Ecology 67:419–425. [13]

Crawford-Sidebotham, T. J. 1972. The role of slugs and snails in the maintenance of the cyanogenesis polymorphism of *Lotus corniculatus* and *Trifolium repens*. Heredity 28:405–411. [10]

Crawley, M. J. 1983. Herbivory: The dynamics of animal-plant interactions. University of California Press, Berkeley. [1, 2, 10, 13, 17]

———. 1985. Reduction of oak fecundity by low-density herbivore populations. Nature 314:163–164. [13]

———. 1987. The effects of insect herbivores on the growth and reproductive performance of English oak. In V. Labeyrie, G. Fabres, and D. Lachaise, eds., Insects-plants, 307–311. Proceedings of the Sixth International Symposium on Insect-Plant Relationships. Dr. W. Junk, Dordrecht. [13]

———. 1989. Insect herbivores and plant population dynamics. Annual Review of Entomology 34:531–564. [13, 17]

Crawley, M. J., and M. Akhteruzzaman. 1988. Individual variation in the phenology of oak trees and its consequences for herbivorous insects. Functional Ecology 2:409–415. [11]

Crawley, M. J., and M. Nachapong. 1985. The establishment of seedlings from primary and regrowth seeds of ragwort (*Senecio jacobaea*). Journal of Ecology 73:255–261. [13]

Crawley, M. J., and R. Pattrasudhi. 1988. Interspecific competition between insect herbivores: Asymmetric competition between cinnabar moth and the ragwort seed-head fly. Ecological Entomology 13:243–249. [11]

Crespi, B. J., and F. L. Bookstein. 1989. A path-analytical model for the measurement of selection on morphology. Evolution 43:18–28. [7]

Crill, P. 1977. An assessment of stabilizing selection in crop variety development. Annual Review of Phytopathology 15:185–202. [5]

Cromartie, W. J. 1981. The environmental control of insects using crop diversity. In D. Pimentel, ed., CRC handbook of pest management in agriculture, 1:223–225. CRC Press, Boca Raton, FL. [5]

Crow, J. F. 1986. Basic concepts in population, quantitative, and evolutionary genetics. W. H. Freeman, New York. [3, 7]

Crow, J. F., and T. Nagylaki. 1976. The rate of change of a character correlated with fitness. American Naturalist 110:207–213. [3]

Crowley, P. H. 1979. Predator-mediated coexistence: An equilibrium interpretation. Journal of Theoretical Biology 80:129–144. [5]

Cruickshank, I. A. M., and D. R. Perrin. 1964. Pathological function of phenolic compounds in plants. In J. B. Harbourne, ed., Biochemistry of phenolic compounds, 511–544. Academic Press, New York. [3]

Cubit, J. D. 1974. Interaction of seasonally changing physical factors and grazing affecting high intertidal communities on a rocky shore. Ph.D. diss., University of Oregon, Eugene. [10]

Cullen, J. M., P. F. Kable, and M. Catt. 1973. Epidemic spread of a rust imported for biological control. Nature 244:462–464. [15, 18]

Cummins, J. E., and A. W. Day. 1977. Genetic and cell cycle analysis of a smut fungus (*Ustilago violacea*). Methods in Cell Biology 15:445–469. [18]

Curtis, C. F., L. M. Cook, and R. J. Wood. 1978. Selection for and against insecticide resistance and possible methods of inhibiting the evolution of resistance in mosquitos. Ecological Entomology 3:273–287. [17]

Cuthbert, F. P., Jr., and B. W. Davis, Jr. 1970. Resistance in sweet potatoes to damage by soil insects. Journal of Economic Entomology 63:360–363. [2, 11]

Cuthbert, F. P., Jr., R. L. Fery, and O. L. Chambliss. 1974. Breeding for resistance to cowpea curculio in southern peas. HortScience 9:69–70. [2]

DaCosta, C. P., and C. M. Jones. 1971. Cucumber beetle resistance and mite susceptibility controlled by the bitter gene in *Cucumis sativus* L. Science 172:1145–1146. [2, 4, 11, 17]

Daday, H. 1954a. Gene frequencies in wild populations of *Trifolium repens* L. I. Distribution by latitude. Heredity 8:61–78. [17]

———. 1954b. Gene frequencies in wild populations of *Trifolium repens* L. II. Distributed by altitude. Heredity 8:377–384. [17]

———. 1958. Gene frequencies in wild populations of *Trifolium repens*. L. III. World distribution. Heredity 12:169–184. [17]

———. 1965. Gene frequencies in wild populations of *Trifolium repens* L. IV. Mechanisms of natural selection. Heredity 20:355–365. [10, 17]

Dahms, R. G. 1948. Comparative tolerance of small grains to greenbugs from Oklahoma and Mississippi. Journal of Economic Entomology 41:825–826. [5]

Dahms, R. G., T. H. Johnston, A. M. Schlehuber, and E. A. Wood, Jr. 1955. Reaction of small-grain varieties and hybrids to greenbug attack. Oklahoma Agricultural Experiment Station Technical Bulletin T-55. [5]

Dahms, R. G., R. O. Snelling, and F. A. Fenton. 1936. Effect of different varieties of sorghum on biology of the chinch bug. Journal of Economic Entomology 28:160–161. [9]

Dale, J. E. 1959. Some effects of the continuous removal of floral buds on the growth of the cotton plant. Annals of Botany 23:636–649. [13]

Damman, H. 1987. Leaf quality and enemy avoidance by the larvae of a pyralid moth. Ecology 68:88–97. [11, 12]

———. 1989. Facilitative interactions between two lepidopteran herbivores of *Asimina*. Oecologia 78:214–219. [6, 11]

Danell, K., T. Elmqvist, L. Ericson, and A. Salomonson. 1985. Sexuality in willows and preference by bark-eating voles: Defense or not? Oikos 44:82–90. [13, 16]

Danell, K., and K. Huss-Danell. 1985. Feeding by insects and hares on birches earlier affected by moose browsing. Oikos 44:75–81. [11]

Dannell, K., K. Huss-Danell, and R. Bergstrom. 1985. Interactions between browsing moose and two species of birch in Sweden. Ecology 66:1867–1878. [13]

Daubeny, H. A. 1973. Haida red raspberry. Canadian Journal of Botany 53:345–346. [2]

Davis, J. 1967. Some effects of deer browsing on chamise sprouts after fire. American Midland Naturalist 77:234–238. [13]

Davis, B. N. K. 1973. The Hemiptera and Coleoptera of stinging nettle (*Urtica dioica* L.) in East Anglia. Journal of Applied Ecology 10:213–237. [10]

————. 1983. Insects on nettles. Cambridge University Press, Cambridge. [10]

Davis, D. W., W. Randle, and J. V. Groth. 1988. Some sources of partial resistance to common leaf rust (*Puccinia sorghi*) in maize and strategy for screening. Maydica 23:1–13. [8]

Davy, A. J. 1980. Biological flora of the British Isles: *Deschampsia caespitosa* (L.) Beauv. Journal of Ecology 68:1075–1096. [10]

Dawson, E. Y. 1966. Cacti in the Galapagos Islands, with special reference to the relationship with tortoises. In R. I. Bowman, ed., The galapagos, 204–209. University of California Press, Berkeley. [13]

Day, P. R. 1974. Genetics of host-parasite interaction. W. H. Freeman, San Francisco. [4, 14, 15, 18]

————. 1978. The genetic basis of epidemics. In J. G. Horsfall and E. B. Cowling, eds., Plant disease 2:263–285. Academic Press, New York. [14]

Day, P. R., J. A. Barrett, and M. S. Wolfe. 1983. The evolution of host-parasite interaction. In T. Kosuge, C. P. Meredity, and A. Hollaender, eds., Genetic engineering of plants: An agricultural perspective, 419–430. Plenum Press, New York. [15]

de Araújo, A. M. 1976. The relationship between altitude and cyanogenesis in white clover (*Trifolium repens* L.). Heredity 37:291–293. [17]

DeBarr, G. L. 1969. The damage potential of a flower thrip in slash pine seed orchards. Journal of Forestry 67:326–327. [13]

DeBarr, G. L., E. P. Merkel, C. H. O'Gwynn, and M. H. Zoerb, Jr. 1972. Differences in insect infestation in slash pine seed orchards due to phorate treatment and clonal variation. Forest Science 18:56–64. [13]

Delph, L. F. 1986. Factors regulating fruit and seed production in the desert annual *Lesquerella gordonii*. Oecologia 69:471–474. [13]

Delwiche, P. A., and P. H. Williams. 1981. Thirteen marker genes in *Brassica nigra*. Journal of Heredity 72:289–290. [15]

Dement, W. A., and H. A. Mooney. 1974. Seasonal variation in the production of tannins and cyanogenic glucosides in the chaparral shrub, *Heteromeles arbutifolia*. Oecologia 15:65–76. [16]

Dempster, E. R. 1955. Maintenence of genetic heterogeneity. Cold Spring Harbor Symposium on Quantitative Biology 20:25–32. [7]

den Boer, P. J. 1968. Spreading of risk and stabilization of animal numbers. Acta Biotheoretica 18:165–194.

Denno, R. F., S. Larsson, and K. L. Olmstead. 1990. Role of enemy-free space and plant quality in host-plant selection by willow beetles. Ecology 71:124–137. [6, 11]

Denno, R. F., and M. S. McClure, eds. 1983. Variable plants and herbivores in natural and managed systems. Academic Press, New York. [2, 10]

De Nooij, M. P. 1988. The role of weevils in the infection process of the fungus *Phomopsis subordinaria* in *Plantago lanceolata*. Oikos 52:51–58. [14]

De Nooij, M. P., and J. M. M. Van Dame. 1988a. Variation in host susceptibility among and within populations of *Plantago lanceolata* L. infected by the fungus *Phomopsis subordinaria* (Desm.) Trav. Oecologia 75:535–538. [14]

———. 1988b. Variation in pathogenicity among and within populations of the fungus *Phomopsis subordinaria* infecting *Plantago lanceolata*. Evolution 42:1166–1171. [8]

De Nooij, M. P., and H. A. Van der Aa. 1987. *Phomopsis subordinaria* and associated stalk disease in natural populations of *Plantago lanceolata*. Canadian Journal of Botany 65:2318–2325. [14]

de Ponti, O. M. B. 1983. Resistance to insects promotes the stability of integrated pest control. In F. Lamberti, T. M. Waller, and N. A. van der Graaff, eds., Durable resistance in crops, 211–225. Plenum, New York. [2]

dePonti, O. M. B., and F. Garretsen. 1980. Resistance in *Cucumis sativus* L. to *Tetranychus urticae* Koch. 7. The inheritance of resistance and the relation between these characters. Euphytica 29:513–523. [2, 4]

dePonti, O. M. B., G. G. Kennedy, and F. Gould. 1983. Different resistance of nonbitter cucumbers to *Tetranychus urticae* in the Netherlands and the U.S.A. Cucurbit Genetics Cooperative. [2]

De Steven, D. 1981. Predispersal seed predation in a tropical shrub (*Mabea occidentalis:* Euphorbiaceae). Biotropica 13:146–150. [13]

———. 1983. Reproductive consequences of insect seed predation in *Hamamelis virginiana*. Ecology 64:89–98. [13]

De Steven, D., and F. E. Putz. 1984. Impact of mammals on early recruitment of a tropical canopy tree, *Dipteryx panamensis*, in Panama. Oikos 43:207–216. [13]

———. 1985. Mortality rates of some rain forest palms in Panama. Principes 29: 162–165.

Dethier, V. G. 1954. Evolution of feeding preferences in phytophagous insects. Evolution 8:33–54. [5]

———. 1980. Food-aversion learning in two polyphagous caterpillars, *Diacrisia virginica* and *Estigmene congrua*. Physiological Entomology 5:321–325. [4]

Detling, J. K., M. I. Dyer, and D. T. Winn. 1979. Net photosynthesis, root respiration, and regrowth of *Bouteloua gracilis* following simulated grazing. Oecologia 41:127–134. [13]

Detling, J. K., and E. L. Painter. 1983. Defoliation responses of western wheatgrass populations with diverse histories of prairie dog grazing. Oecologia 57:65–71. [13]

Deverall, B. J. 1972. Phytoalexins. In J. B. Harbourne, ed., Phytochemical ecology, 217–234. Academic Press, New York. [3]

———. 1977. Defence mechanisms of plants. Cambridge University Press, Cambridge. [14]

Devlin, B., K. Roeder, and N. C. Ellstrand. 1988. Fractional paternity assignment: Theoretical development and comparison to other methods. Theoretical and Applied Genetics 76:369–380. [3]

Devonshire, A. L. 1975. Studies of the acetylcholinesterase from houseflies (*Musca domestica* L.) resistant and susceptible to organophosphorus insecticides. Biochemical Journal 149:463–469. [17]

Diamond, J., and T. J. Case, eds. 1986. Community ecology. Harper and Row, New York. [11]

Dickerson, G. E. 1955. Genetic slippage in response to selection for multiple objectives. Cold Spring Harbor Symposium on Quantitative Biology 21:213–224. [3]

Dickey, J. L., and M. Levy. 1979. Development of powdery mildew (*Erysiphe polygoni*) on susceptible and resistant races of *Oenothera biennis*. American Journal of Botany 66:1114–1117. [15]

Diehl, S. R., and G. L. Bush. 1984. An evolutionary and applied perspective of insect biotypes. Annual Review of Entomology 29:471–504. [2, 5, 6]

Dillon, P. M., S. Lowrie, and D. McKey. 1983. Disarming the "Evil Woman": Petiole constriction by a sphingid larva circumvents mechanical defenses of its host plant, *Cnidoscolus urens* (Euphorbiaceae). Biotropica 15:112–116. [10]

Dimock, E. J., II, R. R. Silen, and V. E. Allen. 1976. Genetic resistance in Douglas-fir to damage by snowshoe hare and black-tailed deer. Forest Science 22:106–121. [10, 13]

Dimock, M. B. 1981. Chemical and physiological studies of antibiosis in the wild tomato *Lycopersicon hirsutum* f. *glabratum* to the tomato fruitworm *Heliothis zea*. M.S. thesis, North Carolina State University, Raleigh. [2]

Dimock, M. B., and W. M. Tingey. 1985. Resistance in *Solanum* spp. to the Colorado potato beetle: Mechanisms, genetic resources, and potential. In D. N. Ferro and R. H. Voss, eds., Proceedings of the symposium on the ecology and management of the Colorado potato beetle, 79–104. Seventeenth International Congress of Entomology, Massachusetts Agriculture Experimental Station Research Bulletin no. 704. [2]

Dinoor, A. 1970. Sources of oat crown rust resistance in hexaploid and tetraploid wild oats in Israel. Canadian Journal of Botany 48:153–161. [14]

Dinoor, A., and N. Eshed. 1984. The role and importance of pathogens in natural plant communities. Annual Review of Phytopathology 22:443–466. [14, 15, 18]

———. 1987. The analysis of host and pathogen populations in natural ecosystems. In M. S. Wolfe and C. E. Caten, eds., Populations of plant pathogens: Their dynamics and genetics, 75–88. Blackwell Scientific Publications. Oxford. [8, 14]

Dinus, R. J. 1974. Knowledge about natural ecosystems as a guide to disease control in managed forests. Proceedings of the American Phytopathological Society 1:184–190. [14]

Dirzo, R. 1980. Experimental studies on slug-plant interactions. I. The acceptability of thirty plant species to the slug *Agriolimax carunae*. Journal of Ecology 68:981–998. [10]

———. 1984. Herbivory: A phytocentric overview. In R. Dirzo and J. Sarukhán, eds., Perspectives in plant population biology, 141–165. Sinauer, Sunderland, MA. [10, 13, 17]

———. 1985. Metabolitos secundarios en las plantas: Atributos panglossianos o de valor adaptativo? Ciencia 36:137–145. [13]

Dirzo, R., and J. L. Harper. 1982a. Experimental studies on slug-plant interactions. III. Differences in the acceptability of individual plants of *Trifolium repens* to slugs and snails. Journal of Ecology 70:101–118. [9, 10, 13]

———. 1982b. Experimental studies on slug-plant interactions. IV. The performance of cyanogenic and acyanogenic morphs of *Trifolium repens* in the field. Journal of Ecology 70:119–138. [9, 10, 13, 14, 17]

Dixon, A. F. G. 1971a. The role of aphids in wood formation. I. The effect of the sycamore aphid, *Drepanosiphum platanoidis* (Schr.) (Aphididae), on the growth of sycamore, *Acer pseudoplatanus* L. Journal of Applied Ecology 8:165–179. [13]

———. 1971b. The role of aphids in wood formation. II. The effect of the lime aphid, *Eucallipterus tiliae* L. (Aphididae), on the growth of lime *Tilia* x *vulgaris* Hayne. Journal of Applied Ecology 8:393–399. [13]

Dixon, R. A., J. A. Bailey, J. N. Bell, G. P. Bolwell, C. L. Cramer, K. Edwards, M. A. M. S. Hamden, C. J. Lamb, M. P. Robbins, T. B. Ryder, and W. Schuch. 1986. Rapid changes in gene expression in response to microbial elicitation. Philosophical Transactions of the Royal Society of London. B314:411–426. [16]

Doherty, H. M., R. R. Selvendran, and D. J. Bowles. 1988. The wound response of tomato plants can be inhibited by aspirin and related hydroxy-benzoic acids. Physiological and Molecular Plant Pathology 33:377–384. [16]

Dolinger, P. M., P. R. Ehrlich, W. L. Fitch, and D. E. Breedlove. 1973. Alkaloid and predation patterns in Colorado lupine populations. Oecologia 13:191–204. [4, 13]

Donnelly, E. D., and W. B. Anthony. 1970. Effect of genotype and tannin on dry matter digestibility in sericea lespedeza. Crop Science 10:200–202. [10]

———. 1973. Relationship of sericea lespedeza leaf and stem tannin to forage quality. Agronomy Journal 65:993–994. [10]

———. 1983. Breeding low-tannin sericea. III. Variation in forage quality factors among lines. Crop Science 23:982–984. [10]

Donoghue, M. J. 1989. Phylogenies and the analysis of evolutionary sequences, with examples from seed plants. Evolution 43:1137–1156. [17]

Doolittle, S. P. 1954. The use of wild *Lycopersicon* species for tomato disease control. Phytopathology 44:409–414. [14]

Dorschner, K. W., J. D. Ryan, R. C. Johnson, and R. D. Eikenbary. 1987. Modification of host nitrogen levels by the greenbug (Homoptera: Aphididae): Its role in resistance of winter wheat to aphids. Environmental Entomology 16:1007–1011. [11]

Doss, R. P., C. H. Shanks, Jr., J. D. Chamberlain, and J. K. L. Garth. 1987. Role of leaf hairs in resistance of a clone of beach strawberry, *Fragaria chiloensis*, to feeding by adult black vine weevil, *Otiorhynchus sulcatus* (Coleoptera: Curculionidae). Environmental Entomology 16:764–768. [13]

Dover, B. A., R. Noblet, R. F. Moore, and B. M. Shepard. 1987. Development and emergence of *Pediobius foveolatus* from Mexican beetle larvae fed foliage from *Phaseolus lunatus* and resistant and susceptible soybeans. Journal of Agricultural Entomology 4:271–279. [12]

Dowd, P. F. 1989. Fusaric acid: A secondary fungal metabolite that synergizes the toxicity of cooccurring host allelochemicals to the corn earworm, *Heliothis zea* (Lepidoptera). Journal of Chemical Ecology 15:249–254. [3]

Dreyer, D. L., K. C. Jones, and R. J. Molyneaux. 1985. Feeding deterrency of some pyrrolizidine, indolizidine, and quinolizidine alkaloids towards pea aphid (*Acyrthosiphon pisum*) and evidence for phloem transport of indolizidine alkaloid swainsonine. Journal of Chemical Ecology 11:1045–1051. [13]

Dritschilo, W., J. Krummel, D. Nafus, and D. Pimental. 1979. Herbivorous insects colonising cyanogenic and acyanogenic *Trifolium repens*. Heredity 42:49–56. [13]

Duffey, S. S. 1980. Sequestration of plant natural products by insects. Annual Review of Entomology 25:447–477. [12, 17]

———. 1986. Plant glandular trichomes: Their partial role in defense against insects.

In B. Juniper and Sir R. Southwood, eds., Insects and the plant surface, 151–172. Edward Arnold, London. [13]

Duffey, S. S., and K. A. Bloem. 1986. Plant defense-herbivore-parasite interactions and biological control. In M. Kogan, ed., Ecological theory and integrated pest management practice, 135–183. John Wiley and Sons, New York. [12]

Duffey, S. S., K. A. Bloem, and B. C. Campbell. 1986. Consequences of sequestration of plant natural products in plant-insect-parasitoid interactions. In D. J. Boethel and R. D. Eikenbary, eds., Interaction of host plant resistance and parasitoids and predators of insects, 31–60. Halsted Press, New York. [12]

Duggan, A. E. 1985. Pre-dispersal seed predation by *Anthocharis cardamines* (Pieridae) in the population dynamics of the perennial *Cardamine pratensis* (Brassicaceae). Oikos 44:99–106. [13]

Dumas, B. A., and A. J. Mueller. 1986. Distribution of biotype C and E greenbugs on wheat in Arkansas. Journal of Entomological Science 21:38–42. [5]

Duncan, D. P., and A. C. Hodson. 1958. Influence of the forest tent caterpillar upon the aspen forests of Minnesota. Forest Science 4:71–93. [13]

Dunn, J. A., and D. P. Kempton. 1972. Resistance to attack by *Brevicoryne brassicae* among plants of Brussels sprouts. Annals of Applied Biology 72:1–11. [5]

Dwinell, L. D. 1985. Oak hosts affect pathogenic variability of *Fusiform* rust fungus. In J. Barrows-Broaddus and H. R. Powers, eds., Proceedings of the Rusts of Hard Pines Working Conference, 251–258. Georgia Center for Continuing Education, University of Georgia, Athens. [8]

Dyer, M. I., J. K. Detling, D. C. Coleman, and D. W. Hilbert. 1982. The role of herbivores in grasslands. In J. R. Estes, R. J. Tyrl, and J. N. Brunken, eds., Grasses and grasslands: Systematics and ecology, 255–295. University of Oklahoma Press, Norman. [10]

Eastop, V. F. 1973. Biotypes of aphids. Bulletin of the Entomological Society of New Zealand. 2:40–51. [5]

Eck, H. V., F. C. Gilmore, D. B. Ferguson, and G. C. Wilson. 1975. Heritability of nitrate reductase and cyanide levels in seedlings of grain sorghum cultivars. Crop Science 15:421–423. [13]

Edmunds, G. F., and D. N. Alstad. 1978. Coevolution in insect herbivores and conifers. Science 199:941–945. [9, 11, 13]

———. 1981. Responses of black pineleaf scales to host plant variability. In R. F. Denno and H. Dingle, eds., Insect life history patterns: Habitat and geographic variation, 29–38. Springer-Verlag, New York. [9]

Edson, J. L. 1985. The influences of predation and resource subdivision on the coexistence of goldenrod aphids. Ecology 66:1736–1743. [5, 11]

Edwards, J. 1984. Spatial pattern of clonal structure of the perennial herb, *Aralia nudicaulis* L. (Araliaceae). Bulletin of the Torrey Botanical Club 111:28–33. [13]

———. 1985. Effects of herbivory by moose on flower and fruit production of *Aralia nudicaulis*. Journal of Ecology 73:861–868. [13]

Edwards, M. D., C. W. Stuber, and J. F. Wendel. 1987. Molecular-marker-facilitated investigations of quantitative-trait loci in maize. I. Numbers, genomic distribution, and types of gene action. Genetics 116:113–125. [3]

Edwards, P. J., and S. D. Wratten. 1980. Ecology of insect-plant interactions. Studies in Biology 121. Edward Arnold, London. [13]

Edwards, P. J., S. D. Wratten, and S. Greenwood. 1986. Palatibility of British trees to insects: Constitutive and induced defences. Oecologia 69:316–319. [16]

Efron, B. 1979. Bootstrap methods: Another look at the jackknife. Annals of Statistics 7:1–26. [3]

Efron, B., and G. Gong. 1983. A leisurely look at the bootstrap, the jackknife, and cross-validation. American Statistician 37:36–48. [3]

Ehleringer, J. R., I. Ullmann, O. L. Lange, G. D. Farquhar, I. R. Cowan, E.-D. Schulze, and H. Ziegler. 1986. Mistletoes: A hypothesis concerning morphological and chemical avoidance of herbivory. Oecologia 70:234–237. [13]

Ehrlich, P. R., and P. H. Raven. 1964. Butterflies and plants: A study in coevolution. Evolution 18:586–608. [2, 4, 10, 13, 17]

Eisen, E. J., and A. M. Saxton. 1983. Genotype by environment interactions and genetic correlations involving two environmental factors. Theoretical and Applied Genetics 67:75–86. [3]

Eisenbach, J., and T. E. Mittler. 1987a. Extra-nuclear inheritance in a sexually produced aphid: The ability to overcome host plant resistance by biotype hybrids of the greenbug. *Schizaphis graminum.* Experientia 43:332–334. [5, 6]

————. 1987b. Polymorphism of biotypes E and C of the aphid *Schizaphis graminum* (Homoptera: Aphididae) in response to different scotophases. Environmental Entomology 16:519–523. [6]

Elden, T. C., J. H. Elgin, and J. L. Soper. 1986. Inheritance of pubescence in selected clones from two alfalfa populations and relationship to potato leafhopper resistance. Crop Science 26:1143–1146. [2]

El-Heineidy, A. H., P. Barbosa, and P. Gross. 1988. Influence of dietary nicotine on the fall armyworm, *Spodoptera frugiperda,* and its parasitoid, the ichneumonid wasp, *Hyposoter annulipes.* Entomologia Experimentalis et Applicata 46:227–232. [12]

Ellingboe, A. H. 1985. Prospects for using recombinant DNA technology to study race-specific interactions between host and parasite. In J. V. Groth and W. R. Bushness, eds., Genetic basis of biochemical mechanisms of plant disease, 103–125. American Phytopathological Society Press, St. Paul. [14]

Ellis, W. M., R. J. Keymer, and D. A. Jones. 1977a. The defensive function of cyanogenesis in natural populations. Experientia 33:309–311. [10]

————. 1977b. The effect of temperature on the polymorphism of cyanogenesis in *Lotus corniculatus* L. Heredity 38:339–347. [17]

————. 1977c. On the polymorphism of cyanogenesis in *Lotus corniculatus* L. VIII. Ecological studies in Anglesey. Heredity 39:45–65. [17]

Ellison, L. 1960. Influence of grazing on plant succession of rangelands. Botanical Review 26:1–78. [10, 13]

Ellison, R. L. and J. N. Thompson. 1987. Variation in seed and seedling size: The effects of seed herbivores on *Lomatium grayi* (Umbelliferae). Oikos 49:269–280. [13]

Ellstrand, N. C., and J. Antonovics. 1985. Experimental studies on the evolutionary significance of sexual reproduction. II. A test of the density-dependent selection hypothesis. Evolution 39:657–666. [18]

Elmqvist, T., L. Ericson, K. Danell, and A. Salomonson. 1987. Flowering, shoot production, and vole bark herbivory in a boreal willow. Ecology 68:1623–1629. [13]

———. 1988. Latitudinal sex ratio variation in willows, *Salix* spp., and gradients in vole herbivory. Oikos 51:259–266. [13]

Elmqvist, T., and H. Gardfjell. 1988. Differences in response to defoliation between males and females of *Silene dioica*. Oecologia 77:225–230. [13, 16]

Elzen, G. W., H. J. Williams, and S. B. Vinson. 1983. Response by the parasitoid *Campoletis sonorensis* (Hymenoptera: Ichneumonidae) to chemicals (synomones) in plants: Implications for host habitat location. Environmental Entomology 12:1872–1876. [12]

———. 1984. Isolation and identification of cotton synomones mediating searching behavior by parasitoid *Campoletis sonorensis*. Journal of Chemical Ecology 10:1251–1264. [12]

———. 1986. Wind tunnel flight responses by hymenopterous parasitoid *Campoletis sonorensis* to cotton cultivars and lines. Entomologia Experimentalis et Applicata 42:285–289. [12]

Emmelin, N., and W. Feldberg. 1947. The mechanism of the sting of the common nettle (*Urtica urens*). Journal of Physiology (London) 106:440–445. [10]

Endler, J. A. 1986. Natural selection in the wild. Princeton University Press, Princeton, NJ. [1, 5, 7, 13, 14, 17]

Ennos, R. A. 1981a. Detection of selection in populations of white clover (*Trifolium repens* L.). Biological Journal of the Linnean Society 15:75–82. [13]

———. 1981b. Manifold effects of the cyanogenic loci in white clover. Heredity 46:127–132. [17]

———. 1983. Maintenance of genetic variation in plant populations. Evolutionary Biology 16:129–155. [15]

Erdman, M. D. 1983. Nutrient and cardenolide composition of unextracted and solvent-extracted *Calotropis procera*. Journal of Agricultural and Food Chemistry 31:509–513. [16]

Eshed, N., and A. Dinoor. 1981. Genetics of pathogenicity in *Puccinia coronata:* The host range among grasses. Phytopathology 71:156–163. [15]

Estes, J. A., and P. D. Steinberg. 1988. Predation, herbivory, and kelp evolution. Paleobiology 14:19–36. [17]

Evans, E. W., C. C. Smith, and R. P. Gendron. 1989. Timing of reproduction in a prairie legume: Seasonal impacts of insects consuming flowers and seeds. Oecologia 78:220–230. [13]

Eyjolfsson, R. 1970. Recent advances in the chemistry of cyanogenic glycosides. Fortschritte der Chemie Organischer Naturstoffe 28:74–108. [17]

Ezcurra, E., J. C. Gomez, and J. Becerra. 1987. Diverging patterns of host use by phytophagous insects in relation to leaf pubescence in *Arbutus xalapensis* (Ericaceae). Oecologia 72:479–480. [13]

Faeth, S. E. 1896. Indirect interactions between temporally separated herbivores mediated by the host plant. Ecology 67:479–494. [6, 11]

———. 1987. Community structure and folivorous insect outbreaks: The roles of vertical and horizontal interactions. In P. Barbosa and J. C. Schultz, eds., Insect outbreaks, 135–171. Academic Press, New York. [1, 11]

———. 1988. Plant-mediated interactions between seasonal herbivores: Enough for evolution or coevolution? In K. C. Spencer, ed., Chemical mediation of coevolution, 391–414. American Institute of Biological Sciences, Washington, DC. [1, 11, 17]

Faeth, S. E., S. Mopper, and D. Simberloff. 1981. Abundances and diversity of leaf-mining insects on three oak host species: Effects of host-plant phenology and nitrogen content of leaves. Oikos 37:238–251. [6, 11]

Fagerström, T. 1989. Anti-herbivory chemical defense in plants: A note on the concept of cost. American Naturalist 133:281–287. [17]

Fagerström, T., S. Larsson, and O. Tenow. 1987. On optimal defense in plants. Functional Ecology 1:73–81. [17]

Falconer, D. S. 1952. The problem of environment and selection. American Naturalist 86:293–298. [3, 7]

———. 1953. Selection for large and small size in mice. Journal of Genetics 51:470–501. [3]

———. 1981. Introduction to quantitative genetics. 2d ed., Longman, London. [2, 3, 6, 7, 11, 14, 17]

Farentinos, R. C., P. J. Capretta, R. E. Kepner, V. M. Littlefield. 1891. Selective herbivory in tassel-eared squirrels: Role of monoterpenes in ponderosa pines chosen as feeding trees. Science 213:1273–1275. [10, 13]

Farrar, R. R., J. D. Barbour, and G. G. Kennedy. 1989. Quantifying food consumption and growth in insects. Annals of the Entomological Society of America 82:593–598. [4]

Feder, J. L., C. A. Chilcote, and G. L. Bush. 1988. Genetic differentiation between sympatric host races of the apple maggot fly Rhagoletis pomonella. Nature 336:61–64. [4, 6]

Federation of British Plant Pathologists. 1973. A guide to the use of terms in plant pathology. Phytopathological Papers, no. 17. Commonwealth Mycological Institute, Kew. [14]

Feeny, P. P. 1968. Effect of oak leaf tannins on larval growth of the winter moth Operophtera brumata. Journal of Insect Physiology 14:805–817. [7]

———. 1970. Seasonal changes in oak leaf tannins and nutrients as a cause of spring feeding by winter moth caterpillars. Ecology 51:565–581. [17]

———. 1975. Biochemical coevolution between plants and their insect herbivores. In L. H. Gilbert and P. H. Raven, eds., Coevolution of animals and plants, 3–19. University of Texas Press, Austin. [17]

———. 1976. Plant apparency and chemical defense. In J. W. Wallace and R. L. Mansell, eds., Biochemical interaction between plants and insects, 1–40. Plenum, New York. [2, 10, 11, 13, 16, 17]

———. 1977. Defensive ecology of the Cruciferae. Annals of the Missouri Botanical Garden 64:221–234. [4]

Feeny, P., K. Sachdeve, L. Rosenberry, and M. Carter. 1989. Luteolin 7-0-(6'-0-malonyl)-beta-D-glucoside and trans-chlorogenic acid: Oviposition stimulants for the black swallowtail butterfly. Phytochemistry 27:3439–3448. [4]

Fehr, W. R. 1987. Principles of cultivar development. Vol. 1. Macmillan, New York. [2]

Felsenstein, J. 1976. The theoretical population genetics of variable selection and migration. Annual Review of Genetics 10:253–280. [5]

————. 1988. Sex and the evolution of recombination. In R. E. Michod and B. R. Levin, eds., The evolution of sex, 74–86. Sinauer, Sunderland, MA. [15]

Fenner, M. 1987. Seedlings. New Phytologist 106 (Suppl.): 35–47. [13]

Ferguson, J. E., and R. L. Metcalf. 1985. Cucurbitacins: Plant-derived defense compounds for diabroticites (Coleoptera: Chrysomelidae). Journal of Chemical Ecology 11:311–318. [17]

Fernandes, G. W., and T. G. Whitham. 1989. Selective fruit abscission by *Juniperus monosperma* as an induced defense against predators. American Midland Naturalist 121:389–392. [13]

Fernando, R. L., S. A. Knights, and D. Gianola. 1984. On a method of estimating the genetic correlation between characters measured in different experimental units. Theoretical and Applied Genetics 67:175–178. [3]

Fery, R. L., and F. P. Cuthbert, Jr. 1972. Association of plant density, cowpea curculio damage, and *Choanephora* pod rot in southern peas. Journal of the American Society of Horticultural Science 97:800. [2]

————. 1974a. Effect of plant density on fruitworm damage in tomato. HortScience 9:140–141. [2]

————. 1974b. Resistance of tomato cultivars to the fruitworm, *Heliothis zea* (Boddie). HortScience 9:469–470. [2]

————. 1975. Inheritance of pod resistance to cowpea curculio infestation in southern peas. Journal of Heredity 66:43–44. [2]

————. 1978. Inheritance and selection of nonpreference resistance to the cowpea curculio in the southern bean (*Vigna unguiculata* L. Walp.). Journal of the American Society for Horticultural Science 103:370–372. [2]

Fery, R. L., and G. G. Kennedy. 1987. Genetic analysis of 2-tridecanone concentration, leaf trichome characteristics, and tobacco hornworm resistance in *Lycopersicon*. Journal of the American Society for Horticultural Science 112:886–891. [2]

Field, C., and H. A. Mooney. 1986. The photosynthesis-nitrogen relationship in wild plants. In T. J. Givnish, ed., On the economy of plant form and function, 25–55. Cambridge University Press, Cambridge. [16]

Filip, G. M., and C. A. Parker. 1987. Simultaneous infestation by dwarf mistletoe and western spruce budworm decrease growth of Douglas fir in the Blue Mountains of Oregon. Forest Science 33:767–773. [13]

Finch, S., and T. H. Jones. 1989. An analysis of the deterrent effect of aphids on cabbage root fly (*Delia radicum*) egg-laying. Ecological Entomology 14:387–391. [6]

Fish, J. 1978. Resistance of *Sorghum bicolor* to *Rhopalosiphum maidis* and *Peregrinus maidis* as affected by differences in growth stage of the host. Entomologia Experimentalis et Applicata. 23:227–236. [2]

Fisher, R. A. 1918. The correlation between relatives on the supposition of Mendelian inheritance. Transactions of the Royal Society of Edinburgh 52:399–433. [3]

————. 1959. The genetical theory of natural selection, 2d ed. Dover, New York. [7]

Fleming, R. 1980. Selection pressures and plant pathogens: Robustness of the model. Phytopathology 70:175–178. (Errata: Phytopathology 71 (1981): 268) [17]

Flemion, F., and B. McNear. 1951. Reduction of vegetative growth and seed yield in umbelliferous plants by *Lygus oblineatus*. Contributions of the Boyce Thompson Institute 16:279–283. [4]

Flor, H. H. 1942. Inheritance of pathogens in *Melamspora lini*. Phytopathology 32:653–659. [2]

———. 1955. Host-parasite interaction in flax rust: Its genetics and other implications. Phytopathology 45:680–685. [5, 8, 15]

———. 1956. The complementary genetic systems in flax and flax rust. Advances in Genetics 8:29–54. [2, 14, 15, 17]

———. 1971. Current status of the gene-for-gene concept. Annual Review of Phytopathology 9:272–296. [17]

Forde, M. B. 1964. Inheritance patterns of turpentine composition in *Pinus attenuata* × *radiata* hybrids. New Zealand Journal of Botany 4:53–59. [2]

Forno, I. W., and J. L. Semple. 1987. Response of *Salvinia molesta* to insect damage: Changes in nitrogen, phosphorous, and potassium content. Oecologia 73:71–74. [13]

Forsberg, J. 1987. Size discrimination among conspecific host plants in two pierid butterflies, *Pieris napi* L. and *Pontia daplidice* L. Oecologia 72:52–57. [13]

Foster, G. S., R. K. Campbell, and W. T. Adams. 1984. Heritability, gain, and C effects in rooting of western hemlock cuttings. Canadian Journal of Forest Research 14:628–638. [3]

Foster, S. A. 1986. On the adaptive value of large seeds for tropical moist forest trees: A review and synthesis. Botanical Review 52:260–299. [13]

Foster, W. A. 1984. The distribution of the sea-lavender aphid *Staticobium staticis* on a marine saltmarsh and its effect on host plant fitness. Oikos 42:97–104. [13]

Foulds, W. 1977. The physiological response to moisture supply of cyanogenic and acyanogenic phenotypes of *Trifolium repens* L. and *Lotus corniculatus* L. Heredity 39:219–234. [10, 17]

Foulds, W., and J. P. Grime. 1972a. The influence of soil moisture on the frequency of cyanogenic plants in populations of *Trifolium repens* and *Lotus corniculatus*. Heredity 28:143–146. [13, 17]

———. 1972b. The response of cyanogenic and acyanogenic phenotypes of *Trifolium repens* to soil moisture supply. Heredity 28:181–187. [13, 16, 17]

Foulds, W., and L. Young. 1977. Effect of frosting, moisture stress, and potassium cyanide on the metabolism of cyanogenic and acyanogenic phenotypes of *Lotus corniculatus* L. and *Trifolium repens* L. Heredity 38:19–24. [13, 17]

Fowler, N. L., and M. D. Rausher. 1985. Joint effects of competitors and herbivores on growth and reproduction in *Aristolochia reticulata*. Ecology 66:1580–1587. [11, 13]

Fowler, S. V., and J. H. Lawton. 1985. Rapidly induced defenses and talking trees: The devil's advocate position. American Naturalist 126:181–195. [16]

Fox, L. R. 1981. Defense and dynamics in plant-herbivore systems. American Zoologist 21:853–864. [10, 12, 17]

———. 1988. Diffuse coevolution within complex communities. Ecology 69:906–907. [13]

Fox, L. R., and B. J. Macauley. 1977. Insect grazing on *Eucalyptus* in response to variation in leaf tannins and nitrogen. Oecologia 29:145–162. [13]

Fox, L. R., and P. A. Morrow. 1981. Specialization: Species property or local phenomenon? Science 211:1466–1470. [4]

———. 1983. Estimates of damage by herbivorous insects on *Eucalyptus* trees. Australian Journal of Ecology 8:139–147. [13]

———. 1986. On comparing herbivore damage in Australian and north temperate systems. Australian Journal of Ecology 11:387–393. [4, 13]

Fraenkel, G. S. 1959. The raison d'être of secondary plant substances. Science 129:1466–1470. [4, 10, 13]

Frank, S. A., and M. Slatkin. 1990. Evolution in a variable environment. American Naturalist 136:244–260. [7]

Frankel, O. H., and A. H. D. Brown. 1984. Current plant genetic resources: A critical appraisal. In O. H. Frankel and A. H. D. Brown, eds., Genetics: New frontiers, vol. 4. Oxford and IBH Publishing Company, New Delhi. [2]

Franklin, R. T. 1970. Insect influences on the forest canopy. In D. E. Reichle, ed., Analysis of temperate forest ecosystems, 86–99. Springer-Verlag, Berlin. [13]

Fraser, R. S. S. 1985. Mechanisms of resistance to plant diseases. Dr. W. Junk Publishers, Dordrecht. [2]

Frazer, B. D. 1972. Population dynamics and the recognition of biotypes in the pea aphid. Canadian Entomologist 104:1729–1733. [6]

Frazer, B. D., and N. Gilbert. 1976. Coccinellids and aphids: A quantitative study of the impact of adult ladybirds (Coleoptera: Coccinellidae) preying on field populations of pea aphids (Homoptera: Aphididae). Journal of the Entomological Society of British Columbia 73:33–56. [5]

Freeland, W. J. 1974. Vole cycles: another hypothesis. American Naturalist 108:238–245. [10]

Freeland, W. J., and D. H. Janzen. 1974. Strategies in herbivory by mammals: The role of plant secondary compounds. American Naturalist 108:269–289. [10]

Frey, K. J., and J. A. Browning. 1971. Association between genetic factors for crown rust resistance and yield in oats. Crop Science 11:757–760. [15]

Frey, K. J., J. A. Browning, and M. D. Simons. 1973. Management of host resistance genes to control diseases. Zeitschrift für Pflanzenkrankheiten und Pflanzenschutz. 80:160–180. [9]

———. 1977. Management systems for host genes to control disease loss. Annals of the New York Academy of Science 287:255–274. [5]

Fritz, R. S. 1990a. Effects of genetic and environmental variation on resistance of willow to sawflies. Oecologia 82:325–332. [2, 6, 11, 13]

———. 1990b. Variable competition between insect herbivores on genetically variable host plants. Ecology 71:2008–2011. [6, 11]

Fritz, R. S., W. S. Gaud, C. F. Sacchi, and P. W. Price. 1987a. Patterns of intra- and interspecific association of gall-forming sawflies in relation to shoot size on their willow host plant. Oecologia 73:159–169. [9]

———. 1987b. Variation in herbivore density among host plants and its consequences for community structure: Field studies on willow sawflies. Oecologia 72:577–588. [2, 11, 13]

Fritz, R. S., and J. Nobel. 1989. Plant resistance, plant traits, and host plant choice of the leaf-folding sawfly on the arroyo willow. Ecological Entomology 14:393–401. [11, 13]

———. 1990. Host plant variation in mortality of the leaf-folding sawfly on the arroyo willow. Ecological Entomology 15:25–35. [11, 12]

Fritz, R. S., and P. W. Price. 1988. Genetic variation among plants and insect community structure: Willows and sawflies. Ecology 69:845–856. [1, 2, 3, 6, 9, 11, 13, 17]

———. 1990. A field test of interspecific competition on oviposition of gall-forming sawfly species on willow. Ecology 71:99–106. [11]

Fritz, R. S., C. F. Sacchi, and P. W. Price. 1986. Competition versus host plant phenotype in species composition: Willow sawflies. Ecology 67:1608–1618. [1, 6, 9, 11, 17]

Frost, H. B. 1948. Genetics and breeding. In H. J. Webber and L. D. Batchelor, eds., The citrus industry, 817–913. University of California Press, Berkeley. [15]

Fry, J. D. 1990. Trade-offs in fitness on different hosts: Evidence from a selection experiment with the phytophagous mite *Tetranychus urticae* Koch. American Naturalist 136:569–580. [6]

Fry, J. D. 1992. The mixed-model analysis of variance applied to quantitative genetics: Biological meaning of the parameters. Evolution (in press). [3]

Futuyma, D. J. 1983a. Evolutionary interactions among herbivorous insects and plants. In D. J. Futuyma and M. Slatkin, eds., Coevolution, 207–231. Sinauer, Sunderland, MA. [5, 10, 12, 13, 17]

———. 1983b. Selective factors in the evolution of host choice by phytophagous insects. In S. Ahmad, ed., Herbivorous insects: Host-seeking behavior and mechanisms, 227–244. Academic Press, New York. [5, 6]

———. 1986. The role of behavior in host associated divergence in herbivorous insects. In M. Huettel, ed., Evolutionary genetics of invertebrate behavior, 295–302. Plenum Press, New York. [6]

———. 1987. The role of behavior in host associated divergence in herbivorous insects. In M. Huettel, ed., Evolutionary genetics of invertebrate behavior. Plenum, New York. [7]

Futuyma, D. J., R. F. Cort, and I. van Noordwijk. 1984. Adaptations to host plants in the fall cankerworm (*Alsophila pometaria*) and its bearing on the evolution of host affiliation in phytophagous insects. American Naturalist 123:287–296. [5]

Futuyma, D. J., and F. Gould. 1979. Association of plants and insects in a deciduous forest. Ecological Monographs 49:33–60. [6]

Futuyma, D. J., and G. C. Mayer. 1980. Non-allopatric speciation in animals. Systematic Zoology 29:254–271. [6]

Futuyma, D. J., and G. Moreno. 1988. The evolution of ecological specialization. Annual Review of Ecology and Systematics 19:207–234. [4, 6]

Futuyma, D. J., and S. C. Peterson. 1985. Genetic variation in the use of resources by insects. Annual Review of Entomology 30:217–238. [2, 5, 6]

Futuyma, D. J., and T. E. Philippi. 1987. Genetic variation and covariation in responses to host plants by *Alsophila pometaria* (Lepidoptera: Geometridae). Evolution 41:269–279. [3, 4, 5, 6]

Futuyma, D. J., and M. Slatkin, eds. 1983. Coevolution. Sinauer Associates, Sunderland, MA. [11]

Futuyma, D. J., and S. S. Wasserman. 1981. Food plant specialization and feeding efficiency in the tent caterpillars, *Malacosoma disstria* Hubner and *M. americanum* (F.). Entomologia Experimentalis et Applicata 30:106–110. [5, 7]

Fyfe, J. L., and N. Gilbert. 1963. Partial diallel crosses. Biometrics 19:278–286. [3]

Gabriel, W. 1988. Quantitative genetic models for parthenogenetic species. In G. de Jong, ed., Population genetics and evolution, 73–82. Springer-Verlag, Berlin. [6]

Gaines, S. D. 1985. Herbivory and between-habitat diversity: The differential effectiveness of defenses in a marine plant. Ecology 66:473–485. [10]

Gaines, S. D., and J. Lubchenco. 1982. A unified approach to marine plant-herbivore interactions. II. Biogeography. Annual Review of Ecology and Systematics 13:111–138. [17]

Gall, L. F. 1987. Leaflet position influences caterpillar feeding and development. Oikos 49:172–179. [11]

Gallagher, K. D. 1988. Effects of host plant resistance on the microevolution of the rice brown planthopper, *Nilaparvata lugens* (Stål) (Homoptera: Delphacidae). Ph.D. diss. University of California, Berkeley. [5]

Gallun, R. L. 1972. Genetic interrelationships between host plants and insects. Journal of Environmental Quality 1:259–265. [17]

Gallun, R. L. 1977. Genetic basis of Hessian fly epidemics. Annals of the New York Academy of Sciences 287:223–229. [2, 5]

Gallun, R. L., and J. H. Hatchett. 1968. Interrelationships between races of Hessian fly, *Mayetiola destructor* (Say), and resistance in wheat. In Proceedings of the Third International Wheat Genetics Symposium, 258–262. Australian Academy of Science, Canberra. [17]

Gallun, R. L., and G. S. Khush. 1980. Genetic factors affecting expression and stability of resistance. In F. G. Maxwell and P. R. Jennings, eds., Breeding plants resistant to insects, 63–86. John Wiley and Sons, New York. [2, 4, 5, 11]

Gallun, R. L., K. J. Starks, and W. D. Guthrie. 1975. Plant resistance to insects attacking cereals. Annual Review of Entomology 20:337–357. [13]

Gange, A. C., and V. K. Brown. 1989. Effects of root herbivory by an insect on a foliar-feeding species, mediated through changes in the host plant. Oecologia 81:38–42. [6]

Ganskopp, D. 1988. Defoliation of Thurber needlegrass: Herbage and root responses. Journal of Range Management 41:472–476. [13]

Garner, F. H. 1963. The palatability of herbage plants. Journal of the British Grassland Society 18:79–89. [10]

Garrish, R. S., and T. D. Lee. 1989. Physiological integration in *Cassia fasciculata* Michx.: Inflorescence removal and defoliation experiments. Oecologia 81:279–284. [13]

Gaynor, D. L., and W. F. Hunt. 1983. The relationship between nitrogen supply, endophytic fungus, and Argentine stem weevil resistances in ryegrass. Proceedings of the New Zealand Grassland Association 44:257–263. [2]

Gebhardt, M. D., and S. C. Stearns. 1988. Reaction norms for developmental time and weight at eclosion in *Drosophila mercatroum*. Journal of Evolutionary Biology 1:335–354. [7]

Geiger, H. H., and M. Heun. 1989. Genetics of quantitative resistance to fungal diseases. Annual Review of Phytopathology 27:317–341. [3]

Gentile, A. G., and A. K. Stoner. 1968. Resistance in *Lycopersicon* and *Solanum* species to the potato aphid. Journal of Economic Entomology 61:1152–1154. [2]

George, J. A., and T. R. Davidson. 1963. Pollen transmission of necrotic ring spot and sour cherry yellow virus from tree to tree. Canadian Journal of Plant Science 43:276–286. [18]

Georgiadis, N. J., and S. J. McNaughton. 1988. Interactions between grazers and a cyanogenic grass, *Cynadon plectostachyus*. Oikos 51:343–350. [10, 13]

Gerechter-Amitai, Z. K., and R. W. Stubbs. 1970. A valuable source of yellow rust resistance in Israeli populations of wild emmer, *Triticum dicoccoides*. Euphytica 19:12–21. [15]

Gergerich, R. C., and H. A. Scott. 1988. The enzymatic function of ribonuclease determines plant virus transmission by leaf-feeding beetles. Phytopathology 78:270–272. [14]

Gershenzon, J. 1984. Changes in the levels of plant secondary metabolites under water and nutrient stress. In C. Steelink and S. A. Loewin, eds., Phytochemical adaptations to stress. Recent Advances in Phytochemistry 18:273–320. [13]

Getz, W. M., and J. Pickering. 1983. Epidemic models: Thresholds and population regulation. American Naturalist 121:892–898. [18]

Gibberd, R., P. J. Edwards, and S. D. Wratten. 1988. Wound-induced changes in the acceptability of tree-foliage to Lepidoptera: Within-leaf effects. Oikos 51:43–47. [16]

Gibbs, D. 1974. Chemotaxonomy of flowering plants. McGill-Queen's University Press, Montreal. [10, 17]

Gibbs, J. N. 1978. Intercontinental epidemiology of Dutch elm disease. Annual Review of Phytopathology 16:287–307. [18]

Gibson, L., and M. Visser. 1982. Interspecific competition between two field populations of grass-feeding bugs. Ecological Entomology 7:61–67. [11]

Gibson, R. W. 1979. The geographical distribution, inheritance, and pest-resisting properties of sticky-tipped foliar hairs on potato species. Potato Research 22:223–236. [2]

Giertych, M. 1970. The influence of defoliation on flowering in pine *Pinus sylvestris*. Aboretum Kornickie 15:93–98. [13]

Giese, R. L., and K. H. Knauer. 1977. Ecology of the walking stick. Forest Science 23:45–63. [13]

Gilbert, L. E. 1971. Butterfly-plant coevolution: Has *Passiflora adenopoda* won the selectional race with heliconiine butterflies? Science 172:585–586. [13]

———. 1975. Ecological consequences of a coevolved mutualism between butterflies and plants. In L. E. Gilbert and P. H. Raven, eds., Coevolution of animals and plants, 210–240. University of Texas Press, Austin. [13]

Gill, B. S., W. J. Raupp, H. C. Sharma, L. E. Browder, J. H. Hatchett, T. L. Harvey, J. G. Moseman, and J. G. Waines. 1986. Resistance of *Aegilops squarrosa* to wheat leaf rust, wheat powdery mildew, greenbug, and Hessian fly. Plant Disease 70:553–556. [11]

Gillespie, J. 1973. Natural selection for within-generation variance in offspring number. Genetics 76:601–606. [7]

———. 1974 Polymorphism in patchy environments. American Naturalist 108:145–151. [7]

Gillespie, J. H. 1975. Natural selection for resistance to epidemics. Ecology 56:493–495. [15]

Gillespie, J. H., and M. Turelli. 1989. Genotype-environment interaction in the maintenance of polygenic variation. Genetics 121:129–138. [7]

Gilreath, M. E., G. S. McCutcheon, G. R. Carner, and S. G. Turnipseed. 1986. Pathogen incidence in noctuid larvae from selected soybean genotypes. Journal of Agricultural Entomology 3:213–226. [12]

Giroux, J. F., and J. Bedard. 1987. Effects of simulated feeding by snow geese on *Scirpus americanus* rhizome. Oecologia 74:137–143. [13]

Gladstones, J. S. 1970. Lupins as crop plants. Field Crop Abstracts 23:123–148. [10]

Glander, K. E. 1981. Feeding patterns in mantled howling monkeys. In A. C. Kamil and T. D. Sargent, eds., Foraging behavior, 231–257. Garland Publishing, New York. [10]

Glesener, R. R., and D. Tilman. 1978. Sexuality and the components of environmental uncertainty: Clues from geographical pathogenesis in terrestrial animals. American Naturalist 112:659–673. [18]

Glover, D. V., and E. H Stanford. 1966. Tetrasomic inheritance of resistance in alfalfa to the pea aphid. Crop Science 6:161–165. [2]

Goldschmidt, V. 1928. Vererbungsversuche mit den biologischen Arten des Antherenbrandes (*Ustilago violacea* Pers.). Zeitschrift für Botanik 21:1–90. [18]

Gonzalez-Candelas, F., and L. D. Mueller. 1989. A two-locus, two-allele model of natural selection in alternating environments: Implications for phenotypic plasticity. (manuscript) [7]

Goodnight, C. J. 1987. On the effect of founder events on epistatic genetic variance. Evolution 41:80–91. [3]

––––––. 1988. Epistasis and the effect of founder events on the additive genetic variance. Evolution 42:441–454. [3]

Gori, D. F. 1983. Post-pollination phenomena and adaptive floral changes. In C. E. Jones and R. J. Little, eds., Handbook of experimental pollination biology, 31–45. Van Nostrand Reinhold, New York. [15]

Gould, F. 1978. Resistance of cucumber varieties to *Tetranychus urticae:* Genetic and environmental determinants. Journal of Economic Entomology 71:680–683. [4]

––––––. 1979. Rapid host range evolution in a population of the phytophagous mite *Tetranychus urticae* Koch. Evolution 33:791–802. [3, 6, 7]

––––––. 1983. Genetics of plant-herbivore systems: Interactions between applied and basic studies. In R. F. Denno and M. S. McClure, eds., Variable plants and herbivores in natural and managed systems, 599–653. Academic Press, New York. [2, 5, 6, 9, 11, 17]

––––––. 1984. Role of behavior in the evolution of insect adaptation to insecticides and resistant host plants. Bulletin of the Entomological Society of America 30:33–41. [4, 6, 17]

––––––. 1986a. Simulation models for predicting durability of insect-resistant germ plasm: A deterministic diploid, two-locus model. Environmental Entomology 15:1–10. [5, 9]

––––––. 1986b. Simulation models for predicting durability of insect-resistant germ plasm: Hessian fly (Diptera: Cecidomyiidae)-resistant winter wheat. Environmental Entomology 15:11–23. [5, 9]

––––––. 1988. Genetics of pairwise and multispecies plant-herbivore coevolution. In

K. C. Spencer, ed., Chemical mediation of coevolution, 13–55. American Institute of Biological Sciences, Washington, DC. [10, 11, 17]

Gould, S. J., and R. C. Lewontin. 1979. The spandrels of San Marco and the Panglossian paradigm: A critique of the adaptationist programme. Proceedings of the Royal Society of London, B205:581–598. [10]

Gould, S. J., and Vrba, E. S. 1982. Exaptation: A missing term in the science of form. Paleobiology 8:4–15. [10]

Gouyon, P. H., P. Fort, and G. Caraux. 1983. Selection of seedlings of *Thymus vulgaris* by grazing slugs. Journal of Ecology 71:299–306. [10]

Grant, J. F., and M. Shepard. 1985. Influence of three soybean genotypes on development of *Voria ruralis* (Diptera: Tachinidae) and on foliage consumption by its host, the soybean looper (Lepidoptera: Noctuidae). Florida Entomologist 68:672–677. [12]

Grant, W. M., and S. A. Archer. 1983. Calculation of selection coefficients against unnecessary genes for virulence from field data. Phytopathology 73:547–557. [5]

Gray, A. J., M. G. Drury, and A. F. Raybould. 1990. *Spartina* and ergot fungus *Claviceps purpurea:* A singular contest? In J. J. Burdon and S. R. Leather, eds., Pests, pathogens, and plant communities, 63–79. Blackwell Scientific Publications, Oxford. [14]

Graybill, F. A. 1976. Theory and application of the linear model. Duxbury Press, Boston, MA. [3, 5]

Greaves, R. 1966. Insect defoliation of eucalypt regrowth in the Florentine Valley, Tasmania. Appita 19:119–126. [13]

———. 1967. The influence of insects on the productive capacity of Australian forests. Australian Forest Research 3:36–45. [13]

Green, H. F. 1986. Does aggregation prevent competitive exclusion? A response to Atkinson and Shorrocks. American Naturalist 128:301–304. [5]

Green, M. B., and P. A. Hedin. 1986. Natural resistance of plants to pests: Roles of allelochemicals. ACS Symposium Series 296. American Chemical Society, Washington, DC. [2]

Greenwood, R. M., and I. A. E. Atkinson. 1977. Evolution of divaricating plants in New Zealand in relation to moa browsing. Proceedings of the New Zealand Ecological Society 24:21–33. [13]

Gregory, P., W. M. Tingey, D. A. Ave, and P. Bouthyette. 1986. Potato glandular trichomes: A phytochemical defense mechanism against insects. In M. B. Green and P. A. Hedin, eds., Natural resistance of plants to pests, 160–167. ACS Symposium Series 296. American Chemical Society, Washington, DC. [2]

Gregory, R. A., and P. M. Wargo. 1986. Timing of defoliation and its effect on bud development, starch reserves, and sap sugar concentration in sugar maple. Canadian Journal of Forest Research 16:10–17. [13]

Greig-Smith, P. W., and M. F. Wilson. 1985. Influences of seed size, nutrient composition, and phenolic content on the preferences of bullfinches feeding in ash trees. Oikos 44:47–54. [13]

Gressel, J. 1987. Biotechnologically conferring herbicide resistance in crops: The present realities. In L. van Vloten-Doting, G. S. P. Groot, and T. C. Hall, eds. Molecular form and function of the plant genome, 489–504. NATO ASI Series 83. Plenum Press, New York. [17]

Griffing, B. 1956a. Concept of general and specific combining ability in relation to diallel crossing systems. Australian Journal of Biological Sciences 9:463–493. [3]

———. 1956b. A generalized treatment of the use of diallel crosses in quantitative inheritance. Heredity 10:31–50. [3]

Griffiths, E., and S. D. Wratten. 1979. Intra- and inter-specific differences in cereal aphid low-temperature tolerance. Entomologia Experimentalis et Applicata 26: 161–167. [6]

Grime, J. P., G. M. Blythe, and J. D. Thornton. 1970. Food selection by the snail *Cepaea nemoralis* L. In Animal populations in relation to their food resources, 73–99. British Ecological Society Symposium no. 10. [10]

Grime, J. P., S. F. MacPherson-Stewart, and R. S. Dearman. 1968. An investigation of leaf palatability using the snail *Cepaea nemoralis* L. Journal of Ecology 56:405–420. [10]

Gross, R. S., and P. A. Werner. 1983. Relationships among flowering phenology, insect visitors, and seed-set of individuals: Experimental studies on four co-occurring species of goldenrod (*Solidago:* Compositae). Ecological Monographs 53:95–117. [13]

Groth, J. V. 1976. Multilines and "super races": A simple model. Phytopathology. 66:937–939. [5, 8]

———. 1988. *Uromyces appendiculatus,* rust of Phaseolus beans. In D. S. Ingram and P. H. Williams, eds., Advances in plant pathology. Vol. 6, Genetics of plant pathogenic fungi, ed. G. S. Sidhu, 389–400. Academic Press, London. [8]

Groth, J. V. and H. M. Alexander. 1988. Genetic divergence in spatially separated pathogen populations. In M. J. Jeger, ed., Spatial components of plant disease epidemics, 165–181. Prentice-Hall, New Jersey. [8]

Groth, J. V., and C. O. Person. 1977. Genetic interdependence of host and parasite in epidemics. Annals of the New York Academy of Science 287:97–106. [5]

Groth, J. V., and A. P. Roelfs. 1982. The effect of sexual and asexual reproduction on race abundance in cereal rust fungus populations. Phytopathology 72:1503–1507. [5]

———. 1987a. The analysis of virulence diversity in populations of plant pathogens. In M. S. Wolfe and C. E. Caten, eds., Populations of plant pathogens: Their dynamics and genetics, 633–674. Blackwell Scientific Publications, Oxford. [8]

———. 1987b. The concept and measurement of phenotypic diversity in *Puccinia graminis* on wheat. Phytopathology 77:1395–1399. [8]

Grubb, P. J. 1986. Sclerophylls, pachyphylls, and pycnophylls: The nature and significance of hard leaf surfaces. In B. E. Juniper and T. R. E. Southwood, eds., Insects and the plant surface, 137–150. Edward Arnold, London. [10]

Guenther, E. 1948. The essential oils. D. van Nostrand Co., New York. [4]

Gulmon, S. L., and C. C. Chu. 1981. The effects of light and nitrogen on photosynthesis, leaf characteristics and dry matter allocation in the chaparral shrub, *Diplacus aurantiacus*. Oecologia 49:207–212. [17]

Gulmon, S. L. and H. A. Mooney. 1986. Costs of defense and their effects on plant productivity. In T. J. Givnish, ed., On the economy of plant form and function, 681–698. Cambridge University Press, Cambridge. [13, 16, 17]

Gutierrez, A. P. 1986. Analysis of the interaction of host plant resistance, phytophagous and entomophagous species. In D. J. Boethel and R. D. Eikenbary, eds., Interaction of host plant resistance and parasitoids and predators of insects, 198–215. Halsted Press, New York. [12]

Gutierrez, A. P., J. U. Baumgaertner, and C. G. Summers. 1984. Multitrophic models of predator-prey energetics. Canadian Entomologist 116:923–963. [5]

Gwynne, M. O., and R. H. V. Bell. 1968. Selection of vegetation components by grazing ungulates in the Serengeti National Park. Nature 220:390–393. [10]

Hackerott, H. L. 1972. Registration of KS30 sorghum germplasm. Crop Science 12:719. [5]

Hagerman, A. E. 1988. Extraction of tannin from fresh and preserved leaves. Journal of Chemical Ecology 14:453–461. [4]

Hairston, N. G., F. E. Smith, and L. B. Slobodkin. 1960. Community structure, population control, and competition. American Naturalist 94:421–425. [10, 11]

Haldane, J. B. S. 1949. Disease and evolution. La Ricerca Scientifica 19 (Suppl.) 68–76. [14, 15]

Hamilton, W. D. 1980. Sex versus non-sex versus parasite. Oikos 35:282–290. [14, 15, 18]

Hamilton, W. D., R. Axelrod, and R. Tanese. 1990. Sexual reproduction as an adaptation to resist parasites (a review). Proceedings of the National Academy of Sciences 87:3566–3573. [14]

Hamm, J. J., and B. R. Wiseman. 1986. Plant resistance and nuclear polyhedrosis virus for suppression of the fall armyworm (Lepidoptera: Noctuidae). Florida Entomologist 69:541–549. [12]

Hammond, A. M., and T. N. Hardy. 1988. Quality of diseased plants as hosts for insects. In E. A. Heinricks, ed., Plant stress–insect interactions, 381–432. J. Wiley and Sons, New York. [2]

Hammond, K., and J. W. James. 1970. Genes of large effect and the shape of the distribution of a quantitative character. Australian Journal of Biological Sciences 23:867–876. [3]

Hamrick, J. L., and L. R. Holden. 1979. Influence of microhabitat heterogeneity on gene frequency distribution and gametic phase disequilibrium in *Avena barbata*. Evolution 33:521–533. [15]

Hanks, L. M., and R. F. Denno. 1989. The role of demic adaptation in colonization and spread of scale insect populations. In K. C. Kim, ed., Evolution of insect pests: The pattern of variations. John Wiley, New York. [9]

Hanny, B. W. 1980. Gossypol, flavonoid, and condensed tannin content of cream and yellow anthers of five cotton (*Gossypium hirsutum* L.) cultivars. Journal of Agriculture and Food Chemistry 28:504–506. [16]

Hanny, B. W., J. C. Bailey, and W. R. Meredith, Jr. 1979. Yellow cotton pollen suppresses growth of larvae of tobacco budworm. Environmental Entomology 8:706–707. [16]

Hanover, J. W. 1966a. Genetics of terpenes. I. Gene control of monoterpene levels in *Pinus monticola* Dougl. Heredity 21:73–84. [2, 13, 16, 17]

———. 1966b. Inheritance of 3-carene concentration in *Pinus monticola*. Forest Science 12:447–450. [2]

Hanski, I. 1981. Coexistence of competitors in patchy environment with and without predation. Oikos 37:306–312. [5]

———. 1983. Coexistence of competitors in patchy environment. Ecology 64:493–500. [5]

Hanson, H. C., L. D. Love, and M. S. Morris. 1931. Effects of different systems of grazing by cattle upon a western wheat-grass type of range. Colorado Agricultural Experiment Station Bulletin 377:1–82. [13]

Harberd, D. J. 1961. Observations on population structure and longevity of *Festuca rubra*. New Phytologist 60:184–206. [15]

———. 1967. Observations on natural clones in *Holcus mollis*. New Phytologist 66:401–408. [15]

Harborne, J. B. 1972. Phytochemical ecology. Proceedings of the Phytochemical Society Symposium, Royal Holloway College, Englefield Green, Surrey. Academic Press, London. [2]

———. 1988. Introduction to ecological biochemistry. Academic Press, London. [2]

Hardy, T. N., K. Clay, and A. M. Hammond, Jr. 1985. Fall armyworm (Lepidoptera: Noctuidae): A laboratory bioassay and larval preference study for the fungal endophyte of perennial ryegrass. Journal of Economic Entomology 15:1083–1089. [2, 10]

Hare, J. D. 1980. Variation in fruit size and susceptibility to seed predation among and within populations of the cocklebur, *Xanthium strumarium* L. Oecologia 46:217–222. [2, 9, 11, 13]

———. 1983. Manipulation of host suitability for herbivore pest management. In R. F. Denno and M. S. McClure, eds., Variable plants and herbivores in natural and managed systems, 655–680. Academic Press, New York. [2, 12]

Hare, J. D. 1987. Growth of *Leptinotarsa decemlineata* in response to simultaneous variation in protein and glycoalkaloid concentration. Journal of Chemical Ecology 13:39–46.

Hare, J. D., and J. A. Dodds. 1987. Survival of the Colorado potato beetle on virus-infected tomato in relation to plant nitrogen and alkaloid content. Entomologia Experimentalis et Applicata 43:31–36.

Hare, J. D., and D. J. Futuyma. 1978. Different effects of variation in *Xanthium strumarium* L. (Compositae) on two insect seed predators. Oecologia 37:109–112. [11, 13]

Hare, J. D., and G. G. Kennedy. 1986. Genetic variation in plant-insect associations: Survival of *Leptinotarsa decemlineata* populations on *Oolanum carolinense*. Evolution 40:1031–1043. [2, 6]

Harlan, J. R. 1976. Diseases as a factor in plant evolution. Annual Review of Phytopathology 14:31–51. [15, 18]

———. 1979. Origins of agriculture and crop evolution. In M. K. Harris, ed., Biology and breeding for resistance to arthropods and pathogens in agricultural plants, 1–8. Texas A & M University Publication MP 145. College Station, TX. [2]

Harlan, J. R., and K. J. Starks. 1980. Germplasm resources and needs. In F. G. Maxwell and P. R. Jennings, eds., Breeding plants resistant to insects. John Wiley and Sons, New York. [2]

Harley, J. L. 1969. The biology of mycorrhiza. Leonard Hill, London. [17]

Harley, K. L. S., and A. J. Thorsteinson. 1967. The influence of plant chemicals on the feeding behavior, development, and survival of the two-striped grasshopper, *Melanoplus bivittatus* (Say), Acrididae: Orthoptera. Canadian Journal of Zoology 45:305–319. [13]

Harper, J. L. 1967. A Darwinian approach to plant ecology. Journal of Ecology 55:247–270. [18]

———. 1977. Population biology of plants. Academic Press, New York. [1, 2, 10, 13, 14, 17, 18]

———. 1982. After description. In E. I. Newman, ed., The plant community as a working mechanism, 11–25. Blackwell, Oxford. [10, 14, 18]

———. 1990. Pests, pathogens, and plant communities: An introduction. In J. J. Burdon and S. R. Leather, eds., Pests, pathogens, and plant communities, 3–14. Blackwell Scientific Publications, Oxford. [14]

Harris, M. K. 1975. Allopatric resistance: Searching for sources of insect resistance for use in agriculture. Environmental Entomology 4:661–669. [2]

———. 1982. Genes for resistance to insects in agriculture with a discussion of host-parasite interaction in *Carya*. In H. M. Heybrook, B. R. Stephan, and K. Von Weissenberg, eds. Resistance to diseases and pests in forest trees, 72–83. Proceedings of the 3rd international workshop Wageningen (Netherlands). Centre for Agricultural Publishing and Documentation, Wageningen. [2]

———, ed. 1979. Biology and breeding for resistance to anthropods and pathogens in agricultural plants. Texas A&M University Publication MP 1451. College Station, TX. [2]

Harris, M. K., and R. A. Frederiksen. 1984. Concepts and methods regarding host plant resistance to arthropods and pathogens. Annual Review of Phytopathology 22:247–272. [2, 4, 11]

Harris, M. K., and C. E. Rogers, eds. 1988. The entomology of indigenous and naturalized systems in agriculture. Westview Press, Boulder. [2]

Harris, W., and J. L. Brock. 1972. Effect of Porina caterpillar (*Wiseana* spp.) infestations on yield and competitive interactions of ryegrass and white clover varieties. New Zealand Journal of Agricultural Research 15:723–740. [9]

Harrison, E. 1987. Treefall gaps versus forest understory as environments for a defoliating moth on a tropical forest shrub. Oecologia 72:65–68. [13]

Harrison, S., and R. Karban. 1986. Effects of an early-season folivorous moth on the success of a later-season species, mediated by a change in the quality of a shared host, *Lupinus arboreus* Sims. Oecologia 69:354–359. [6, 11]

Harry, I. B., and D. D. Clarke. 1986. Race-specific resistance in groundsel (*Senecio vulgaris*) to the powdery mildew *Erysiphe fischeri*. New Phytologist 103:167–175. [8, 14, 15]

———. 1987. The genetics of race-specific resistance in groundsel (*Senecio vulgaris*) to the powdery mildew fungus *Erysiphe fischeri*. New Phytologist 107:715–723. [14, 15]

Hartl, D. L. 1980. Principles of population genetics. Sinauer Associates, Sunderland, MA. [2]

Hartley, S. E. 1988. The inhibition of phenolic biosynthesis in damaged and undamaged birch foliage and its effect on insect herbivores. Oecologia 76:65–70. [16]

Hartnett, D. C. 1989. Density- and growth stage-dependent responses to defoliation in two rhizomatous grasses. Oecologia 80:414–420. [13]

Hartnett, D. C., and W. G. Abrahamson. 1979. The effects of stem gall insects on the life history patterns of *Solidago canadensis*. Ecology 60:910–917. [13]

Hartnett, D. C., and F. A. Bazzaz. 1984. Leaf demography and plant-insect interactions: Goldenrods and phloem-feeding aphids. American Naturalist 124:137–142. [16]

Harvell, C. D. 1986. The ecology and evolution of inducible defenses in a marine bryozoan: Cues, costs, and consequences. American Naturalist 128:810–823. [10]

Harvey, T. L., and H. L. Hackerott. 1969a. Plant resistance to a greenbug biotype injurious to sorghum. Journal of Economic Entomology 62:1271–1274. [5]

———. 1969b. Recognition of greenbug biotypes injurious to sorghum. Journal of Economic Entomology 62:776–779. [6]

Harvey, T. L., H. L. Hackerott, F. L. Sorensen, R. H. Painter, E. E. Ortman, and D. C. Peters. 1960. The development and performance of Cody alfalfa, a spotted alfalfa aphid resistant variety. Kansas Agricultural Experiment Station Technical Bulletin 114:1–26. [2]

Haskins, F. A., and H. J. Gorz. 1986. Relationship between contents of leucoanthocyanidin and dhurrin in sorghum leaves. Theoretical and Applied Genetics 73:2–3. [16]

Hassell, M. P. 1978. The dynamics of arthropod predator-prey systems. Monographs in Population Biology 13. Princeton University Press, Princeton. [12]

Hassell, M. P., and R. M. Anderson. 1984. Host susceptibility as a component in host parasitoid systems. Journal of Animal Ecology 53:611–621. [12]

Hassell, M. P., and R. M. May 1988. Spatial heterogeneity and the dynamics of parasitoid-host systems. Annales Zoologici Fennici 25:55–61. [11]

Hassell, M. P., and J. K. Waage. 1984. Host-parasitoid population interactions. Annual Review of Ecology and Systematics 29:89–114.[11]

Hastings, A. 1978. Spatial heterogeneity and the stability of predator-prey systems: predator-mediated coexistence. Theoretical Population Biology 14:380–395. [5]

Hastings, A., and C. L. Hom. 1989. Pleiotropic stabilizing selection limits the number of polymorphic loci to at most the number of characters. Genetics 122:459–464. [17]

Hastings, A., and C. L. Wolin. 1989. Within-patch dynamics in a metapopulation. Ecology 70:1261–1266. [5]

Haukioja, E., and S. Hanhimaki. 1985. Rapid wound-induced resistance in white birch (*Betula pubescens*) foliage to the geometrid *Epirrita autumnata:* A comparison of trees and moths within and outside the outbreak range of the moth. Oecologia 65:223–228. [3]

Haukioja, E., and S. Neuvonen. 1985. Induced long-term resistance of birch foliage against defoliators: Defensive or incidental? Ecology 66:1303–1308. [11, 13, 16]

Haukioja, E., and P. Niemelä. 1977. Retarded growth of a geometrid larva after mechanical damage to leaves of its host tree. Annales Zoologici Fennici 14:48–52. [16]

Haukioja, E., P. Niemelä, and S. Siren. 1985. Foliage phenols and nitrogen in relation to growth, insect damage, and ability to recover after defoliation, in the mountain birch *Betula pubescens* spp. *tortuosa*. Oecologia 65:214–222. [13]

Haukioja, E., J. Soumela, and S. Neuvonen. 1985. Long-term inducible resistance in birch foliage: Triggering cues and efficacy on a defoliator. Oecologia 65:363–369. [11]

Havel, J. 1986. Predator-induced defenses: A review. In W. C. Kerfoot and A. Sih, eds., Predation: Direct and indirect effects on aquatic communities. University Press of New England, Hanover, NH. [10]

Hawkins, B. A. 1988. Foliar damage, parasitoids, and indirect competition: A test using herbivores of birch. Ecological Entomology 13:301–308. [11]

Hay, M. E. 1981. Spatial patterns of grazing intensity on a Caribbean barrier reef: Herbivory and algal distribution. Aquatic Botany 11:97–109. [17]

———. 1984. Patterns of fish and urchin grazing on Caribbean coral reefs: Are previous results typical? Ecology 65:446–454. [17]

Hay, M. E., J. E. Duffy, C. A. Pfister, and W. Fenical. 1987. Chemical defense against different marine herbivores: Are amphipods insect equivalents? Ecology 68:1567–1580. [10]

Hay, M. E., and W. Fenical. 1988. Marine plant-herbivore interactions: The ecology of chemical defense. Annual Review of Ecology and Systematics 19:111–145. [10, 17]

Hay, M. E., V. J. Paul, S. M. Lewis, K. Gustafson, J. Tucker, and R. N. Trindell. 1988. Can tropical seaweeds reduce herbivory by growing at night?: Diel patterns of growth, nitrogen content, herbivory, and chemical versus morphological defenses. Oecologia 75:233–245. [13, 17]

Hayman, B. I. 1954a. The analysis of variance of diallel tables. Biometrics 10:235–244. [3]

———. 1954b. The theory and analysis of diallel crosses. Genetics 39:789–809. [3]

———. 1958. The theory and analysis of diallel crosses. II. Genetics 43:63–85. [3]

Heath, M. C. 1987. Evolution of plant resistance and susceptibility to fungal invaders. Canadian Journal of Plant Pathology 9:389–397. [15]

Hedrick, P. W. 1985. Genetics of populations. Jones and Bartlett Publishers, Boston. [3]

———. 1986. Genetic polymorphism in heterogeneous environments: A decade later. Annual Review of Ecology and Systematics 17:535–566. [5, 15]

Hedrick, P. W., M. E. Ginevan, and E. P. Ewing. 1976. Genetic polymorphism in heterogeneous environments. Annual Review of Ecology and Systematics 7:1–32. [5, 15]

Hedrick, P. W., S. Jain, and L. Holden. 1978. Multilocus systems in evolution. Evolutionary Biology 11:101–184. [15]

Hefendehl, F. W., and M. J. Murray. 1976. Genetic aspects of the biosynthesis of natural odors. Lloydia 39:39–52. [13]

Heichel, G. H., and N. C. Turner, 1976. Phenology and leaf growth of defoliated hardwood trees. In J. Anderson and H. Kaya, eds., Perspectives in forest entomology, 31–40. Academic Press, New York. [13]

———. 1983. CO_2 assimilation of primary and regrowth foliage of red maple (*Acer rubrum* L.) and red oak (*Quercus rubra* L.): Adaptation to defoliation. Oecologia 57:14–19. [13]

———. 1984. Branch growth and leaf numbers of red maple (*Acer rubrum* L.) and red oak (*Quercus rubra* L.): Response to defoliation. Oecologia 62:1–6. [13]

Heinricks, E. A., ed. 1988. Plant stress–insect interactions. J. Wiley and Sons, New York. [2]

Heithaus, E. R. 1981. Seed predation by rodents on three ant-dispersed plants. Ecology 62:136–145. [13]

Hendrix, S. D. 1979. Compensatory reproduction in a biennial herb following insect defoliation. Oecologia 42:107–118. [13, 16]

———. 1984. Variation in seed weight and its effect on germination in *Pastinaca sativa* L. (Umbelliferae). American Journal of Botany 71:795–802. [13]

———. 1988. Herbivory and its impact on plant reproduction. In J. Lovett Doust and L. Lovett Doust, eds., Plant reproductive ecology: Patterns and strategies, 246–263. Oxford Press, Oxford. [13]

Hendrix, S. D., and E. J. Trapp. 1981. Plant-herbivore interactions: Insect induced changes in host plant sex expression and fecundity. Oecologia 49:119–122. [13]

———. 1989. Floral herbivory in *Pastinaca sativa:* Do compensatory responses offset reductions in fitness? Evolution 43:891–895. [4, 13]

Hepting, G. H. 1974. Death of the American chestnut. Journal of Forest History 18:60–67. [15]

Herrera, C. M. 1982. Defense of ripe fruits from pests: Its significance in relation to plant disperser interactions. American Naturalist 120:218–241. [13]

———. 1984. Selective pressures on fruit seediness: Differential predation by fly larvae on the fruits of *Berberis hispanica*. Oikos 42:166–170. [13]

———. 1985. Grass/grazer radiation: An interpretation of silica-body diversity. Oikos 45:446–447. [13]

Hewett, E. W. 1977. Some effects of infestation on plants: A physiological viewpoint. New Zealand Entomologist 6:235–243. [13]

Hill, W. G. 1970. Design of experiments to estimate heritability by regression of offspring on selected parents. Biometrics 26:566–571. [3]

Hill, W. G., and R. Thompson. 1977. Design of experiments to estimate offspring-parent regression using selection parents. Animal Production 24:163–168. [3]

Hirschberg, J., A. B. Yehuda, I. Pecker, and N. Ohad. 1987. Mutations resistant to photosystem II herbicides. In D. von Wettstein and N. H. Chua, eds., Plant molecular biology, 357–366. NATO ASI Series, Life Sciences vol. 140. Plenum Press, New York. [17]

Hladik, C. M. 1978. Adaptive strategies of primates in relation to leaf-eating. In G. G. Montgomery, ed., Arboreal folivores, 373–395. Smithsonian Institution, Washington, DC. [10]

Hockey, P. A. R., and G. M. Branch. 1984. Oystercatchers and limpets: Impact and implications. A preliminary assessment. Ardea 72:199–206. [17]

Hocking, B. 1970. Insect interactions with the swollen thorn acacias. Transactions of the Royal Entomological Society of London 122:211–255. [10]

Hodgkinson, K. C., and H. G. Baas Becking. 1977. Effect of defoliation on root growth of some arid zone perennial plants. Australian Journal of Agricultural Research 29:31–41. [13]

Hodgkinson, K. C., N. G. Smith, and G. E. Miles. 1972. The photosynthetic capacity of stubble leaves and their contribution to growth of the lucerne plant after high level cutting. Australian Journal of Agricultural Research 23:225–238. [13]

Hodgson, E. 1985. Microsomal mono-oxygenases. In G. A. Kerkut and L. I. Gilbert, eds., Comprehensive insect physiology, biochemistry, and pharmacology, 225–321. Pergamon Press, New York. [4]

Hoekstra, R. F., R. Bijlsma, and A. J. Dolman. 1985. Polymorphism from environmental heterogeneity: Models are only robust if the heterozygote is close in fitness to the favoured homozygote in each environment. Genetical Research, Cambridge 45:299–314. [5]

Hoffman, G. D., and P. B. McEvoy. 1986. Mechanical limitations on feeding by meadow spittlebugs Philaenus spumarius (Homoptera: Cercopidae) on wild and cultivated host plants. Ecological Entomology 11:415–426. [6]

Hoffmann, A. A., P. A. Parsons, and K. M. Nielson. 1984. Habitat selection: Olfactory response of Drosophila melanogaster depends on resources. Heredity 53:139–143. [6]

Holmes, R. D., and K. Jepson-Innes. 1989. A neighborhood analysis of herbivory in Bouteloua gracilis. Ecology 70:971–976. [13]

Holt, J. S., S. R. Radosevich, and A. J. Stemler. 1983. Differential efficiency of photosynthetic oxygen evolution in flashing light in triazine-resistant and triazine-susceptible biotypes of Senecio vulgaris L. Biochimica et Biophysica Acta 722:245–255. [17]

Holt, R. 1977. Predation, apparent competition, and the structure of prey communities. Theoretical Population Biology 12:197–229. [11]

———. 1984. Spatial heterogeneity, indirect interactions, and the coexistence of prey species. American Naturalist 124:377–406. [11]

———. 1987. Prey communities in patchy environments. Oikos 50:276–290. [11]

Hooker, A. L., and K. M. S. Saxena. 1971. Genetics of disease resistance in plants. Annual Review of Genetics 5:407–424. [8, 15]

Horber, E. 1980. Types and classification of resistance. In F. G. Maxwell and P. R. Jennings, eds., Breeding plants resistant to insects, 15–21. Wiley, New York. [17]

Horn, H. S., and R. H. MacArthur. 1972. Competition between fugitive species in a harlequin environment. Ecology 53:749–752. [5]

Horrill, J. C., and A. J. Richards. 1986. Differential grazing by the mollusc Arion hortensis Fer. on cyanogenic and acyanogenic seedlings of the white clover, Trifolium repens L. Heredity 56:277–281. [10, 13]

Horvitz, C. C., C. Turnbill, and D. J. Harvey. 1987. The biology of the immature stages of Eurybia elvina (Lepidoptera: Riodinidae), a myrmecophilous metalmark butterfly. Annals of the Entomological Society of America 80:513–519. [13]

Howard, D. J., and R. G. Harrison. 1984. Habitat segregation in ground crickets: The role of interspecific competition and habitat selection. Ecology 65:69–76. [11]

Howe, H., and J. Smallwood. 1982. Ecology of seed dispersal. Annual Review of Ecology and Systematics 13:201–228. [13]

Howe, H. F., E. W. Schupp, and L. C. Westley. 1985. Early consequence of seed dispersal for a neotropical tree (Virola surinamensis). Ecology 66:781–791. [13]

Hruska, A. J. 1988. Cyanogenic glucosides as defense compounds: A review of the evidence. Journal of Chemical Ecology 14:2213–2217. [10, 17]

Hsiang, T., and B. J. Van der Kamp. 1985. Variation of rust virulence and host resistance of *Melampsora* on black cottonwood. Canadian Journal of Plant Pathology 7:247–252. [8]

Hughes, M. A., and J. D. Stirling. 1982. A study of dominance and the locus controlling cyanoglucoside production in *Trifolium repens*. Euphytica 3:477–483. [17]

Hughes, M. A., J. D. Stirling, and D. B. Colinge. 1984. The inheritance of cyanoglucoside content in *T. repens*. Biochemical Genetics 22:139–151. [17]

Hughes, R. D., and M. A. Bryce. 1984. Biological characterization of two biotypes of pea aphid, one susceptible and the other resistant to fungal pathogens, coexisting on lucerne in Australia. Entomologia Experimentalis et Applicata 36:225–229. [6]

Hughes, R. D., and M. A. Hughes. 1988. Temporary loss of antibiosis in plants of a lucerne cultivar selected for resistance to spotted alfalfa aphid. Entomologia Experimentalis et Applicata 49:75–82. [6]

Huheey, J. E. 1984. Warning coloration and mimicry. In W. J. Bell and R. Carde, eds., Chemical ecology of insects, 257–297. Sinauer Associates, Sunderland, MA. [12]

Hunter, M. D. 1987. Opposing effects of spring defoliation on late season oak caterpillars. Ecological Entomology 12:373–382. [11]

Hurlbert, S. H. 1984. Pseudoreplication and the design of ecological field experiments. Ecological Monographs 54:187–211. [9]

Hutchinson, G. E. 1951. Copepodology for the ornithologist. Ecology 32:571. [5]

Hutson, V., and R. Law. 1981. Evolution of recombination in populations experiencing frequency-dependent selection with time delay. Proceedings of the Royal Society of London B213:345–359. [15]

Huxley, C. R. 1986. Evolution of benevolent ant-plant relationships. In B. E. Juniper and T. R. E. Southwood, eds., Insects and the plant surface, 257–282. Edward Arnold, London. [10]

Hyder, D. N. 1972. Defoliation in relation to vegetative growth. In V. B. Youngner and C. M. McKell, eds., The biology and utilization of grasses, 304–317. Academic Press, New York. [13]

Ikeda, T., R. Matsumura, and D. M. Benjamin. 1977. Chemical basis for feeding adaptation of pine sawflies, *Neodiprion rugifrons* and *Neodiprion swainei*. Science 197:497–498. [13]

Ilott, T. W., M. E. Durgan, and R. W. Michelmore. 1987. Genetics of virulence in California populations of *Bremia lactucae* (lettuce downy mildew). Phytopathology 77:1381–1386. [8]

Ingham, R. E., and J. K. Detling. 1986. Effects of defoliation and nematode consumption on growth and leaf gas exchange in *Bouteloua curtipendula*. Oikos 46:23–28. [13]

Ingram, G. B., and J. T. Williams. 1988. *In situ* conservation of wild relatives of crops. In J. W. Holden and J. T. Williams, eds., Crop genetic resources: Conservation and evaluation. George Allen and Unwin, London. [2]

Inouye, D. W. 1982. The consequences of herbivory: A mixed blessing for *Jurinea mollis* (Asteraceae). Oikos 39:267–272. [13]

Inouye, D. W., and O. R. Taylor, Jr. 1979. A template region plant-ant-seed predator system: Consequence of extra floral nectar secretion by *Helianthella quinquenervis*. Ecology 60:1–7. [13]

Irving, R. S., and R. P. Adams. 1973. Genetic and biosynthetic relationships on mono-terpenes. Recent Advances in Phytochemistry 6:187–214. [2]

Isenhour, D. J., and B. R. Wiseman. 1989. Parasitism of the fall armyworm (Lepidop-tera: Noctuidae) by *Campoletis sonorensis* (Hymenoptera: Ichneumonidae) as af-fected by host feeding on silks of *Zea mays* L. cv. Zapalote Chico. Environmental Entomology 18:394–397. [12]

Isenhour, D. J., B. R. Wiseman, and J. R. Layton. 1989. Enhanced predation by *Orius insidiosus* (Hemiptera: Anthocoridae) on larvae of *Heliothis zea* and *Spodoptera frugiperda* (Lepidoptera: Noctuidae) caused by prey feeding on resistant corn geno-types. Environmental Entomology 18:418–422. [12]

Islam, Z., and M. J. Crawley. 1983. Compensation and regrowth in ragwort (*Senecio jacobaea*) attacked by cinnabar moth (*Tyria jacobaeae*). Journal of Ecology 71:829–843. [13]

Isman, M. G., S. S. Duffey, and G. G. E. Scudder. 1977. Variation in cardenolide con-tent of the lygaeid bugs, *Oncopeltus fasciatus* and *Lygaeus kalmii kalmii* and of their milkweed hosts (*Asclepias* spp.) in central California. Journal of Chemical Ecology 3:613–624. [16]

Ives, A. R. 1988. Covariance, coexistence, and the population dynamics of two com-petitors using a patchy resource. Journal of Theoretical Biology 133:345–361. [1, 5, 11]

Ives, A. R., and R. M. May. 1985. Competition within and between species in a patchy environment: Relations between microscopic and macroscopic models. Journal of Theoretical Biology 115:65–92. [5, 11]

Ivins, J. D. 1952. The relative palatability of herbage plants. Journal of the British Grassland Society 7:43–54. [10]

Jablonski, D., and R. A. Lutz. 1983. Larval ecology of marine benthic invertebrates: Paleobiological implications. Biological Reviews 58:21–89. [7]

Jackston, T. A. 1980. The effect of defoliation on yield of radish. Annals of Applied Biology 94:415–419. [13]

Jaenike, J. 1978. An hypothesis to account for the maintenance of sex within popula-tions. Evolutionary Theory 3:191–194. [14, 15]

———. 1982. Environmental modification of oviposition behavior in *Drosophila*. American Naturalist 119:784–802. [9, 14]

———. 1985. Genetic and environmental determinants of food preference in *Droso-phila tripunctata*. Evolution 39:362–369. [7]

———. 1986a. Genetic complexity of host-selection behavior in *Drosophila*. Evolution 39:1295–1302. [6]

———. 1986b. Parasite pressure and the evolution of amanitin tolerance in *Drosophila*. Evolution 39:1295–1302. [6]

———. 1987. Genetics of oviposition-site preference in *Drosophila tripunctata*. Hered-ity 59:363–369. [6]

———. 1990. Host specialization in phytophagous insects. Annual Review of Ecology and Systematics 21:243–273. [6]

Jain, S. K., and K. L. Mehra, eds. 1982. Conservation of tropical plant resources. Pro-ceedings of the Regional Workshop of Conservation of Tropical Plant Resources in South-East Asia, New Delhi. [2]

James, W. C. 1974. Assessment of plant diseases and losses. Annual Review of Phytopathology 12:27–48. [14]

James, W. C., and P. S. Teng. 1979. The quantification of production constraints associated with plant disease. Applied Biology 4:201–267. [14]

James, W. O. 1950. Alkaloids in the plant. In R. H. F. Manske and H. F. Holes, eds., The alkaloids: Chemistry and physiology, 1:16–90. Academic Press, New York. [13]

Jameson, D. A. 1963. Responses of individual plants to harvesting. Botanical Review 29:532–594. [13]

Janzen, D. H. 1966. Coevolution of mutualism between ants and acacias in Central America. Evolution 20:249–275. [10, 13]

———. 1969. Seed-eaters versus seed size, number, toxicity and dispersal. Evolution 23:1–27. [13]

———. 1970. Herbivores and the number of tree species in tropical forests. American Naturalist 104:501–528. [3, 13]

———. 1971. Seed predation by animals. Annual Review of Ecology and Systematics. 2:465–492. [10, 13]

———. 1972a. Association of a rainforest palm and seed-eating beetles in Puerto Rico. Ecology 53:258–261. [10]

———. 1972b. Protection of *Barteria* (Passifloraceae) by *Pachysima* ants (Pseudomyrmecinae) in a Nigerian rain forest. Ecology 531:885–892. [10]

———. 1973a. Comments on host-specificity of tropical herbivores and its relevance to species richness. In V. H. Heywood, ed., Taxonomy and ecology. Systematics Association Special Volume 5:201–211. Academic Press, London. [17]

———. 1973b. Community structure of secondary compounds in plants. Pure and Applied Chemistry 34:529–538. [17]

———. 1973c. Dissolution of mutualism between *Cecropia* and its *Azteca* ants. Biotropica 5:15–28. [13, 17]

———. 1974. Tropical blackwater rivers, animals, and mast fruiting by the Dipterocarpaceae. Biotropica 6:69–103. [17]

———. 1975a. Behavior of *Hymenaea courbaril* when its predispersal seed predator is absent. Science 189:145–147. [13, 17]

———. 1975b. Ecology of plants in the tropics. Studies in Biology 58. Edward Arnold, London. [13]

———. 1975c. Interactions of seeds and their insect predators/parasitoids in a tropical deciduous forest. In P. W. Price, ed., Evolutionary strategies of parasitic insects and mites, 154–186. Plenum Press, New York. [10]

———. 1975d. Intra- and interhabitat variations in *Guazuma ulmfolia* (Sterculiaceae) seed predation by *Amblycerus cistelinus* (Bruchidae) in Costa Rica. Ecology 56:1009–1013. [13]

———. 1976a. Effect of defoliation on fruitbearing branches of Kentucky coffee tree, *Gymnocladus dioicus* (Leguminosae). American Midland Naturalist 95:474–478. [13]

———. 1976b. Why bamboos wait so long to flower. Annual Review of Ecology and Systematics 7:347–391. [13]

———. 1977. Intensity of predation on *Pithecellobium saman* (Leguminosae) seeds by

Merobruchus columbinus and *Stator limbatus* (Bruchidae) in Costa Rican deciduous forest. Tropical Ecology 18:162–176. [13]

———. 1978a. The ecology and evolutionary biology of seed chemistry as relates to seed predation. In J. B. Harborne, ed., Biochemical aspects of plant and animal coevolution, 163–206. Academic Press, New York. [16]

———. 1978b. Reduction of seed predation on *Bauhinia pauletia* (Leguminosae) through habitat destruction in a Costa Rican deciduous forest. Brenesia 14–15:325–335. [13]

———. 1978c. Seeding patterns of tropical trees. In P. B. Tomlinson and M. H. Zimmerman, eds., Tropical trees as living systems, 128–183. Cambridge University Press, Cambridge. [13]

———. 1979. New horizons in the biology of plant defenses. In G. A. Rosenthal and D. H. Janzen, eds., Herbivores: Their interaction with secondary plant metabolites, 331–350. Academic Press, New York. [10, 13, 16, 17, 18]

———. 1980a. Specificity of seed-attacking beetles in a Costa Rican deciduous forest. Journal of Ecology 68:929–952. [10]

———. 1980b. When is it coevolution? Evolution 34:611–612. [4, 10, 11]

———. 1982. Cenizero tree (Leguminosae) *Pithecellobium saman* delayed fruit development in Costa Rican deciduous forest. American Journal of Botany 69:1269–1276. [13]

———. 1984. A host plant is more than its chemistry. Illinois Natural History Survey Bulletin 33:141–174. [13]

Janzen, D. H., C. A. Ryan, I. E. Liener, and G. Pearce. 1986. Potentially defensive proteins in mature seeds of 59 species of tropical Leguminosae. Journal of Chemical Ecology 12:1469–1480. [16]

Jarosz, A. M. 1984. Ecological and evolutionary dynamics of *Phlox-Erysiphe cichoracearum* interactions. Ph.D. diss., Purdue University. [14, 15]

Jarosz, A. M., and J. J. Burdon. 1988. The effect of small-scale environmental changes on disease incidence and severity in a natural plant-pathogen interaction. Oecologia 75:278–281. [15]

———. 1990. Predominance of a single major gene for resistance to *Phakopsora pachyrhizi* in a population of *Glycine argyrea*. Heredity 64:347–353. [8, 14]

Jayakar, S. C. 1970. A mathematical model for interaction of gene frequencies in a parasite and its host. Theoretical Population Biology 1:140–164. [5, 15, 17]

Jefferies, M. J., and J. H. Lawton. 1984. Enemy free space and the structure of ecological communities. Biological Journal of the Linnean Society 23:269–286. [6, 11]

Jeger, M. J. 1987. Modelling the dynamics of pathogen populations. In M. S. Wolfe and C. E. Caten, eds., Populations of plant pathogens: Their dynamics and genetics. Oxford: Blackwell. [5]

Jeger, M. J., E. Griffiths, and D. G. Jones. 1981. Disease progress of non-specialized fungal pathogens in intraspecific mixed strands of cereal cultivars. I. Models. Annals of Applied Biology 98:187–198. [5]

Jennersten, O. 1983. Butterfly visitors as vectors of *Ustilago violacea* spores between caryophyllaceous plants. Oikos 40:125–130. [15]

———. 1988. Insect dispersal of fungal disease: Effects of *Ustilago* infection on pollinator attraction in *Viscaria vulgaris*. Oikos 51:163–170. [14]

Jennings, C. W., W. A. Russell, and W. D. Guthrie. 1974a. Genetics of resistance in maize to first and second brood European corn borer. Crop Science 14:394–398. [2]

————. 1974b. Genetics of resistance in maize to second brood European corn borer. Iowa State Journal of Research 48:267–280. [2]

Jennings, T. J., and P. J. Barkham. 1975. Food of slugs in mixed deciduous woodlands. Oikos 26:211–221. [10]

Jensen, H. F. 1952. Intra-varietal diversification in oat breeding. Agronomy Journal 44:30–34. [5]

Jermy, T. 1976. Insect-host plant relationship: Coevolution or sequential evolution? In T. Jermy, ed., The host plant in relationship to insect behavior and reproduction. Plenum Press, New York. [2, 4]

————. 1984. Evolution of insect-host plant relationships. American Naturalist 124:609–630. [4, 6, 12, 13]

Jermy, T., F. E. Hanson, and V. G. Dethier. 1968. Induction of specific food preference in lepidopterous larvae. Entomologia Experimentalis et Applicata 11:211–230. [9]

Jesiotr, L. J. 1979. The influence of the host plant on the reproduction potential of the twospotted spider mite, *Tetranychus urticae* Koch (Acarina: Tetranychidae). II. Responses of the field population feeding on roses and beans. Ekologia Polska 27:351–355. [2]

————. 1980. The influence of host plants on the reproduction potential of the twospotted spider mite, *Tetranychus urticae* Koch (Acarina: Tetranychidae). IV. Changes within different populations affected by a new species of host plant. Ekologia Polska 28:633–647. [2]

Jogia, M. K., A. R. E. Sinclair, and R. J. Andersen. 1989. An antifeedant in balsam poplar inhibits browsing by snowshoe hares. Oecologia 79:189–192. [13]

Johnson, A. E., R. J. Molyneux, and G. B. Merrill. 1895. Chemistry of toxic range plants. Variation in pyrrolizidine alkaloid content of *Senecio*, *Amsinckia*, and *Crotalaria* species. Journal of Agricultural and Food Chemistry 33:50–55. [16]

Johnson, A. L., and B. H. Beard. 1977. Sunflower moth damage and inheritance of the phytomelanin layer in sunflower achenes. Crop Science 17:369–372. [2]

Johnson, C. D., and C. N. Slobodchikoff. 1979. Coevolution of *Cassia* (Leguminosae) and its seed beetle predators (Bruchidae). Environmental Entomology 8:1059–1064. [13]

Johnson, H. B. 1975. Plant pubescence: An ecological perspective. Botanical Review 41:233–258. [10]

Johnson, J. C., M. T. Nielsen and G. B. Collins. 1988. Inheritance of glandular trichomes in tobacco. Crop Science 28:241–244. [2]

Johnson, M. W., S. C. Welter, N. C. Toscano, I. P. Ting, and J. T. Trumble. 1983. Reduction of tomato leaflet photosynthesis rates by mining activity of *Liriomyza sativae* (Diptera: Agromyzidae). Journal of Economic Entomology 76:1061–1063. [13]

Johnson, R. 1983. Genetic background of durable resistance. In F. Lamberti, J. M. Waller, and N. A. van der Graaff, eds. Durable resistance in crops, 5–23. Plenum, New York. [5]

————. 1984. A critical analysis of durable resistance. Annual Review of Phytopathology 22:309–330. [8]

Jokela, J. J. 1966. Incidence and heritability of *Melampsora* rust in *Populus deltoides*

Bartr. In H. D. Gerhold, E. J. Schreiner, R. E. McDermott, and J. A. Winieski, eds., Breeding pest-resistant trees, 111–117. Pergamon Press, Oxford [3]

Jones, A., J. M. Schalk, and P. D. Dukes. 1979. Heritability estimates for resistances in sweet potato to soil insects. Journal of the American Society for Horticultural Science 104:424–426. [11]

Jones, C. G., and J. S. Coleman. 1988. Leaf disc size and insect feeding preference: Implications for assays and studies on induction of plant defense. Entomologia Experimentalis et Applicata 47:167–172. [4]

Jones, D. 1983. The influence of host density and gall shape on the survivorship of *Diastrophus kincaidii* Gill. (Hymenoptera: Cynipidae). Canadian Journal of Zoology 61:2138–2142. [11]

Jones, D. A. 1962. Selective eating of the acyanogenic form of the plant *Lotus corniculatus* L. by various animals. Nature 193:1109–1110. [13]

———. 1966. On the polymorphism of cyanogenesis in *Lotus corniculatus:* Selection by animals. Canadian Journal of Genetics and Cytology 8:556–567. [10, 13]

———. 1971. Chemical defense mechanisms and genetic polymorphism. Science 173:945. [10]

———. 1972. Cyanogenic glycosides and their function. In J. B. Harborne, ed., Phytochemical ecology, 103–124. Academic Press, New York. [10, 17]

———. 1973. Coevolution and cyanogenesis. In V. H. Heywood, ed., Taxonomy and ecology, 213–242. Academic Press, London. [17]

———. 1977. On the polymorphism of cyanogenesis in *Lotus corniculatus* L. VII. The distribution of the cyanogenic form in Western Europe. Heredity 39:27–44. [17]

———. 1979. Chemical defense: Primary or secondary function? American Naturalist 113:445–451. [17]

Jones, D. A., and A. D. Ramnani. 1985. Altruism and movement of plants. Evolutionary Theory 7:143–148. [10]

Jones, L. S., F. N. Briggs, and R. A. Blanchard. 1950. Inheritance of resistance to the pea aphid in alfalfa hybrids. Hilgardia 20:9–17. [2]

Jong, G. de. 1989. Phenotypically plastic characters in isolated populations. An A. Fontdevila, ed., Evolutionary biology of transient unstable populations. Springer-Verlag, Heidelberg. [7]

Jong, G. de. 1990. Quantitative genetics of reaction norms. Journal of Evolutionary Biology 3:447–468. [7]

Journet, A. R. P. 1980. Intraspecific variation in food plant favourability to photophagous insects: Psyllids on *Eucalyptus blakelyi* M. Ecological Entomology 5:249–261. [11]

Julien, M. H., and A. S. Bourne. 1986. Compensatory branching and changes in nitrogen content in the aquatic weed *Salvinia molesta* in response to disbudding. Oecologia 70:250–257. [13]

Juniper, B. E., and T. R. E. Southwood, eds. 1986. Insects and the plant surface. Edward Arnold, London. [10]

Jürges, K., and G. Röbbelen. 1980. Möglichkeiten einer auslese auf glucosinolat-gehalt in köhlruben (*Brassica napus* var. *napobrassica* [L.] Rchb). Zeitschrift für Pflanzenzuchtung 85:265–274. [16]

Juvik, J. A., and M. A. Stevens. 1982a. Inheritance of foliar tomatine content in tomatoes. Journal of the American Society for Horticultural Science 107:1061–1065. [2]

———. 1982b. Physiological mechanisms of host-plant resistance in the genus *Lycopersicon* to *Heliothis zea* and *Spodoptera exigua*, two insect pests of the cultivated tomato. Journal of the American Society for Horticultural Science 107:1065–1069. [11]

Kakes, P. 1987. On the polymorphism for cyanogenesis in natural populations of *Trifolium repens* L. in the Netherlands. I. Distribution of the genes *Ac* and *Li*. Acta Botanica Neerlandica 36:59–69. [17]

———. 1989. An analysis of the costs and benefits of the cyanogenic system in *Trifolium repens* L. Theoretical and Applied Genetics 77:111–118. [10, 13, 16, 17]

Kalisz, S. 1986. Variable selection on the timing of germination in *Collinsia verna* (Scrophulariaceae). Evolution 40:479–491. [3]

Kalloo, M. K. Banerjee, R. K. Kashyap, and A. K. Yadav. 1989. Genetics of resistance to fruit borer, *Heliothis armigera* Hubner, in *Lycopersicon*. Plant Breeding 102:173–175. [4]

Kampmeijer, P., and J. C. Zadoks. 1977. EPIMUL, a simulator of foci and epidemics in mixtures of resistant and susceptible plants, mosaics and multilines. Centre for Agricultural Publishing and Documentation, Wageningen, Netherlands. [5]

Karban, R. 1980. Periodical cicada nymphs impose periodical oak tree wood accumulation. Nature 287:326–327. [13]

———. 1982. Experimental removal of seventeen-year cicada nymphs and growth of host apple trees. Journal of the New York Entomological Society 90:74–81. [13]

———. 1985. Resistance against spider mites in cotton induced by mechanical abrasion. Entomologia Experimentalis et Applicata 37:137–141. [2]

———. 1986. Interspecific competition between folivorous insects on *Erigeron glaucus*. Ecology 67:1063–1072. [1, 11]

———. 1987a. Effects of clonal variation of the host plant, interspecific competition, and climate on the population size of a folivorous thrips. Oecologia 74:298–303. [6, 9, 11, 13]

———. 1987b. Herbivory dependent on plant age: A hypothesis based on acquired resistance. Oikos 48:336–337. [11]

———. 1989a. Community organization of folivores of *Erigeron glaucus:* Effects of competition, predation, and host plant. Ecology 70:1028–1039. [9, 11]

———. 1989b. Fine-scale adaptation of herbivorous thrips to individual host plants. Nature 340:60–61. [6, 9]

Karban, R., R. Adamchek, and W. C. Schnathorst. 1987. Induced resistance and interspecific competition between spider mites and a vascular wilt fungus. Science 235:678–680. [11]

Karban, R., and J. R. Carey. 1984. Induced resistance of cotton seedlings to mites. Science 225:53–54. [2]

Kareiva, P. 1982a. Exclusion experiments and the competitive release of insects feeding on collards. Ecology 63:696–704. [11]

———. 1982b. Insects and adaptations. Book Review. Science 215:658–659. [9]

———. 1983. Influence of vegetation texture on herbivore populations: Resource concentration and herbivore movement. In R. F. Denno and M. S. McClure, eds., Variable plants and herbivores in natural and managed systems, 259–290. Academic Press, New York. [3, 5, 11]

———. 1986. Patchiness, dispersal, and species interactions: Consequences for communities of herbivorous insects. In J. Diamond and T. J. Case, eds., Community ecology, 192–206. Harper and Row, New York. [11]

Karlin, S., and J. McGregor. 1974. Towards a theory of the evolution of modifier genes. Theoretical Population Biology 5:59–103. [17]

Kartohardjono, A., and E. A. Heinrichs. 1984. Populations of the brown planthopper, *Nilaparvata lugens* (Stål) (Homoptera: Delphacidae) and its predators on rice varieties with different levels of resistance. Environmental Entomology 13:359–365. [12]

Kaufmann, W. G., and R. V. Flanders. 1985. Effects of variably resistant soybean and lima bean cultivars on *Pediobius foveolatus* (Hymenoptera: Eulophidae), a parasitoid of the Mexican bean beetle, *Epilachna varivestis* (Coleoptera: Coccinellidae). Environmental Entomology 14:678–682. [12]

Kea, W. C., S. G. Turnipseed, and G. R. Carner. 1978. Influence of resistant soybeans on the susceptibility of lepidopterous pests to insecticides. Journal of Economic Entomology 71:58–60. [12]

Kearsey, M. J. 1965. Biometrical analysis of a random mating population: A comparison of five experimental designs. Heredity 20:205–235. [3]

———. 1967. The genetic architecture of body weight and egg hatchability in *Drosophila melanogaster*. Genetics 56:23–37. [3]

Kearsley, M. J. C., and T. G. Whitham. 1989. Developmental changes in resistance to herbivory: Implications for individuals and populations. Ecology 70:422–434. [1, 11]

Keeler, K. H. 1977. The extrafloral nectaries of *Ipomoea carnea* (Convolvulaceae). American Journal of Botany 64:1182–1188. [17]

———. 1980. The extrafloral nectaries of *Ipomoea leptophylla* (Convolvulaceae). American Journal of Botany 67:216–222. [17]

Keeler, R. F., K. R. Van Kampen, and L. F. James, eds. 1978. Effects of poisonous plants on livestock. Academic Press, New York. [10]

Kelley, S., J. Antonovics, and A. Schmitt. 1988. A test of the short-term advantage of sexual reproduction. Nature 331:714–716. [18]

Kemp, R. F. O. 1986. Do fungal species really exist? A study of basidiomycete species with special reference to those in *Coprinus* section *Lanatuli*. Bulletin of the British Mycology Society 19:34–39. [8]

Kemp, W. B. 1937. Natural selection within plant species as exemplified in a permanent pasture. Journal of Heredity 28:329–333. [13]

Kempthorne, L. 1956. The theory of the diallel cross. Genetics 41:451–459. [3]

Kempthorne, O. 1969. An introduction to genetic statistics. Iowa State University Press, Ames. [3]

Kempthorne, O., and R. N. Curnow. 1961. The partial diallel cross. Biometrics 17:229–250. [3]

Kendall, W. A., and R. T. Sherwood. 1975. Palatability of leaves of tall fescue and reed canarygrass and some of their alkaloids to meadow voles. Agronomy Journal 67:667–671. [10]

Kennedy, G. G. 1978. Recent advances in insect resistance of vegetable and fruit crops in North America: 1966–1977. Bulletin of the Entomological Society of America 24:375–384. [4]

———. 1986. Consequences of modifying biochemically mediated insect resistance in

Lycopersicon species. In M. B. Green and P. A. Hedin, eds., Natural resistance of plants to pests, 130–141. ACS Symposium Series 296. American Chemical Society, Washington, DC. [2]

Kennedy, G. G., F. Gould, O. M. B. dePonti, and R. E. Stinner. 1987. Ecological, agricultural, genetic, and commercial considerations in the deployment of insect-resistant germplasm. Environmental Entomology 16:327–338. [2, 5, 9]

Kennedy, G. G., and A. N. Kishaba. 1976. Bionomics of *Aphis gossypii* on resistant and susceptible cantaloupe. Environmental Entomology 5:357–361. [2]

Kennedy, G. G., A. N. Kishaba, and G. W. Bohn. 1975. Response of several pest species to *Cucumis melo* L. lines resistant to *Aphis gossypii* Glover. Environmental Entomology 4:653–657. [12]

Kennedy, G. G., and D. C. Margolies. 1985. Mobile anthropod pests: Management in diversified agroecosystems. Bulletin of the Entomological Society of America 35:21–27. [2]

Kennedy, G. G., D. L. McLean, and M. G. Kinsey. 1978. Probing behavior of *Aphis gossypii* on resistant and susceptible muskmelon. Journal of Economic Entomology 71:13–16. [2]

Kennedy, G. G., J. Nienhuis, and T. Helentjaris. 1987. Mechanisms of anthropod resistance in tomatoes. In D. J. Nevins and R. A. Jones, eds., Tomato biotechnology, 145–154. Vol. 4 of Plant biology. Alan R. Liss, New York. [2]

Kennedy, G. G., G. A. Schaefers, and D. K. Ourecky. 1973. Resistance in red raspberry to *Amphorophora agathonica* Hottes and *Aphis rubicola* Oestlund. HortScience 8:311–313. [2]

Kennedy, G. G., C. E. Sorenson, and R. L. Fery. 1985. Mechanisms of resistance to Colorado potato beetle in tomato. In D. N. Ferro and R. H. Voss, eds., Proceedings of the Symposium on the Colorado Potato Beetle, 107–116. Seventeenth International Congress of Entomology. Massachusetts Agricultural Experiment Station Research Bulletin 704. [2]

Kennedy, G. G., R. T. Yamamoto, M. B. Dimock, W. G. Williams, and J. Bordner. 1981. Effect of daylength and light intensity on 2-tridecanone levels and resistance in *Lycopersicon hirsutum* f. *glabratum* to *Manduca sexta*. Journal of Chemical Ecology 7:707–716. [2]

Kennedy, J. S. 1950. Host-finding and host alternation in aphids. Eighth International Congress of Entomology, Stockholm, 423–426. [6]

———. 1951. A biological approach to plant viruses. Nature (London) 168:890–894. [2]

Kenneth, R. G., and J. Palti. 1984. The distribution of downy and powdery mildews and of rusts over tribes of Compositae (Asteraceae). Mycologia 76:705–718. [15]

Kermack, W. O., and A. G. McKendrick. 1927. Contributions to the mathematical theory of epidemics. Journal of the Royal Statistical Society 115:700–721. [18]

Kerns, D. L., D. C. Peters, and G. J. Puterka. 1987. Greenbug biotype and grain sorghum seed sale surveys in Oklahoma, 1986. Southwestern Entomologist. 12:237–243. [5]

Kerns, D. L., G. J. Puterka, and D. C. Peters. 1989. Intrinsic rate of increase for greenbug (Homoptera: Aphididae) biotypes E, F, G, and H on small grain and sorghum varieties. Environmental Entomology 18:1074–1078. [6]

Kettlewell, H. B. D. 1956. Further selection experiments on industrial melanism in the Lepidoptera. Heredity 10:287–301. [7]

Keymer, R. J., and W. M. Ellis. 1978. Experimental studies on plants of *Lotus corniculatus* L. from Anglesey polymorphic for cyanogenesis. Heredity 40:189–206. [10, 17]

Khan, M. A., J. Antonovics, and A. D. Bradshaw. 1976. Adaptation to heterogeneous environments. III. The inheritance of response to spacing in flax and linseed (*Linum usitatissimum*). Australian Journal of Agricultural Research 27:649–659. [7]

Killebrew, J. F., K. W. Roy, G. W. Lawrence, K. S. McLean, and H. H. Hodges. 1988. Greenhouse and field evaluation of *Fusarium solani* pathogenicity to soybean seedlings. Plant Disease 72:1067–1070. [8]

Kindler, S. D., and S. M. Spomer. 1986. Biotypic status of six greenbug (Homoptera: Aphididae) isolates. Environmental Entomology 15:567–572. [5]

Kindler, S. D., S. M. Spomer, T. L. Harvey, R. L. Burton, and K. J. Starks. 1984. Status of biotype-E greenbugs (Homoptera: Aphididae) in Kansas, Nebraska, Oklahoma, and northern Texas during 1980–1981. Journal of the Kansas Entomological Society 57:155–158. [5]

Kinghorn, B. P. 1987. The nature of 2-locus epistatic interactions in animals: Evidence from Sewell Wright's guinea pig data. Theoretical and Applied Genetics 73:595–604. [3]

Kingsley, P., J. M. Scriber, C. R. Grau, and P. A. Delwiche. 1983. Feeding and growth performance of *Spodoptera eridania* (Noctuidae: Lepidoptera) on "vernal" alfalfa, as influenced by *Verticillium* wilt. Protection Ecology 5:127–134. [3]

Kinloch, B. B., and R. W. Stonecypher. 1969. Genetic variation in susceptibility to fusiform rust in seedlings from a wild population of loblolly pine. Phytopathology 59:1246–1255. [14]

Kinsman, S. 1982. Herbivore responses to *Oenothera biennis* (Onagraceae): Effects of the host plant's size, genotype, and resistant conspecific neighbors. Ph.D. diss., Cornell University, Ithaca, NY. [2, 9, 11, 17]

Kinsman, S., and W. J. Platt. 1984. The impact of an herbivore upon *Mirabilis hirsuta*, a fugitive prairie plant. Oecologia 65:2–6. [13]

Kiraly, Z. 1980. Defenses triggered by the invader: Hypersensitivity. In J. G. Horsfall and E. B. Cowling, eds., Plant disease, 5:201–224. Academic Press, New York. [14]

Kirk, R. E. 1982. Experimental design: Procedures for the behavioral sciences, 2d ed. Brooks/Cole, Monterey, CA. [9]

Kirk, W. D. 1987. How much pollen can thrips destroy? Ecological Entomology 12:31–40. [13]

Kishaba, A. N., G. W. Bohn, and H. H. Toba. 1976. Genetic aspects of antibiosis to *Aphis gossypii* in *Cucumis melo* from India. Journal of the American Society for Horticultural Science 101:557–561. [2]

Kiyosawa, S. 1972. Theoretical comparison between mixture and rotation cultivations of disease resistant varieties. Annals of the Phytopathological Society of Japan 38:52–59. [5]

Kiyosawa, S., and M. Shiyomi. 1972. A theoretical evaluation of the effect of mixing resistant variety with susceptible variety for controlling plant diseases. Annals of the Phytopathological Society of Japan 38:41–51. [5]

Kjellson, R. 1985. Seed fate in a population of *Carex poluifera* L. II. Seed predation and its consequences for dispersal and seed banks. Oecologia 67:424–429. [13]

Klenke, J. R. 1985. Recurrent selection for European corn borer resistance in a maize synthetic. Ph.D. diss., Iowa State University, Ames. (Abstract) [17]

Klinkhamer, P. G. L., T. J. De Jong, and E. van der Meigjden. 1988. Production, dispersal, and predation of seeds in the biennial *Cirsium vulgare*. Journal of Ecology 76:403–414. [13]

Klinkowski, M. 1970. Catastrophic plant diseases. Annual Review of Phytopathology 8:37–60. [18]

Klun, J. A., W. D. Guthrie, A. R. Hallauer, and W. A. Russell. 1970. Genetic nature of concentration of 2,4-dihyroxy-7-methoxy-2H-1, 4-benzoxazine-3 (4H)-one and resistance to the European corn borer in a diallel set of eleven maize inbred lines. Crop Science 10:87–90. [2, 13]

Klun, J. A., and J. F. Robinson. 1969. Concentration of two, 1,4-benzoxazinanes in dent corn at various stages of development of the plant and its relation to resistance of the host plant to the European corn borer. Journal of Economic Entomology 62:214–220. [2]

Knapp, S. J. 1986. Confidence intervals for heritability for two-factor mating design in single environment linear models. Theoretical and Applied Genetics 72:587–591. [3]

Knapp, S. J., and W. C. Bridges. 1987. Confidence interval estimators for heritability for several mating and experimental designs. Theoretical and Applied Genetics 73:759–763. [3]

―――. 1988. Parametric and jackknife confidence interval estimators for two-factor mating design genetic variance ratios. Theoretical and Applied Genetics 76:385–392. [3]

Knapp, S. J., W. W. Stroup, and W. M. Ross. 1985. Exact confidence intervals for heritability on a progeny mean basis. Crop Science 25:192–194. [3]

Kogan, M. 1982. Plant resistance in pest management. In R. L. Metcalf and W. H. Luckman, eds., Introduction to insect pest management, 2d ed., 93–134. John Wiley and Sons, New York. [12]

―――. 1986. Bioassays for measuring quality of insect food. In T. Miller and J. Miller, eds., Insect-plant interactions, 155–190. Springer-Verlag, New York. [4]

Kogan, M., and J. Paxton. 1983. Natural inducers of plant resistance. In P. A. Hedin, ed., Plant resistance to insects, 153–171. ACS Symposium Series 208, American Chemical Society, Washington, DC. [2]

Kohn, L. A. P., and W. R. Atchley. 1988. How similar are genetic correlation structures? Evolution 42:467–481. [7]

Kooistra, E. 1971. Red spider mite tolerance in cucumber. Euphytica 20:47–50. [4]

Koptur, S. 1979. Facultative mutualism between weedy vetches bearing extrafloral nectaries and weedy ants in California. American Journal of Botany 66:1016–1020. [10]

Krischik, V. A., and R. F. Denno. 1983. Individual, population, and geographic patterns in plant defense. In R. F. Denno and M. S. McClure, eds., Variable plants and herbivores in natural and managed systems, 463–512. Academic Press, New York. [2, 13, 16, 17]

Kuhlman, E. G., and H. R. Powers, Jr. 1988. Resistance responses in half-sib loblolly pine progenies after inoculation with *Cronartium quercuum* f. sp. *fusiforme*. Phytopathology 78:484–487. [8]

Kulman, H. M. 1971. Effects of insect defoliation on growth and mortality of trees. Annual Review of Entomology 16:289–324. [1, 13]

Kvenberg, J. E., and P. A. Jones. 1974. Comparison of alate offspring produced by two biotypes of greenbug. Environmental Entomology 3:407–408. [6]

Labeyrie, V. 1987. Towards a synthetic approach to insect-plant relationships. In V. Labeyrie, G. Fabres, and D. Lachaise, eds., Insects-plants, 3–7. Proceedings of the Sixth International Symposium on Insect-Plant Relationships. Dr. W. Junk, Dordrecht. [13]

Labeyrie, V., and M. Hossaert. 1985. Ambiguous relations between *Bruchus affinis* and the *Lathyrus* group. Oikos 44:107–113. [13]

Ladygina, E. Y., V. A. Makarova, and N. S. Ignat'eva. 1970. Morphological and anatomical description of *Pastinaca sativa* fruit and localization of furanocoumarins in them. Farmatsiya (Moscow) 19:29–35 (in Russian) (Chemical Abstracts 74:61588x, 1970). [16]

Lamb, R. J., and P. A. MacKay. 1979. Variability in migratory tendency within and among natural populations of the pea aphid, *Acyrthosiphon pisum*. Oecologia 39:289–299. [6]

Lambert, L., and T. C. Klein. 1984. Influence of three soybean plant genotypes and their F1 intercrosses on the development of five insect species. Journal of Economic Entomology 77:622–625. [11]

Lamberti, F., J. M. Waller, and N. A. van de Graaff, eds. 1983. Durable resistance in crops. Plenum, New York. [5, 13]

Lampert, E. P., D. L. Haynes, A. J. Sawyer, D. P. Jokinen, S. G. Wellso, R. L. Gallun, and J. J. Roberts. 1983. Effects of regional releases of resistant wheats on the population dynamics of the cereal leaf beetle (Coleoptera: Chrysomelidae). Annals of the Entomological Society of America 76:972–980. [12]

Lampitt, L. H., J. H. Bushill, H. S. Rooke, and E. M. Jackson. 1943. Solanine: Glycoside of the potato. II. Its distribution in the potato plant. Journal of the Society of Chemical Industry (London) 62:48–51. [16]

Lande, R. 1976. The maintenance of genetic variability by mutation in a polygenic character with unlinked loci. Genetical Research 26:221–235. [6, 7, 17]

———. 1979. Quantitative genetic analysis of multivariate evolution, applied to brain:body size allometry. Evolution 33:402–416. [3]

———. 1981. The minimum number of genes contributing to quantitative variation between and within populations. Genetics 99:541–553. [3]

Lande, R., and S. J. Arnold. 1983. The measurement of selection on correlated characters. Evolution 37:1210–1226. [3, 4, 7, 13]

Landsberg, J., and C. Ohmart. 1989. Levels of insect defoliation in forests: Patterns and concepts. Trends in Ecology and Evolution 4:96–100. [13]

Langenheim, J. H. 1984. The roles of plant secondary chemicals in wet tropical ecosystems. In E. Medina, H. A. Mooney, and C. Vazquez-Yanes, eds., Physiological ecology of plants of the wet tropics, 189–208. Dr. W. Junk, The Hague. [13]

Langenheim, J. H., and G. D. Hall. 1983. Sesquiterpene deterrence of leaf-tying lepidopteran, *Stenoma ferrocanella*, on *Hymenaea stignocarpa* in central Brazil. Biochemical Systematics and Ecology 11:29–36. [13]

Langenheim, J. H., S. P. Arrhenius, and J. C. Nascimento. 1981. Relationship of light intensity to leaf resin composition and yield in the tropical genera *Hymenaea* and *Copaifera*. Biochemical Systematics and Ecology 9:27–37. [13]

Langenheim, J. H., D. E. Lincoln, W. H. Stubblebine, and A. C. Gabrielli. 1982. Evolutionary implications of leaf resin pocket patterns in the tropical tree *Hymenaea* (Caesalpinioideae: Leguminosae). American Journal of Botany 69:595–607. [13]

Langenheim, J. H., W. L. Stubblebine, D. E. Lincoln, and C. E. Foster. 1978. Implications of variation in resin composition among organs, tissues, and populations in the tropical legume *Hymenaea.* Biochemical Systematics and Ecology 6:299–313. [16]

Larsson, S., C. Bjorkman, and R. Gref. 1986. Responses of *Neodiprion sertifer* (Hym., Diprionidae) larvae to variation in needle resin acid concentration in Scots pine. Oecologia 70:77–84. [9]

Larsson, S., and C. P. Ohmart. 1988. Leaf age and larval performance of the leaf beetle *Paropsis atomaria*. Ecological Entomology 13:19–24. [11]

Larsson, S., A. Wiren, L. Lundgren, and T. Ericsson. 1986. Effects of light and nutrient stress on leaf phenolic chemistry in *Salix dasyclados* and susceptibility to *Galerucella lineola*. Oikos 47:205–210. [13]

Latch, G. C. M., W. F. Hunt, and D. R. Musgrave. 1985. Endophytic fungi affect growth of perennial ryegrass. New Zealand Journal of Agricultural Research 28:165–168. [17]

Latch, G. C. M., and J. A. Lancashire. 1970. The importance of some effects of fungal disease on pasture yield and composition. Proceedings of the Eleventh International Grassland Congress, 688–691.

Latch, G. C. M., L. R. Potter, and B. F. Tyler. 1987. Incidence of endophytes in seeds from collections of *Lolium* and *Festuca* species. Annals of Applied Biology 111:59–64. [10]

Lauenstein, V. G. 1980. Zum suchverhalten von *Anthochoris nemorum* L. (Het. Anthochridae). Zeitschrift für Angewandte Entomologie 89:428–442. [12]

Lauter, D. J., and D. N. Munns. 1986. Water loss via the glandular trichomes of chickpea (*Cicer arietinum* L.). Journal of Experimental Botany 37:640–649. [17]

Lawton, J. H. 1986. The effects of parasitoids on phytophagous insect communities. In J. Waage and D. Greathead, eds., Insect parasitoids, 265–287. Academic Press, London. [11]

Lawton, J. H., and M. P. Hassell. 1984. Interspecific competition in insects. In C. B. Huffaker and R. L. Rabb, eds., Ecological entomology, 451–495. John Wiley and Sons, New York. [11]

Lawton, J. H., and S. McNeill. 1979. Between the devil and the deep blue sea: On the problem of being a herbivore. In R. M. Anderson, B. D. Turner, and L. R. Taylor, eds., Population dynamics, 223–244. Blackwell, London. [11, 12]

Lawton, J. H., and D. R. Strong, Jr. 1981. Community patterns and competition in folivorous insects. American Naturalist 118:317–338. [11]

Leach, J. G. 1940. Insect transmission of plant diseases. McGraw-Hill, New York. [18]

Leather, S. R., A. D. Watt, and G. I. Forrest. 1987. Insect-induced chemical changes in young lodgepole pine (*Pinus contorta*): The effect of previous defoliation on oviposition, growth, and survival of the pine beauty moth, *Panolis flammea*. Ecological Entomology 12:275–281. [11]

Lee, D. W., and J. B. Lowry. 1980. Young-leaf anthocyanin and solar ultraviolet. Biotropica 12:75–76. [17]

Lee, J. A. 1968. Genetical studies concerning the distribution of trichomes on the leaves of *Gossypium hirsutum* L. Genetics 60:575. [2]

———. 1971. Some problems in breeding smooth-leaved cottons. Crop Science 11:448–450. [2]

———. 1973. The inheritance of gossypol level in *Gossypium* II: Inheritance of seed gossypol in two strains of cultivated *Gossypium barbadense* L. Genetics 75:259–264. [2]

Lee, T. D., and F. A. Bazzaz. 1980. Effects of defoliation and competition on growth and reproduction in the annual plant *Abutilon theophrasti*. Journal of Ecology 68:813–821. [13, 16]

Legg, P. D., D. F. Matzinger, and T. J. Mann. 1965. Genetic variation and covariation in a *Nicotiana tabacum* L. synthetic two generations after synthesis. Crop Science 5:30–33. [17]

Lemke, C. A., and M. A. Mutschler. 1984. Inheritance of glandular trichomes in crosses between *Lycopersicon esculentum* and *Lycopersicon pennellii*. Journal of the American Society for Horticultural Science 109:592–596. [2]

Lenski, R. E. 1988a. Experimental studies of pleiotropy and epistasis in *Escherichia coli*. I. Variation in competitive fitness among mutants resistant to virus T4. Evolution 42:425–432. [3, 17]

———. 1988b. Experimental studies of pleiotropy and epistasis in *Escherichia coli*. II. Compensation for maladaptive effects associated with resistance to virus T4. Evolution 42:433–440. [3, 17]

Leonard, K. J. 1969a. Factors affecting rates of stem rust increase in mixed plantings of susceptible and resistant oat varieties. Phytopathology 59:1845–1850. [5]

———. 1969b. Selection in heterogeneous populations of *Puccinia graminis* f. sp. *avenae*. Phytopathology 59:1851–1857. [5]

———. 1977. Selection pressures and plant pathogens. Annals of the New York Academy of Sciences 287:207–222. [5, 15, 17, 18]

Leonard, K. J., and R. J. Czochor. 1978. In response to "Selection pressures and plant pathogens: Stability of equilibria." Phytopathology 68:971–973. [5, 17]

———. 1980. Theory of genetic interactions among populations of plants and their pathogens. Annual Review of Phytopathology 18:237–258. [5, 15, 17, 18]

Leonard, K. J., and W. E. Fry. 1986. Plant disease epidemiology. Vol. 1, Population dynamics and management. Macmillan, New York. Vol. 2, Genetics resistance and management. McGraw-Hill, New York. [2, 18]

Le Pelley, R. H. 1932. *Lygus simonyi* Reut. (Hem. Capsid.), a pest of coffee in Kenya colony. Bulletin of Entomological Research 23:85–100. [13]

Leppik, E. E. 1970. Gene centers of plants as sources of disease resistance. Annual Review of Phytopathology 8:323–344. [2, 14]

Lessells, C. M. 1985. Parasitoid foraging: Should parasitism be density dependent? Journal of Animal Ecology 54:27–41. [11]

Leung, H., and M. Taga. 1988. *Magnaporthe grisea* (Pyricularia) species: The blast fungus. In D. S. Ingram and P. H. Williams, eds., Advances in plant pathology. Vol. 6, Genetics of plant pathogenic fungi, ed. G. S. Sidhu, 175–188. Academic Press, London. [8]

Levene, H. 1953. Genetic equilibrium when more than one ecological niche is available. American Naturalist 87:331–373. [6]

Levin, D. A. 1972. Pollen exchange as a function of species proximity in *Phlox*. Evolution 26:257–258. [18]

————. 1973. The role of trichomes in plant defense. Quarterly Review of Biology 48:3–15. [10, 12, 13]

————. 1975. Pest pressure and recombination systems in plants. American Naturalist 109:437–451. [15, 18]

————. 1976a. Alkaloid-bearing plants: An ecogeographic perspective. American Naturalist 110:261–284. [13]

————. 1976b. The chemical defenses of plants to pathogens and herbivores. Annual Review of Ecology and Systematics 7:121–159. [2, 10, 13]

Levin, D. A., and H. W. Kerster. 1974. Gene flow in seed plants. Evolutionary Biology 7:139–220. [15, 18]

Levin, S. A. 1976. Population dynamic models in heterogeneous environments. Annual Review of Ecology and Systematics 7:287–310. [2, 5]

————. 1983. Some approaches to the modelling of coevolutionary interactions. In M. H. Nitecki, ed., Coevolution, 21–65. University of Chicago Press, Chicago. [15, 17]

Levin, S. A., and J. D. Udovic. 1977. A mathematical model of coevolving populations. American Naturalist 111:657–675. [17, 18]

Levins, R. 1966. The strategy of model building in population biology. American Scientist 54:421–431. [14]

————. 1968. Evolution in changing environments. Princeton University Press, Princeton, NJ. [7]

Levins, R., and E. Culver. 1971. Regional coexistence of species and competition between rare species. Proceedings of the National Academy of Sciences. 68:1246–1248. [5]

Lewis, A. 1984. Plant quality and grasshopper feeding: Effects of sunflower condition on preference and performance in *Melanoplus differentialis*. Ecology 65:836–843. [13]

Lewis, A. C., and H. F. van Emden. 1986. Assays for insect feeding. In J. R. Miller and T. A. Miller, eds., Insect-plant interactions, 95–119. Springer-Verlag, Berlin. [17]

Lewis, D. H. 1988. Evolutionary aspects of mutualistic associations between fungi and photosynthetic organisms. In A. D. M. Rayner, C. M. Brasier, and D. Moore, eds., Evolutionary biology of the fungi, 161–178. Cambridge University Press, Cambridge. [17]

Lewis, G. C., and R. O. Clements. 1986. A survey of ryegrass endophyte (*Acremonium loliae*) in the U.K. and its apparent ineffectuality on a seedling pest. Journal of Agricultural Science (Cambridge) 107:633–638. [10]

Lewis, J. W. 1981. On the coevolution of pathogen and host. I. General theory of discrete time coevolution. Journal of Theoretical Biology 93:927–951. [15]

Lewis, W. M. 1987. The cost of sex. In S. C. Stearns, ed., The evolution of sex and its consequences, 33–57. Birkhauser Verlag, Basel. [15]

Lewontin, R. C. 1974a. The analysis of variance and the analysis of causes. American Journal of Human Genetics 26:400–411. [3]

————. 1974b. The genetic basis of evolutionary change. Columbia University Press, New York. [7]

————. 1983. Gene, organism, and environment. In D. S. Bendal, ed., Evolution from molecules to man. Columbia University Press, New York. [7]

Lieberei, R., B. Biehl, A. Giesemann, and N. T. V. Junqueira. 1989. Cyanogenesis inhibits active defense reactions in plants. Plant Physiology 90:33–36. [17]

Lieberman, D., and M. Lieberman. 1984. The causes and consequences of synchronous flushing in a dry tropical forest. Biotropica 16:193–201. [1, 13, 17]

Lightfoot, D. C., and W. G. Whitford. 1989. Interplant variation in creosote bush foliage characteristics and canopy anthropods. Oecologia 81:166–175. [4, 13]

Lin, J., M. H. Dickson, and C. J. Eckenrode. 1984. Resistance of Brassica lines to the diamondback moth (Lepidoptera: Yponomeutidae) in the field, and inheritance of resistance. Journal of Economic Entomology. 77:1293–1296. [4]

Lincoln, D. E., and J. H. Langenheim. 1976. Geographic patterns of monoterpenoid variation in Satureja douglasii. Biochemical Systematics and Ecology 4:237–248. [2]

————. 1978. Effect of light and temperature on monoterpenoid yield and composition. Biochemical Systematics and Ecology 6:21–32. [2]

————. 1979. Variation of Satureja douglasii monoterpenoids in relation to light intensity and herbivory. Biochemical Systematics and Ecology 7:289–298. [2, 3, 13]

————. 1981. A genetic approach to monoterpenoid compositional variation in Satureja douglasii. Biochemical Systematics and Ecology 9:153–160. [2]

Lincoln, D. E., and H. A. Mooney. 1984. Herbivory on Diplacus aurantiacus shrubs in sun and shade. Oecologia 64:173–177. [3, 13]

Linde, D. C., J. V. Groth, and A. P. Roelfs. 1990. Comparison of isozyme and virulence diversity patterns in the bean rust fungus Uromyces appendiculatus. Phytopathology 80:141–147. [8]

Lindroth, R. L. 1988a. Adaptations of mammalian herbivores to plant chemical defenses. In K. C. Spencer, ed., Chemical mediation of coevolution, 414–445. Academic Press, San Diego. [10]

————. 1988b. Hydrolysis of phenolic glycosides by midgut beta-glucosidases in Papilio glaucus subspecies. Insect Biochemistry 18:784–792. [4]

Lindroth, R. L., and M. S. Pajutee. 1987. Chemical analysis of phenolic glycosides: Art, facts, and artifacts. Oecologia 74:144–148. [4]

Linhart, D. E., and H. A. Mooney. 1984. Herbivory on Diplacus aurantiacus shrubs in sun and shade. Oecologia 64:173–176. [13]

Linhart, Y. B. 1989. Interactions between genetic and ecological patchiness in forest trees and their dependent species. In J. H. Bock and Y. B. Linhart, eds., Evolutionary ecology of plants, 392–340. Westview Press, Boulder. [13]

Linhart, Y. B., J. B. Mitton, K. B. Sturgeon, and M. L. Davis. 1981. Genetic variation in space and time in a population of ponderosa pine. Heredity 46:407–426. [13]

Linhart, Y. B., M. A. Snyder, and S. A. Habeck. 1989. The influence of animals on genetic variability within ponderosa pine stands, illustrated by the effects of Abert's squirrel and porcupine. In A. Teacle, W. W. Lovington, and R. H. Hamre, eds., Multiresource management of ponderosa pine forests, 141–148. United States Department of Agriculture Forest Service General Technical Report RM-185. [13]

Llewellyn, M., and V. K. Brown. 1985. The effect of hostplant species on adult weight and the reproductive potential of aphids. Journal of Animal Ecology 54:639–650. [4]

Lockwood, J., T. Sparks, and R. Story. 1984. Evolution of insect resistance to insecticides: A reevaluation of the roles of physiology and behavior. Bulletin of the Entomological Society of America 30:41–51. [4]

Lodge, R. W. 1962. Autecology of *Cynosurus cristatus* L. II. Ecotypic variation. Journal of Ecology 50:75–86. [13]

Loehle, C. 1988. Tree life history strategies: The role of defenses. Canadian Journal of Forest Research 18:209–222. [13, 17]

Loehle, C., and J. H. K. Pechmann. 1988. Evolution: The missing ingredient in systems ecology. American Naturalist 132:884–899. [18]

Loeschcke, V. 1987. Genetic constraints on adaptive evolution. Springer-Verlag, Berlin. [3]

Lofsvold, D. 1986. Quantitative genetics of morphological differentiation in *Peromyscus*. I. Tests of homogeneity of genetic covariance structure among species and subspecies. Evolution 40:559–573. [7]

Łomnicki, A. 1974. Evolution of herbivore-plant, predator-prey, and parasite-host systems: A theoretical model. American Naturalist 108:167–180. [17]

Long, B. J., G. M. Dunn, J. S. Bowman, and D. J. Routley. 1976. Relationship of hydroxamic acid concentration (DIHHOA) to resistance to the corn leaf aphid. Crop Science 17:55–58. [2]

Lookado, S. E., and A. J. Pollard. 1991. Chemical contents of the stinging trichomes of *Cnidoscolus texanus*. Journal of Chemical Ecology 17:1909–1916. [10]

Louda, S. M. 1982. Limitation of the recruitment of the shrub *Haplopappus squarrosus* (Asteraceae) by flower- and seed-feeding insects. Journal of Ecology 70:43–53. [13]

———. 1983. Seed predation and seedling mortality in the recruitment of a shrub, *Haplopappus venetus* (Asteraceae), along a climatic gradient. Ecology 64:511–521. [13]

———. 1984. Herbivore effect on stature, fruiting, and leaf dynamics of a native crucifer. Ecology 65:1379–1386. [10, 13]

———. 1986. Insect herbivory in response to root-cutting and flooding stress on a native crucifer under field conditions. Acta Oecologia, Oecologia Generalis 7:37–53. [3, 13]

———. 1987. Variation in methylglucosinolate and insect damage to *Cleome serrulata* (Capparaceae) along a natural soil moisture gradient. Journal of Chemical Ecology 13:569–581. [13]

Louda, S. M., N. Huntly, and P. M. Dixon. 1987. Insect herbivory across a sun/shade gradient: Response to experimentally-induced in situ plant stress. Acta Oecologia 8:357–363. [3, 6, 11]

Louda, S. M., K. H. Keeler, and R. D. Holt. 1989. Herbivore influences on plant performance and competitive interactions. In J. B. Grace and D. Tilman, eds., Perspectives in plant competition. 413–444. Academic Press, New York. [13]

Louda, S. M., and J. E. Rodman. 1983a. Concentration of glucosinolates in relation to habitat and insect herbivory for the native crucifer *Cardamine cordifolia*. Biochemical Systematics and Ecology 11:199–207. [4, 13]

———. 1983b. Ecological patterns in the glucosinolate content of a native mustard, *Cardamine cordifolia*, in the Rocky Mountains. Journal of Chemical Ecology 9:397–422. [6, 16]

Louda, S. M., and P. H. Zedler. 1985. Predation in insular plant dynamics: An experimental assessment of postdispersal fruit and seed survival. Enewetak Atoll, Marshall Island. American Journal of Botany 72:438–445. [13]

Lovegrove, B. G., and J. U. M. Jarvis. 1986. Coevolution between mole-rats (Bathyergidae) and a geophyte, *Micranthus* (Iridaceae). Cimbebasia 8:79–85. [13]

Lovett Doust, J., and L. Lovett Doust. 1985. Sex ratios, clonal growth, and herbivory in *Rumex acetosella*. In J. White, ed., Studies in plant demography: A festschrift for John L. Harper, 327–341. Academic Press, New York. [13]

Lowe, H. J. B. 1984. A behavioral difference among clones of the grain aphid *Sitobion avenae*. Ecological Entomology 9:119–122. [6]

Lowe, H. J. B., and L. R. Taylor. 1964. Population parameters, wing production, and behavior in red and green *Acyrthosiphon pisum* (Harris) (Homoptera: Aphididae). Entomologia Experimentalis et Applicata 48:117–125. [6]

Lowe, S., and F. E. Strong. 1963. The unsuitability of some viruliferous plants as hosts for the green peach aphid, *Myzus persicae*. Journal of Economic Entomology 56:307–309. [2]

Lowman, M. D. 1984. An assessment of techniques for measuring herbivory: Is rainforest defoliation more intense than we thought? Biotropica 16:264–268. [13]

———. 1985. Temporal and spatial variability in insect grazing of the canopies of five Australian rainforest tree species. Australian Journal of Ecology 10:7–24. [10, 11]

Lowman, M. D., and J. D. Box. 1983. Variation in leaf toughness and phenolic content among five species of Australian rain forest trees. Australian Journal of Ecology 8:17–25. [10]

Lubbers, A. E., and M. J. Lechowicz. 1989. Effects of leaf removal on reproduction vs. belowground storage in *Trillium grandiflorum*. Ecology 70:85–96. [13]

Lubchenco, J., and J. Cubit. 1980. Heteromorphic life histories of certain marine algae as adaptations to variations in herbivory. Ecology 61:676–687. [13]

Lubchenco, J, and S. D. Gaines. 1981. A unified approach to marine plant-herbivore interactions. I. Populations and communities. Annual Review of Ecology and Systematics 12:405–437. [17]

Luckwill, L. C. 1943. The genus *Lycopersicon*: An historical, biological, and taxonomic survey of the wild and cultivated tomatoes. Aberdeen University Studies 120. [2]

Lüdtke, M., and H. Hahn. 1953. Über den Linamaringehalt gesunder und von *Colletotrichum lini* befallener junger Leinpflanzen. Biochemische Zeitschrift 324:433–442. [17]

Luig, N. H., and I. A. Watson. 1977. The role of barley, rye, and grasses in the 1973–

42 wheat stem rust epiphytotic in southern and eastern Australia. Proceedings of the Linean Society of New South Wales 101:65–76. [5]

Lukefahr, M. J., and J. E. Houghtaling. 1969. Resistance of cotton strains with high gossypol content to *Heliothis* spp. Journal of Economic Entomology 62:588–591. [12, 16]

Lukefahr, M. J., J. E. Houghtaling, and H. M. Graham. 1971. Suppression of *Heliothis* populations with glabrous cotton strains. Journal of Economic Entomology 64:486–488. [12]

Lukens, R. J., and R. Mullany. 1972. The influence of shade and wet soil on southern corn leaf blight. Plant Disease Reporter 56:203–206.

Lundstroem, A. N. 1887. Pflanzenbiologische Studien. II. Die Anpassung der Pflanzen an Thiere. Nova Acta Uppsala, ser. 3, XIII. [13]

Lyman, J. M., and C. Cardona. 1982. Resistance in lima beans to a leafhopper, *Eupersca kraemeri*. Journal of Economic Entomology 75:281–286. [2]

Lynch, M., and W. Gabriel. 1987. Environmental tolerance. American Naturalist 129:283–303. [7]

Lynch, S. P., and R. A. Martin. 1987. Cardenolide content and thin-layer chromatography profiles of monarch butterflies, *Danaus plexippus* L., and their larval host-plant milkweed, *Asclepias viridis* Walt., in northwestern Louisiana. Journal of Chemical Ecology 13:47–70. [12]

MacClement, W. D., and M. G. Richards. 1956. Virus in wild plants. Canadian Journal of Botany 34:793–799. [14]

Macedo, C. A., and J. H. Langenheim. 1989. A further investigation of leaf sesquiterpene variation in relation to herbivory in two Brazilian populations of *Copaifera langsdorfii*. Biochemical Systematics and Ecology 17:207–216. [13]

MacGarvin, M., J. H. Lawton, and P. A. Heads. 1986. The herbivorous insect communities of open and woodland bracken: Observations, experiments and habitat manipulations. Oikos 47:135–148. [13]

Mack, R. N., and D. A. Pyke. 1984. The demography of *Bromus tectorum:* The role of microclimate, grazing, and disease. Journal of Ecology 72:731–748. [13]

Mack, T. P., and Z. Smilowitz. 1980. Soluble protein electrophoretic patterns from two biotypes of *Myzus persicae* (Sulzer). American Potato Journal 57:365–369. [5]

MacKay, P. A. 1989. Clonal variation in sexual morph production in *Acyrthosiphon pisum* (Homoptera: Aphididae). Environmental Entomology 18:558–562. [5]

MacKay, P. A., and R. J. Lamb. 1988. Genetic variation in asexual populations of two aphids in the genus *Acyrthosiphon*, from an Australian lucerne field. Entomologia Experimentalis et Applicata 48:117–125. [6]

MacLean, D. A., and D. P. Ostaff. 1989. Patterns of balsam fir mortality by an uncontrolled spruce budworm outbreak. Canadian Journal of Forest Research 19:1087–1095. [13]

Maddox, G. D., and N. Cappuccino. 1986. Genetic determination of plant susceptibility to an herbivorous insect depends on environmental context. Evolution 40:863–866. [2, 3, 6, 9, 11, 13, 17]

Maddox, G. D., and R. B. Root. 1987. Resistance to sixteen diverse species of herbivorous insects within a population of goldenrod, *Solidago altissima:* Genetic variation and heritability. Oecologia 72:8–14. [2, 4, 6, 9, 11, 13]

———. 1990. Structure of the selective encounter between goldenrod (*Solidago altissima*) and its diverse insect fauna. Ecology 71:2115–2124. [6, 11, 17]

Maga, J. A. 1980. Potato glycoalkaloids. CRC Critical Reviews in Food Science 12:371–405. [16]

Maher, E. P., and M. A. Hughes. 1973. Studies on the nature of *Li* locus in *Trifolium repens*. Biochemical Genetics 8:113–126. [17]

Mahmoud, A., J. P. Grime, and S. B. Furness. 1975. Polymorphisms in *Arrhenatherum elatius* (L.) Beauv. ex J. & C. Presl. New Phytologist 75:269–275. [13]

Mandel, J. 1954. Chain block designs with two-way elimination of heterogeneity. Biometrics 10:251–272. [3]

Manly, B. F. J. 1985. The statistics of natural selection on animal populations. Chapman and Hall, New York. [7]

Markkula, M. 1963. Studies on the pea aphid, *Acyrthosiphon pisum* (Hom., Aphididae), with special reference to the difference in the biology of the green and red forms. Annales Agriculturae Fenniae 2 (Suppl. 1). [6]

Marquis, R. J. 1984. Leaf herbivores decrease fitness of a tropical plant. Science 226:537–539. [2, 6, 10, 13, 16]

———. 1987. Variacion en la herbivoria foliar y su importancia selectiva en *Piper arieianum* (Piperaceae). Revista de Biologia Tropical 35 (Suppl.): 133–149. [13]

———. 1988a. Intra-crown variation in leaf herbivory and seed production in striped maple, *Acer pennsylvanicum* L. (Aceraceae). Oecologia 77:51–55. [13, 16]

———. 1988b. Phenological variation in the neotropical understory shrub *Piper arieianum:* Causes and consequences. Ecology 69:1552–1565. [13]

———. 1990. Genotypic variation in leaf damage in *Piper arieianum* (Piperaceae) by a multi-species assemblage of herbivores. Evolution 44:104–120. [6, 9, 11, 13]

———. 1991a. A bite is a bite is a bite? Constraints on response to folivory in *Piper arieianum* (Piperaceae). Ecology, in press. [13]

———. 1991b. Herbivore fauna of *Piper* (Piperaceae) in a Costa Rican wet forest: Diversity specificity and impact. In P. W. Price, T. M. Lewinsohn, G. W. Fernandes, and W. W. Benson, eds., Plant-Animal interactions: Evolutionary ecology in tropical and temperate regions, 179–208. J. Wiley and Sons, New York. [13]

Marquis, R. J., and H. E. Braker. 1987. Influence of method of presentation on results of plant-host preference tests with two species of grasshopper. Entomologia Experimentalis et Applicata 44:59–63. [4]

———. 1992. Plant-herbivore interactions at La Selva: Diversity, specificity, and impact. In L. M. McDade et al., eds., La Selva. University of Chicago Press, Chicago. In Press. [13]

Marshall, D. R. 1977. The advantages and hazards of genetic homogeneity. Annals of the New York Academy of Sciences 287:1–20. [5, 9]

———. 1989. Modelling the effects of multiline varieties on the population genetics of plant pathogens. In K. J. Leonard and W. E. Fry, eds., Plant disease epidemiology. Vol. 2, Genetics, resistance, and management, 284–317. McGraw-Hill, New York. [5]

Marshall, D. R., and J. J. Burdon. 1981. Multiline varieties and disease control. 3. Combined use of overlapping and disjoint gene sets. Australian Journal of Biological Sciences 34:81–95. [5]

Marshall, D. R., J. J. Burdon, and W. J. Müller. 1986. Multiline varieties and disease control. 6. Effects of selection at different stages of pathogen life cycle on the evolution of virulence. Theoretical and Applied Genetics 71:801–809. [5]

Marshall, D. R., and A. J. Pryor. 1978. Multiline varieties and disease control. I. The "dirty crop" approach with each component carrying a unique single resistance gene. Theoretical and Applied Genetics 51:177–184. [8]

———. 1979. Multiline varieties and disease control. II. The "dirty crop" approach with each component carrying two or more genes for resistance. Euphytica 28:145–159. [8]

Marshall, D. R., and B. S. Weir. 1987. Multiline varieties and disease control. V. The "dirty crop" approach with complex mixtures of genotypes based on overlapping gene sets. Theoretical and Applied Genetics 69:463–474. [18]

Marten, G. C., R. F. Burns, A. B. Simons, and F. J. Wooding. 1973. Alkaloids and palatability of *Phalaris arundinacea* L. grown in diverse environments. Agronomy Journal 65:199–201. [10]

Martin, J. S., and M. M. Martin. 1982. Tannin assays in ecological studies: Lack of correlation between phenolics, proanthocyanidins, and protein-precipitating constituents in mature foliage of six oak species. Oecologia 54:205–211. [4]

Martin, M. M., and J. S. Martin. 1984. Surfactants: Their role in preventing the precipitation of proteins by tannins in insect guts. Oecologia 61:342–345. [13]

Martin, R. A., and S. P. Lynch. 1988. Cardenolide content and thin-layer chromatography profiles of monarch butterflies, *Danaus plexippus* L. and their larval hostplant milkweed, *Asclepias asperula* subsp. *capricornu* (Woods.), in north central Texas. Journal of Chemical Ecology 14:295–318. [12]

Maschinski, J., and T. G. Whitham. 1989. The continuum of plant responses to herbivory: The influence of plant association, nutrient availability, and timing. American Naturalist 134:1–19. [13, 16]

Mason, C. F. 1970. Food, feeding rates, and assimilation in woodland snails. Oecologia 4:358–373. [10]

Mather, K., and J. L. Jinks. 1982. Biometrical genetics: The study of continuous variation. 3d ed. Chapman and Hall, London. [7, 17]

Mathre, D. E. 1978. Disrupted reproduction. In J. G. Horsfall and E. B. Cowling, eds., Plant disease, 3:257–278. Academic Press, New York. [15]

Matsumura, F. 1975. Toxicology of insecticides. Plenum Press, New York. [4]

Mattson, W. J. 1978. The role of insects in the dynamics of cone production of red pine. Oecologia 33:327–349. [13]

———. 1980. Herbivory in relation to plant nitrogen content. Annual Review of Ecology and Systematics 11:119–161. [3]

Mattson, W. J., and N. D. Addy. 1975. Phytophagous insects as regulators of forest primary production. Science 190:512–522. [10]

Mattson, W. J., and R. A. Haack. 1987a. The role of drought in outbreaks of planteating insects. Bioscience 37:110–118. [6, 13]

———. 1987b. The role of drought stress in provoking outbreaks of phytophagous insects. In P. Barbosa and J. Schultz, eds., Insect outbreaks: Ecological and evolutionary perspectives, 365–407. Academic Press, Orlando, FL. [11]

Mattson, W. J., and J. M. Scriber. 1987. Nutritional ecology on insect folivores of woody plants: Nitrogen, water, fiber, and mineral considerations. In F. Slansky, Jr., and J. G. Rodriguez, eds., Nutritional ecology of insects, mites, spiders, and related invertebrates. Wiley-Interscience, New York. [7]

Matzinger, D. F., and T. J. Mann. 1964. Genetic studies on associations between nicotine content and yield of flue-cured tobacco. In Proceedings of the Third World Tobacco Science Congress, 357–365. Salisbury, Southern Rhodesia, February 18–26, 1963. [17]

————. 1971. Inheritance and relationships among plant characters and smoke constituents in flue-cured tobacco. In Proceedings of the Fifth International Tobacco Science Congress, 68–76. Hamburg, Germany, September 14–19, 1970. [17]

Matzinger, D. F., T. J. Mann, and C. C. Cockerham. 1972. Recurrent family selection and correlated response in *Nicotiana tabacum* L. I. 'Dixie Bright 244' × 'Coker 139.' Crop Science 12:40–43. [13, 17]

Matzinger, D. F., T. J. Mann, and H. F. Robinson. 1960. Genetic variability in flue-cured varieties of *Nicotiana tabacum*. I. Hicks Broadleaf × Coker 139. Agronomy Journal 52:8–11. [17]

Matzinger, D. F., E. A. Wernsman, and W. W. Weeks. 1989. Restricted index selection for total alkaloids and yield in tobacco. Crop Science 29:74–77. [16, 17]

Maxwell, F. G., and P. R. Jennings, eds. 1980. Breeding plants resistant to insects. John Wiley and Sons, New York. [2, 5, 9, 11, 13]

May, B., and F. R. Holbrook. 1978. Absence of genetic variability in the green peach aphid, *Myzus persicae* (Hemiptera: Aphididae). Annals of the Entomological Society of America 71:809–812. [5]

May, R. M., and R. M. Anderson. 1979. Population biology of infectious diseases. II. Nature 280:455–461. [18]

————. 1983. Epidemiology and genetics in the coevolution of parasites and hosts. Proceedings of the Royal Society of London, series B219:281–313. [18]

May, R. M., and A. P. Dobson. 1986. Population dynamics and the rate of evolution of pesticide resistance. In National Research Council, Pesticide resistance: Strategies and tactics for management, 170–193. National Academy Press, Washington, DC. [17]

Maynard Smith, J. 1966. Sympatric speciation. American Naturalist 100:637–650. [6, 18]

Maynard Smith, J., and R. Hoekstra. 1980. Polymorphisms in a varied environment: How robust are the models? Genetical Research (Cambridge) 35:45–57. [5]

Mazanec, Z. 1966. The effect of defoliations by *Didymuria violescens* (Phasmatidae) on the growth of alpine ash. Australian Forestry 30:125–130. [13]

————. 1974. Influence of jarrah leaf miner on the growth of jarrah. Australian Forestry 37:32–42. [13]

McCarty, E. C., and R. Price. 1942. Growth and carbohydrate content of important mountain forage plants in central Utah as affected by clipping and grazing. United States Department of Agriculture Technical Bulletin no. 818. United States Department of Agriculture, Washington, DC. [13]

McClure, M. S. 1980a. Competition between exotic scale species: Scale insects on hemlock. Ecology 61:1391–1401. [11]

————. 1980b. Foliar nitrogen: A basis for host suitability for elongate hemlock scale, *Florinia externa* (Homoptera: Diaspididae). Ecology 61:72–79. [13]

McClure, M. S., and P. W. Price. 1975. Competition and coexistence among sympatric *Erythroneura* leafhoppers (Homoptera: Cicadellidae) on American sycamore. Ecology 56:1388–1397. [11]

McCrea, K. D., and W. G. Abrahamson. 1985. Evolutionary impacts of the goldenrod ball gallmaker on *Solidago altissima* clones. Oecologia 68:20–22. [13]

———. 1987. Variation in herbivore infestation: Historical vs. genetic factors. Ecology 68:822–827. [3, 11, 9, 13]

McCutcheon, G. S., and S. G. Turnipseed. 1981. Parasites of lepidopterous larvae in insect resistant and susceptible soybeans in South Carolina. Environmental Entomology 10:69–74. [12]

McDermitt, D. K., and R. S. Loomis. 1981. Elemental composition of biomass and its relation to energy content, growth efficiency, and growth yield. Annals of Botany 48:275–290. [17]

McDonald, B. A., and J. P. Martinez. 1990. DNA restriction fragment length polymorphisms among *Mycosphaerella graminicola* (Anamorph *Septoria tritici*) isolates collected from a single wheat field. Phytopathology 80:1368–1373. [8]

McDonald, B. A., J. M. McDermott, S. B. Goodwin, and R. W. Allard. 1989. The population biology of host-pathogen interactions. Annual Review of Phytopathology 27:77–94. [5]

McDonald, G. I. 1981. Differential defoliation of neighboring Douglas-fir trees by western spruce budworm. United States Department of Agriculture Research Note INT-306. [13]

McKenzie, H. 1964. Inheritance of sawfly reaction and stem solidness in spring wheat crosses: Sawfly reactions. Canadian Journal of Plant Science 45:583–589. [2]

McKenzie, J. A., J. M. Dearn, and M. J. Whitten. 1980. Genetic basis of resistance to diazinon in Victorian populations of the Australian sheep blowfly, *Lucilia cuprina*. Australian Journal of Biological Sciences 33:85–95. [17]

McKenzie, J. A., and A. Purvis. 1984. Chromosomal localisation of fitness modifiers of diazinon resistance genotypes of *Lucilia cuprina*. Heredity 53:625–634. [3]

McKenzie, J. A., M. J. Whitten, and M. A. Adena. 1982. The effect of genetic background on the fitness of diazinon resistance genotypes of the Australian sheep blowfly, *Lucilia cuprina*. Heredity 49:1–9. [3, 17]

McKey, D. 1974a. Adaptive patterns in alkaloid physiology. American Naturalist 108:305–320. [10, 13, 16, 17]

———. 1974b. Ant-plant eating by Colobus monkeys. Biotropica 6:269–270. [10]

———. 1979. The distribution of secondary compounds within plants. In G. A. Rosenthal and D. H. Janzen, eds., Herbivores: Their interaction with secondary plant metabolites, 55–133. Academic Press, New York. [10, 13, 16, 17]

McLain, D. K. 1981. Resource partitioning by three species of hemipteran herbivores on the basis of host plant density. Oecologia 48:414–417. [13]

———. 1984. Host plant density and territorial behavior of the seed bug, *Neacoryphus bicrucis* (Hempitera: Lygaeidae). Behavioral Ecology and Sociobiology 14:181–187. [13]

McMichael, S. C. 1960. Combined effects of glandless genes gl_2 and gl_3 on pigment glands in the cotton plant. Agronomy Journal 52:385–386. [2]

———. 1969. Selection for glandless seeded cotton plants. Crop Science 9:518–520. [2]

McMurtry, J. A., and E. H. Stanford. 1960. Observations of feeding habits of the spotted alfalfa aphid on resistant and susceptible alfalfa plants. Journal of Economic Entomology 53:714–717. [2]

McNaughton, S. J. 1979a. Grassland-herbivore dynamics. In A. R. E. Sinclair and M. Norton-Griffiths, eds., Serengeti: Dynamics of an ecosystem, 82–103. University of Chicago Press, Chicago. [10]

———. 1979b. Grazing as an optimization process: Grass-ungulate relationships in the Serengeti. American Naturalist 113:691–703. [10]

———. 1983a. Compensatory plant growth as a response to herbivory. Oikos 40:329–336. [10, 13, 16]

———. 1983b. Physiological and ecological implications of herbivory. In O. L. Lange, P. S. Nobel, E. B. Osmond, and H. Zieger, eds., Physiological plant ecology III, 657–677. Encyclopedia of Plant Physiology, n.s. 12C. Springer-Verlag, Berlin. [10, 13]

———. 1984. Grazing lawns: Animals in herds, plant form, and coevolution. American Naturalist 124:863–886. [10, 13]

———. 1986. On plants and herbivores. American Naturalist 128:765–770. [10, 13]

McNaughton, S. J., and F. S. Chapin III. 1985. Effects of phosphorous nutrition and defoliation on C_4 graminoids from the Serengeti Plains. Ecology 66:1616–1629. [13]

McNaughton, S. J., and N. J. Georgiadis. 1986. Ecology of African grazing and browsing mammals. Annual Review of Ecology and Systematics 17:39–65. [10]

McNaughton, S. J., and Tarrants, J. L. 1983. Grass leaf silicification: Natural selection for an inducible defense against herbivores. Proceedings of the National Academy of Sciences U.S.A. 80:790–791. [10, 13]

McNaughton, S. J., J. L. Tarrants, M. M. McNaughton, and R. H. Davis. 1985. Silica as a defense against herbivory and a growth promoter in African grasses. Ecology 66:528–535. [13]

McNaughton, S. J., L. L. Wallace, and M. B. Coughenour. 1983. Plant adaptation in an ecosystem context: Effects of defoliation, nitrogen, and water on growth of an African C_4 sedge. Ecology 64:307–318. [13]

McNeill, J. 1977. The biology of Canadian weeds. 25. *Silene alba* (Miller) E. H. L. Krause. Canadian Journal of Plant Science 57:1103–1114. [18]

McPheron, B. A., D. C. Smith, and S. H. Berlocher. 1988. Genetic differences between host races of *Rhagoletis pomonella*. Nature 336:64–66. [4, 6]

Meade, T. 1989. The effect of host plant genetic variation on the susceptibility of two polyphagous lepidopteran herbivores to *Bacillus thuringiensis* var. kurstaki. M.S. thesis, University of California, Riverside. [12]

Meagher, T. R. 1986. Analysis of paternity with a natural population of *Chamaelirium luteum*. I. Identification of male parentage. American Naturalist 128:199–215. [3, 18]

Meagher, T. R., and E. Thompson. 1987. Analysis of parentage for naturally established seedlings of *Chamaelirium luteum* (Liliaceae). Ecology 68:803–812. [3]

Mebrahtu, T. 1985. Efficiency of different selection and screening methods to identify productive soybean breeding lines resistant to Mexican bean beetle. Ph.D. diss., University of Maryland, College Park. (Abstract) [17]

Mehlenbacher, S. A., R. L. Plaisted, and W. M. Tingey. 1983. Inheritance of glandular trichomes in crosses with *Solanum berthaultii*. American Potato Journal 60:699–708. [2]

——. 1984. Heritability of trichome density and droplet size in interspecific crosses and hybrids and relationship to aphid resistance. Crop Science 24:320–322. [2]

Mendoza, A., D. Pinero, and J. Sarukhan. 1987. Effects of experimental defoliation on growth, reproduction, and survival of *Astrocaryum mexicanum*. Journal of Ecology 75:545–554. [13]

Metcalf, R. L., A. M. Rhodes, R. A. Metcalf, J. Ferguson, E. R. Metcalf, and P. Y. Lu. 1982. Curcurbitacin contents and diabroticite feeding upon *Curcurbita* species. Environmental Entomology 11:931–937. [4]

Michels, G. J., Jr., T. J. Kring, R. W. Behle, A. C. Bateman, and N. M. Heiss. 1987. Development of greenbug (Homoptera: Aphididae) on corn: Geographic variations in host-plant range of biotype E. Journal of Economic Entomology 80:394–397. [5]

Mihaliak, C. A., and D. E. Lincoln. 1985. Growth pattern and carbon allocation to volatile leaf terpenes under nitrogen-limiting conditions in *Heterotheca subaxillaris* (Asteraceae). Oecologia 66:423–426. [3, 16]

——. 1989. Plant biomass partitioning and chemical defense: Response to defoliation and nitrate limitation. Oecologia 80:122–126. [13]

Miles, J. W., and J. M. Lenné. 1984. Genetic variation within a natural *Stylosanthes guianensis, Colletotrichum gloeosporioides* host-pathogen population. Australian Journal of Agricultural Research 35:211–218. [8, 14,]

Miller, J. S., and P. Feeny. 1983. Effects of benzyllisoquinoline alkaloids on the larvae of polyphagous lepidoptera. Oecologia 58:332–339. [13]

Miller, R. F., and G. B. Donart. 1981. Response of *Muhlenbergia porteri* Scribn. to season of defoliation. Journal of Range Management 34:91–94. [13]

Miller, R. G. 1969. Jackknifing variances. Annals of Mathematical Statistics 39:567–582. [3]

Miller, R. G. 1974a. The jackknife: A review. Biometrika 61:1–15. [3]

Miller, R. G. 1974b. An unbalanced jackknife. Annals of Statistics 2:880–891. [3]

Miller, T., and J. Miller, eds. 1986. Insect-plant interactions. Springer Series in Experimental Entomology, Springer Verlag, New York. [4]

Milner, R. J. 1985. Distribution in time and space of resistance to the pathogenic fungus *Erynia neoaphidis* in the pea aphid *Acyrthosiphon pisum*. Entomologia Experimentalis et Applicata 37:235–240. [6]

Milton, K. 1979. Factors influencing leaf choice by howler monkeys: A test of some hypotheses of food selection by generalist herbivores. American Naturalist 114:362–378. [10]

Misaghi, I. J. 1982. Physiology and biochemistry of plant-pathogen interactions. Plenum Press, New York. [14]

Mitchell-Olds, T., and J. J. Rutledge. 1986. Quantitative genetics in natural plant populations: A review of the theory. American Naturalist 127:379–402. [3, 4, 14]

Mitchell-Olds, T., and R. G. Shaw. 1987. Regression analysis of natural selection: Statistical inference and biological interaction. Evolution 41:1149–1161. [3, 7]

Mittelbach, G. G,. and K. L. Gross. 1984. Experimental studies of seed predation in old fields. Oecologia 65:7–13. [13]

Mitter, C., and D. J. Futuyma. 1983. An evolutionary-genetic view of host-plant utilization by insects. In R. F. Denno and M. S. McClure, eds., Variable plants and herbivores in natural and managed systems, 427–460. Academic Press, New York. [5, 10]

Mode, C. J. 1958. A mathematical model for the co-evolution of obligate parasites and their hosts. Evolution 12:158–165. [5, 15, 17]

———. 1960. A model of a host-pathogen system with particular reference to the rusts of cereals. In O. Kempthorne, ed., Biometrical genetics, 84–97. Proceedings of the International Union of the Biological Sciences, ser. B, no. 38.[5]

———. 1961. A generalized model of a host-pathogen system. Biometrics 17:386–404. [5]

Moesta, P., and H. Grisebach. 1982. L-2-Aminooxy-3-phenylpropionic acid inhibits phytoalexin accumulation in soybean with concomitant loss of resistance against *Phytophthora megasperma* f. sp. *glycinea.* Physiological Plant Pathology 21:65–70. [16]

Moffatt, J. M., W. D. Worrall, and R. L. Reinisch. 1983. Biotype constitution of the greenbug (*Schizaphis graminum*) in the Texas Rolling Plains. Agronomy Abstracts. 1983 Annual Meeting American Society of Agronomy, Madison, WI. p. 73. [5]

Molau, U., B. Eriksen, and J. T. Knudsen. 1989. Predispersal seed predation in *Bartsia alpina.* Oecologia 81:181–185. [13]

Mole, S., J. C. Rogler, C. J. Morell, and L. G. Butler. 1990. Herbivore growth reduction by tannins: Use of Waldbauer ratio techniques and manipulation of salivary protein production to elucidate mechanisms of action. Biochemical Systematics and Ecology 18:183–198. [10]

Mole, S., J. A. M. Ross, and P. G. Waterman. 1988. Light-induced variation in phenolic levels in foliage of rain-forest plants. I. Chemical changes. Journal of Chemical Ecology 14:1–21. [16]

Mooney, H. A., and C. Chu. 1974. Seasonal carbon allocation in *Heteromeles arbutifolia,* a California evergreen shrub. Oecologia 14:295–306. [16]

Mooney, H. A., and S. L. Gulmon. 1982. Constraints on leaf structure and function in reference to herbivory. BioScience 322:198–206. [16, 17]

Mooney, H. A., S. L. Gulmon, and N. D. Johnson. 1983. Physiological constraints on plant chemical defenses. In P. A. Hedin, ed., Plant resistance to insects, 21–36. ACS Symposium Series 208. [16]

Moore, L. R. 1978. Seed predation in the legume *Crotalaria.* I. Intensity and variability of seed predation in native and introduced populations of *C. pallidi* Ait. Oecologia 34:185–202. [13]

Moore, L. V., J. H. Myers, and R. Eng. 1988. Western tent caterpillars prefer the sunny side of the tree, but why? Oikos 51:321–326. [13]

Mopper, S., T. G. Whitham, and P. W. Price. 1990. Plant phenotype and interspecific competition between insects determine sawfly performance and density. Ecology 71:2135–2144. [6, 11]

Moran, N. 1981. Intraspecific variability in herbivore performance and host quality: A field study of *Uroleucon caligatum* (Homoptera: Aphididae) and its *Solidago* hosts (Asteraceae). Ecological Entomology 6:301–306. [2, 5, 6, 7, 9, 11, 13]

———. 1986. Morphological adaptation to host plants in *Uroleucon* (Homoptera: Aphididae). Evolution 40:1044–1050. [6]

Moran, N., and W. D. Hamilton. 1980. Low nutritive quality as defense against herbivores. Journal of Theoretical Biology 86:247–254. [10]

Moran, N. A. 1988. The evolution of host-plant alteration in aphids: Evidence for specialization as a dead end. American Naturalist 132:681–706. [7]

Moran, N. A., and T. G. Whitham. 1990. Interspecific competition between root-feeding and leaf-galling aphids mediated by host-plant resistance. Ecology 71: 1050–1058. [6, 11]

Morgan, A. C. 1910. An observation upon the toxic effect of the food of the host upon its parasites. Proceedings of the Entomological Society of Washington 12:72. [12]

Morishima, H., and H. J. Oka. 1979. Genetic diversity in rice populations of Nigeria: Influence of community structure. Agro-Ecosystems 5:263–269. [18]

Morrow, P. A., and L. R. Fox. 1980. Effects of variation in *Eucalyptus* essential oil yield on insect growth and grazing damage. Oecologia 45:209–219. [13]

———. 1989. Estimates of pre-settlement insect damage in Australian and North American forests. Ecology 70:1055–1060. [13]

Morrow, P. A., and V. C. LaMarche, Jr. 1978. Tree ring evidence for chronic insect suppression of productivity in subalpine *Eucalyptus*. Science 201:1244–1246. [13]

Mortlock, R. P. 1982. Regulatory mutations and the development of new metabolic pathways by bacteria. Evolutionary Biology 14:205–268. [17]

Moseman, J. G., E. Nevo, M. A. El Morshidy, and D. Zohary. 1984. Resistance of *Triticum dicoccoides* to infection with *Erysiphe graminis tritici*. Euphytica 33:41–47. [14]

Mothes, K. 1976. Secondary plant substances as materials for chemical high quality breeding in higher plants. In J. Wallace and R. L. Mansell, eds., Biochemical interactions between plants and insects, 385–405. Recent Advances in Phytochemistry 10:385–405. Plenum Press, New York. [17]

Muhammed, A., R. Aksel, and R. C. Borstel, eds. 1976. Genetic diversity in plants. International Symposium on Genetic Control of Diversity in Plants. Plenum Press, New York. [2]

Mulkern, G. B. 1967. Food selection by grasshoppers. Annual Review of Entomology 12:59–78. [10]

Muller, C. H. 1969. The "co-" in coevolution. Science 164:197–198. [17]

Müller, H., C. S. A. Stinson, K. Marquardt, and D. Schroeder. 1989. The entomofaunas of roots of *Centaurea maculosa* Lam., *C. diffusa* Lam., and *C. vallesiaca* Jordan in Europe. Journal of Applied Entomology 107:83–95. [13]

Mundt, C. C. 1989. Modeling disease increase in host mixtures. In K. J. Leonard and W. E. Fry, eds., Plant disease epidemiology. vol. 2, Genetics, resistance, and management, 150–181. McGraw-Hill, New York. [5]

Mundt, C. C., and L. S. Brophy. 1988. Influence of number of host genotype units on the effectiveness of host mixtures for disease control: A modeling approach. Phytopathology 78:1087–1094. [5]

Mundt, C. C., and J. A. Browning. 1985. Genetic diversity and cereal rust management. In A. P. Roelfs and W. R. Bushness, eds., The cereal rusts, 2:527–560. Academic Press, Orlando. [14]

Mundt, C. C., and K. J. Leonard. 1986. Analysis of factors affecting disease increase and spread in mixtures of immune and susceptible plants in computer-simulated epidemics. Phytopathology 76:832–840. [5]

Mundt, C. C., K. J. Leonard, W. M. Thal, and J. H. Fulton. 1986. Computerized simulation of crown rust epidemics in mixtures of immune and susceptible oat plants with different genotype unit area and spatial distributions of initial disease. Phytopathology 76:590–598. [5]

Murfet, I. C. 1977. Environmental interaction and the genetics of flowering. Annual Review of Plant Physiology 28:253–278. [13]

Murray, M. J. 1960a. The genetic basis for the conversion of menthone to menthol in Japanese mint. Genetics 47:925–929. [2]

———. 1960b. The genetic basis for a third ketone group in *Mentha spicata* L. Genetics 47:931–937. [2]

Murray, M. J., D. E. Lincoln, and F. W. Hefendehl. 1980. Chemogenetic evidence supporting multiple allele control of the biosynthesis of (−)-menthone and (+)-isomenthone stereoisomers in *Mentha* species. Phytochemistry 19:2103–2110. [2]

Murray, R. D. H., J. Mendez, and S. A. Brown. 1982. The natural coumarins. John Wiley and Sons, Chichester, England. [16]

Myers, N. 1980. The sinking ark: A new look at the problem of disappearing species. Pergamon Press, Oxford. [2]

———. 1981. Conversion rates in tropical moist forest: Review of a recent survey. In F. Mergen, ed., Tropical forests: Utilization and conservation, 48–66. Yale School of Forestry and Environmental Studies, New Haven. [2]

Myint, M. M., H. R. Rapusas, and E. A. Heinrichs. 1986. Integration of varietal resistance and predation for the management of *Nephotettix virescens* (Homoptera: Cicadellidae) populations on rice. Crop Protection 5:259–265. [12]

Nachit, M., and W. Feucht. 1983. Inheritance of phenolic compounds, indoles, and growth vigour in *Prunus* crosses (Cherries). Zeitschrift für Pflanzenzuchtungg 90:166–171. [16]

Nagylaki, T. 1977. Selection in one- and two-locus systems. Springer-Verlag, Berlin. [7]

Namkoong, G. 1983. Preserving natural diversity. In C. M. Schonewald-Cox, S. M. Chambers, B. MacBride, and W. L. Thomas, eds., Genetics and conservation. Benjamin Cummings, Menlo Park. [2]

Nash, A. F., and R. G. Gardner. 1988. Heritability of tomato early blight resistance derived from *Lycopersicon hirsutum* P.I. 126–445. Journal of the American Society for Horticultural Science 113:264–268. [3]

Nass, H. G. 1972. Cyanogenesis in *Sorghum bicolor, Sorghum sudanense, Lotus,* and *Trifolium repens*: A review. Crop Science 12:503–506. [17]

Nath, P., and C. V. Hall. 1965. The genetic basis for cucumber beetle resistance in *Cucurbita pepo* L. Journal of the American Society for Horticultural Science 86:442–445. [2]

National Research Council. 1969. Insect pest management and control, Vol. 3, Principles of plant and animal pest control. National Academy of Sciences, Publ. 1695. Washington, DC. [2]

Neal, J. J. 1987. Metabolic costs of mixed-function oxidase induction in *Heliothis zea*. Entomologia Experimentalis et Applicata 43:175–179. [4]

Neher, D. A., C. K. Augspurger, and H. T. Wilkinson. 1987. Influence of age structure of plant populations on damping-off epidemics. Oecologia 74:419–424. [15]

Nelson, C. J., J. N. Seiber, and L. P. Brower. 1981. Seasonal and intraplant variation of cardenolide content in the California milkweed, *Asclepias eriocarpa,* and implications for plant defense. Journal of Chemical Ecology 7:981–1010. [16]

Nelson, R. R. 1972. *Stabilizing racial populations of plant pathogens by use of resistance genes.* Journal of Environmental Quality 1:220–227. [5]

———. 1973. Breeding plants for disease resistance. Pennsylvania State University Press, University Park. [5, 14]

———. 1978. Genetics of horizontal resistance to plant diseases. Annual Review of Phytopathology 16:359–378. [8]

———. 1979. The evolution of parasitic fitness. In J. B. Horsfall and E. B. Cowling, eds., Plant disease: An advanced treatise, 4:23–46. Academic Press, New York. [18]

Neuvonen, S., and K. Danell. 1987. Does browsing modify the quality of birch foliage for *Epirrita autumnata* larvae: Oikos 49:156–160. [11]

Neuvonen, S., and E. Haukioja. 1984. Low nutritive quality as defence against herbivores: Induced responses in birch. Oecologia 63:71–74. [9, 13]

Nevo, E., Z. Gerechter-Amitai, A. Beiles, and E. M. Golenberg. 1986. Resistance of wild wheat to stripe rust: Predictive method by ecology and allozyme genotypes. Plant Systematics and Evolution 153:13–30. [15]

Nevo, E., J. G. Moseman, A. Beiles, and D. Zohary. 1984. Correlation of ecological factors and allozymic variations with resistance to *Erysiphe graminis hordei* in *Hordeum spontaneum* in Israel: Patterns and applications. Plant Systematics and Evolution 145:79–96. [15]

Newton, I. 1967. The feeding ecology of the bullfinch (*Pyrrhula pyrrhula* L.) in southern England. Journal of Animal Ecology 36:721–744. [13]

Ng, D. 1988. A novel level of interactions in plant-insect systems. Nature 334:611–612. [6, 7]

Nielsen, M. T., G. A. Jones, and G. B. Collins. 1982. Inheritance pattern for secreting and nonsecreting glandular trichomes in tobacco. Crop Science 22:1051–1053. [2]

Nielsen, M. T., P. D. Legg, and C. C. Litton. 1985. Effects of two introgressed disease resistance factors on agronomic characteristics and certain chemical components in burley tobacco. Crop Science 25:698–701. [16]

Nielson, M. W., and H. Don. 1974a. A new virulent biotype of the spotted alfalfa aphid in Arizona. Journal of Economic Entomology 67:64–66. [6]

———. 1974b. Probing behavior of biotypes of the spotted alfalfa aphid on resistant and susceptible alfalfa clones. Entomologia Experimentalis et Applicata 67: 477–486.

Nielson, M. W., and W. F. Lehman. 1980. Breeding approaches in alfalfa. In F. G. Maxwell and P. R. Jennings, eds., Breeding plants resistant to insects, 277–312. J. Wiley and Sons, New York. [2]

Niemelä, P. 1987. Does leaf morphology of some plants mimic caterpillar damage? Oikos 50:256–257. [11]

Niemezyk, J. D. 1980. New evidence indicates greenbug overwinters in North. Weeds, Trees, and Turf. 19:64. [5]

Nienhuis, J., T. Helentjaris, M. Slocum, B. Ruggers, and A. Schaefer. 1987. Restriction fragment length polymorphism analysis of loci associated with insect resistance in tomato. Crop Science 27:797–803. [2, 3]

Nitao, J. K. 1988. Artificial defloration and furanocoumarin induction in *Pastinaca sativa* (Umbelliferae). Journal of Chemical Ecology 14:1515–1522. [16]

———. 1989. Enzymatic adaptation in a specialist herbivore for feeding on furanocoumarin-containing plants. Ecology 70:629–635. [4]

Nitao, J. K., and A. R. Zangerl. 1986. Floral development and chemical defense allocation in wild parsnip (*Pastinaca sativa*). Ecology 68:521–529. [16]

Nordlund, D. A., R. L. Jones, and W. J. Lewis, eds. 1981. Semiochemicals: Their role in pest control. John Wiley and Sons, New York. [12]

Nordlund, D. A., W. J. Lewis, and M. A. Altieri. 1988. Influences of plant-produced allelochemicals on the host/prey selection behavior of entomophagous insects. In P. Barbosa and D. K. Letourneau, eds., Novel aspects of insect-plant interactions, 65–90. John Wiley and Sons, New York. [12]

Nordlund, D. A., and C. E. Sauls. 1981. Kairomones and their use for management of entomophagous insects. XI. Effect of host plants on kairomonal activity of frass from *Heliothis zea* larvae for the parasitoid *Microplitis croceipes*. Journal of Chemical Ecology 7:1057–1061. [12]

Norris, D. M., and M. Kogan. 1980. Biochemical and morphological bases of resistance. In F. G. Maxwell and P. R. Jennings, eds., Breeding plants resistant to insects, 23–61. John Wiley and Sons, New York. [2, 11]

Notteghem, J.-L. 1985. Definition of a strategy for the use of resistance through the genetic analysis of host-pathogen relationships: The rice *Pyricularia oryzae* couple. Agronomic Tropicale. Serie Riz et Riziculture et Cultures Vivrieres Tropicales 40:129–147. [3]

Nowak, R. S., and M. M. Caldwell. 1984. A test of compensatory photosynthesis in the field: Implications for herbivory tolerance. Oecologia 61:311–318. [13]

Núñez-Farfán, J., and R. Dirzo. 1988. Within-gap spatial heterogeneity and seedling performance in a Mexican tropical forest. Oikos 51:274–284. [13]

Oates, J. D., J. J. Burdon, and J. B. Brouwer. 1983. Interactions between *Avena* and *Puccinia* species. II. The pathogens: *Puccinia coronata* and *P. graminis*. Journal of Applied Ecology 20:585–596. [15]

Obrycki, J. J. 1986. The influence of foliar pubescence on entomophagous species. In D. J. Boethel and R. D. Eikenbary, eds., Interactions of plant resistance and parasitoids and predators of insects, 61–83. Ellis Horwood, Chicester. [11, 12]

Obrycki, J. J., and M. J. Tauber. 1984. Natural enemy activity on glandular pubescent potato plants in the greenhouse: An unreliable predictor of effects in the field. Environmental Entomology 13:679–683. [12]

O'Dowd, D. J., and A. M. Gill. 1984. Predator satiation and site alteration following fire: Mass reproduction of alpine ash (*Eucalyptus delegatensis*) in southwestern Australia. Ecology 65:1052–1066. [13]

O'Dowd, D. J., and M. E. Hay. 1980. Mutualism between harvester ants and a desert ephemeral: Seed escape from rodents. Ecology 61:531–540. [3]

Oelrichs, P. B., and Robertson, P. A. 1970. Purification of pain-producing substances from *Dendrocnide* (*Laportea*) *moroides*. Toxicon 8:89–90. [10]

Oesterheld, M., and S. J. McNaughton. 1988. Intraspecific variation in the response of *Themeda triandra* to defoliation: The effect of time of recovery and growth rates on compensatory growth. Oecologia 77:181–186. [13]

Old, K. M., M. J. Dudzinski, and J. C. Bell. 1988. Isozyme variability in field populations of *Phytophthora cinnamomi* in Australia. Australian Journal of Botany 36:355–360. [8]

Olesen, P. O. 1978. On cyclosis and topophysis. Silvae Genetica 27:173–178. [3]

Olney, H. P. 1968. Growth substances from *Veratrum tenuipetalum*. Plant Physiology 43:293–302. [1, 17]

Olson, B. E., and J. H. Richards. 1988. Tussock regrowth after grazing: Intercalary meristem and axillary bud activity of tillers of *Agropyron desertorum*. Oikos 51:374–382. [13]

Oluomi-Sadeghi, H., C. G. Helm, M. Kogan, and D. F. Schoeneweiss. 1988. Effect of water stress on abundance of twospotted spider mite on soybeans under greenhouse conditions. Entomologia Experimentalis et Applicata 48:85–90. [13]

Onuf, C. P. 1978. Nutritive value as a factor in plant-insect interactions with an emphasis on field studies. In G. G. Montgomery, ed., The ecology of arboreal folivores, 85–96. Smithsonian Institution Press, Washington, DC. [13]

Onuf, C. P., J. M. Teal, and I. Valiela. 1977. Interactions of nutrients, plant growth, and herbivory in a mangrove system. Ecology 58:514–526. [13]

Onukogu, R. A., W. D. Guthrie, W. A. Russel, G. L. Reed, and J. C. Robbins. 1978. Location of genes that condition resistance in maize to sheath-collar feeding by second generation European corn borer. Journal of Economic Entomology 71: 1–4. [2]

Orians, G. H., and D. H. Janzen. 1974. Why are embryos so tasty? American Naturalist 108:581–592. [17]

Orr, D. B., and D. J. Boethel. 1985. Comparative development of *Copidosoma truncatellum* (Hymenoptera: Encyrtidae) and its host, *Pseudoplusia includens* (Lepidoptera: Noctuidae), on resistant and susceptible soybean genotypes. Environmental Entomology 14:612–616. [12]

———. 1986. Influence of plant antibiosis through four trophic levels. Oecologia 70:242–249. [12]

Orr, D. B., D. J. Boethel, and W. A. Jones. 1985. Biology of *Telenomus chloropus* (Hemiptera: Pentatomidae) reared on resistant and susceptible soybean genotypes. Canadian Entomologist 117:1137–1142. [12]

Ortega, A., S. K. Vasal, J. Milan, and C. Hershey. 1980. Breeding for insect resistance in maize. In F. G. Maxwell and P. R. Jennings, eds., Breeding plants resistant to insects, 371–419. J. Wiley, New York. [13]

Ortman, E. E., and D. C. Peters. 1980. Introduction to F. G. Maxwell and P. R. Jennings, eds., Breeding plants resistant to insects, 3–13. John Wiley and Sons, New York. [11]

Ortman, E. E., E. L. Sorensen, R. H. Painter, T. L. Harvey, and H. L. Hackerott. 1960. Selection and evaluation of pea aphid–resistant alfalfa plants. Journal of Economic Entomology 53:881–887. [2]

Østergaard, H. 1983. Predicting development of epidemics on cultivar mixtures. Phytopathology 73:166–172. [5]

Ostry, M. E., and H. S. McNabb, Jr. 1985. Susceptibility of *Populus* species and hybrids to disease in the north central United States. Plant Disease 69:755–757. [8]

Otte, D. 1975. Plant preference and plant succession. Oecologia 18:129–144. [10]

Overman, J. L., and L. E. MacCarter. 1972. Evaluating seedlings of cantaloupe for varietal nonpreference-type resistance to *Diabrotica* spp. Journal of Economic Entomology 65:1140–1144. [2]

Owen, D. F. 1978. The effect of a consumer, *Phytomyza ilicis,* on seasonal leaf-fall in the holly, *Ilex aquifolium.* Oikos 31:268–271. [13]

———. 1980. How plants may benefit from the animals that eat them. Oikos 35:230–235. [10]

Owen, D. F., and R. G. Wiegert. 1976. Do consumers maximize plant fitness? Oikos 27:488–492. [10]

———. 1981. Mutualism between grasses and grazers: An evolutionary hypothesis. Oikos 36:376–378. [10, 13]

———. 1982a. Beating the walnut tree: More on grass/grazer mutualism. Oikos 39:115–116. [10]

———. 1982b. Grasses and grazers: Is there a mutualism? Oikos 38:258–259. [10]

———. 1984. Aphids and plant fitness: 1984. Oikos 43:403. [10]

Paige, K. N., and T. G. Whitham. 1987. Overcompensation in response to mammalian herbivory: The advantage of being eaten. American Naturalist 129:407–416. [10, 13]

Painter, E. L., and J. K. Detling. 1981. Effects of defoliation on net photosynthesis and regrowth of western wheatgrass. Journal of Range Management 34:68–71. [13]

Painter, R. H. 1930. The biological strains of Hessian fly. Journal of Economic Entomology 21:322–326. [4]

———. 1951. Insect resistance in crop plants. Macmillan, New York. [2, 4, 5, 9, 17]

———. 1958. Resistance of plants to insects. Annual Review of Entomology 3:267–290. [17]

Pair, S. D., B. R. Wiseman, and A. N. Sparks. 1986. Influence of four corn cultivars on fall armyworm (Lepidoptera: Noctuidae) establishment of parasitization. Florida Entomologist 69:566–570. [12]

Pallant, D. 1969. The food of the grey field slug *Agriolimax reticulatus* (Muller) in woodland. Journal of Animal Ecology 38:391–397. [10]

Palmer, A. R., and C. Strobeck. 1986. Fluctuating asymmetry: Measurement, analysis, patterns. Annual Review of Ecology and Systematics 17:391–421. [7]

Panda, N. 1979. Principles of host-plant resistance to insects. Allanheld, Osmum and Co., New York. [2]

Pandey, M. C., and R. D. Wilcoxson. 1970. The effect of light and physiological races on *Leptosphaerulina* leaf spot of alfalfa and selection for resistance. Phytopathology 60:1456–1462. [3]

Papaj, D. R., and M. D. Rausher. 1983. Individual variation in host location by phytophagous insects. In S. Ahmad, ed., Herbivorous insects: Hostseeking behavior and mechanisms, 77–124. Academic Press, New York. [4]

Park, T. 1954. Experimental studies of interspecific competition. II: Temperature, humidity, and competition in two species of *Tribolium.* Physiological Zoology 27:177–238. [11]

Parker, M. A. 1985. Local population differentiation for compatibility in an annual legume and its host-specific fungal pathogen. Evolution 39:713–723. [8, 11, 14, 15]

———. 1986. Individual variation in pathogen attack and differential reproductive success in the annual legume, *Amphicarpaea bracteata.* Oecologia 69:253–259. [14, 15]

————. 1987. Pathogen impact on sexual vs. asexual reproductive success in *Arisaema triphyllum*. American Journal of Botany 74:1758–1763. [15]

————. 1988a. Disequilibrium between disease-resistance variants and allozyme loci in an annual legume. Evolution 42:239–247. [14, 15]

————. 1988b. Genetic uniformity and disease resistance in a clonal plant. American Naturalist 132:538–549. [15]

————. 1988c. Polymorphism for disease resistance in the annual legume *Amphicarpaea bracteata*. Heredity 60:27–31. [14, 15]

————. 1989. Disease impact and local genetic diversity in the clonal plant *Podophyllum peltatum*. Evolution 43:540–547. [15]

————. 1990. The pleiotropy theory for polymorphism of disease resistance genes in plants. Evolution 44:1872–1875. [15]

————. 1991. Nonadaptive evolution of disease resistance in an annual legume. Evolution 45:1209–1217. [15]

Parker, M. A., and R. B. Root. 1981. Insect herbivores limit habitat distribution of a native composite, *Michaeranthera canescens*. Ecology 62:1390–1392. [13]

Parker, M. A., and A. G. Salzman. 1985. Herbivore exclosure and competitor removal: Effects on juvenile survivorship and growth in the shrub *Gutierrezia microcephala*. Journal of Ecology 73:903–913. [13]

Parlevliet, J. E., and J. C. Zadoks. 1977. The integrated concept of disease resistance: A new view including horizontal and vertical resistance in plants. Euphytica 26:5–21. [8, 18]

Parr, J. C., and R. Thurston. 1972. Toxicity of nicotine in synthetic diets to larvae of the tobacco hornworm. Annals of Entomological Society of America 65:1185–1188. [13]

Pasek, J. E., and M. E. Dix. 1988. Insect damage to conelets, second-year cones, and seeds of ponderosa pine in southeastern Nebraska. Journal of Economic Entomology 81:1681–1690. [13]

Pashley, D. 1988. Quantitative genetics, development, and physiological adaptation in host strains of fall armyworm. Evolution 42:93–102. [4]

Pasteels, J. M., J. C. Gregoire, and M. Rowell-Rahier. 1983. The chemical ecology of defense in anthropods. Annual Review of Entomology 28:263–289. [12]

Pasteels, J. M., M. Rowell-Rahier, and M. J. Raupp. 1988. Plant-derived defense in chrysomelid beetles. In P. Barbosa and D. K. Letourneau, eds., Novel aspects of insect-plant interactions, 235–272. John Wiley and Sons, New York. [12]

Paterniani, E. 1969. Selection for reproductive isolation between two populations of maize, *Zea mays* L. Evolution 23:534–547. [13]

Paterson, A. H., E. S. Lander, J. D. Hewitt, S. Peterson, S. E. Lincoln, and S. D. Tanksley. 1988. Resolution of quantitative traits into mendelian factors by using a complete linkage map of restriction fragment length polymorphisms. Nature 335:721–726. [3]

Pathak, M. D., and R. H. Painter. 1958. Differential amounts of material taken up by four biotypes of corn leaf aphid from resistant and susceptible sorghums. Annals of the Entomological Society of America 51:250–254. [6]

Paul, N. D., and P. G. Ayres. 1986a. The impact of pathogen (*Puccinia lagenophorae*) on populations of groundsel (*Senecio vulgaris*) overwintering in the field. I. Mortality, vegetative growth, and the development of size hierarchies. Journal of Ecology 74:1069–1084. [14]

———. 1986b. The impact of a pathogen (*Puccinia lagenophorae*) on populations of groundsel (*Senecio vulgaris*) overwintering in the field. II. Reproduction. Journal of Ecology 74:1085–1094. [14]

———. 1987. Survival, growth, and reproduction of groundsel (*Senecio vulgaris*) infected by rust (*Puccinia lagenophorae*) in the field during summer. Journal of Ecology 75:61–71. [14]

Paul, N. D., P. G. Ayres, and L. E. Wyness. 1989. An assessment of fungicides for chemical exclusion experiments. Functional Ecology 3:759–769. [14]

Paul, V. J., and K. L. Van Alstyne. 1988. Chemical defense and chemical variation in some tropical Pacific species of *Halimeda* (Halimedaceae: Chlorophyta). Coral Reefs 6:263–269. [10]

Paulissen, M. A. 1987. Exploitation by, and effects of, caterpillar grazers on the annual, *Rudbeckia hirta* (Compositae). American Midland Naturalist 117:439–441. [13]

Pease, C. M., and J. J. Bull. 1988. A critique of methods for measuring life history trade-offs. Journal of Evolutionary Biology 1:293–303. [3, 4, 17]

Peckarsky, B. L., and S. I. Dodson. 1980. An experimental analysis of biological factors contributing to stream community structure. Ecology 61:1283–1290. [11]

Pedersen, P. N. 1960. Methods of testing the pseudo-resistance of barley to infection by loose smut, *Ustilago nuda*. Acta Agriculturae Scandinavica 10:312–332. [15]

Penning de Vries, F. W. T. 1974. Substrate utilization and respiration in relation to growth and maintenance in higher plants. Netherlands Journal of Agricultural Science 22:40–44. [17]

———. 1975a. The cost of maintenance processes in plant cells. Annals of Botany 39:77–92. [16]

———. 1975b. Use of assimilates in higher plants. In J. P. Cooper, ed., Photosynthesis and productivity in different environments, 459–480. Cambridge University Press, Cambridge. [17]

Penning de Vries, F. W. T., A. H. M. Brunsting, and H. H. van Laar. 1974. Products, requirements, and efficiency of biosynthesis: A quantitative approach. Journal of Theoretical Biology 45:339–377. [17]

Pereira, A. S. Rodrigues. 1978. Effects of leaf removal on yield components in sunflower. Netherlands Journal of Agricultural Science 26:133–144. [13]

Perrin, R. M. 1977. Pest management in multiple cropping systems. Agro-Ecosystems 3:93–118. [5]

Perrin, R. M., and M. L. Phillips. 1978. Some effects of mixed cropping on the population dynamics of insect pests. Entomologia Experimentalis et Applicata 24:385–393. [5]

Perry, D. A., and G. B. Pitman. 1983. Genetic and environmental influences in host resistance to herbivory: Douglas-fir and the Western Spruce budworm. Zeitschrift für angewandte Entomologie 96:217–229. [9, 13]

Person, C. 1959. Gene-for-gene relationships in host:parasite systems. Canadian Journal of Botany 37:1101–1130. [8]

———. 1966. Genetic polymorphism in parasitic systems. Nature 212:266–267. [15]

Person, C. O., D. J. Sambroski, and R. Rohringer. 1952. The gene for gene concept. Nature 194:561–562. [2]

Peterson, G. C., J. W. Johnson, G. L. Teetes, and D. T. Rosenow. 1984. Registration of Tx2783 greenbug resistant sorghum germplasm line. Crop Science. 24:390. [5]

Phillips, P. A., and M. M. Barnes. 1975. Host race formation among sympatric apple, walnut, and plum populations of the codling moth, *Laspeyresia pomonella*. Annals of the Entomological Society of America 68:1053–1060. [6]

Phillips, P. C., and S. J. Arnold. 1989. Visualizing multivariate selection. Evolution 43:1209–1222. [3]

Pielou, E. C. 1966. The measurements of diversity in different types of biological collections. Journal of Theoretical Biology 13:131–144. [9]

Pillemer, E. A., and W. M. Tingey. 1976. Hooked trichomes: A physical plant barrier to a major agricultural pest. Science 193:482–484. [2]

Pilson, D. 1990a. Control of insect distribution patterns I. Preference and performance. (Manuscript) [6]

———. 1990b. Control of insect distribution patterns II. Indirect interactions between herbivores. (Manuscript) [6]

———. 1992. Relative resistance of goldenrod to aphid attack: Changes through the growing season. Evolution (in press). [6]

Pilson, D., and M. D. Rausher. 1990. Clonal variation in aphid host-use characters and golden-rod resistance. (Manuscript) [6]

Pimentel, D. 1976. World food crisis: Energy and pests. Bulletin of the Entomological Society of America 22:20–26. [16]

———. 1988. Herbivore population feeding pressure on plant hosts: Feedback evolution and host conservation. Oikos 53:289–302. [17]

Pimentel, D., and A. G. Wheeler, Jr., 1973. Influence of alfalfa resistance on a pea aphid population and its associated parasites, predators, and competitors. Environmental Entomology 2:1–11. [12]

Pitelka, L. F., and J. W. Ashmun. 1985. Physiology and integration of ramets in clonal plants. In J. B. C. Jackson, L. W. Buss, and R. E. Cook, eds., Population biology and evolution of clonal organisms, 399–434. Yale University Press, New Haven. [13]

Plapp, F. W., Jr. 1976. Biochemical genetics of insecticide resistance. Annual Review of Entomology 21:179–197. [17]

———. 1986. Genetics and biochemistry of insecticide resistance in anthropods: Prospects for the future. In Committee on Strategies for the Management of Pesticide Resistance Populations, Pesticide resistance: Strategies and tactics for management, 74–86. National Academy Press, Washington, DC. [17]

Platt, A. W. 1941. The influence of some environmental factors on the expressions of the solid stem character in certain wheat varieties. Scientific Agriculture 22:216–223. [9]

Plaut, H. N. 1965. On the phenology and control value of *Stethorus punctillum* Weise as a predator of *Tetranychus cinnabarinus* Boisd. in Israel. Entomophaga 10:133–137. [12]

Pollard, A. J. 1986. Variation in *Cnidoscolus texanus* in relation to herbivory. Oecologia 70:411–413. [10]

Pollard, A. J., and D. Briggs. 1982. Genecological studies of *Urtica dioica* L. I. The nature of intraspecific variation in *U. dioica*. New Phytologist 92:453–470. [10]

————. 1984. Genecological studies of *Urtica dioica* L. III. Stinging hairs and plant-herbivore interactions. New Phytologist 97:507–522. [10]

Polley, H. W., and J. K. Detling. 1988. Herbivory tolerances of *Agropyron smithii* populations with different grazing histories. Oecologia 77:261–267. [13]

Pope, D. D. 1982. Biometrical analysis of pathogenicity in the *Ustilago hordei-Hordeum vulgate* host-parasite system. M.S. thesis, University of British Columbia, Vancouver. [8]

Populer, C. 1978. Changes in host susceptibility with time. In J. G. Horsfall and E. B. Cowling, eds., Plant disease 2:239–262. Academic Press, New York. [15]

Porter, K. B., G. L. Peterson, and O. Vise. 1982. A new greenbug biotype. Crop Science 22:847–850. [5, 6]

Potter, D. A., and T. W. Kimmerer. 1988. Do holly leaf spines really deter herbivory? Oecologia 75:216–221. [10]

————. 1989. Inhibition of herbivory on young holly leaves: Evidence for the defensive role of saponins. Oecologia 78:322–329. [13]

Powell, J. E., and L. Lambert. 1984. Effects of three resistant soybean genotypes on development of *Microplitis croceipes* and leaf consumption by its *Heliothis* spp. hosts. Journal of Agricultural Entomology 1:169–176. [12]

Power, A. G. 1988. Leafhopper response to genetically diverse maize stands. Entomologia Experimentalis et Applicata 49:213–219. [5, 9]

Power, M. E., and W. J. Matthews. 1983. Algae-grazing minnows (*Campostoma anomalum*), piscivorous bass (*Micropterus* spp.), and the distribution of attached algae in a small, prairie-margin stream. Oecologia 60:328–332. [17]

Powers, H. R., Jr., and F. R. Matthews. 1980. Comparison of six geographic sources of loblolly pine for fusiform rust resistance. Phytopathology 70:1141–1143. [14]

Powers, H. R., Jr., R. A. Schmidt, and G. A. Snow. 1981. Current status and management of fusiform rust on southern pines. Annual Review of Phytopathology 19:353–371. [8]

Powers, L. 1950. Determining scales and the use of transformations in studies of weight per locule of tomato fruit. Biometrics 6:145–163. [3]

Preszler, R. W., and P. W. Price. 1988. Host quality and sawfly populations: A new approach to life table analysis. Ecology 69:2012–2020. [11]

Price, P. W. 1980. Evolutionary biology of parasites. Princeton University Press, Princeton, NJ. [10]

————. 1981. Semiochemicals in evolutionary time. In D. A. Nordlund, R. L. Jones, and W. L. Lewis, eds., Semiochemicals: Their role in pest control, 251–279. John Wiley and Sons, New York. [1, 11, 12]

————. 1983a. Alternative paradigms in community ecology. In P. W. Price, C. N. Slobodchikoff, and W. S. Gaud, eds., A new ecology: Novel approaches to interactive systems, 353–383. John Wiley and Sons, New York. [17]

————. 1983b. Hypotheses on organization and evolution in herbivorous insect communities. In R. F. Denno and M. S. McClure, eds., Variable plants and herbivores in natural and managed systems, 559–598. Academic Press, New York. [6, 11]

————. 1986. Ecological aspects of host plant resistance and biological control: Interactions among three trophic levels. In D. J. Boethel and R. D. Eikenbary, eds., Interactions of plant resistance and parasitoids and predators of insects, 11–30. Ellis Horwood, Chicester. [11, 12]

————. 1987. The role of natural enemies in insect populations. In P. Barbosa and J. C. Schultz, eds., Insect outbreaks, 287–313. Academic Press, New York. [11]

————. 1988. Inversely density-dependent parasitism: The role of plant refuges for hosts. Journal of Animal Ecology 57:89–96. [11]

Price, P. W., C. E. Bouton, P. Gross, B. A. McPheron, J. N. Thompson, and A. E. Weis. 1980. Interactions among three trophic levels: Influence of plants on interactions between insect herbivores and natural enemies. Annual Review of Ecology and Systematics 11:41–65. [1, 3, 11, 12]

Price, P. W., and K. M. Clancy. 1986a. Interactions among three trophic levels: Gall size and parasitoid attack. Ecology 67:1593–1600. [11, 12]

————. 1986b. Multiple effects of precipitation on *Salix lasiolepis* and populations of the stem-galling sawfly, *Euura lasiolepis*. Ecological Research 1:1–14. [6, 11]

Price, P. W., and T. P. Craig. 1984. Life history, phenology, and survivorship of stem-galling sawfly, *Euura lasiolepis* (Hymenoptera: Tenthredinidae), on the arroyo willow, *Salix lasiolepis,* in northern Arizona. Annals of the Entomological Society of America 77:712–719. [11]

Price, P. W., H. Roininen, and J. Tahvanainen. 1987. Plant age and attack by the bud galler, *Euura mucronata*. Oecologia 73:334–337. [11]

Price, P. W., and M. F. Willson. 1979. Abundance of herbivores on six milkweed species in Illinois. American Midland Naturalist 101:76–86. [13]

Price, T., M. Kirkpatrick, and S. J. Arnold. 1988. Directional selection and the evolution of breeding date in birds. Science 240:798–799. [3]

Primack, R. B., and J. Antonovics. 1981. Experimental ecological genetics in *Plantago*. Components of seed yield in the ribwort plantain *Plantago lanceolata* L. Evolution 35:1069–1079. [14]

Prokopy, R. J., S. R. Diehl, and S. S. Cooley. 1988. Behavioral evidence for host races in *Rhagoletis pomonella* flies. Oecologia 76:138–147. [4, 6]

Pulliam, H. R. 1975. Diet optimization with nutrient constraints. American Naturalist 109:765–768. [10]

Pullin, A. S., and J. E. Gilbert. 1989. The stinging nettle, *Urtica dioica,* increases trichome density after herbivore and mechanical damage. Oikos 54:275–280. [10]

Puranik, S. B., and D. E. Mathre. 1971. Biology and control of ergot on male sterile wheat and barley. Phytopathology 61:1075–1080. [14]

Puterka, G. J., and D. C. Peters. 1988. Rapid technique for determining greenbug (Homoptera: Aphididae) biotypes B, C, E, and F. Journal of Economic Entomology 81:396–399. [5]

————. 1989. Inheritance of greenbug, *Schizaphis graminum* (Rodani) virulence to Gb2 and Gb3 resistance genes in wheat. Genome 32:109–114. [5]

Puterka, G. J., D. C. Peters, D. Kerns, J. E. Slosser, L. Bush, D. W. Worrall, and R. W. McNew. 1988. Designation of two new greenbug (Homoptera: Aphididae) biotypes G and H. Journal of Economic Entomology 81:1754–1759. [5]

Puterka, G. J., J. E. Slosser, and E. C. Gilmore. 1892. Biotype C and E greenbugs: Distribution in the Texas Rolling Plains and damage to four small grain varieties. Southwestern Entomologist 7:4–8. [5]

Putman, W. L. 1955. Bionomics of *Stethorus punctillum* Weise (Coleoptera: Coccinellidae) in Ontario. Canadian Entomologist 87:9–33. [12]

Quiring, D. T., and J. N. McNeil. 1987. Foraging behavior of a Dipteran leaf miner on exploited and unexploited hosts. Oecologia 73:7–15. [11]

Quisumbing, A. R. 1975. Host plant resistance in cucumbers: Influence of plot size and seeding rate in screening for field resistance to cucumber beetles. Ph.D. diss., North Carolina State University, Raleigh. [2]

Radwan, M. A. 1972. Differences between Douglas-fir genotypes in relation to browsing preference by black-tailed deer. Canadian Journal of Forest Research 2:250–255. [10]

Radwan, M. A., G. L. Crouch, C. A. Harrington, and W. D. Ellis. 1982. Terpenes of ponderosa pine and feeding preferences by pocket gophers. Journal of Chemical Ecology 8:241–253. [16]

Rafes, P. M. 1970. Estimation of the effects of phytophagous insects on forest production. In D. E. Reichle, ed., Analysis of temperate forest ecosystems, 100–106. Ecological Studies, Analysis and Synthesis, vol. 1. Springer-Verlag, Berlin. [13]

Rai, J. P. N., and R. S. Tripathi. 1985. Effect of herbivory by the slug, *Mariaella dussumieri*, and certain insects on growth and competitive success of two sympatric annual weeds. Agriculture, Ecosystems, and Environment 13:125–137. [13]

Raman, K. V., W. M. Tingey, and P. Gregory. 1979. Potato glycoalkaloids: Effects on survival and feeding behavior of the potato leafhopper. Journal of Economic Entomology 72:337–341. [13]

Ramey, H. H. 1962. Genetics of plant pubescence in upland cotton. Crop Science 2:269. [2]

Randolph, P. A., J. C. Randolph, and C. A. Barlow. 1975. Age-specific energetics of the pea aphid, *Acyrthosiphon pisum*. Ecology 56:359–369. [5]

Ratcliffe, R. H., and J. J. Murray. 1983. Selection for greenbug (Homoptera: Aphididae) resistance in Kentucky bluegrass cultivars. Journal of Economic Entomology 76:1221–1224. [5]

Rathcke, B. J. 1976. Competition and coexistence within a guild of herbivorous insects. Ecology 57:76–87. [11]

Raupp, M. J., and R. F. Denno. 1983. Leaf age as a predictor of herbivore distribution and abundance. In R. F. Denno and M. S. McClure, eds., Variable plants and herbivores in natural and managed systems, 91–124. Academic Press, New York. [11, 16]

Rausher, M. D. 1980. Host abundance, juvenile survival, and oviposition preference in *Battus philenor*. Evolution 34:342–355. [4]

———. 1983. Ecology of host-selection behavior in phytophagous insects. In R. F. Denno and M. S. McClure, eds., Variable plants and herbivores in natural and managed systems, 223–257. Academic Press, New York. [5, 6, 9]

———. 1984a. The evolution of habitat preference in subdivided populations. Evolution 38:596–608. [6, 7, 18]

———. 1984b. Tradeoffs in performance on different hosts: Evidence from within- and between-site variation in the beetle *Deloyala guttata*. Evolution 38:582–595. [3, 4, 5, 6, 7]

———. 1988a. The evolution of habitat preference. III. The evolution of avoidance and adaptation. In K. C. Kim, ed., Evolution of insect pests: The pattern of variations. Wiley, New York. [6]

———. 1988b. Is coevolution dead? Ecology 69:898–901. [4, 6, 13]

————. 1992. The measurement of selection on quantitative traits: Biases due to environmental covariances between traits and fitness. Evolution in press.

Rausher, M. D., and P. Feeny. 1980. Herbivory, plant density, and plant reproduction success: The effect of *Battus philenor* on *Aristolochia reticulata*. Ecology 61:905–917. [13, 14]

Rausher, M. D., D. A. Mckay, and M. C. Singer. 1981. Pre- and post-alighting host discrimination by *Euphydryas editha* butterflies: The behavioral mechanisms causing clumped distributions of egg clusters. Animal Behavior 29:1220–1228. [13]

Rausher, M. D., and E. L. Simms. 1989. The evolution of resistance to herbivory in *Ipomoea purpurea*. I. Attempting to detect selection. Evolution 43:563–572. [3, 4, 6, 7, 13, 14]

Raymond, W. F. 1969. The nutritive value of forage crops. Advances in Agronomy 21:1–108. [10]

Read, D. J. 1968. Some aspects of the relationship between shade and fungal pathogenicity in an epidemic disease of pines. New Phytologist 67:39–48. [3]

Reader, R. 1979. Impact of leaf-feeding insects on three bog ericads. Canadian Journal of Botany 57:2107–2112. [13]

Real, L. A. 1980. Fitness, uncertainty, and the role of diversification in evolution and behavior. American Naturalist 115:623–628. [18]

Rees, S. B, and J. B. Harborne. 1985. The role of sesquiterpene lactones and phenolics in the chemical defense of the chicory plant. Phytochemistry 24:2225–2231. [16]

Rehr, S. S., E. A. Bell, D. H. Janzen, and P. P. Feeny. 1973. Insecticidal amino acids in legume seeds. Biochemical Systematics 1:63–67. [17]

Rehr, S. S., P. P. Feeny, and D. H. Janzen. 1973. Chemical defenses in Central American non-ant acacias. Journal of Animal Ecology 42:405–416. [17]

Rehr, S. S., D. H. Janzen, and P. P. Feeny. 1973. L-dopa in legume seeds: A chemical barrier to insect attack. Science 181:81–82. [17]

Reichardt, P. B., J. P. Bryant, T. P. Clausen, and G. D. Wieland. 1984. Defense of winter-dormant Alaska paper birch against snowshoe hares. Oecologia 65:58–69. [13]

Reichman, O. J. 1979. Desert granivore foraging and its impact on seed densities and distributions. Ecology 60:1085–1092. [13]

————. 1988. Comparison of the effects of crowding and pocket gopher disturbance on mortality, growth, and seed production of *Berteroa incana*. American Midland Naturalist 120:58–69. [13]

Reichman, O. J., and J. U. M. Jarvis. 1989. The influence of three sympatric species of fossorial mole-rats (Bathyergidae) on vegetation. Journal of Mammology 70:763–771. [13]

Reichman, O. J., and S. C. Smith. 1985. Impact of pocket gopher burrows on overlying vegetation. Journal of Mammology 66:720–725. [13]

Renwick, J. A. A. 1988. Comparative mechanisms of host selection by insects attacking pine trees and crucifers. In K. Spencer, ed., Chemical mediation of coevolution, 303–316. Academic Press, New York. [4]

Rhoades, D. F. 1977. Integrated antiherbivore, antidesiccant, and ultraviolet screening properties of creosote-bush resin. Biochemical Systematics and Ecology 5:281–290. [1, 17]

————. 1979. Evolution of plant chemical defenses against herbivory. In G. A. Rosenthal and D. H. Janzen, eds., Herbivores: Their interaction with secondary plant metabolites, 3–54. Academic Press, New York. [10, 13, 16, 17]

————. 1983. Herbivore population dynamics and plant chemistry. In R. F. Denno and M. S. McClure, eds., Variable plants and herbivores in natural and managed systems, 155–222. Academic Press, New York. [3, 10, 13, 17]

————. 1985. Offensive-defensive interactions between herbivores and plants: Their relevance in herbivore population dynamics and ecological theory. American Naturalist 125:205–238. [10, 17]

Rhoades, D. F., and R. G. Cates. 1976. Toward a general theory of plant antiherbivore chemistry. Recent Advances in Phytochemistry 10:168–213. [2, 10, 13, 17]

Rhomberg, L. R., S. Joseph, and R. S. Singh. 1985. Seasonal variation and clonal selection in cyclically parthenogenetic rose aphids (*Macrosiphym rosae*). Canadian Journal of Genetics and Cytology 27:224–232. [6]

Rice, R. L., D. E. Lincoln, and J. H. Langenheim. 1978. Palatability of monoterpenoid compositional types of *Satureja douglasii* to a generalist molluscan herbivore, *Ariolimax dolichophallus*. Biochemical Systematics and Ecology 6:45–53. [13]

Rice, W. R. 1983a. Parent-offspring pathogen transmission: A selective agent promoting sexual reproduction. American Naturalist 121:187–203. [14, 15]

————. 1983b. Sexual reproduction: An adaptation reducing parent-offspring contagion. Evolution 37:1317–1320. [9]

————. 1989. Analyzing tables of statistical tests. Evolution 43:223–225. [3]

Richards, J. H. 1984. Root growth response to defoliation in two *Agropyron* bunchgrasses: Field observations with an improved root periscope. Oecologia 64:21–25. [13]

Rick, C. M. 1986. Tomato mutants: Freaks, anomalies, and breeders' resources. HortScience 21:918–919. [2]

Ricklefs, R. E. 1973. Ecology. Chiron Press, Portland, OR. [10]

Rickson, F. R. 1977. Progressive loss of ant-related traits in *Cecropia peltata* on selected Caribbean islands. American Journal of Botany 64:585–592. [13]

Riemenschneider, D. E. 1990. Susceptibility of intra- and inter-specific hybrid poplars to *Agrobacterium tumefaciens* strain C58. Phytopathology 80:1099–1102. [8]

Risch, S. 1985. Effects of induced chemical changes on interpretation of feeding preference tests. Entomologia Experimentalis et Applicata 39:81–84. [4]

Risch, S. J. 1981. Insect herbivore abundance in tropical monocultures and polycultures: An experimental test of two hypotheses. Ecology 62:1325–1340. [5]

Risch, S. J., D. Andow, and M. A. Altieri. 1983. Agroecosystem diversity and pest control: Data, tentative conclusions, and new research directions. Environmental Entomology 12:625–629. [5]

Ritland, K. 1990. Gene identity and the genetic demography of plant populations. In A. H. Brown, M. T. Clegg, A. L. Kahler, and B. S. Weir, eds. Plant population genetics, breeding, and genetic resources, 181–190. Sinauer, Sunderland, MA. [3]

Roberts, K. L., R. Thurston, and G. A. Jones. 1981. Density of types of trichomes on leaf lamina of various tobacco cultivars. Tobacco Science 25:68–69. [2]

Robertson, A. 1956. The effect of selection against extreme deviants based on deviation or on homozygosis. Journal of Genetics 54:236–248. [17]

————. 1959. The sampling variance of the genetic correlation coefficient. Biometrics 15:469–485. [3]

————. 1967. The nature of quantitative genetic variation. In R. A. Brink, ed., Heritage from Mendel, 265–280. University of Wisconsin Press, Madison. [3]

Robinson, R. A. 1976. Plant pathosystems. Springer, New York. [5, 8, 17]

Robinson, T. 1974. Metabolism and function of alkaloids in plants. Science 184:430–435. [3, 17]

————. 1979. The evolutionary ecology of alkaloids. In G. A. Rosenthal and D. H. Janzen, eds., Herbivores: Their interaction with secondary plant metabolites, 413–448. Academic Press, London. [4, 10]

Roe, R., and B. E. Mottershead. 1962. Palatability of *Phalaris arundinacea* L. Nature 193:255–256. [10]

Roelfs, A. P. 1985. Race specificity and methods of study. In W. R. Bushnell and A. P. Roelfs, eds., The cereal rusts, 1:1565–1592. Academic Press, Orlando. [14]

Roelfs, A. P., and J. V. Groth. 1980. A comparison of virulence phenotypes in wheat stem rust populations reproducing sexually and asexually. Phytopathology 70:855–862. [8]

Roff, D. A., and T. A. Mousseau. 1987. Quantitative genetics and fitness: Lessons from *Drosophila*. Heredity 58:103–118. [3]

Rogers, D. J., and M. J. Sullivan. 1986. Nymphal performance of *Geocoris punctipes* (Hemiptera: Lygaeidae) on pest-resistant soybeans. Environmental Entomology 15:1032–1036. [12]

Roitberg, B. D., and J. H. Myers. 1978. Adaptation of alarm pheromone responses of the pea aphid *Acyrthosiphon pisum* (Harris). Canadian Journal of Zoology 56:103–108. [6]

————. 1979. Behavioral and psychological adaptations of pea aphids to high ground temperatures and predator disturbance. Canadian Entomologist 111:515–519. [6]

Roland, J., and J. H. Myers. 1987. Improved insect performance from host-plant defoliation: Winter moth on oak and apple. Ecological Entomology 12:409–414. [11]

Room, P. M. 1988. Effects of temperature, nutrients, and a beetle on branch architecture of the floating weed *Salvinia molesta* and simulations of biological control. Journal of Ecology 76:826–848. [13]

Root, R. B. 1973. Organization of a plant-anthropod association in simple and diverse habitats: The fauna of collards (*Brassica oleracea*). Ecological Monographs 43:95–124. [4, 5, 11]

Rose, M. R. 1984. Genetic variance in *Drosophila* life history: Untangling the data. American Naturalist 123:565–569. [3]

Rose, M. R., P. M. Service, and E. W. Hutchinson. 1987. Three approaches to trade-offs in life-history evolution. In V. Loeschcke, ed. Genetic constraints on adaptive evolution, 91–105. Springer-Verlag, Berlin. [7]

Rosenberg, A. 1978. The supervenience of biological concepts. Philosophy of Science 45:368–386. [18]

Rosenthal, G. A. 1977. The biological effects and mode of action of L-canavanine, a structural analogue of L-arginine. Quarterly Review of Biology 52:155–178. [17]

————. 1986. The chemical defenses of higher plants. Scientific American 254:76–81. [17]

Rosenthal, G. A., and D. H. Janzen, eds. 1979. Herbivores: Their interaction with secondary plant metabolites. Academic Press, New York. [2, 10]

Ross, M. D., and W. T. Jones. 1983. A genetic polymorphism for tannin production in

Lotus corniculatus and its relationship to cyanide polymorphism. Theoretical and Applied Genetics 64:263–268. [16]

Rossiter, M. C., J. C. Schultz, and I. T. Baldwin. 1988. Relationships among defoliation, red oak phenolics, and gypsy moth growth and reproduction. Ecology 69:267–275. [4]

Rothschild, M., T. Reichstein, J. Von Euw, R. Aplin, and R. R. M. Harman. 1970. Toxic Lepidoptera. Toxicon 8:293–299. [10]

Roughgarden, J. 1979. Theory of population genetics and evolutionary ecology: An introduction. Macmillan Publishing Co., New York. [14]

Roughgarden, J., and J. Diamond. 1986. The role of species interaction in community ecology. In J. Diamond and T. J. Case, eds., Community ecology, 333–343. Harper and Row, New York. [11]

Roundy, B. A., and G. B. Ruyle. 1989. Effects of herbivory on twig dynamics of a sonoran desert shrub *Simmondsia chinensis* (Link) Schn. Journal of Applied Ecology 26:701–710. [13]

Rousi, M. 1988. Resistance breeding against voles in birch: Possibilities for increasing resistance by provenance transfers. European and Mediterranean Plant Protection Organization Bulletin 18:257–263. [17]

―――. 1989. Susceptibility of winter-dormant *Pinus sylvestris* families to vole damage. Scandinavian Journal of Forest Research 4:149–161. [17]

Roux, J. B. 1960. La selection de cotoniers sans gossypol. Coton et Fibres Tropicales 15:1–14. [2]

Rowell-Rahier, M. 1984. The presence or absence of phenolglycosides on *Salix* (Salicaceae) leaves and the level of dietary specialisation of some of their herbivorous insects. Oecologia 62:26–30. [11]

Rowell-Rahier, M., and J. M. Pasteels. 1986. Economics of chemical defense in Chrysomelinae. Journal of Chemical Ecology 12:1189–1203. [11]

Ruberson, J. R., M. J. Tauber, C. A. Tauber, and W. M. Tingey. 1989. Interactions on three trophic levels: *Edovum puttleri* Grissell (Hymenoptera: Eulophidae), the Colorado potato beetle, and insect-resistant potatoes. Canadian Entomologist 121:841–851. [12]

Ruess, R. W. 1984. Nutrient movement and grazing: Experimental effects of clipping and nitrogen source on nutrient uptake in *Kyllingia nervosa*. Oikos 43:183–188. [13]

―――. 1988. The interaction of defoliation and nutrient uptake in *Sporobolus kentrophyllus,* a short grass species from the Serengeti Plains. Oecologia 77:550–556. [13]

Reuss, R. W., S. J. McNaughton, and M. B. Coughenour. 1983. The effects of clipping, nitrogen source, and nitrogen concentration on the growth responses and nitrogen uptake of an East African sedge. Oecologia 59:253–261. [13]

Russell, E. P. 1989. Enemies hypothesis: A review of the effect of vegetational diversity on predatory insects and parasitoids. Environmental Entomology 18:590–599. [5]

Russell, G. E. 1978. Plant breeding for pest and disease resistance. Butterworth and Co. London. [5]

Russo, S., and P. Tricerri. 1976. Selection for blast (*Pyricularia oryzae*) resistance in rice by natural infection test. Genetica Agraria 30:143–152. [3]

Ryan, C. A. 1979. Proteinase inhibitors. In G. A. Rosenthal and D. H. Janzen, eds.,

Herbivores: Their interaction with secondary plant metabolites, 599–618. Academic Press, New York. [2]

Sacchi, C. F., P. W. Price, T. P. Craig, and J. K. Itami. 1988. Impact of shoot galler attack on sexual reproduction in the arroyo willow. Ecology 69:2021–2029. [13]

Sagar, G. R. 1970. Factors controlling the size of plant populations. Proceedings of the Tenth British Weed Control Conference, 965–979. British Crop Protection Council, Brighton. [14]

Saghai Maroof, M. A., R. K. Webster, and R. W. Allard. 1983. Evolution of resistance to scald, powdery mildew, and new blotch in barley composite cross II populations. Theoretical and Applied Genetics 66:279–283. [15]

Salim, M., and E. A. Heinrichs. 1986. Impact of varietal resistance in rice and predation on the mortality of *Sogatella furcifera* (Horvath) (Homoptera: Delphacidae). Crop Protection 5:395–399. [12]

Salisbury, E. 1961. Weeds and aliens. Collins, London. [10]

Salto, C. E. 1976. Pulgon verde de los cereales. I. Presencia de dos biotipos en Argentina. Chacra Experimental de Barrow Publication Tecnica no. 14. [5]

Salto, C. E., R. D. Eikenbary, and K. J. Starks. 1983. Compatibility of *Lysiphlebus testaceipes* (Hymenoptera: Braconidae) with greenbug (Homoptera: Aphididae) biotypes "C" and "E" reared on susceptible and resistant oat varieties. Environmental Entomology 12:603–604. [12]

Sands, D. C., and J. L. McIntyre. 1977. Possible methods to control pear blast, caused by *Pseudomonas syringae*. Plant Disease Reporter 61:311–312. [15]

SAS Institute. 1985. SAS user's guide: Statistics. 5th ed. Cary, NC. [3]

Sato, Y., and N. Ohsaki. 1987. Host-habitat location by *Apanteles glomeratus* and effect of food-plant exposure on host-parasitism. Ecological Entomology 12:291–297. [6, 11]

Sauer, C. 1952. Agricultural origins and dispersals. American Geographical Society, New York. [4]

Schaeffer, L. R., J. W. Wilton, and R. Thompson. 1978. Simultaneous estimation of variance and covariance components from multivariate mixed model equations. Biometrics 34:199–208. [3]

Schafer, J. F. 1971. Tolerance to plant disease. Annual Review of Phytopathology 9:235–252. [14]

Scharpf, R. F., and Koerber, T. W. 1986. Destruction of shoots, flowers, and fruits by dwarf mistletoe and grasshoppers in California. Canadian Journal of Forest Research 16:166–168. [13]

Scheffé, H. 1959. The analysis of variance. Wiley, New York. [3]

Scheiner, S. M. 1989. Variable selection along an environmental gradient. Evolution 43:548–562. [7]

Scheiner, S. M., and C. J. Goodnight. 1984. The comparison of phenotypic plasticity and genetic variation in populations of the grass *Danthonia spicata*. Evolution 38:845–855. [7]

Scheiner, S. M., and R. F. Lyman. 1989. The genetics of phenotypic plasticity I. Heritability. Journal of Evolutionary Biology 2:95–107. [7, 11]

———. 1991. The genetics of phenotypic plasticity II. Response to selection. Journal of Evolutionary Biology 3:23–50. [7]

Schemske, D. W. 1984. Population structure and local selection in *Impatiens pallida* (Balsaminaceae), a selfing annual. Evolution 38:817–832. [15]

Schlichting, C. 1986. The evolution of phenotypic plasticity in plants. Annual Review of Ecology and Systematics 17:667–693. [7]

Schluter, D. 1988. Estimating the form of natural selection on a quantitative trait. Evolution 42:849–861. [3, 7]

Schmaulhausen, I. I. 1949. Factors of evolution: The theory of stabilizing selection. Blankston, Philadelphia. [7]

Schmidt, D. J., and J. C. Reese. 1986. Sources of error in nutritional index studies of insects on artificial diet. Journal of Insect Physiology 32:193–198. [4]

Schmidt, R. A. 1978. Diseases in forest ecosystems: The importance of functional diversity. In J. B. Horsfall and E. B. Cowling, eds., Plant disease: An advanced treatise 2:287–315. Academic Press, New York. [14, 18]

Schmitt, J., and J. Antonovics. 1986. Experimental studies of the evolutionary significance of sexual reproduction. VI. Effect of neighbor relatedness and aphid infestation on seedling performance. Evolution 40:7–11. [18]

Schoener, T. W. 1983. Field experiments on interspecific competition. American Naturalist 122:240–285. [11]

———. 1988. Leaf damage in island buttonwood, *Conocarpus erectus:* Correlations with pubescence, island area, isolation, and the distribution of major carnivores. Oikos 53:253–266. [13]

Schowalter, T. D., W. W. Hargrove, and D. A. Crossley, Jr. 1986. Herbivory in forested ecosystems. Annual Review of Entomology 31:177–196. [13]

Schowalter, T. D., and M. I. Haverty. 1989. Influence of host genotype on Douglas-fir seed losses to *Contarinia oregonensis* (Diptera: Cecidomyiidae) and *Megastigmus spermotrophus* (Hymenoptera: Torymidae) in western Oregon. Environmental Entomology 18:94–97. [11, 13]

Schroth, M. N., S. V. Thomson, D. C. Hildebrand, and W. J. Moller. 1974. Epidemiology and control of fire blight. Annual Review of Phytopathology 12:389–412. [15, 18]

Schultz, D. E., and D. C. Allen. 1977. Effects of defoliation by *Hydria prunivorata* on the growth of black cherry. Environmental Entomology 6:276–283. [13]

Schultz, J. C. 1983a. Habitat selection and foraging tactics of caterpillars in heterogeneous trees. In R. F. Denno and M. S. McClure, eds., Variable plants and herbivores in natural and managed systems, 61–91. Academic Press, New York. [3, 11]

———. 1983b. Impact of variable plant defensive chemistry on susceptibility of insects to natural enemies. In P. A. Hedin, ed., Plant resistance to insects, 37–54. American Chemical Society, Washington, DC. [11]

———. 1988a. Many factors influence the evolution of herbivore diets, but plant chemistry is central. Ecology 69:896–897. [13]

———. 1988b. Plant responses induced by herbivores. Trends in Ecology and Evolution 3:45–49. [13]

Schultz, J. C., and I. T. Baldwin. 1982. Oak leaf quality declines in response to defoliation by gypsy moth larvae. Science 217:149–151. [3, 11]

Schupp, E. W. 1988. Factors affecting post-dispersal seed survival in a tropical forest. Oecologia 76:525–530. [13]

Schuster, M. F., and M. Calderon. 1986. Interactions of host plant resistant genotypes and beneficial insects in cotton ecosystems. In D. J. Boethel and R. D. Eikenbary, eds., Interaction of host plant resistance and parasitoids and predators of insects, 84–97. Halsted Press, New York. [12]

Schuster, M. F., P. D. Calvin, and W. C. Langston. 1983. Interaction of high tannin with bollworm control by Pydrin and Dipel. Proceedings of the 1983 Beltwide Cotton Production Research Conference, 72–73. National Cotton Council of America, Memphis. [12]

Scott, A. W., and R. D. B. Whalley. 1984. The influence of intensive sheep grazing on genotypic differentiation in *Danthonia linkii, D. richardsonii* and *D. racemosa* on the New England Tablelands. Australian Journal of Ecology 9:419–429. [13]

Scott, D. R. 1970. *Lygus* bugs feeding on developing carrot seed: Plant resistance to that feeding. Journal of Economic Entomology, 63:959–961. [2]

———. 1977. Selection for *Lygus* bug resistance in carrot. HortScience 12:452. [2]

Scott, G. E., F. F. Dicke, and G. R. Peshe. 1966. Location of genes conditioning resistance in corn to leaf feeding of the European corn borer. Crop Science 6:400–446. [2]

———. 1967. Effect of second brood European corn borer infestation on 45 single cross corn hybrids. Crop Science 7:229–230. [2]

Scott, J. A. 1986. The butterflies of North America: A natural history and field guide. Stanford University Press, Stanford, CA. [6]

Scriber, J. M. 1977. Limiting effects of low leaf-water content on the nitrogen utilization, energy budget, and larval growth of *Hylaphora cecropia* (Lepidoptera: Saturniidae). Oecologia 28:269–287. [7, 13]

———. 1979. Effects of leaf-water supplementation upon post-ingestive nutritional indices of forb-, vine-, and tree-feeding Lepidoptera. Entomologia Experimentalis et Applicata 25:240–252. [7]

———. 1981. Sequential diets, metabolic costs, and growth of *Spodoptera eridania* (Lepidoptera: Noctuidae) feeding upon dill, lima bean, and cabbage. Oecologia 51:175–180. [4]

———. 1983. Evolution of feeding specialization, physiological efficiency, and host races in selected *Papilionidae* and *Saturniidae*. In R. F. Denno and M. S. McClure, eds., Variable plants and herbivores in natural and managed systems, 373–412. Academic Press, New York. [7]

Scriber, J. M., and P. P. Feeny. 1979. Growth of herbivorous caterpillars in relation to feeding specialization and to the growth form of their food plants. Ecology 60:829–850. [4, 7]

Scriber, J. M., and F. Slansky, Jr. 1981. Nutritional ecology of immature insects. Annual Review of Entomology 26:183–211. [7]

Scurlock, J. H., R. G. Mitchell, and K. K. Ching. 1982. Insects and other factors affecting noble fir seed production at two sites in Oregon. Northwest Science 56:101–107. [13]

Searle, S. R. 1971. Linear models. Wiley, New York. [3]

Seastedt, T. R., D. A. Crossley, Jr., and W. W. Hargrove. 1983. The effects of low-level consumption by canopy arthropods on the growth and nutrient dynamics of black locust and red maple trees in the southern Appalachians. Ecology 64:1040–1048. [13]

Sebesta, E. E., and E. A. Wood, Jr. 1978. Transfer of greenbug resistance from rye to wheat with X-rays. Agronomy Abstracts 70:61–62. [5]

Sedcole, J. R. 1978. Selection pressures and plant pathogens: Stability of equilibria. Phytopathology. 68:967–970. [5]

Segal, A., J. Manisterski, G. Fischbeck, and I. Wahl. 1980. How plant populations defend themselves in natural ecosystems. In J. G. Horsfall and E. B. Cowling, eds., Plant disease. vol. 5, How plants defend themselves, 75–102. Academic Press, New, York. [8, 14, 15]

Seger, J. 1988. Dynamics of some simple host-parasite models with more than two genotypes in each species. Philosophical Transactions of the Royal Society of London B319:541–555. [15]

Seger, J., and H. J. Brockmann. 1987. What is bet hedging? Oxford Survey of Evolutionary Biology 4:182–211. [7]

Seger, J., and W. D. Hamilton. 1988. Parasites and sex. In R. E. Michod and B. R. Levin, eds., The evolution of sex, 176–193. Sinauer, Sunderland, MA. [14, 15]

Seif el Din, A., and M. Obeid. 1971. Ecological studies of the vegetation of the Sudan. IV. The effect of simulated grazing on the growth of *Acacia senegal* (L.) Willd. seedlings. Journal of Applied Ecology 8:211–216. [10]

Seigler, D., and P. W. Price. 1976. Secondary compounds in plants: Primary functions. American Naturalist 110:101–105. [16, 17]

Seigler, D. S. 1977. Primary roles for secondary compounds. Biochemical Systematics and Ecology 5:195–199. [17]

Service, P. 1984a. The distribution of aphids in response to variation among individual host plants. Ecological Entomology 9:321–328. [6]

————. 1984b. Genotypic interactions in an aphid-host plant relationship: *Uroleucon rudbeckiae* and *Rudbeckia laciniata*. Oecologia (Berlin) 61:271–276. [2, 5, 6, 9, 11, 13]

Service, P. M., and R. Lenski. 1982. Aphid genotype, plant phenotypes, and genetic diversity: A demographic analysis of experimental data. Evolution 36:1276–1282. [6, 7]

Service, P. M., and M. R. Rose. 1985. Genetic covariation among life-history components: The effect of novel environments. Evolution 39:943–945. [6, 7, 17]

Settle, W. H., and L. T. Wilson. 1990. Invasion by the variegated leafhopper and biotic interactions: Parasitism, competition, and apparent competition. Ecology 71:1461–1470. [11]

Shah, M. A. 1982. The influence of plant surfaces on the searching behavior of coccinellid larvae. Entomologia Experimentalis et Applicata 31:377–380. [12]

Shanahan, G. J. 1967. The sheep blowfly's tolerance to insecticides. Agricultural Gazette of New South Wales 78:444–445. [17]

Sharma, G. C., and C. V. Hall. 1971. Cucurbitacin B and total sugar inheritance in *Cucurbita pepo* L. related to spotted cucumber beetle feeding. Journal of the American Society for Horticultural Science 96:750–754. [2]

Shaw, D. S. 1987. The breeding system of *Phytophthora infestans:* The role of the A2 mating type. In P. R. Day and C. J. Gellis, eds., Genetics and plant pathogenesis, 161–174. Blackwell Scientific Publications, Oxford. [8]

————. 1988. The *Phytophthora* species. In D. S. Ingram and P. H. Williams, eds.,

Advances in plant pathology. Vol. 6, Genetics of plant pathogenic fungi, ed. G. S. Sidhu, 27–51. Academic Press, London. [8]

Shaw, R. G. 1986. Response to density in a wild population of the perennial herb *Salvia lyrata*: Variation among families. Evolution 40:492–505. [18]

———. 1987. Maximum-likelihood approaches applied to quantitative genetics of natural populations. Evolution 41:812–826. [3]

Shea, M. M., and M. A. Watson. 1989. Patterns of leaf and flower removal: Their effect on fruit growth in *Chamaenerion angustifolium* (fireweed). American Journal of Botany 76:884–890. [13]

Shehata, M. A., D. W. Davis, and F. L. Pflager. 1983. Breeding for resistance to *Aphanomyces euteiches* root rot and *Rhizoctonia solani* stem rot in peas (*Pisum sativum*). Journal of the American Society for Horticultural Science 108:1080–1085. [3]

Shelly, T. E., M. D. Greenfield, and K. R. Downum. 1987. Variation in host plant quality: Influences on the mating system of a desert grasshopper. Animal Behaviour 35:1200–1209. [11]

Shelton, A. M., R. F. Becker, and J. T. Andaloro. 1983. Varietal resistance to onion thrips (*Thrips tabaci*) (Thysanoptera: Thripidae) in processing cabbage. Journal of Economic Entomology 76:85–86. [2]

Shepherd, R. J. 1972. Transmission of viruses through seed and pollen. In C. I. Kado and H. O, Agrawal, eds., Principles and techniques of plant virology, 267–292. Van Nostrand-Reinhold, Princeton, NJ. [3]

Sholes, O. D. V. 1981. Herbivory by species of *Trirhabda* (Coleoptera: Chrysomelidae) on *Solidago altissima* (Asteraceae): Variation between years. Proceedings of the Entomological Society of Washington 83:274–282. [13]

Sholes, O. D. V., and S. W. Beatty. 1987. Influence of host phenology and vegetation on the abundance of *Tamalia coweni* galls (Homoptera: Aphididae) on *Arctostaphylos insularis* (Ericaceae). American Midland Naturalist 118:198–204. [13]

Shorrocks, B., W. Atkinson, and P. Charlesworth. 1979. Competition on a divided and ephemeral resource. Journal of Animal Ecology 48:899–908. [11]

Shorrocks, B., and J. Rosewell. 1988. Aggregation does prevent competitive exclusion: A response to Green. American Naturalist 131:765–771. [5]

Shrimpton, A. E., and A. Robertson. 1988. The isolation of polygenic factors controlling bristle score in *Drosophila melanogaster*. II. Distribution of third chromosomal bristle effects within chromosome sections. Genetics 118:445–459. [3]

Sidhu, G. S., ed. 1988. Genetics of plant pathogenic fungi. Vol. 6 of Advances in plant pathology, ed. D. S. Ingram and P. H. Williams. Academic Press, London. [8, 14]

Sidhu, G. S., and J. M. Webster. 1977. Genetics of simple and complex host-parasite interactions. In Induced mutations against plant diseases, 59–79. International Atomic Energy Agency, Vienna. [14]

Silen, R. R., W. H. Randall, and N. L. Mandell. 1986. Estimates of genetic parameters for deer-browsing of Douglas-fir. Forest Science 32:178–184. [13]

Silvertown, J. W. 1980. The evolutionary ecology of mast seeding in trees. Biological Journal of the Linnean Society 14:235–250. [13]

Simberloff, D., and P. Stiling. 1987. Larval dispersal and survivorship in a leaf-mining moth. Ecology 68:1647–1657. [11]

Simmonds, N. W. 1962. Variability in crop plants: Its use and conservation. Biological Reviews 37:422–465. [9]

Simms, E. L. 1990. Examining selection on the multivariate phenotype: Plant resistance to herbivores. Evolution 44:1177–1188. [3]

Simms, E. L., and M. D. Rausher. 1987. Costs and benefits of plant defense to herbivory. American Naturalist 130:570–581. [2, 3, 4, 6, 9, 10, 11, 13, 16, 17]

———. 1989. The evolution of resistance to herbivory in Ipomoea purpurea. II. Natural selection by insects and costs of resistance. Evolution 43:573–585. [3, 4, 10, 11, 13, 16, 17]

Simons, A. B., and G. C. Marten. 1971. Relationship of indole alkaloids to palatability of Phalaris arundinacea L. Agronomy Journal 63:915–919. [10]

Simons, M. D. 1979. Influence of genes for resistance to Puccinia coronata from Avena sterilis on yield and rust reaction of cultivated oats. Phytopathology 69:450–452. [15]

Simpson, D. M. 1947. Fuzzy leaf in cotton and its association with short lint. Journal of Heredity 38:153–156. [2]

Sinclair, A. R. E. 1975. The resource limitation of trophic levels in tropical grassland ecosystems. Journal of Animal Ecology 44:497–520. [10]

Sinden, S. L., J. M. Shalk and A. K. Stoner. 1978. Effects of daylength and maturity of tomato plants on tomatine content and resistance to the Colorado potato beetle. Journal of the American Society of Horticultural Science 103:596–600. [2]

Sinden, S. L., L. L. Sanford, and R. E. Webb. 1984. Genetic and environmental control of potato glycoalkaloids. American Potato Journal 61:141–156. [2]

Singer, M. C. 1988. The definition and measurement of oviposition preference in plant-feeding insects. In J. R. Milles and T. A. Miller, eds., Plant insect interactions. Springer-Verlag, New York. [7]

Singer, M. C., D. Ng, and C. D. Thomas. 1988. Heritability of oviposition preference and its relationship to offspring performance within a single insect population. Evolution 42:977–985. [4, 6, 7]

Singer, M. C., C. D. Thomas, H. L. Billington, and C. Parmesan. 1989. Variation among conspecific insect populations in the mechanistic basis of diet breadth. Animal Behaviour 37:751–759. [6]

Singh, I. D., and J. B. Weaver, Jr. 1972. Studies on the heritability of gossypol in leaves and flower buds of Gossypium. Crop Science 12:294–297. [2]

Singh, P., and B. M. Sharma. 1977. Prospects of cultivation of henbane. In C. K. Atal and B. M. Kapur, eds., Cultivation and utilization of medicinal and aromatic plants, 39–43. Regional Research Lab, Jammu Tawi. [16]

Singh, R. S., and R. H. Painter. 1964. Effect of temperature and host plants on progeny production of four biotypes of corn leaf aphid, Rhopalosiphon maidis. Journal of Economic Entomology 57:348–350. [6]

Sites, R. W., and S. A. Phillips, Jr. 1989. Root-feeding insects of Senecio riddellii in eastern New Mexico and northwestern Texas. Journal of Range Management 42:404–406. [13]

Sizer, C. E., J. A. Maga, and C. J. Craven. 1980. Total glycoalkaloids in potatoes and potato chips. Journal of Agricultural and Food Chemistry 28:578–579. [16]

Slansky, F., Jr., and P. Feeny. 1977. Stabilization of nitrogen accumulation by larvae of the cabbage butterfly on wild and cultivated food plants. Ecological Monographs 47:209–228. [7]

Slansky, F., and J. Rodriguez. 1987. Nutritional ecology of insects, mites, spiders, and related invertebrates. J. Wiley and Sons, New York. [4]

Slatkin, M. 1974. Competition and regional coexistence. Ecology 55:128–134. [5]

———. 1978. Spatial patterns in polygenic characters. Journal of Theoretical Biology 70:213–228. [17]

———. 1987. Heritable variation and heterozygosity under a balance between mutations and stabilizing selection. Genetical Research 50:53–62. [3]

Slatkin, M. and R. Lande. 1976. Niche width in a fluctuating environment-density independent model. American Naturalist 110:31–55.

Smedegaard-Petersen, V., and O. Stolen. 1981. Effect of energy-requiring defense reactions on yield and grain quality in a powdery mildew–resistant barley cultivar. Phytopathology 71:396–399. [16]

Smiley, J. 1978. Plant chemistry and the evolution of host specificity: New evidence from *Heliconius* and *Passiflora*. Science 201:745–747. [5]

———. 1986. Ant constancy at *Passiflora* extrafloral nectaries: Effects on caterpillar survival. Ecology 67:516–521. [13]

Smiley, J. T., J. M. Horn, and N. E. Rank. 1985. Ecological effects of salicin at three trophic levels: New problems from old adaptations. Science 229:649–651. [11, 12]

Smith, A. E., and D. M. Secoy. 1981. Plants used for agricultural pest control in western Europe before 1850. Chemistry and Industry January 3, 1981, 12–17. [4]

Smith, C. C. 1970. The coevolution of pine squirrels (*Tamiasciurus*) and conifers. Ecological Monographs 3:349–371. [13]

Smith, C. M. 1989. Plant resistance to insects: A fundamental approach. J. Wiley and Sons, New York. [2, 4, 5]

Smith, C. M., J. L. Frazier, and W. E. Knight. 1976. Attraction of clover head weevil, *Hypera meles*, to flower bud volatiles of several species of *Trifolium*. Journal of Insect Physiology 22:1517–1521. [2]

Smith, C. M., R. F. Wilson, and C. A. Brim. 1979. Feeding behavior of Mexican bean beetle on leaf extracts of resistant and susceptible soybean genotypes. Journal of Economic Entomology 72:374–377. [2]

Smith, D. C. 1988. Heritable divergence of *Rhagoletis pomonella* host races by seasonal asynchrony. Nature 336:66–67. [4, 6]

Smith, R. H., and M. H. Bass. 1972. Relation of artificial pod removal to soybean yields. Journal of Economic Entomology 65:606–608. [13]

Smith, T. J., III, H. T. Chan, C. C. McIvor, and M. B. Robblee. 1989. Comparisons of seed predation in tropical tidal forests from three continents. Ecology 70:146–151. [13]

Snaydon, R. W. 1980. Plant demography in agricultural systems. In O. T. Solbrig, ed., Demography and evolution in plant populations, 131–160. University of California Press, Berkeley. [2]

Snaydon, R. W., and M. S. Davies. 1972. Rapid population differentiation in a mosaic environment. II. Morphological variation in *Anthoxanthum odoratum*. Evolution 26:390–405. [14]

Snyder, J. C., and C. D. Carter. 1984. Leaf trichomes and resistance of *Lycopersicon hirsutum* and L. *esculentum* to spider mites (*Tetranychus urticae*). Journal of the American Society for Horticultural Science 109:837–843. [2]

Sokal, R. R., and F. J. Rohlf. 1981. Biometry. 2d ed. W. H. Freeman and Co., New York. [3]

Solomon, B. P. 1981. Response of a host-specific herbivore to resource density, relative abundance, and phenology. Ecology 62:1205–1214. [13]

———. 1983. Compensatory production in *Solanum carolinense* following attack by a host-specific herbivore. Journal of Ecology 71:681–690. [13]

Son, C. L., and S. E. Smith. 1988. Mycorrhizal growth responses: Interactions between photo irradiance and phosphorus nutrition. New Phytologist 108:305–314. [17]

Soper, J. F., M. S. McIntosh, and T. C. Elden. 1984. Diallele analysis of potato leafhopper resistance among selected alfalfa clones. Crop Science 24:667–670. [2, 4]

Sorenson, C. E., R. L. Fery, and G. G. Kennedy. 1989. Relationship between Colorado potato beetle (Coleoptera: Chrysomelidae) and tobacco hornworm (Lepidoptera: Sphingidae) resistance in *Lycopersicon hirsutum* f. *glabratum*. Journal of Economic Entomology 82:1743–1748. [2]

Sork, V. L. 1987. Effects of predation and light on seedling establishment in *Gustavia superba*. Ecology 68:1341–1350. [13]

Sork, V. L., and D. H. Boucher. 1977. Dispersal of sweet pignut hickory in a year of low fruit production, and the influence of predation by a curculionid beetle. Oecologia 28:289–299. [13]

Sosa, O. 1981. Biotypes J and L of the Hessian fly discovered in an Indiana wheat field. Journal of Economic Entomology 74:180–182. [5]

Sosulski, F. W., E. A. Paul, and W. L. Hutcheon. 1963. The influence of soil moisture, nitrogen fertilization, and temperature on quality and amino acid composition of thatcher wheat. Canadian Journal of Soil Science 43:219–228. [3]

Southwood, T. R. E. 1972. The insect/plant relationship: An evolutionary perspective. In H. F. van Emden, ed., Insect/plant relationships, 3–30. Blackwell Scientific Publications, Oxford. [13]

———. 1977. Habitat, the template for ecological strategies. Journal of Animal Ecology 46:337–365. [7]

———. 1986. Plant surfaces and insects: An overview. In B. Juniper and Sir R. Southwood, eds., Insects and the plant surface, 1–22. Edward Arnold, London. [10, 13]

———. 1987. Plant variety and its interaction with herbivorous insects. In V. Labeyrie, G. Fabres, and D. Lachaise, eds., Insects-plants, 61–19. Proceedings of the Sixth International Symposium on Insect-Plant Relationships. Dr. W. Junk Publishers, Dordrecht. [13]

Southwood, T. R. E., V. K. Brown, and P. M. Reader. 1986. Leaf palatability, life expectancy, and herbivore damage. Oecologia 70:544–548. [13]

Southwood, T. R. E., and P. M. Reader. 1976. Population census data and key factor analysis for the vibumum whitefly, *Aleurotrachelus jelinekii* (Frauenf.), on three bushes. Journal of Animal Ecology 45:313–325. [9, 11]

Spangelo, L. P. S., C. S. Hsu, S. O. Fejer, and R. Watkins. 1970. Combining ability analysis and interrelationships between thorniness and yield traits in gooseberry. Canadian Journal of Plant Science 50:439–444. [10]

Spears, E. E., Jr., and P. G. May. 1988. Effect of defoliation on gender expression and fruit set in *Passiflora incarnata*. American Journal of Botany 75:1842–1847. [13]

Spencer, K. C. 1988a. Chemical mediation of coevolution in the *Passiflora-Heliconius* interaction. In K. C. Spencer, ed., Chemical mediation of coevolution, 167–240. Academic Press, New York. [4, 13]

―――. ed. 1988b. Chemical mediation of coevolution. Academic Press, San Diego. [10, 13]

Sprague, G. F., and R. G. Dahms. 1972. Development of crop resistance to insects. Journal of Environmental Quality 1:28–34. [13]

SPSS, Inc. 1986. SPSS-X user's guide, 2d ed. McGraw-Hill Book Co., New York. [3]

Stadler, E., and H. R. Buser. 1984. Defense chemicals in leaf surface wax synergistically stimulate oviposition by a phytophagous insect. Experientia 40:1157–1159. [4]

Stahl, E. 1888. Pflanzen und Schnecken. Biologische Studie über die Schutzmittel der Pflanzen gegen Schneckenfrass. Jenaische Zeitschrift für Medizin und Naturwissenschaft 22:557–684. [4, 10]

Stakman, E. C., W. Q. Loegering, R. C. Cassell, and L. Hines. 1943. Population trends of physiological races of *Puccinia graminis tritici* in the United States for the period 1930 to 1941. Phytopathology 33:884–898. [15]

Stalter, R., and J. Serrao. 1983. The impact of defoliation by gypsy moths on the oak forest at Greenbrook Sanctuary, New Jersey. Bulletin of the Torrey Botanical Club 110:526–529. [13]

Stamp, N. E. 1984. Effect of defoliation by checkerspot caterpillars (*Euphydryas phaeton*) and sawfly larvae (*Macrophya nigra* and *Tenthredo grandis*) on their host plants (*Chelone* spp.). Oecologia 63:275–280. [13]

Stanton, M. L. 1983. Spatial patterns in the plant community and their effects upon insect search. In S. Ahmad, ed., Herbivorous insects: Host-seeking behavior and mechanisms, 125–157. Academic Press, New York. [13]

Stanton, N. L. 1983. The effect of clipping and phytophagous nematodes on net primary production of blue grama, *Bouteloua gracilis*. Oecologia 40:249–257. [13]

Stark, R. W. 1965. Recent trends in forest entomology. Annual Review of Entomology 10:303–324. [13]

Starks, K. J., and R. L. Burton. 1977. Greenbug: Determining biotypes, culturing, and screening for plant resistance, with notes on rearing parasitoids. Technical Bulletin no. 1556. Agricultural Research Service. United States Department of Agriculture. [5]

Starks, K. J., R. L. Burton, and O. G. Merkle. 1983. Greenbugs (Homoptera: Aphididae) plant resistance in small grains and sorghum to biotype E. Journal of Economic Entomology 76:877–880. [5]

Starks, K. J., R. Muniappan, and R. D. Eikenbary. 1972. Interaction between plant resistance and parasitism against the greenbug on barley and sorghum. Annals of the Entomological Society of America 65:650–655. [11, 12]

Starks, K. J., and D. J. Schuster. 1976. Greenbug: Effects of continuous culturing on resistant sorghum. Environmental Entomology 5:720–723. [5]

Starks, K. J., E. A. Wood, and G. L. Teetes. 1973. Effects of temperature on the preference of two greenbug biotypes for sorghum selections. Environmental Entomology 2:351–354. [6]

Stearns, S. C. 1989. The evolutionary significance of phenotypic plasticity. BioScience 39:436–446. [7]

Stebbins, G. L. 1981. Coevolution of grasses and herbivores. Annals of the Missouri Botanical Garden 68:75–86. [13]

Stenseth, N. C. 1983. Grasses, grazers, mutualism, and coevolution: A comment about handwaving in ecology. Oikos 41:152–153. [10]

Stephenson, A. G. 1980. Fruit set, herbivory, fruit reduction, and the fruiting strategy of *Catalpa speciosa* (Bignoniaceae). Ecology 61:57–64. [13]

Stern, K., and L. Roche. 1974. Genetics of forest ecosystems. Springer-Verlag, Berlin. [13]

Steuber, C. W., M. D. Edwards, and J. F. Wendel. 1987. Molecular marker-facilitated investigations of quantitative trait loci in maize. II. Factors influencing yield and its components. Crop Science 27:639–648. [3, 13]

Steward, J. L., and K. H. Keeler. 1988. Are there trade-offs among antiherbivore defenses in *Ipomoea* (Convolvulaceae)? Oikos 53:79–86. [17]

Stewart, I. 1985. Identification of caffeine in citrus flowers and leaves. Journal of Agricultural and Food Chemistry 33:1163–1165. [16]

Stickler, F. C., and A. W. Pauli. 1961. Leaf removal in grain sorghum. I. Effects of certain defoliation treatments on yield and components of yield. Agronomy Journal 53:99–102. [13]

Stiling, P. D. 1980. Competition and coexistence among *Eupteryx* leafhoppers (Hemiptera: Cicadellidae) occurring on stinging nettles (*Urtica dioica*). Journal of Animal Ecology 49:793–805. [11]

———. 1987. The frequency of density dependence in insect host-parasitoid systems. Ecology 68:844–856. [11]

Stiling, P. D., and D. R. Strong, Jr. 1984. Experimental density manipulation of stem-boring insects: Some evidence for interspecific competition. Ecology 65:1683–1685. [11]

Stinner, R. E. 1979. Biological monitoring essentials in studying wide-area movement. In R. L. Rabb and G. G. Kennedy, eds., Movement of highly mobile insects: Concepts and methodology in research, 199–211. University Graphics, North Carolina State University, Raleigh. [2]

Stipanovic, R. D. 1983. Function and chemistry of plant trichomes in insect resistance. In P. A. Hedin, ed., Mechanisms of plant resistance to insects, 69–100. American Chemical Society Symposium Series 208. American Chemistry Society, Washington, DC. [13]

Stovall, M. E., and K. Clay. 1988. The effect of the fungus *Balansia cyperi* on growth and reproduction of purple nutsedge, *Cyperus rotundus*. New Phytologist 109:351–359. [17]

St. Pierre, R. G. 1989. Magnitude, timing, and causes of immature fruit loss in *Amelanchier alnifolia* (Rosaceae). Canadian Journal of Botany 67:726–731. [13]

Strauss, S. Y. 1987. Direct and indirect effects of host-plant fertilization on an insect community. Ecology 68:1670–1678. [11]

Strauss, S. Y. 1988. Determining the effects of herbivory using naturally damaged plants. Ecology 69:1628–1630. [13]

Strong, D. R., Jr. 1982a. Harmonious coexistence of hispine beetles on *Heliconia* in experimental and natural communities. Ecology 63:1039–1049. [11]

————. 1982b. Potential interspecific competition and host specificity: Hispine beetles on *Heliconia*. Ecological Entomology 7:217–220. [11]

Strong, D. R., Jr., J. H. Lawton, and T. R. E. Southwood. 1984. Insects on plants: Community patterns and mechanisms. Harvard University Press, Cambridge. [6, 10, 11, 12, 13, 16, 17]

Stubblebine, W., J. H. Langenheim, and D. Lincoln. 1978. Vegetative response to photoperiod in the tropical leguminous tree *Hymenaea courbaril* L. Biotropica 10:18–29. [13]

Sturgeon, K. B. 1979. Monoterpene variation in ponderosa pine xylem resin related to western pine beetle predation. Evolution 33:803–814. [4, 13]

Sullivan, T. P., W. T. Jackson, J. Pojar, and A. Banner. 1986. Impact of feeding damage by the porcupine on western hemlock-Sitka spruce forests of north-central British Columbia. Canadian Journal of Forest Research 16:642–647. [13]

Sullivan, T. P., and D. S. Sullivan. 1986. Impact of feeding damage by snowshoe hares on growth rates of juvenile lodgepole pine in central British Columbia. Canadian Journal of Forest Research 16:1145–1149. [13]

Sullivan, T. P., and A. Wyse. 1987. Impact of red squirrel feeding damage on spaced stands of lodgepole pine in the Caribou Region of British Columbia. Canadian Journal of Forest Research 17:666–674. [13]

Summers, C. G. 1988. Cultivar and temperature influence on development, survival, and fecundity in four successive generations of *Acryrthosiphon kondoi* (Homoptera: Aphididae). Journal of Economic Entomology 81:515–521. [6]

Sutton, R. D. 1984. The effect of host plant flowering on the distribution and growth of hawthorn psyllids (Homoptera: Psylloidea). Journal of Animal Ecology 53:37–50. [13]

Suzuki, T., and G. R. Waller. 1987. Purine alkaloids in tea seeds during germination. In G. R. Waller ed., Allelochemicals: Role in agriculture and forestry, 289–294. American Chemical Society, Washington, DC. [16]

Swain, T. 1977. Secondary compounds as protection agents. Annual Review of Plant Physiology 28:479–501. [2]

————. 1979. Tannins and lignins. In G. A. Rosenthal and D. H. Janzen, eds., Herbivores: Their interaction with secondary plant metabolites, 657–682. Academic Press, New York. [4, 10]

Tabashnik, B. E., and B. A. Croft. 1982. Managing pesticide resistance in crop-anthropod complexes: Interactions between biological and operational factors. Environmental Entomology 11:1137–1144. [17, 18]

Tabashnik, B. E., and F. Slansky, Jr. 1987. Nutritional ecology of forb-chewing insects. In F. Slansky, Jr., and J. G. Rodriquez, eds., Nutritional ecology of insects, mites, spiders, and related invertebrates. Wiley-Interscience, New York. [7, 17]

Tahvanainen, J., E. Helle, R. Julkunen-Tiitto, and A. Lavola. 1985. Phenolic compounds of willow bark as deterrents against feeding by mountain hare. Oecologia 65:319–323. [10]

Takada, H. 1979. Characteristics of forms of *Myzus persicae* (Sulzer) (Homoptera: Aphididae) distinguished by colour and esterase differences, and their occurrence in populations on different host plants in Japan. Applied Entomology and Zoology 14:370–375. [6]

Tallamy, D. W. 1985. Squash beetle feeding behavior: An adaptation against induced cucurbit defenses. Ecology 66:1574–1579. [11, 16]

Tallamy, D. W., and V. A. Krischik. 1989. Variation and function of cucurbitacins in *Cucurbita*: An examination of current hypotheses. American Naturalist 133:766–786. [13]

Tamaki, G., B. Annis, L. Fox, R. K. Gupta, and A. Meszleny. 1982. Comparison of yellow holocyclic and green anholocyclic strains of *Myzus persicae* (Sulzer): Low temperature adaptability. Environmental Entomology 11:231–233. [6]

Tanksley, S. D., M. F. Herculano, and C. M. Rick. 1982. Use of naturally-occurring enzyme variation to detect and map genes controlling quantitative traits in an interspecific backcross of tomato. Heredity 49:11–25. [3]

Taylor, C. E., and G. P. Georghiou. 1979. Suppression of insecticide resistance by alteration of gene dominance and migration. Journal of Economic Entomology 72:105–109. [17]

Taylor, O. R., Jr., and D. W. Inouye. 1985. Synchrony and periodicity of flowering in *Frasera speciosa* (Gentianaceae). Ecology 66:521–527. [13]

Teetes, G. L., C. A. Schaefer, J. R. Gipson, R. C. McIntyre, and E. E. Latham. 1975. Greenbug resistance to organophosphorus insects on the Texas High Plains. Journal of Economic Entomology 68:214–216. [5]

Tempel, A. S. 1983. Bracken fern (*Pteridium aquilinum* Kuhn) and nectar-feeding ants: A non-mutualistic interaction. Ecology 64:1411–1422. [13]

Templeton, Y. E., D. O. LeBeest, and R. J. Smith. 1979. Biological weed control with mycoherbicides. Annual Review of Phytopathology 17:301–310. [18]

Tester, C. F. 1977. Constituents of soybean cultivars differing in insect resistance. Phytochemistry 16:1899–1901. [17]

Thomas, C. D., D. Ng, M. C. Singer, J. L. B. Mallet, C. Parmesan, and H. L. Billington. 1987. Incorporation of a European weed into the diet of a North American herbivore. Evolution 41:892–901. [6]

Thomas, C. E. 1986. Butterfly larvae reduce host plant survival in vicinity of alternative host species. Oecologia 70:113–117. [13]

Thomas, L. P., and M. A. Watson. 1988. Leaf removal and the apparent effects of architectural constraints on development in *Capsicum annuum*. American Journal of Botany 75:840–843. [13]

Thompson, E. A., and R. G. Shaw. 1990 Pedigree analysis for quantitative traits: Variance components without matrix inversion. Biometrics 46:399–413. [3]

Thompson, J. N. 1978. Within-patch structure and dynamics in *Pastinaca sativa* and resource availability to a specialized herbivore. Ecology 59:443–448. [13]

——. 1982. Interaction and coevolution. John Wiley and Sons, New York. [10]

——. 1983. Selection pressures on phytophagous insects feeding on small host plants. Oikos 40:438–444. [10]

——. 1985. Postdispersal seed predation in *Lomatium* spp. (Umbelliferae): Variation among individuals and species. Ecology 66:1608–1616. [13]

——. 1988a. Coevolution and alternative hypotheses on insect/plant interactions. Ecology 69:893–895. [6]

——. 1988b. Evolutionary ecology of the relationship between oviposition preference and performance of offspring in phytophagous insects. Entomologia Experimentalis et Applicata 47:3–14. [6, 13]

————. 1988c. Evolutionary genetics of oviposition preference in swallowtail butterflies. Evolution 42:1223–1234. [7]

————. 1988d. Variation in interspecific interactions. Annual Review of Ecology and Systematics 19:65–87. [11]

————. 1989. Concepts of coevolution. Trends in Ecology and Evolution 4:179–183. [18]

Thompson, J. N., and P. W. Price. 1977. Plant plasticity, phenology, and herbivore dispersion: Wild parsnip and the parsnip webworm. Ecology 58:1112–1119. [13]

Thompson, J. N., W. Wehling, and R. Podolsky. 1990. Evolutionary genetics of host use in swallowtail butterflies. Nature 344:148–150. [4]

Thompson, J. N., and M. R. Willson. 1979. Evolution of temperate fruit/bird interactions: Phonological strategies. Evolution 33:973–982. [13]

Thorpe, K. W., and P. Barbosa. 1986. Effects of consumption of high and low nicotine tobacco by *Manduca sexta* (Lepidoptera: Sphingidae) on survival of gregarious endoparasitoid *Cotesia congregata* (Hymenoptera: Braconidae). Journal of Chemical Ecology 12:1329–1337. [12]

Thottappilly, G., J. H. Tsai, and J. E. Bath. 1972. Differential aphid transmission of two bean yellow virus strains and comparative transmission by biotypes and stages of the pea aphid. Annals of the Entomological Society of America 65:912–915. [6]

Thurston, E. L. 1974. Morphology, fine structure, and ontogeny of the stinging emergence of *Urtica dioica*. American Journal of Botany 61:809–817. [10]

————. 1976. Morphology, fine structure, and ontogeny of the stinging emergence of *Tragia ramosa* and *T. saxicola* (Euphorbiaceae). American Journal of Botany 63:710–718. [10]

Thurston, E. L., and Lersten, N. R. 1969. The morphology and toxicology of plant stinging hairs. Botanical Review 35:393–412. [10]

Till, I. 1987. Variability of expression of cyanogenesis in white clover (*Trifolium repens* L.). Heredity 59:265–271. [13]

Till-Bottraud, I., P. Kakes, and B. Dommée. 1988. Variable phenotypes and stable distribution of the cyanotypes of *Trifolium repens* in Southern France. Acta Oecologia, Oecologia Plantarum 9:393–404. [17]

Tilman, D. 1978. Cherries, ants and tent caterpillars: Timing of nectar production in relation to susceptibility of caterpillars to ant predation. Ecology 59:686–692. [13]

Tingey, W. M. 1984. Glycoalkaloids as pest resistance factors. American Potato Journal 61:157–167. [2]

————. 1986. Techniques for evaluating plant resistance to insects. In J. R. Miller and T. A. Miller, eds., Insect-plant interactions, 251–284. Springer-Verlag, Berlin. [17]

Tingey, W. M., R. L. Plaisted, J. E. Laubengayer, and S. A. Mehlenbacher. 1982. Green peach aphid resistance by glandular trichomes in *Solanum tuberosum* × S. *berthaultii* hybrids. American Potato Journal 59:241–251. [2]

Tingey, W. M., and S. R. Singh. 1980. Environmental factors influencing the magnitude and expression of resistance. In F. G. Maxwell and P. R. Jennings, eds., Breeding plants resistant to insects, 87–113. John Wiley and Sons, New York. [2, 6, 9, 11]

Tiritilli, M. E., and J. N. Thompson. 1988. Variation in swallowtail/plant interactions: Host selection and the shapes of survivorship curves. Oikos 53:153–160. [13]

Toldine, T. E. 1984. Relation between DIMBOA content and *Helminthosporium turcicum* resistance in maize. (In Hungarian with English abstract). Novenytermeles 33:213–218. [2]

Tomiuk, J. 1987. The neutral theory and enzyme polymorphism in populations of aphid species. In J. Holman, J. Pelikan, A. F. G. Dixon, and L. Weismann, eds., Population structure, genetics and taxonomy of aphids and thysanoptera, 45–62. SPB Academic Publishing. The Hague, Netherlands. [5]

Tooby, J. 1982. Pathogens, polymorphism, and the evolution of sex. Journal of Theoretical Biology 97:557–576. [15]

Tookey, H. L., C. H. VanEtten, and M. E. Daxenbichler. 1980. In I. E. Liener, ed., Toxic constituents of plant foodstuffs, 2d ed., 103–142. Academic Press, New York. [16]

Treacy, M. F., G. R. Zummo, and J. H. Benedict. 1985. Interactions of host-plant resistance in cotton with predators and parasites. Agriculture, Ecosystems, and Environment 13:151–157. [12]

Trenbath, B. R. 1975. Diversify or be damned? Ecologist 5:76–83. [5]

———. 1977. Interactions among diverse hosts and diverse parasites. Annals of the New York Academy of Sciences 287:124–150. [5]

———. 1984. Gene introduction strategies for the control of crop diseases. In G. R. Conway, ed., Pest and pathogen control: Strategic, tactical, and policy models, 142–168. John Wiley and Sons, New York. [5]

Tscharntke, T. 1987. Growth regulation of *Phragmites australis* by the gallmidge *Giraudiella inclusa*. In V. Labeyrie, G. Fabres, and D. Lachaise, eds., Insects-plants, 201–205. Proceedings of the Sixth International Symposium on Insect-Plant Relationships. Dr. W. Junk Publisher, Dordrecht. [13]

———. 1989. Attack by a stem-boring moth increases susceptibility of *Phragmites australis* to gall-making by a midge: Mechanisms and effects on midge population dynamics. Oikos 55:93–100. [6]

Tseng, T., W. D. Guthrie, W. A. Russell, J. C. Robbins, J. R. Coats, and J. J. Tollefson. 1984. Evaluation of two procedures to select for resistance to the European corn borer in a synthetic cultivar of maize. Crop Science 24:1129–1133. [2]

Tsingalia, M. H. 1989. Variation in seedling predation and herbivory in *Prunus africana* in the Kakamega Forest, Kenya. African Journal of Ecology 27:207–217. [13]

Tsukamoto, M. 1983. Methods of genetic analysis of insecticide resistance. In G. P. Georghiou and T. Saito, eds., Pest resistance to pesticides, 71–98. Plenum Press, New York. [17]

Tuite, J., and G. H. Foster. 1979. Control of storage diseases of grain. Annual Review of Phytopathology 17:343–366. [14]

Tuomi, J., P. Niemelä, F. S. Chapin III, J. P. Bryant, and S. Siren. 1988. Defensive responses of trees in relation to their carbon/nutrient balance. In W. J. Mattson, J. Levieux, and C. Bernard-Dagan, eds., Mechanisms of woody plant defenses against insects: Search for Pattern, 57–72. Springer-Verlag, New York. [13]

Tuomi, J., P. Niemelä, E. Haukioja, and S. Neuvonen. 1984. Nutrient stress: An explanation for plant anti-herbivore responses to defoliation. Oecologia 61:208–210. [11]

Tuomi, J., P. Niemelä, I. Jussila, T. Vuorisalo, and V. Jormalainen. 1989. Delayed budbreak: A defensive response of mountain birch to early-season defoliation? Oikos 54:87–91. [13]

Tuomi, J., S. Nisula, T. Vuorisalo, P. Niemelä, and V. Jormalainen. 1988. Reproductive effort of short shoots in silver birch (*Betula pendula* Roth). Experientia 44:540–541. [13]

Tuomi, J., T. Vuorisalo, P. Niemelä, S. Nisula, and V. Jormalainen. 1988. Localized effects of branch defoliations on weight gain of female inflorescences in *Betula pubescens*. Oikos 51:327–330. [13]

Turchin, P. 1990. Rarity of density dependence or population regulation with lags? Nature 344:660–663. [6]

Turelli, M. 1984. Heritable genetic variation via mutation-selection balance: Lerch's zeta meets the abdominal bristle. Theoretical Population Biology 75:401–413. [7]

———. 1988. Population genetic models for polygenic variation and evolution. In B. S. Weir, E. J. Eisen, M. M. Goodman, and G. Namkoong, eds., Proceedings of the Second International Conference on Quantitative Genetics, 601–618. Sinauer, Sunderland, MA. [3]

Turner, H. N., and S. S. Y. Young. 1969. Quantitative genetics in sheep breeding. Cornell University Press, Ithaca, NY. [3]

Turner, M. E., J. C. Stephens, and W. W. Anderson. 1982. Homozygosity and patch structure in plant populations as a result of nearest-neighbor pollination. Proceedings of the National Academy of Sciences USA 79:203–207. [15]

Ueckert, D. N. 1979. Impact of a white grub (*Phyllophaga crinita*) on a shortgrass community and evaluation of selected rehabilitation practices. Journal of Range Management 32:445–448. [13]

Ueckert, D. N., and Hansen, R. M. 1971. Dietary overlap of grasshoppers on sandhill rangeland in northeastern Colorado. Oecologia 8:276–295. [10]

Ueda, N., and H. Takada. 1977. Differential relative abundance of green-yellow and red forms of *Myzus persicae* (Sulzer) (Homoptera: Aphididae) according to host plant and season. Applied Entomology and Zoology 12:124–133. [6]

Unruh, T. R., and R. F. Luck. 1987. Deme formation in scale insects: A test with the pinyon needle scale and a review of other evidence. Ecological Entomology 12:439–449. [9, 11]

Uphof, J. C. T. 1962. Plant hairs. In W. Zimmerman and P. G. Ozenda, eds., Encyclopedia of plant anatomy, vol. 4. Gebruder Bortraeger, Berlin. [10]

Van Alphen, J. J. M., and L. E. M. Vet. 1986. An evolutionary approach to host finding and selection. In J. Waage and D. Greathead, eds., Insect parasitoids, 23–61. Academic Press, London. [11]

Van Alstyne, K. L. 1988. Herbivore grazing increases polyphenolic defenses in the intertidal brown alga *Fucus distichus*. Ecology 69:655–663. [10]

Vandenberg, P., and D. F. Matzinger. 1970. Genetic diversity and heterosis in *Nicotiana*. III. Crosses among tobacco introductions and flue-cured varieties. Crop Science 10:437–440. [16, 17]

van der Meijden, E., A. M. van Zoelen, and L. L. Soldaat. 1989. Oviposition by the cinnabar moth, *Tyria jacobaeae*, in relation to nitrogen, sugars, and alkaloids of ragwort, *Senecio jacobaea*. Oikos 54:337–344. [13]

van der Meijden, E., M. Wijn, and H. J. Verkaar. 1988. Defence and regrowth: Alternative plant strategies in the struggle against herbivores. Oikos 51:355–363. [13]

Vanderplank, J. E. *See* Van der Plank

Van der Plank, J. E. 1963. Plant diseases: Epidemic and control. Academic Press, New York. [5, 14]

————. 1968. Disease resistance in plants. Academic Press, New York. [5, 8]

————. 1975. Principles of plant infection. Academic Press, New York. [15]

————. 1983. Durable resistance in crops: Should the concept of physiological races die? In F. Lamberti, J. M. Waller, and N. A. Van der Graaff, eds., Durable resistance in crops, 41–44. Plenum, New York. [5]

————. 1984. Disease resistance in plants. Academic Press, Orlando. [2]

Van Dijk, P., J. E. Parlevliet, G. H. J. Kema, A. C. Zevan, and R. W. Stubbs. 1988. Characterization of the durable resistance to yellow rust in old winter wheat cultivars in the Netherlands. Euphytica 38:149–158. [14]

Van Dyne, G. M. 1968. Measuring quantity and quality of the diet of large herbivores. In F. B. Golley and H. K. Buechner, eds., A practical guide to the study of the productivity of large herbivores. Blackwell, Oxford. [10]

Van Emden, H. F. 1966. Plant insect relationships and pest control. World Review of Pest Control 5:115–123. [12]

————. 1978. Insects and secondary plant substances: An alternative viewpoint with special reference to aphids. In J. B. Harbourne, ed., Biochemical aspects of plant and animal co-evolution, 309–323. Academic Press, New York. [12]

————. 1986. The interaction of plant resistance and natural enemies: Effects on populations of sucking insects. In D. J. Boethel and R. D. Eikenbary, eds., Interactions of plant resistance and parasitoids and predators of insects, 138–150. John Wiley and Sons, New York. [5]

————., ed. 1972. Insect-plant relationships. Blackwell, Oxford. [10]

Van Ginkel, M., and A. L. Scharen. 1988. Host-pathogen relationships of wheat and *Septoria tritici*. Phytopathology 78:762–766. [8]

Van Tooren, B. F. 1988. The fate of seeds after dispersal in chalk grassland: The role of the bryophyte layer. Oikos 53:41–48. [13]

Verkaar, H. J. 1986. When does grazing benefit plants? Trends in Ecology and Evolution 1:168–169. [13]

————. 1987. Population dynamics: The influence of herbivory. New Phytologist 106:49–60. [13]

Via, S. 1984. The quantitative genetics of polyphagy in an insect herbivore. II. Genetic correlations in larval performance within and among host plants. Evolution 38:896–905. [3, 4, 5, 6, 7]

————. 1986. Genetic covariance between oviposition preference and larval performance in an insect herbivore. Evolution 40:778–785. [6]

————. 1987. Genetic constraints on the evolution of phenotypic plasticity. In V. Leoschcke, ed. Genetic constraints on adaptive evolution, 47–71. Springer-Verlag, Berlin. [7]

————. 1989. Field estimation of variation in host plant use between local populations of pea aphids from two crops. Ecological Entomology 14:357–364. [5, 6]

————. 1990. Ecological genetics and host adaptation in herbivorous insects: The experimental study of evolution in natural and agricultural systems. Annual Review of Entomology 35:421–446. [6, 7]

Via, S., and R. Lande. 1985. Genotype-environment interaction and the evolution of phenotypic plasticity. Evolution 39:505–523. [3, 7]

————. 1987. Evolution of genetic variability in a spatially heterogeneous environment: Effects of genotype-environment interaction. Genetical Research 49:147–156. [3]

Vinson, S. B. 1976. Host selection by insect parasitoids. Annual Review of Entomology 21:109–138. [11]

————. 1981. Habitat location. In D. A. Nordlund, R. L. Jones, and W. J. Lewis, eds., Semiochemicals: Their role in pest control, 51–78. John Wiley and Sons, New York. [11, 12]

————. 1984. Parasitoid-host relationship. In W. J. Bell and R. Carde, eds., Chemical ecology of insects, 205–233. Sinauer Associates, Sunderland, MA. [12]

Vinson, S. B., and P. Barbosa. 1987. Interrelationships of nutritional ecology of parasitoids. In F. Slansky, Jr., and J. G. Rodriquez, eds., Nutritional ecology of insects, mites, spiders, and related invertebrates, 673–695. John Wiley and Sons, New York. [12]

Vinson, S. B., and G. F. Iwantsch. 1980. Host suitability for insect parasitoids. Annual Review of Entomology 25:397–419. [12]

Von Caemmerer, S., and G. D. Farquhar. 1984. Effects of partial defoliation, changes of irradiance during growth, short-term water stress, and growth at enhanced $p(CO_2)$ on the photosynthetic capacity of leaves of *Phaseolus vulgaris* L. Planta 160:320–329. [13]

von Schonborn, A. 1966. The breeding of insect-resistant forest trees in central and northwestern Europe. In H. D. Gerhold, E. J. Schreiner, R. E. McDermott, and J. A. Winieski, eds., Breeding pest-resistant trees, 25–27. Pergamon Press, Oxford. [11]

Waddington, C. H. 1957. The strategy of the genes: A discussion of some aspects of theoretical biology. Allen and Unwin, London. [7]

Wahl, I. 1970. Prevalence and geographic distribution of resistance to crown rust in *Avena sterilis*. Phytopathology 60:746–749. [14, 15]

Wahl, I., Y. Anikster, J. Manisterski, and A. Segal. 1984. Evolution at the center of origin. In W. R. Bushness and A. P. Roelfs, eds., The cereal rusts, 1:39–77. Academic Press, Orlando. [14]

Wahl, I., N. Eshed, A. Segal, and Z. Sobel. 1978. Significance of wild relatives of small grains and other wild grasses in cereal powdery mildews. In D. M. Spencer, ed., The powdery mildews, 83–100. Academic Press, London. [14]

Wainhouse, D., and R. S. Howell. 1983. Intraspecific variation in beech scale populations and in susceptibility of their host *Fagus sylvatica*. Ecological Entomology 8:351–359. [2, 9, 11]

Waldbauer, G. P. 1968. The consumption and utilization of food by insects. Advances in Insect Physiology 5:229–288. [4]

Wallace, J. W., and R. L. Mansell, eds. 1976. Biochemical interactions between plants and insects. Plenum Press, New York. [10]

Wallace, L. L. 1987. Effects of clipping and soil compaction on growth, morphology, and mycorrhizal colonization of *Schizachrium scoparium*, a C_4 bunch grass. Oecologia 72:423–428. [13]

Wallace, L. L., S. J. McNaughton, and M. B. Coughenour. 1984. Compensatory photosynthetic responses of three African graminoids to different fertilization, watering, and clipping regimes. Botanical Gazette 145:151–156. [10, 13]

———. 1985. Effects of clipping and four levels of nitrogen on the gas exchange, growth, and production of two East African graminoids. American Journal of Botany 72:222–230. [13]

Waller, D. A., and C. G. Jones. 1989. Measuring herbivory. Ecological Entomology 14:479–481. [17]

Waller, G. R., and E. K. Nowacki. 1978. Alkaloid biology and metabolism in plants. Plenum Press, New York. [2, 4, 17]

Waloff, N., and O. W. Richards. 1977. The effect of insect fauna on growth mortality and natality of broom, *Sarothamnus scoparius*. Journal of Applied Ecology 14:787–798. [13]

Walsh, B. 1867. The apple worm and the apple maggot. Journal of Horticulture 2:338–343. [4]

Wareing, P. F., M. M. Khalifa, and K. J. Treharne. 1968. Rate-limiting processes in photosynthesis at saturating light intensities. Nature 220:453–457. [13]

Warning, W. C. 1934. Anatomy of the vegetative organs of the parsnip. Botanical Gazette 96:44–72. [16]

Wasserman, S. S., and D. J. Futuyma. 1981. Evolution of host plant utilization in cowpea weevil, *Callosobruchus maculatus* Fabricus (Coleoptera: Bruchidae). Evolution 35:605–617. [7]

Waterman, P. G., J. A. M. Ross, and D. B. McKey. 1984. Factors affecting levels of some phenolic compounds, digestibility, and nitrogen content of the mature leaves of *Barteria fistulosa* (Passifloraceae). Journal of Chemical Ecology 10:387–401. [13]

Watkinson, A. R. 1986. Plant population dynamics. In M. J. Crawley, ed., Plant ecology, 137–184. Blackwell Scientific, Oxford. [13]

Watson, I. A. 1970. Changes in virulence and population shifts in plant pathogens. Annual Review of Phytopathology 8:209–230. [9]

Watson, M. A., and B. B. Casper. 1984. Morphogenetic constraints on patterns of carbon distribution in plants. Annual Review of Ecology and Systematics 15:233–258. [13]

Watt, A. D., and A. F. G. Dixon. 1981. The role of cereal growth stages and crowding in the induction of alatae in *Sitobion avenae* and its consequences for population growth. Ecological Entomology 6:441–447. [6]

Way, M. J. 1988. Entomology of wheat. In M. K. Harris and C. E. Rogers, eds., The entomology of indigenous and naturalized systems in agriculture, 183–206. Westview Press, Boulder. [2]

Weber, G. 1985a. Genetic variability in host plant adaptation of the green peach aphid, *Myzus persicae*. Entomologia Experimentalis et Applicata 38:49–56. [5, 6]

———. 1985b. On the ecological genetics of *Sitobion avenae* (F.) (Hemiptera: Aphididae). Zeitschrift für angewandt Entomologie 100:100–110. [5, 6]

Webster, J. A., and C. Inayatullah. 1985. Aphid biotypes in relation to plant resistance: A selected bibliography. Southwestern Entomologist. 10:116–125. [5]

Webster, J. A., C. Inayatullah, and O. G. Merkle. 1986. Susceptibility of 'Largo' wheat to biotype B greenbug (Homoptera: Aphididae). Environmental Entomology 15:700–702. [5]

Webster, J. A., and K. J. Starks. 1984. Sources of resistance in barley to two biotypes of the greenbug, *Schizaphis graminum* (Rondani), Homoptera: Aphididae. Protection Ecology 6:51–55. [5]

Weckerly, F. W., D. W. Sugg, and R. D. Semlitsch. 1989. Germination success of acorns (*Quercus*): Insect predation and tannins. Canadian Journal of Forest Research 19:811–815. [13]

Weir, B. S., and C. C. Cockerham. 1969. Group inbreeding with two linked loci. Genetics 63:711–742. [3]

Weis, A. E., and W. G. Abrahamson. 1985. Potential selective pressures by parasitoids on a plant-herbivore interaction. Ecology 66:1261–1269. [12]

———. 1986. Evolution of host-plant manipulation by gall makers: Ecological and genetic factors in the *Solidago-Eurosta* system. American Naturalist 127:681–695. [7, 11, 12, 13]

Weis, A. E., W. G. Abrahamson, and K. D. McCrea. 1985. Host gall size and oviposition success by the parasitoid *Eurytoma gigantea*. Ecological Entomology 10:341–348. [7, 11]

Weis, A. E., and W. L. Gorman. 1990. Measuring selection on reaction norms: An exploration of the *Eurosta-Solidago* system. Evolution 44:820–831. [6, 7]

Weis, A. E., K. D. McCrea, and W. G. Abrahamson. 1989. Can there be an escalating arms race without coevolution?: Implications of a host-parasitoid simulation. Evolutionary Ecology 3:361–370. [7]

Weller, J. I. 1987. Mapping and analysis of quantitative trait loci in *Lycopersicon* (tomato) with the aid of genetic markers using approximate maximum likelihood methods. Heredity 59:413–421. [3]

Welz, G., and J. Kranz. 1987. Effects of recombination on races of a barley powdery mildew population. Plant Pathology 36:107–113. [8]

Weseloh, R. M. 1981. Host location by parasitoids. In D. A. Nordlund, R. L. Jones, and W. J. Lewis, eds., Semiochemicals: Their role in pest control, 79–95. John Wiley and Sons, New York. [11]

Weste, G. 1974. *Phytophthora cinnamomi:* The cause of severe disease in certain native communities in Victoria. Australian Journal of Botany 22:1–8. [15]

Westoby, M. 1974. An analysis of diet selection by large generalist herbivores. American Naturalist 108:290–304. [10]

———. 1989. Selective forces exerted by vertebrate herbivores on plants. Trends in Ecology and Evolution 4:115–117. [13]

White, J. 1979. The plant as a metapopulation. Annual Review of Ecology and Systematics 10:109–145. [10]

White, J., M. Lloyd, and R. Karban. 1982. Why don't periodical cicadas (*Magicicada* spp.) normally live in coniferous forests? Environmental Entomology 11:475–482. [9]

White, J. F. 1987. The widespread distribution of endophytes in the Poaceae. Plant Disease 71:340–342. [10]

White, T. C. R. 1971. Lerp insects (Homoptera: Psyllidae) on red gum (*E. camuldulensis*) in South Australia. South Australian Naturalist 8:273–293. [11]

Whitham, T. G. 1983. Host manipulation of parasites: Within-plant variation as a defense against rapidly evolving pests. In R. F. Denno and M. S. McClure, eds.,

Variable plants and herbivores in natural and managed systems, 15–41. Academic Press, New York. [9, 11]

Whitham, T. G., and S. Mopper. 1985. Chronic herbivory: Impacts on architecture and sex expression of pinyon pine. Science 228:1089–1091. [13]

Whitman, D. W. 1988. Allelochemical interactions among plants, herbivores, and their predators. In P. Barbosa and D. K. Letourneau, eds., Novel aspects of insect-plant interactions, 11–64. John Wiley and Sons, New York. [12]

Whitman, R. J. 1973. Herbivore feeding and cyanogenesis in *Trifolium repens* L. Heredity 30:241–244. [10]

Whittaker, J. B. 1982. The effects of grazing by a chrysomelid beetle, *Gastrophysa viridula*, on growth and survival of *Rumex crispus* on a shingle bank. Journal of Ecology 70:291–296. [13]

———. 1984. Responses of sycamore (*Acer pseudoplatanus*) leaves to damage by a typhlocybine leaf hopper, *Ossiannilssonola callosa*. Journal of Ecology 72:455–462. [13]

Whittaker, R. H., and P. P. Feeny. 1971. Allelochemicals: Chemical interactions between species. Science 171:757–770. [1, 2, 10, 17]

Wicklow, D. T. 1988. Metabolites in the coevolution of fungal chemical defense systems. In K. A. Pirozynski and D. L. Hawksworth, eds., Coevolution of fungi with plants and animals, 173–201. Academic Press, New York. [3]

Widstrom, N. W., B. R. Wiseman, and W. W. McMillan. 1984. Patterns of resistance in sorghum to the sorghum midge. Crop Science 24:791–793. [2]

Wiggans, G. R., R. L. Quaas, and L. D. Van Vleck. 1980. Estimating a genetic covariance from least squares solutions. Journal of Dairy Science 63:174–177. [3]

Wilhoit, L. R. 1988. The effects of plant varieties on the evolution of the aphids, *Schizaphis graminum* (Rondani). Ph.D. Diss. University of California, Berkeley. [5]

———. 1991. Modelling the population dynamics of different aphid genotypes in plant variety mixtures. Ecological Modelling. 55:257–283. [5]

Wilhoit, L. R., and T. E. Mittler. 1991. Biotypes and clonal variation in greenbug (Homoptera: Aphididae) populations from a locality in California. Environmental Entomology 20:757–767. [5]

Williams, E. H., and M. D. Bowers. 1987. Factors affecting hostplant use by the montane butterfly *Euphydryas gillettii* (Nymphalidae). American Midland Naturalist 118:153–161. [13]

Williams, G. C. 1975. Sex and evolution. Princeton University Press, Princeton, NJ. [7]

Williams, K. S. 1983. The coevolution of *Euphydryas chalcedona* butterflies and their larval host plants. III. Oviposition behavior and host plant quality. Oecologia 56:336–340. [13]

Williams, K. S., and L. E. Gilbert. 1981. Insects as selective agents on plant vegetative morphology: Egg mimicry reduces egg laying by butterflies. Science 212:467–469. [13]

Williams, K. S., and J. H. Myers. 1984. Previous herbivore attack of red alder may improve food quality for fall webworm larvae. Oecologia 63:166–170. [6, 11]

Williams, W. G., G. G. Kennedy, R. T. Yamamoto, J. D. Thacker, and J. Bordner. 1980. 2-Tridecanone: A naturally occurring insecticide from the wild tomato, *Lycopersicon hirsutum* f. *glabratum*. Science 207:888–889. [2]

Willis, C. L. 1969. Toxic constituents of the stinging nettle. M.S. thesis, Iowa State University, Ames. [10]

Willson, M. F., and P. W. Price. 1980. Resource limitation of fruit and seed production in some *Asclepias* species. Canadian Journal of Botany 58:2229–2233. [13]

Wilson, C. L. 1969. Use of plant-pathogens in weed control. Annual Review of Phytopathology 7:411–434. [18]

Wilson, D. S. 1976. Evolution on the level of communities. Science 192:1358–1360. [18]

Wilson, F. D. 1987. Pink bollworm resistance, lint yield, and earliness of cotton isolines in a resistant genetic background. Crop Science 27:957–960. [16]

Wilson, F. D., and T. N. Shaver. 1973. Glands, gossypol content, and tobacco budworm development in seedlings and floral parts of cotton. Crop Science 13:107–110. [2]

Wilson, J., and G. Shaner. 1989. Individual and cumulative effects of long latent period and low infection-type reactions to *Puccinia recondita* on triticale. Phytopathology 79:101–108. [8]

Wilson, R. L., K. J. Starks, H. Pass, and E. A. Wood. 1978. Resistance in four oat lines to two biotypes of the greenbug. Journal of Economic Entomology 71:886–887. [5]

Windle, P. N., and E. H. Franz. 1979. The effects of insect parasitism on plant competition: Greenbugs and barley. Ecology 60:521–529. [13, 16, 17]

Wink, M. 1983. Inhibition of seed germination by quinolizidine alkaloids. Planta 158:365–368. [17]

———. 1984a. Chemical defense of Leguminosae. Are quinolizidine alkaloids part of the antimicrobial defense system of lupins? Zeitschrift Naturforschung 39C:548–552. [17]

———. 1984b. Chemical defense of lupins. Mollusc-repellent properties of quinolizidine alkaloids. Zeitschrift Naturforschung 39C:553–558. [17]

———. 1985. Chemische Verteidigung del Lupinen: Zur biologischen bedeutung der chinolizidinalkaloide. Plant Systematics and Evolution 150:65–81. [16]

———. 1988. Plant breeding: Importance of plant secondary metabolites for protection against pathogens and herbivores. Theoretical and Applied Genetics 75:225–233. [2]

Wink, M., and L. Witte. 1985a. Quinolizidine alkaloids as nitrogen source for lupin seedlings and cell cultures. Zeitschrift Naturforschung 40C:767–775. [17]

———. 1985b. Quinolizidine alkaloids in *Petteria ramentacea* and the infesting aphids, *Aphis cytisorum*. Phytochemistry 24:2567–2568. [16]

Wippich, C., and M. Wink. 1985. Biological properties of alkaloids: Influence of quinolizidine alkaloids and gramine on the germination and development of powdery mildew, *Erysiphe graminis* f. sp. *hordei*. Experientia 41:1477–1479. [17]

Wise, D. H. 1981. A removal experiment with darkling beetles: Lack of evidence for interspecific competition. Ecology 62:727–738. [11]

Wiseman, B. R., C. V. Hall, and R. H. Painter. 1961. Interactions among cucurbit varieties of feeding response of the striped and spotted cucumber beetles. Proceedings of the American Society for Horticultural Science 78:379–384. [2]

Wöhrmann, K. 1984. Population biology of the rose aphid, *Macrosiphum rosae*. In K.

Wöhrmann and V. Loeschcke, eds., Population biology and evolution, 208–216. Springer-Verlag, Berlin. [5]

Wolda, H., A. Zweep, and K. A. Schuitema. 1971. The role of food in the dynamics of populations of the landsnail *Cepaea nemoralis*. Oecologia 7:361–381. [10]

Woledge, J. 1977. The effects of shading and cutting treatments on the photosynthetic rate of ryegrass leaves. Annals of Botany 41:1279–1286. [13]

Wolfe, M. 1983. Genetic strategies and their value in disease control. In T. Kommedahl and P. H. Williams, eds., Challenging problems in plant health, 461–473. American Phytopathological Society, St. Paul. [14]

Wolfe, M. S. 1985. The current status and prospects of multiline cultivars and variety mixtures for disease resistance. Annual Review of Phytopathology 23:251–273. [5]

Wolfe, M. S., and J. A. Barrett. 1979. Disease in crops: Controlling the evolution of plant pathogens. Journal of the Royal Society of the Arts 127:321–333. [5]

Wolfe, M. S., J. A. Barrett, and J. E. E. Jenkins. 1981. The use of cultivar mixtures for disease control. In J. F. Jenkins and R. T. Plaumb, eds., Strategies for the control of cereal disease, 73–80. Blackwell Scientific Publications, Oxford. [5]

Wolfe, M. S., and C. E. Caten, eds. 1987. Populations of plant pathogens: Their dynamics and genetics. Blackwell Scientific, Oxford. [2]

Wolfe, M. S., and D. R. Knott. 1982. Populations of plant pathogens: Some constraints on analysis of variation in pathogenicity. Plant Pathology 31:79–90. [5]

Wolfe, M. S., P. N. Minchin, and J. A. Barrett. 1984. Some aspects of the development of heterogeneous cropping. In E. J. Gallagher, ed., Cereal production, 95–104. Butterworths, London. [5]

Wolff, K., and W. van Delden. 1987. Genetic analysis of ecological relevant morphological variability in *Plantago lanceolata* L. I. Population characteristics. Heredity 58:183–192. [11]

Woltereck, R. 1909. Weitere experimentelle Untersuchen über Artveränderung, speziell über das Wesen quantitätiver Artunterschiede bei Daphniden. Verhandlungen der Deutschen Zoologischen Gesellschaft 1909:110–172. [7]

Wood, E. A., Jr. 1961. Biological studies of a new greenbug (*Toxoptera graminum*) biotype. Journal of Economic Entomology 54:1171–1173. [5]

Wood, E. A. 1971. Designation and reaction of three biotypes of the greenbug on resistant and susceptible species of sorghum. Journal of Economic Entomology 64:183–185. [6]

Wood, E. A., and K. J. Starks. 1972. Effect of temperature and host plant interaction on the biology of three biotypes of the greenbug. Environmental Entomology 1:230–234. [6]

Wood, R. K. S. 1982. Active defense mechanisms in plants. Plenum Press, New York. [14]

Wood, T. K., and S. I. Guttman. 1982. Ecological and behavioral basis for reproductive isolation in the sympatric *Enchenopa binotata* complex (Homoptera: Membracididae). Evolution 36:233–242. [6]

Woodhead, S. 1983. Surface chemistry of *Sorghum bicolor* and its importance in feeding by *Locusta migratoria*. Physiological Entomology 8:345–352. [10]

Woodhead, S., and E. A. Bernays. 1978. The chemical basis of resistance of *Sorghum bicolor* to attack by *Locusta migratoria*. Entomologia Experimentalis et Applicata 24:123–144. [10]

Woodhead, S., and R. F. Chapman. 1986. Insect behavior and the chemistry of plant surface waxes. In B. E. Juniper and T. R. E. Southwood, eds., Insects and the plant surface, 123–135. Edward Arnold, London. [10]

Woods, D. L., and K. W. Clark. 1979. Palatability of reed canarygrass pasture. Canadian Journal of Plant Science 54:89–91. [13]

Wool, D., S. Bunting, and H. F. van Emden. 1978. Electrophoretic study of genetic variation in British *Myzus persicae* (Sulz.) (Hemiptera, Aphididae). Biochemical Genetics 16:987–1006. [5]

Wool, D., and O. Manheim. 1986. Population ecology of the gall-forming aphid, *Aploneura lentisci* (Pass.) in Israel. Researches in Population Ecology 28:151–162. [11]

Worthen, W. B. 1989. Predator-mediated coexistence in laboratory communities of mycophagous *Drosophila* (Diptera: Drosophilidae). Ecological Entomology 14: 117–126.

Wright, J. W. 1976. Introduction to forest genetics. Academic Press, New York. [9]

Wright, S. 1946. Isolation by distance under diverse systems of mating. Genetics 321:39–59. [15]

———. 1955. Classification of the factors of evolution. Cold Springs Harbor Symposia on Quantitative Biology 20:16–24. [17]

———. 1968 Evolution and the genetics of populations. Vol. 1, Genetic and biometric foundations. University of Chicago Press, Chicago. [3]

———. 1970. Random drift and the shifting balance theory of evolution. In K. Kojima, ed., Mathematical topics in population genetics, 1–31. Springer-Verlag, Berlin. [3]

———. 1978. Evolution and the genetics of populations. Vol. 4, Variability within and among natural populations. University of Chicago Press, Chicago. [2, 3]

Wright, S. J. 1983. The dispersion of eggs by a bruchid beetle among *Scheelea* palm seeds and the effect of distance to the parent palm. Ecology 64:1016–1021. [13]

Wu, C. F. J. 1986. Jackknife, bootstrap, and other resampling methods in regression analysis. Annals of Statistics 14:1261–1295. [3]

Wyatt, I. J. 1970. The distribution of *Myzus persicae* (Sulz.) on year-round chrysanthemums. II. Winter season: The effect of parasitization by *Aphidius matricariae* Hal. Annals of Applied Biology 65:31–41. [12]

Yamada, Y. 1962. Genotype by environment interaction and genetic correlation of the same trait under different environments. Japanese Journal of Genetics 37:498–509. [3]

Yanes, J., Jr., and D. J. Boethel. 1983. Effect of a resistant soybean genotype on the development of the soybean looper (Lepidoptera: Noctuidae) and an introduced parasitoid, *Microplitis demolitor* Wilkinson (Hymenoptera: Braconidae). Environmental Entomology 12:1270–1274. [12]

Youden, W. J., and W. S. Connor. 1953. The chain block design. Biometrics 9:127–140. [3]

Young, H. C., Jr., and J. M. Prescott. 1977. A study of race populations of *Puccinia recondita* f. sp. *tritici*. Phytopathology 67:528–532. [8]

Young, T. P. 1985. *Lobelia telekii* herbivory, mortality, and size at reproduction: Variation with growth rate. Ecology 66:1879–1883. [13]

———. 1987. Increased thorn length in *Acacia depranolobium:* An induced response to browsing. Oecologia 71:436–438. [8, 13, 16]

Zagory, D., and W. J. Libby. 1985. Maturation-related resistance of *Pinus radiata* to western gall rust. Phytopathology 75:1443–1447. [11]

Zahorik, D. M., and K. A. Houpt. 1981. Species differences in feeding strategies, food hazards, and the ability to learn food aversions. In A. C. Kamil and T. D. Sargent, eds., Foraging behavior, 289–310. Garland Publishing, New York. [10]

Zammit, C., and C. Hood. 1986. Impact of flower and seed predators on seed set in two *Banksia* shrubs. Australian Journal of Ecology 11:187–193. [13]

Zammit, C., and W. Westoby. 1988. Pre-dispersal seed losses, and the survival of seeds and seedlings of two serotinous *Banksia* shrubs in burnt and unburnt heath. Journal of Ecology 76:200–214. [13]

Zangerl, A. R. 1986. Leaf value and optimal defense in *Pastinaca sativa* L. (Umbelliferae). American Midland Naturalist 116:432–437. [16, 17]

———. 1990. Furanocoumarin induction in wild parsnip: Evidence for an induced defense against herbivores. Ecology 71:1926–1932. [4, 13, 16]

Zangerl, A. R., and M. R. Berenbaum. 1986. Furanocoumarins in wild parsnip: Effects of photosynthetically active radiation, ultraviolet light, and nutrients. Ecology 68:516–520. [16]

———. 1990. Furanocoumarin induction in wild parsnip: Genetics and populational variation. Ecology 71:1933–1940. [4, 11, 13, 16]

Zangerl, A. R., M. R. Berenbaum, and E. Levine. 1989. Genetic control of seed chemistry and morphology in wild parsnip (*Pastinaca sativa*). Journal of Heredity 80:404–407. [4, 16]

Zangerl, A., J. Nitao, and M. R. Berenbaum. 1991. Parthenocarpic fruits in wild parsnip: Decoy defense against a specialized herbivore. Evolutionary Ecology 5:136–145. [4]

Zavarin, E., and F. W. Cobb. 1970. Oleoresin variability in *Pinus ponderosa*. Phytochemistry 9:2509–2515. [2]

Zavarin, E., W. B. Critchfield, and K. Snajberk. 1969. Turpentine composition of *Pinus contorta* × *Pinus banksiana* hybrids and hybrid derivatives. Journal of Canadian Botany 47:1443–1453. [2]

Zeringue, H. J., Jr. 1987. Changes in cotton leaf chemistry induced by volatile elicitors. Phytochemistry 26:1357–1360. [16]

Zettler, F. W., and T. E. Freeman. 1972. Plant-pathogens as biocontrols of aquatic weeds. Annual Review of Phytopathology 10:455–470. [18]

Zimmer, D. E., and D. Rehder. 1976. Rust resistance of wild *Helianthus* species of the north central United States. Phytopathology 66:208–211. [8, 14, 15]

Zimmer, D. E., J. F. Schafer, and F. L. Patterson. 1963. Mutation for virulence in *Puccinia coronata*. Phytopathology 53:171–176. [8]

Zimmerman, M., and G. H. Pyke. 1988. Experimental manipulations of *Polemonium foliosissimum:* Effects on subsequent nectar production, seed production, and growth. Journal of Ecology 76:777–789. [13]

Zucker, W. V. 1983. Tannins: Does structure determine function? An ecological perspective. American Naturalist 121:335–365. [10, 17]

Zummo, G. R., J. H. Benedict, and J. C. Segers. 1983. No-choice study of plant-insect interactions for *Heliothis zea* (Boddie) (Lepidoptera: Noctuidae) on selected cottons. Environmental Entomology 12:1833–1836. [12]

Contributors

JANIS ANTONOVICS
Department of Botany
Duke University
Durham, NC 27706

HELEN M. ALEXANDER
Departments of Botany and Systematics
    and Ecology
University of Kansas
3038 Haworth
Lawrence, KS 60645-2106

JAMES D. BARBOUR
Department of Entomology
North Carolina State University
Box 7630
Raleigh, NC 27695-7630

FAHKRI A. BAZZAZ
Department of Organismic and
    Evolutionary Biology
Harvard University
16 Divinity Avenue
Cambridge, MA 02138

MAY R. BERENBAUM
Department of Entomology
320 Morrill Hall
University of Illinois
505 South Goodwin
Urbana, IL 61801-3795

BARBARA CHRIST
Department of Plant Pathology
The Pennsylvania State University
University Park, PA 16802

ROBERT S. FRITZ
Department of Biology
Vassar College
Poughkeepsie, NY 12601

JAMES GROTH
Department of Plant Pathology
University of Minnesota
Saint Paul, MN 55108

J. DANIEL HARE
Department of Entomology
University of California
Riverside, CA 92521

RICHARD KARBAN
Department of Entomology
University of California
Davis, CA 95616

GEORGE G. KENNEDY
Department of Entomology
North Carolina State University
Box 7630
Raleigh, NC 27695-7630

ROBERT J. MARQUIS
Department of Biology
University of Missouri
8001 Natural Bridge Road
Saint Louis, MO 63121-4499

MATTHEW A. PARKER
Department of Biological Sciences
State University of New York
Binghamton, NY 13902

DIANA PILSON
Division of Biological Sciences
University of Montana
Missoula, MT 59812

A. JOSEPH POLLARD
Department of Biology
Furman University
Greenville, SC 29613

MARK D. RAUSHER
Department of Zoology
Duke University
Durham, NC 27706

ELLEN L. SIMMS
Department of Ecology and Evolution
1101 East 57th Street
University of Chicago
Chicago, IL 60637

ARTHUR E. WEIS
Department of Ecology and
   Evolutionary Biology
University of California
Irvine, CA 92717

LAWRENCE R. WILHOIT
Department of Entomology
University of California
Davis, CA 95616

ARTHUR R. ZANGERL
Department of Entomology
320 Morrill Hall
University of Illinois
505 South Goodwin
Urbana, IL 61801-3795

Taxonomic Index

569

Subject Index

577